Withdrawn

THE LIVERWORTS OF
BRITAIN AND IRELAND

THE LIVERWORTS OF
BRITAIN AND IRELAND

A.J.E. SMITH

Reader in Botany, School of Biological Sciences
University College of North Wales, Bangor

CAMBRIDGE
UNIVERSITY PRESS

Published by the Press Syndicate of the University of Cambridge
The Pitt Building, Trumpington Street, Cambridge CB2 1RP
40 West 20th Street, New York, NY 10011–4211, USA
10 Stamford Road, Oakleigh, Melbourne 3166, Australia

First published 1990
First paperback edition 1991
Reprinted 1996

Printed in Great Britain at the University Press, Cambridge

British Library cataloguing in publication data

Smith, A.J.E. (Anthony John Edwin)
The liverworts of Britain and Ireland.
1. Great Britain. Liverworts
I. Title
588′.33′0941

Library of Congress cataloguing in publication data

Smith, A.J.E. (Anthony John Edwin), 1935–
The liverworts of Britain and Ireland / A.J.E. Smith.
p. cm.
Bibliography: p.
Includes index.
ISBN 0-521-23834-X (hb)
ISBN 0-521-42473-9 (pb)
1. Liverworts–Great Britain. 2. Liverworts–Ireland. I. Title.
QK558.S65 1989
588′.33′0941–dc20 89-36120 CIP

ISBN 0 521 23834 X hardback
ISBN 0 521 42473 9 paperback

CONTENTS

PREFACE   *page* vii

ACKNOWLEDGEMENTS   ix

INTRODUCTION   1
  Examination of material   1
  Measurements   2
  Use of keys   2
  Nomenclature and taxonomy   4
  Taxonomic categories   4
  Frequency and distribution   5
  Illustrations and descriptions   5

CONSPECTUS OF CLASSIFICATION   6

HEPATICAE   9
  Artificial key to orders (including Anthocerotae)   9
  Calobryales   10
  Jungermanniales   12
  Artificial key to families of Jungermanniales   12
  Metzgeriales   280
  Artificial key to families of Metzgeriales   280
  Sphaerocarpales   308
  Marchantiales   310
  Artificial key to families of Marchantiales   310

ANTHOCEROTAE   334

ADDENDUM   338

GLOSSARY   339

BIBLIOGRAPHY   353

INDEX   355

PREFACE

It is many years since there has been a readily available comprehensive liverwort flora of the British Isles, Symers M. Macvicar's 1926 *The Student's Handbook of British Hepatics* being long out of print as also has been the facsimile edition produced by Wheldon & Wesley in 1960. Since 1926, and particularly during the last 30 years or so, much taxonomic work has been carried out on European hepatics and this, together with recent additions to the British and Irish floras, has rendered Macvicar's *Handbook* very out of date. Although this work has taken eight years to prepare so little is known about many liverwort taxa that it can only be regarded as a guide to British and Irish liverworts rather than a definitive account. A fully detailed flora would take many years to prepare. It must be stressed that this is not a revision of Macvicar's *Handbook* but is a completely new flora embodying recent information and ideas, some of the latter no doubt controversial, on liverwort taxonomy. Although I have sought opinions and guidance from many sources the views expressed here are entirely my own.

Bangor, January 1989   A.J.E. SMITH

ACKNOWLEDGEMENTS

I am greatly indebted to Mr A.R. Perry for the loan of specimens, and the time he took searching for some of these, in the herbaria of the British Bryological Society and the National Museum of Wales at Cardiff. Without access to these specimens my task would have been much more difficult. He also very kindly assisted with checking the proofs.

I am also very grateful to the following individuals for the loan of specimens and/or the provision of information: Dr H. Bischler, Mr A.C. Crundwell, Professor H.J.B. Birks, Professor J.G. Duckett, Dr S.R. Edwards, Mr J.C. Gardiner, Professor P. Greig-Smith, Dr R. Grolle, Dr M.O. Hill, Dr E.W. Jones, Mr D.G. Long, Dr M.E. Newton, Mrs J.A. Paton, Mr D. Synnott, Dr H.L.K. Whitehouse and Mr M. Yeo.

The Directors or Curators of the following institutions have also been most helpful in lending material: Botany School, University of Cambridge; Royal Botanic Garden, Edinburgh; National Botanic Gardens, Dublin; Manchester Museum, University of Manchester.

Dr M.E. Newton and Blackwell Scientific Publications, Osney Mead, Oxford, kindly gave permission to reproduce text and figures on heterochromatin in *Pellia epiphylla* agg. from *J. Bryol.* **14**(2), pp. 219, 226, 228, 1986.

I am indebted to successive heads of the School of Plant Biology for providing facilities, the University College of North Wales for granting me study leave in 1987 and Nerys Owen, Sally Wells and Maria Woolley for typing the manuscript. Finally I would like to thank my wife for putting up with my obsessional interest in bryophytes and tolerating a general lack of enthusiasm for any other activities.

INTRODUCTION

The structure and life-cycle of bryophytes are dealt with in detail in two excellent textbooks and more briefly in several elementary books which are readily available, and hence are not dealt with here. The two textbooks are *Structure and Life of Bryophytes* by E.V. Watson (Hutchinson University Library, London, 1972) and *Introduction to Bryology* by W.B. Schofield (Macmillan, New York, 1985). A useful elementary text is *The Diversity of Green Plants* by P. Bell and C. Woodcock (Edward Arnold, London, 1983). An exhaustive account of all aspects of Hepaticae and Anthocerotae is given by R.M. Schuster in the first volume of his monumental work *The Hepaticae and Anthocerotae of North America east of the Hundredth Meridian* (Columbia University Press, New York, 1966).

EXAMINATION OF MATERIAL

With leafy liverworts fresh or dried specimens may be examined and the only drawback of the latter is loss of colour and oil-bodies; when moistened dried specimens usually regain their original shape. Dried thalloid liverworts and even more so dried hornworts (Anthocerotae) do not recover well when remoistened and are best examined when fresh. A 20% solution of household liquid detergent will help to revivify them but section cutting is still difficult.

Leafy liverworts are best examined by placing two shoots on a slide for microscopic examination, one dorsal side up, the other ventral side up. If the shoots are too large to allow the determination of the shape of leaves *in situ*, leaves can be dissected off, but it should be remembered that decurrent leaf bases frequently remain attached to the stem.

Sections of thalloid species may be cut by placing a portion of thallus between two halves of a length of elder pith (or carrot) or between two pieces of polystyrene. By resting a razor blade flat on one half of pith or polystyrene and cutting successive slices from the specimen by moving the blade in a rotary fashion and very gradually increasing pressure on the blade, at least a few very thin sections will be obtained. Alternatively place a thallus on a slide, cover with a second slide placed at right angles to the thallus, hold down and press the second slide onto the thallus. Place a razor blade at about 10° from the vertical against this slide, then cut slices of thallus, gradually moving the razor blade to the vertical; this will produce excellent sections with a little practice. A good quality double-sided razor blade will produce much better results than a single-sided blade or old fashioned cut-throat razor.

With a number of genera (e.g. *Cephalozia*, *Cephaloziella*) sex determination may be required. As it is essential to avoid breakage of stems careful dissection will be necessary. This may be done by teasing out moist material and carefully extracting stems. Fertile branches or shoot tips are usually recognisable by larger and/or more

1

closely imbricate leaves. In many but not all instances male bracts are saccate at the base and female bracts of a different shape or size from the leaves. Where bracts do not differ from the leaves it may be necessary to use a clearing agent so that gametangia can be seen through the cleared bracts. Gum chloral mountant (G.B.I. Laboratories Ltd, Northgate, Pontefract, W. Yorkshire WF8 1HJ – available through a pharmacist) is an excellent clearing agent. Alternatively, an inflorescence may be gently crushed or torn apart. However, antheridia are often of only short duration but in most species, with a little practice, male bracts where present are readily identified.

MEASUREMENTS

Leaf dimensions. Length of simple and lobed leaves is measured from the middle of the insertion to the extreme apex or a line joining the two longest lobes (Fig. 1). Leaf width is measured across the widest part of the leaf (Fig. 1). Both length and width measurements include teeth where these are present. Underleaves are measured similarly (Fig. 1). Length of conduplicate leaf lobes is measured from the insertion of the keel to the lobe apex and width is measured across the widest part to the keel (Fig. 1).

Cell dimensions. Unless otherwise stated measurements are from the centre of the leaf (midway from the middle of the insertion to the apex) and from middle lamella to middle lamella (Fig. 1). Measurements are from mature leaves and the occasional abnormally small or large cell encountered in liverwort leaves is excluded. Where two species have an overlapping range, e.g. 16–28 μm, 24–36 μm, at least some cells in the range 16–24 μm will be found in one species and 28–36 μm in the second so that cell size in such instances can be used as a discriminating character.

Gemma size. Gemmae are frequently longer than wide or of irregular shape. Measurements are of the longest dimension.

All measurements are from my own observations and do not always agree with those of Macvicar (1926) some of which are the same as Müller's (1954–57) and were presumably taken from the first edition of Müller's *Die Lebermoos Europas* (1906–16).

USE OF KEYS

The characters used to define families, and sometimes to a lesser extent genera, may be ill-defined and difficult to ascertain or are based upon gametophyte reproductive or sporophyte characters that are only very rarely present. Further, the limits of some families and genera and the allocation of certain genera to families is a matter of dispute. This makes the construction of easy to use keys based upon discrete characters almost impossible unless taxa are keyed out almost to the species level (if then).

The key to the families of Jungermanniales and some of the keys to genera and species must be used only as a guide to the allocation of taxa and are not infallible. It is absolutely essential that a careful comparison of material under study is made with descriptions and illustrations before any firm conclusions are reached concerning identity. Where selection of a key character depends upon correct interpretation (i.e. where there may be ambiguity) allowance has been made for this as far as possible. Thus, the detection of underleaves or the determination of the nature of leaf insertion is sometimes difficult and where such characters are used it may be necessary to try both parts of a key dichotomy. It should, however, be borne in mind that the majority of liverworts, and almost all common species, are easily identified once the correct way through the keys concerned has been found.

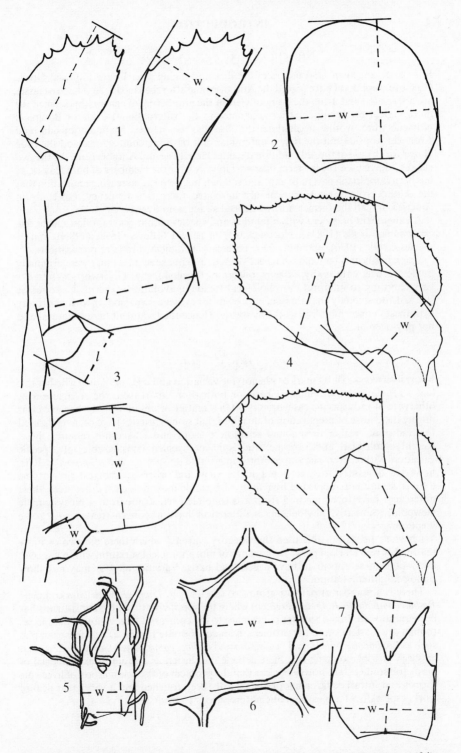

Fig. **1**. Leaf and cell dimensions. 1, *Plagiochila atlantica*; 2, *Southbya nigrella*; 3, *Lejeunea cavifolia*; 4, *Scapania nemorea*; 5, underleaf of *Barbilophozia hatcheri*; 6, mid-leaf cell of *Odontoschisma elongatum*; 7, *Acrobolbus wilsonii*. *l* = length, *w* = width.

NOMENCLATURE AND TAXONOMY

For the arrangement and names of families and higher taxa I have followed Grolle (1983b) except that I have placed the Anthocerotae after the Hepaticae. The two classes are not related and, from the point of view of the numbering of genera, this seemed the more logical proposition. Strict adherence to the International Code of Botanical Nomenclature would necessitate the use of the names Marchantiopsida and Anthocerotopsida but the first is unfamiliar and both are cumbersome and the code allows for the use of the old traditional names Hepaticae and Anthocerotae for the two classes. I have used the lay term liverwort loosely to cover members of both classes, as has been done in the past in Britain and Ireland, but with the increasing realisation that the Anthocerotae and Hepaticae are unrelated there is a tendency, especially in America, to call the former hornworts as distinct from liverworts.

Arrangement of genera within families and species within genera follows, with few exceptions, Grolle (1983a). Nomenclature of genera follows Grolle (1983b) and of species Grolle (1976), updated where necessary as indicated in later publications.

There has been a tendency in recent years to accept as fact ideas that were originally put forward as tentative hypotheses concerning the phylogeny of liverworts. Whilst it is very tempting to interpret morphological trends in evolutionary terms it cannot be stressed too strongly that there is at present no evidence whatsoever to support any hypotheses concerning liverwort phylogeny. Hence the treatment here is natural and not phylogenetic.

TAXONOMIC CATEGORIES

Many liverworts exhibit marked phenotypic variation and it is not known whether this has a genetic or environmental basis or both. For this reason, the recognition or otherwise of intraspecific categories is often a matter of opinion. It has become clear during the course of preparation of this flora that some varieties have been recognised for traditional rather than sound scientific reasons and I have not retained these. Recent cytological and isozyme studies have revealed considerable intraspecific variation within even such apparently invariable species as *Pellia epiphylla*. It is, however, necessary to take a pragmatic view and where cytological or isozyme differences are not supported by readily discernible morphological differences, these differences have to be ignored from a taxonomic even if not from a biosystematic viewpoint. Similarly, phytochemical differences have also been treated with some scepticism.

I have as far as possible used the category 'variety' where there are two or more reasonably well marked peaks in the range of morphological variation within a species and where these appear to have a genetical basis. Varieties may or may not show ecological differentiation.

There is a number of pairs or groups of species (e.g. *Lophozia ventricosa* complex, *Radula complanata*/*R. lindenbergiana*) where the specific status of the constituents has been questioned. In the absence of evidence to the contrary I have retained the species though draw attention to the problem in an appropriate footnote. In a few examples such as *Lophocolea bidentata*/*L. cuspidata* and *Chiloscyphus polyanthos*/*C. pallescens* it has been possible to come to a more definite conclusion as a result of experimental or biometric studies. It is, however, clear that the problem of the delimitation of liverwort species and intraspecific categories has hardly been touched and a very great deal of work, especially of a biosystematic nature, needs to be done.

FREQUENCY AND DISTRIBUTION

Only six categories of frequency have been used: very rare, rare, occasional, frequent, common or very common. Into which of these categories any one species is placed is to some extent a matter of personal opinion and sometimes the selection of which of two categories is arbitrary. Where possible differences in frequency in different parts of the British isles are indicated. Several sources of information have been used: personal observation, distribution maps published in the *Transactions of the British Bryological Society* and *Journal of Bryology*, the *Provisional Atlas of the Bryophytes of the British Isles* (Smith, 1978), the *Distribution of Hepaticae in Scotland* (Macvicar, 1910), *Distribution of Bryophytes in the British Isles* (Corley & Hill, 1981), data accumulated by the Distribution Maps Scheme started by the British Bryological Society in 1960 and local floras.

The figures given at the end of habitat details are the number of vice-counties from which a taxon has been recorded. Thus 17, H3, C means a plant has been record from 17 British, 3 Irish vice-counties and the Channel Islands. The data have been obtained from Corley & Hill (1981) and additions to this published in *Bulletin of the British Bryological Society*, 1982–, and is complete to the end of 1987. A vice-county map is given in Fig. 151.

European distribution has been taken from Düll (1983) and that for Macaronesia from Eggers (1982). Technically Macaronesia includes the Azores, Madeira, the Canary Islands and the Cape Verde Islands. Very few European liverworts occur on the Cape Verde Islands and I have used the term Macaronesia to include only the Azores, Madeira and the Canaries. Where a species does occur on the Cape Verde Islands this is mentioned separately. Extra-European distribution, which in many instances is equivocal because of taxonomic confusion and incomplete knowledge, is taken from Müller (1954, 1957) and Schuster (1966, 1969, 1974, 1980).

Altitudinal range is given where possible but information on this is very incomplete so ranges must be regarded as only approximate. Altitudes are almost completely lacking on herbarium specimens and the only local floras with useful data are those of North Wales and the island of Skye. There is some information in Macvicar (1910).

Frequency of perianths is also given but in many species is almost certainly an overestimate as there is a tendency to collect preferentially material with perianths. With species where determination of sex is necessary this is definitely so.

ILLUSTRATIONS AND DESCRIPTIONS

With only two exceptions (*Fossombronia crozalsii*, *Cephaloziella marcostachya* var. *spiniflora*) all illustrations and descriptions are based upon British and/or Irish material although some of the latter may have been augmented by the examination of European material. Except for critical taxa descriptions are based upon detailed examination of ten randomly selected specimens; abnormal or depauperate specimens have not been included so that descriptions are of 'typical' specimens. Where I was unable to see particular structures and have taken information from other sources this is indicated (e.g. 'Perianth obpyriform, plicate above (Macvicar, 1926)'), otherwise all descriptions are based upon my own observations and measurements. The illustrations were prepared using a Leitz Laborlux microscope with a Leitz drawing apparatus.

CONSPECTUS OF CLASSIFICATION

The arrangement of families and higher taxa follows Grolle (1983b) and that of categories below the rank of family, with one or two minor modifications, that of Grolle (1983a). I have treated the Hepaticae and Anthocerotae as classes within the division Bryophyta. There is an increasing body of evidence that they, like the classes of mosses, have been derived from Rhynalian ancestors. Whether these various classes are directly related or not, their probable common ancestor seems to justify the retention of a single division although some recent authors have raised the Hepaticae and Anthocerotae to divisional rank. Nothing is known of the phylogeny of liverworts or hornworts, although this has been the subject of much speculation, and the following classification should be regarded as natural (or phenetic) and no phylogenetic conclusions should be construed from it.

DIVISION BRYOPHYTA
CLASS I. HEPATICAE
Subclass 1. Jungermanniideae
ORDER 1. CALOBRYALES
FAMILY 1. HAPLOMITRIACEAE
1. HAPLOMITRIUM

ORDER 2. JUNGERMANNIALES
FAMILY 2. LEPICOLEACEAE
2. MASTIGOPHORA

FAMILY 3. HERBERTACEAE
3. HERBERTUS

FAMILY 4. PSEUDOLEPICOLEACEAE
4. BLEPHAROSTOMA

FAMILY 5. TRICHOCOLEACEAE
5. TRICHOCOLEA

FAMILY 6. LEPIDOZIACEAE
6. TELARANEA   8. LEPIDOZIA   9. BAZZANIA
7. KURZIA

FAMILY 7. CALYPOGEIACEAE
10. CALYPOGEIA

FAMILY 8. ADELANTHACEAE
11. ADELANTHUS

FAMILY 9. CEPHALOZIACEAE

12. CEPHALOZIA
13. NOWELLIA
14. CLADOPODIELLA
15. PLEUROCLADULA
16. HYGROBIELLA
17. ODONTOSCHISMA

FAMILY 10. CEPHALOZIELLACEAE

18. CEPHALOZIELLA

FAMILY 11. ANTHELIACEAE

19. ANTHELIA

FAMILY 12. LOPHOZIACEAE

20. TETRALOPHOZIA
21. BARBILOPHOZIA
22. ANASTREPTA
23. LOPHOZIA
24. LEIOCOLEA
25. GYMNOCOLEA
26. EREMONOTUS
27. SPHENOLOBOPSIS
28. ANASTROPHYLLUM
29. TRITOMARIA

FAMILY 13. JUNGERMANNIACEAE

30. JAMESONIELLA
31. MYLIA
32. JUNGERMANNIA
33. NARDIA

FAMILY 14. GYMNOMITRIACEAE

34. MARSUPELLA
35. GYMNOMITRION

FAMILY 15. SCAPANIACEAE

36. DOUINIA
37. DIPLOPHYLLUM
38. SCAPANIA

FAMILY 16. GEOCALYCACEAE

39. LEPTOSCYPHUS
40. LOPHOCOLEA
41. CHILOSCYPHUS
42. HARPANTHUS
43. GEOCALYX
44. SACCOGYNA

FAMILY 17. PLAGIOCHILACEAE

45. PEDINOPHYLLUM
46. PLAGIOCHILA

FAMILY 18. ARNELLIACEAE

47. SOUTHBYA
48. GONGYLANTHUS

FAMILY 19. ACROBOLBACEAE

49. ACROBOLBUS

FAMILY 20. PLEUROZIACEAE

50. PLEUROZIA

FAMILY 21. RADULACEAE

51. RADULA

FAMILY 22. PTILIDIACEAE

52. PTILIDIUM

FAMILY 23. PORELLACEAE

53. PORELLA

FAMILY 24. FRULLANIACEAE

54. FRULLANIA

FAMILY 25. JUBULACEAE

55. JUBULA

FAMILY 26. LEJEUNEACEAE

56. HARPALEJEUNEA
58. LEJEUNEA
61. APHANOLEJEUNEA

57. DREPANOLEJEUNEA   59. COLURA   62. MARCHESINIA
                      60. COLOLEJEUNEA

ORDER 3. METZGERIALES

FAMILY 27. CODONIACEAE

63. FOSSOMBRONIA   64. PETALOPHYLLUM

FAMILY 28. PELLIACEAE

65. PELLIA

FAMILY 29. PALLAVICINIACEAE

66. PALLAVICINIA   67. MOERCKIA

FAMILY 30. BLASIACEAE

68. BLASIA

FAMILY 31. ANEURACEAE

69. ANEURA   70. CRYPTOTHALLUS   71. RICCARDIA

FAMILY 32. METZGERIACEAE

72. METZGERIA   73. APOMETZGERIA

Subclass 2. Marchantiideae

ORDER 4. SPHAEROCARPALES

FAMILY 33. SPHAEROCARPACEAE

74. SPHAEROCARPOS

ORDER 5. MARCHANTIALES

FAMILY 34. TARGIONIACEAE

75. TARGIONIA

FAMILY 35. LUNULARIACEAE

76. LUNULARIA

FAMILY 36. WIESNERELLACEAE

77. DUMORTIERA

FAMILY 37. CONOCEPHALACEAE

78. CONOCEPHALUM

FAMILY 38. AYTONIACEAE

79. REBOULIA

FAMILY 39. MARCHANTIACEAE

80. PREISSIA   81. MARCHANTIA

FAMILY 40. RICCIACEAE

82. RICCIOCARPOS   83. RICCIA

CLASS II. ANTHOCEROTAE

ORDER 6. ANTHOCEROTALES

FAMILY 41. ANTHOCEROTACEAE

84. ANTHOCEROS   85. PHAEOCEROS

CLASS I. HEPATICAE

(MARCHANTIOPSIDA, HEPATICOPSIDA)

Plants thalloid or leafy. Main axis branching laterally, dichotomously, pinnately or irregularly, or ventrally. In thalloid plants, thallus dorsiventral, with or without a well defined midrib, lacking mucilage-filled cavities but sometimes with air-cavities, tissue on dorsal side differentiated or not, central strand of conducting cells usually absent; rhizoids simple and sometimes also tuberculate; ventral scales present or not. Leafy plants erect and isophyllous to dorsiventral and anisophyllous; leaves in three ranks, third rank ventral, usually reduced in size or absent; stem with or without central strand of differentiated cells; rhizoids simple; leaves various in shape, without a midrib. Cells with numerous chloroplasts and usually oil-bodies; apical cells usually protected by slime papillae. Antheridia exogenous in origin, solitary in antheridial cavities or 1–2 in axils of leaf-like bracts, spherical, on long or short stalk. Archegonia flask-shaped on short stalks, consisting of neck, and venter surrounding ventral canal cell and egg cell. In thalloid forms developing sporophyte protected by involucre, pseudoperianth or fleshy calyptra. In leafy forms perianth of 2–3 fused leaves, usually developed following fertilisation, but sometimes lacking; developing sporophyte protected by one or a combination of perianth, fleshy calyptra, perigynium or marsupium. Sporophyte, protected by gametophyte structures until mature, composed of foot, very short seta and capsule, lacking intercalary meristem; mature capsule elevated into air by elongation of seta cells, spherical to shortly cylindrical, wall 1–several cells thick, with or without chloroplasts, stomata lacking, columella lacking, sporogenous tissue giving rise to spores and unicellular, spirally thickened elaters, capsule usually dehiscing by 4 valves. Two subclasses, 6 orders (excluding the Takakiales now placed in a separate class), about 330 extant genera and 5500–6000 species. For a detailed account of the class see Schuster (1966).

ARTIFICIAL KEY TO ORDERS OF LIVERWORTS (INCLUDING ANTHOCEROTAE)

1 Plants leafy, very rarely branching dichotomously, archegonia and capsules terminal on main axis or branch, rhizoids colourless to brown or if purple or violet then leaves not crisped or lettuce-like   2
  Plants thalloid; if leafy then branching dichotomously, leaves usually crisped and lettuce-like, rhizoids usually deep purple or violet and archegonia and sporophytes borne on dorsal side of main axis   3
2 Shoots radially symmetrical with simple leaves in three equal ranks, capsule protected by fleshy calyptra   **1. Calobryales (p. 10)**
  Shoots dorsiventral with two equal ranks of lateral leaves and third rank smaller in size or lacking; if leaves in three equal ranks then leaves bilobed to $\frac{1}{3}$ or more, capsule protected by perianth, perigynium or marsupium
    2. Jungermanniales (p. 12)

3 Thallus cells with single large chloroplast, capsule green, elongate-cylindrical, dehiscing by 2 valves   **6. Anthocerotales (p. 334)**
  Thallus cells with numerous small chloroplasts (lacking in *Cryptothallus*), capsule blackish or brownish at maturity, spherical or shortly cylindrical, dehiscing by 4 valves or irregularly   4
4 Thallus ± circular, to 15 mm diameter, densely covered with numerous fusiform to clavate involucres on dorsal surface, spores in tetrads.
    4. Sphaerocarpales (p. 308)
  Thallus various, without numerous involucres on dorsal surface, spores rarely adhering in tetrads   5
5 Thallus or leafy shoots transluscent, lacking internal differentiation (conducting strand present in Pallaviciniaceae), rhizoids smooth only, capsules not on specially modified branches or embedded in thallus   **3. Metzgeriales (p. 280)**
  Thallus opaque, with epidermal pores (except *Dumortiera*), internal differentiation, smooth and tuberculate rhizoids, capsules borne on modified branches or embedded in thallus   **5. Marchantiales (p. 310)**

Subclass Jungermanniideae

Gametophyte leafy or thalloid, without or with only slight tissue differentiation, air-cavities lacking, rhizoids smooth. Chlorophyllose cells with 2–numerous oil-bodies. Archegonia with 4–5 rows of neck cells, arising at or near stem apex. Sporophyte with seta greatly elongated at maturity. Capsule wall (1–)2–several stratose, usually dehiscing by 4 valves. Spores usually 6–50 μm, rarely in tetrads; spore mother cells 4-lobed before meiosis.

1. CALOBRYALES

A monogeneric order with the characters of the genus. *Takakia* is included in this order by some authorities (e.g. Grolle, 1983b) but it has recently been shown that it is not a liverwort (Crandall-Stotler, *J. Bryol.* **14**, 1–23, 1985).

1. HAPLOMITRIACEAE

HAPLOMITRIUM NEES, *Naturg. Europ. Leberm.* 1: 109. 1833

Stems ± radially symmetrical, arising from horizontal rhizome lacking rhizoids. Leaves in 3 ± similar ranks, insertion transverse, entire or with 2 small lateral lobes. Antheridia scattered round stem. Archegonia naked (perianth lacking) round stem or near apex with stem ceasing growth after fertilisation. Developing sporophyte protected by fleshy calyptra formed partly by upgrowth of archegonial stalk and partly from venter. Capsule shortly cylindrical with unistratose wall dehiscing by a single slit. Elaters bispiral, basal elaterophore present. Seven species.

1. H. hookeri (Sm.) Nees, *Naturg. Europ. Leberm.* 1: 111. 1833   (Fig. 2)

Dioecious. Green or pale green, small tufts or scattered plants. Shoots to about 1 cm in lowland habitats, to 4 cm in montane habitats, stems erect, arising from horizontal, simple or branched rhizome. Lower leaves small, obliquely inserted below, upper larger, transversely inserted, crowded or not, concave, variable in shape, lingulate to

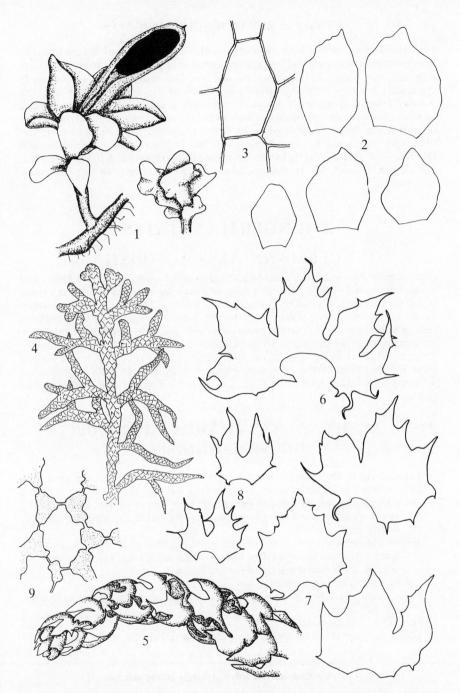

Fig. **2**. 1–3, *Haplomitrium hookeri*: 1, sterile and fertile shoots (× 12); 2, leaves (× 15); 3, mid-leaf cell (× 50). 4–9, *Mastigophora woodsii*: 4, shoot (× 2.5); 5, branch tip (× 23); 6, stem leaves (× 23); 7, branch leaves (× 23); 8, underleaves (× 23); 9, mid-leaf cell (× 450).

ovate-rhomboidal, one or both sides often with a small lobe, apex obtuse to acute, margin entire or obscurely and irregularly toothed; cells thin-walled, decreasing in size from leaf base to apex, 20–32 μm wide in mid-leaf. Gemmae lacking. Capsule shortly cylindrical, protected by fleshy calyptra; spores 19–24 μm. Capsules occational, late summer, autumn, spring. $n = 9$. At low altitudes in dune slacks, sandy soil by water and on tracks and in flushes on heaths, at high altitudes (ascending to 1070 m) on gravelly detritus and in flushes, occasional in the Scottish highlands, very rare elsewhere and usually present in small quantity; W. Cornwall, Brecon, Caernarfon, S. Lancashire and Durham northwards. 37, H11. Belgium, Alps and Carpathians north to Fennoscandia, Spitzbergen, Iceland, W. Himalayas, Maine, New Hampshire, British Columbia, Greenland.

2. JUNGERMANNIALES

(JUNGERMANNIALES ACROGYNAE)

Plants usually dorsiventral with leaves in 2 lateral and 1 ventral ranks, the latter often reduced in size and of different morphology or absent; leaves basically bilobed, arising from 2 meristematic cells, but becoming entire by infilling of sinus or more than 2-lobed by the activity of additional meristematic cells. Stem with apical cell with 3 cutting faces. Rhizoids present. Antheridia produced in axils of usually modified lateral leaves; archegonia terminal on main or branch axis (so that growth is sympodial or acrogynous). Developing sporophyte usually protected by tubular perianth derived from modified fused leaves. Capsule spherical to cylindrical, wall 2–10-stratose, dehiscing by 4 valves. Spores 6–18(–60) μm. About 275 genera and 7000 described species.

ARTIFICIAL KEY TO THE FAMILIES OF JUNGERMANNIALES

1 Leaves divided into uniseriate lobes or segments, or margins with uniseriate cilia 5
  Leaves simple or lobed but lobes not uniseriate nor margins ciliate   2
2 Leaves incubous (i.e. with upper leaf margin lying nearer to stem apex than the
      lower margin so that forward margin of one leaf overlies rear margin of leaf in
      front)   10
  Leaves transversely inserted and orientated or succubous (i.e. with lower margin
      lying nearer to stem apex than upper margin so that rear margin of one leaf
      overlies the forward margin of the leaf behind)   3
3 Leaves simple, margin toothed or not   19
  Leaves variously lobed or conduplicate or at least apex emarginate or conspi-
      cuously lobed   4
4 Leaves in 3 ± equal ranks   25
  Leaves in 2 lateral rows with third row (underleaves) smaller or lacking   27

Leaves with uniseriate lobes or longly ciliate margins

5 Plants with conspicuous purple rhizoids   **Fossombronia fimbriata (p. 287)**
  Rhizoids where present colourless to brownish   6

6 Leaves conduplicate **15. Scapaniaceae (p. 166)**
  Leaves not conduplicate 7
7 Leaves divided into lobes with longly ciliate margins 8
  Leaves divided almost to base into 2–4 uniseriate lobes without marginal cilia 9
8 Plants pale green to creamy, leaf cells without trigones **5. Trichocoleaceae (p. 20)**
  Plants reddish-brown, brown or olive-green, cells with bulging trigones
  22. Ptilidiaceae (p. 243)
9 Underleaves of similar size to leaves, leaf lobes 1 cell wide at base
  4. Pseudolepicoleaceae (p. 18)
  Underleaves smaller than leaves, leaf lobes 1–2 cells wide at base
  6. Lepidoziaceae (p. 21)

Leaves incubous

10 Leaf margin spinosely toothed 11
   Leaf margin entire 12
11 Leaves asymmetrically 3–4-lobed, not conduplicate **2. Lepicoleaceae (p. 15)**
   Leaves conduplicate with small helmet-shaped ventral lobe
   25. Jubulaceae (p. 258)
12 Leaves not conduplicate 13
   Leaves conduplicate with small ventral lobe 14
13 Leaves 3–4-lobed, or with truncate apex with 2–3 teeth and flagelliform shoots
   present **6. Lepidoziaceae (p. 21)**
   Leaves simple or at most bilobed to $\frac{1}{6}$, flagelliform shoots lacking
   7. Calypogeiaceae (p. 35)
14 Ventral lobe joined to lower part of dorsal lobe by conspicuous keel region 15
   Ventral lobe almost free from dorsal lobe, conspicuous keel lacking 16
15 Underleaves absent, keel smooth, perianth dorsiventrally compressed
   21. Radulaceae (p. 236)
   Underleaves present or if absent then keel crenulate with projecting cells, perianth
   not compressed **26. Lejeuneaceae (p. 259)**
16 Ventral lobe helmet-shaped **24. Frullaniaceae (p. 250)**
   Ventral lobe not helmet-shaped 17
17 Underleaves absent **20. Pleuroziaceae (p. 234)**
   Underleaves present 18
18 Dorsal lobe much larger than unmodified ventral lobe **23. Porellaceae (p. 245)**
   Flap-like dorsal lobe much smaller than sac-like, beaked ventral lobe
   Colura (p. 272)

Leaves simple, succubous

19 Underleaves present, conspicuous **16. Geocalycaceae (p. 201)**
   Underleaves absent or minute and not obvious 20
20 Leaves ± opposite **18. Arnelliaceae (p. 230)**
   Leaves alternate 21
21 Leaf margin toothed 22
   Leaf margin entire 23
22 Dorsal margin of leaf inflexed **8. Adelanthaceae (p. 44)**
   Dorsal margin of leaf plane or deflexed **17. Plagiochilaceae (p. 218)**

23 Leaves very obliquely inserted, dorsal margin longly decurrent
                                                17. Plagiochilaceae (p. 218)
    Leaves transversely to obliquely inserted, margin not or only slightly decurrent 24
24 Stems with flagelliform shoots arising from ventral side and/or gemmae at shoot
    tips, leaf cells to 32 μm wide   **9. Cephaloziaceae (p. 46)**
    Stems without flagelliform shoots, gemmae lacking or if present then leaf cells more
    than 50 μm wide   **13. Jungermanniaceae (p. 130)**

Leaves in 3 ± equal ranks, bilobed

25 Cell walls thin, without trigones   **9. Cephaloziaceae (p. 46)**
    Cells with very thick walls or large bulging trigones   26
26 Leaves with flexuose acuminate lobes, leaf with vitta of elongated pitted cells
    extending into lobes, trigones large, bulging   **3. Herbertaceae (p. 16)**
    Leaf lobes acute, appressed, leaf without vitta, cell walls very thick, without
    trigones   **11. Antheliaceae (p. 88)**

Shoots dorsiventral, leaves 2(–5)-lobed

27 Leaves conduplicate   **15. Scapaniaceae (p. 166)**
    Leaves not conduplicate   28
28 Shoots cylindrical or clavate with closely appressed imbricate leaves (*Eremonotus*
    may key out here)   **14. Gymnomitriaceae (p. 153)**
    Shoots not cylindrical or clavate   29
29 Underleaves present on sterile stems   30
    Underleaves absent or minute and subulate on sterile stems   33
30 Plants filiform, leaves to 0.3 mm long, cells 8–16 μm wide or if larger then leaf
    margins toothed   **10. Cephaloziellaceae (p. 72)**
    Plants not filiform, leaves 0.3 mm or more long (except sometimes *Cladopodiella
    francisci*), cells (16–)20–48 μm wide, margins entire   31
31 Dorsal margin of leaf not decurrent   **12. Lophoziaceae (p. 90)**
    Dorsal margin of leaf decurrent   32
32 Underleaves not lobed, not joined at base to lateral leaves, perianth smooth,
    abruptly narrowed to mouth   **12. Lophoziaceae (p. 90)**
    Underleaves bilobed or if simple then joined to lateral leaves at base, perianth 3-
    winged, wide-mouthed   **16. Geocalycaceae (p. 201)**
33 Leaves transversely inserted   34
    Leaves obliquely inserted and succubous   38
34 Leaves to 250 μm long, bilobed to ½ or more   35
    Leaves more than 250 μm long, if less then only bilobed to *ca* ⅓   36
35 At least some branches arising from ventral side of stem, leaf margins entire or
    toothed   **10. Cephaloziellaceae (p. 72)**
    Stems branching dichotomously or all branches arising laterally, leaf margins entire
                                               12. Lophoziaceae (p. 90)
36 Cortical cells of stem translucent, central strand visible by transmitted light and
    ventral flagelliform branches often present or leaf lobes longly acuminate and
    ventral margin inflexed and sac-like   **9. Cephaloziaceae (p. 46)**
    Plants not as above   37
37 Apical leaves or stem tips frequently gemmiferous, leaves symmetrical or

asymmetrical, perianth well developed and protruding beyond female bracts
12. Lophoziaceae (p. 90)
Gemmae or gemmiferous shoot tips lacking, leaves symmetrical, perianth not
developed or not protruding beyond female bracts
14. Gymnomitriaceae (p. 153)
38 Cortical cells of stem translucent, central strand visible by transmitted light
9. Cephaloziaceae (p. 46)
Central strand not visible through translucent cortical cells　　　　　39
39 Leaf lobes acuminate, gemmae lacking　　　　　40
Leaf lobes acute to rounded and gemmae often present, or if lobes acuminate then
abundant reddish-brown gemmae present at tips of upper leaves　　　41
40 Flagelliform shoots commonly present, leaves bilobed to $\frac{1}{3}$, lobes spreading so that
leaves may appear V-shaped, cells 20–28 μm wide
17. Plagiochilaceae (p. 218)
Flagelliform shoots lacking, leaves bilobed $\frac{1}{3}$–$\frac{1}{2}$, not V-shaped, cells 28–32 μm wide
19. Acrobolbaceae (p. 234)
41 Cells at base of leaf lobes 10–20 μm wide, leaves 240–480 μm long, gemmae lacking
13. Jungermanniaceae (p. 130)
Cells at base of leaf lobes 20–44 μm wide or if less then leaves more than 500 μm
long, gemmae often present　　　**12. Lophoziaceae (p. 90)**

2. LEPICOLEACEAE

Stems 1–2-pinnate, ± isophyllous or anisophyllous, mature branches with attenuate
tips. Branches arising laterally and ventrally. Rhizoids arising from leaf bases at tips of
branches. Leaves incubous, overlapping dorsal side of stem, symmetrical or not, 2–4-
lobed, insertion ± transverse; cells with large trigones or thick-walled. Underleaves of
similar size to or smaller than leaves. Gemmae lacking. Perianth ± absent, perigynium
present or not; capsule ovoid. Two genera.

2. MASTIGOPHORA NEES, *Naturg. Europ. Leberm.* 3: 89. 1838

Dioecious. Plants medium-sized to robust. Stems 1–2-pinnate, mature branches with
attenuate apices with small leaves. Rhizoids arising near tips of branches. Leaves
incubous, 2–4-lobed, spinosely toothed; cells with large trigones. Underleaves 2–4-
lobed, spinosely toothed. Male inflorescence on short lateral branch, antheridia 2–3 in
axils of male bracts and underleaves. Female inflorescence terminal on short lateral
branch: perianth rudimentary; capsule ovoid. A relict, disjunct, mainly antipodal
genus of about 10 species.

1. M. woodsii (Hook.) Nees, *Naturg. Europ. Leberm.* 3: 95. 1838　　　(Fig. 2)

Reddish to yellowish-brown, loose patches. Shoots 6–12 cm, stems procumbent,
pinnately to ± bipinnately branched, mature branches with attenuate and sometimes
flagelliform tips. Leaves approximate, imbricate towards stem tips, translucent, very
fragile, overlapping stem on dorsal side, ± transversely inserted, partially embracing
stem, 1.0–1.6 mm long, to twice as wide as long, asymmetrical, 3–4-lobed to $\frac{1}{3}$–$\frac{2}{3}$,
occasionally bilobed or entire, lobes ending in spinose point, margin with sparse
spinose teeth becoming eroded with age, branch leaves smaller than stem leaves, 1–3-
lobed; cells 24–32(–48) μm wide, walls yellowish-brown, trigones very large and
sometimes merging so that walls appear very thick. Underleaves 0.8–1.2 mm long, *ca* as
wide as long, bilobed to *ca* $\frac{2}{3}$, lobes lanceolate, acuminate, margin with sparse spinose

teeth. Gemmae lacking. Capsules unknown. In damp turf, amongst heather, on cliff ledges, in scree, usually in north or east facing sites, 150–1000 m, occasional in W. Scotland and W. Ireland, from Dumbarton north to W. Sutherland, Hebrides, Shetland, Kerry, W. Galway, W. Mayo. 10, H4. Faroes, Himalayas, Yunnan, western N. America.

Only likely to be confused with *Ptilidium ciliare* from which it differs in the mature branches with attenuated apices and the leaf margins spinose-dentate instead of ciliate.

3. HERBERTACEAE

A small family of two genera, the second of which, *Triandrophyllum* Fulf. & Hatch., differs from *Herbertus* in the leaves 2–4-lobed, lacking a vitta, cell walls not pitted, trigones vestigial or absent and perianth longly exserted.

3. HERBERTUS S.F. GRAY, *Nat. Arr. Br. Pl.*: 678, 705. 1821

Dioecious. Shoots normally erect or ascending, often arising from prostrate primary stems, simple or rarely branched, isophyllous or slightly anisophyllous. Rhizoids, if present, at base of erect stems. Leaves ± secund, $\frac{1}{2}$–$\frac{3}{4}$ bilobed, lobes lanceolate, acute to acuminate; median cells elongated with pitted walls, forming a vitta extending into lobes, other cells ± isodiametric, trigones prominent. Male inflorescence bud-like, subterminal, bracts saccate, less deeply bilobed than leaves. Female inflorescence bud-like, terminal, bracts bilobed, more or less concealing perianth. Perianth narrow, split to $\frac{1}{3}$–$\frac{1}{2}$ into 6 laciniate lobes. Capsule globose. An ancient genus of probably 40–50 species with a high degree of endemism, widely distributed on montane islands and colder, unglaciated continental areas away from the tropics.

1　Lateral leaves very asymmetrical and larger than ± symmetrical underleaves
　　　　　　　　　　　　　　　　　　　　　　　　　　　　3. H. borealis
　Lateral leaves only slightly asymmetrical and ± similar to underleaves　　　2
2　Leaves ± suberect when moist, lobes lanceolate and 2.3–4.0 times as long as wide
　　　　　　　　　　　　　　　　　　　　　　　　　　　　1. H. stramineus
　Leaves squarrose-spreading when moist, lobes linear-lanceolate and (3–)4–6 times as
　　　long as wide　　　　　　　　　　　　　　　　　　　　**2. H. aduncus**

1. H. stramineus (Dum.) Lett, *List Brit. Hep.*: 1977. 1902　　　　　　(Fig. 3)
H. aduncus auct.

Olive-green to reddish-brown tufts or patches or scattered shoots amongst other plants. Shoots 3–7 cm, ± erect, with occasional ventral flagelliform branches. Leaves imbricate, ± suberect when moist, hardly altered when dry, ± transversely inserted, lateral leaves ± subfalcate, slightly asymmetrical, underleaves ± straight but otherwise similar, 1.0–1.6 mm long, 0.4–0.8 mm wide, 2–3 times as long as wide, bilobed to $\frac{1}{2}$–$\frac{2}{3}$, lobes lanceolate, acute to acuminate, 2.3–4.0 times as long as wide; vitta cells extending to $\frac{1}{4}$(–$\frac{1}{3}$) way up lobes, other cells ± isodiametric with large trigones, 16–20(–24) μm wide. Perianths unknown. Cliffs and rock ledges, more rarely in turf or dwarf shrub communities, in base-rich habitats from 220–1100 m, rare or occasional in N. Wales and the Lake District, occasional in W. and C. Scotland. 21. S.W. Norway, Faroes, Iceland, Alaska.

2. H. aduncus (Dicks.) S.F. Gray ssp. **hutchinsiae** (Gott.) Schust., *Hep. Anthoc. N. Amer.* 1: 718. 1966　　　　　　　　　　　　　　　　　　　　(Fig. 3)
H. hutchinsiae (Gott.) Evans

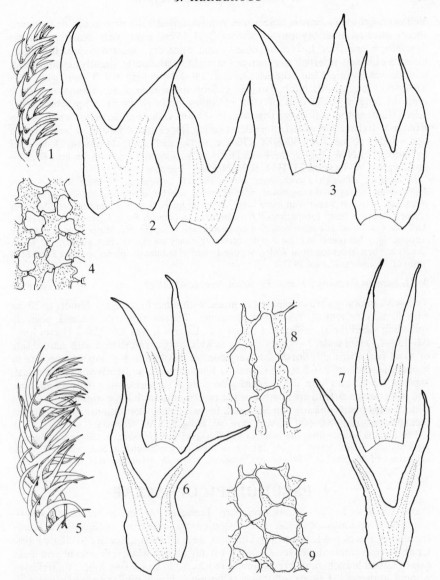

Fig. 3. 1–4, *Herbertus stramineus*: 1, portion of shoot (× 8); 2, lateral leaves (× 33); 3, underleaves (× 33); 4, cells from leaf lobe (× 450). 5–9, *H. aduncus*: 5, part of shoot (× 8); 6, lateral leaves (× 33); 7, underleaves (× 33); 8, vitta cells (× 450); 9, cells from leaf lobe (× 450).

Yellowish-brown to brown, sometimes reddish-tinged tufts or patches or scattered shoots amongst other bryophytes. Shoots (2–)3–15 cm, erect, with occasional ventral flagelliform branches. Leaves flexuose-secund when dry, squarrose-spreading when moist, ± obliquely inserted; stem leaves ± straight to subfalcate, slightly asymmetrical, underleaves ± straight but otherwise similar, 1.0–2.0 mm long, 0.4–0.7 mm wide, 2.2–3.6 times as long as wide, bilobed to $\frac{1}{2}$–$\frac{2}{3}$(–$\frac{3}{4}$), lobes linear-lanceolate, acuminate, (3–)4–6 times as long as wide; cells of vitta extending halfway or more up lobes, other cells ± isodiametric with large trigones, 16–24 μm wide. Fertile plants unknown in Britain or Ireland. On usually north-east facing slopes and scree and in rocky woods and wooded ravines from 30–800(–1100) m, sometimes locally dominant. Occasional to frequent in N.W. Wales, the Lake District, W. Scotland from Arran and Kintyre northwards, W. Ireland. 23, H14. S.W. Norway, Faroes.

 H. aduncus, although a commoner species than *H. stramineus*, is more western in its distribution. It may be distinguished from *H. stramineus* by its often larger size, squarrose-spreading, narrower leaves with more longly tapering lobes. It is also more reddish. Schuster (1966) recognises three subspecies of *H. aduncus*, ssp. *aduncus* from Japan and western N. America, ssp. *hutchinsiae* from oceanic Europe and ssp. *tenuis* (Evans) Miller & Scott from the Appalachians, but points out that it is not certain that they are distinct. Ssp. *tenuis* (as *H. tenuis* Evans) has been recorded from Wales, Scotland and Ireland but the plants so named are small forms of *H. aduncus* (Jones, 1952).

3. H. borealis Crundw., *Trans. Br. bryol. Soc.* 6: 41. 1970   (Fig. 4)

Yellowish to orange-brown patches or mixed with other bryophytes. Shoots to 20 cm with numerous ventral flagelliform branches. Leaves imbricate, secund, scarcely spreading when moist, insertion ± transverse. Lateral leaves 1.4–2.0(–2.1) mm long, 0.6–1.0(–1.1) mm wide, 2–3 times as long as wide, very asymmetrical with ventral side of basal part ± straight, dorsal side very convex, bilobed to $\frac{1}{2}$–$\frac{2}{3}$, lobes lanceolate to broadly lanceolate, 2.0–3.5(–4.3) times as long as wide; underleaves ± symmetrical, smaller, 1.2–1.5 mm long, 0.5–0.7 mm wide; cells of vitta extending halfway up lobes and sometimes nearly to apex, other cells ± isodiametric with large trigones, 17–24 μm wide. Fertile plants unknown in Scotland. In dwarf shrub communities on north-east facing slopes from 400–600 m; very rare, W. Ross. 1. S.W. Norway.

 This plant is distinct from the other two British species of *Herbertus* in colour and the very asymmetrical lateral leaves which are larger than the ± symmetrical underleaves. For the occurrence of this plant in Scotland see Crundwell, *Trans. Br. bryol. Soc.* 6, 41–9, 1970.

4. PSEUDOLEPICOLEACEAE

Plants small to robust. Stems irregularly branched, origin of branches various, isophyllous to ± anisophyllous. Leaf insertion transverse or oblique with leaves succubous, leaves 3–4-lobed to $\frac{1}{3}$ to almost to base into ± equal entire to ciliate lobes. Underleaves ± similar to leaves. Gemmae usually absent. Male inflorescence on main axis or lateral branch, male bracts similar to leaves but less deeply lobed; underleaves without antheridia. Female inflorescence terminal, bracts similar to leaves but with deeper base and segments branching dichotomously. Perianth present, unistratose, shallowly plicate above, contracted to mouth. Capsule ovoid, wall 2–5-stratose. Spores 8–16 μm. A mainly antipodal family of 7 genera.

4. BLEPHAROSTOMA (DUM. EMEND LINDB.) DUM., *Recueil Observ. Jungerm.*: 18. 1835

Monoecious or dioecious. Plants slender. Stem ± isophyllous, in section cortical and inner cells hardly distinct. Leaves transversely inserted. Leaves divided almost to base

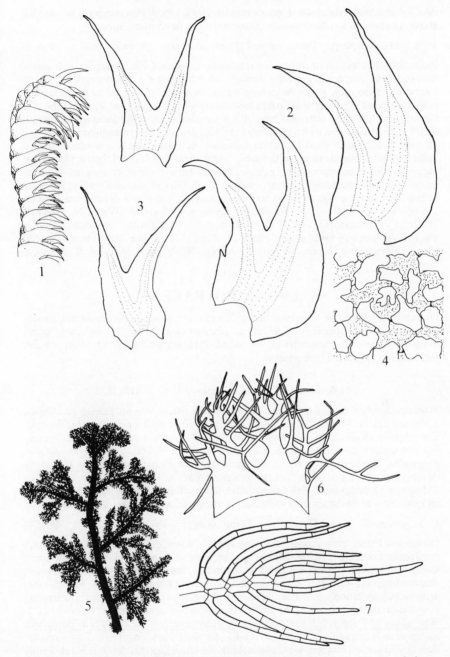

Fig. **4**. 1–4, *Herbertus borealis*: 1 portion of shoot (× 8); 2, lateral leaves (× 33); 3, underleaves (× 33); 4, cells from leaf lobe (× 450). 5–7, *Trichocolea tomentella*: 5, shoot (× 3); 6, leaf (× 15); 7, tip of leaf lobe (× 130).

into 3–4 uniseriate segments. Underleaves usually 3-lobed. Perianth weakly 3-angled above, contracted to ciliate mouth. Three northern hemisphere species.

1. B. trichophyllum (L.) Dum., *Recueil Observ. Jungerm.*: 18. 1835   (Fig. 5)

Paroecious, autoecious or sometimes apparently dioecious. Plants slender, pale green, forming small patches or growing through other bryophytes. Shoots procumbent to ascending, 3–16 mm, stems branching irregularly. Leaves distant to approximate, patent to spreading, divided almost to base into (3–)4 stiff, uniseriate segments 160–620 μm long, each of 7–13 cells, leaf base of 3–4 ± isodiametric cells joined for about half their length, insertion ± transverse; cells at middle of segments rectangular, 27–47 (–53) μm long. Underleaves similar to leaves. Gemmae very rare. Male bracts imbricate, ± similar to leaves. Female bracts with base 2–3 cells deep, 4–5-lobed, lobes similar to leaf segments but dichotomously branched. Perianth subcylindical, contracted to ciliate mouth. Capsules occasional, spring, summer. $n = 9$. On soil, especially where basic, on humus, peat, rotting logs on banks, among rocks, cliffs and in woods in montane areas, ascending to *ca* 1000 m, occasional to frequent in Mid and N. Wales, N. England, Scotland, N. Ireland, rare or very rare elsewhere. 53, H16. From C. Spain, S. Italy and Yugoslavia north to Spitzbergen, Faroes, Iceland, Asia Minor and Caucasus east to China, Siberia and Japan, Azores, E. Africa, N., C. and southern S. America, Greenland.

5. TRICHOCOLEACEAE

Dioecious. Plants small to robust. Stems 1–3-pinnate, stout, branches lateral. Leaves succubous, asymmetrical, 4–5(–9)-lobed, lobes narrow, ciliate. Underleaves resembling leaves, smaller, symmetrical, 2–4-lobed. Perianth lacking, perigynium present, surrounding capsule. Three genera.

5. TRICHOCOLEA DUM., *Comment. Bot.*: 113. 1822

Plants pale green. Stems 1–3-pinnate, paraphyllia often present. Leaves succubous, asymmetrical, 4–7-lobed, lobes longly tapering, often with opposite subulate cilia. Underleaves smaller than lateral leaves, bilobed. Male inflorescence on lateral branch, bracts ± similar to leaves, with 1–2 antheridia per bract, underleaves without antheridia. Archegonia terminal on main axis or branch, becoming lateral by development of innovation, female bracts similar to leaves. Perianth lacking or rudimentary, perigynium developing after fertilisation. Capsule obloid, wall 6–8-stratose. Spores 10–20 μm. About 28 mainly antipodal and tropical American species.

1. T. tomentella (Furh.) Dum., *Jung. Europ. Indig*: 67, 1831   (Fig. 4)

Dioecious. Plants pale green, becoming creamy when dry, in loose fluffy-looking, sometimes extensive patches. Shoots procumbent to suberect, 3–9(–12) cm, stems stout, 2–3-pinnate, with branched paraphyllia, leaves distant to approximate, ± spreading, 0.5–0.9 mm long, divided almost to base into 3–4 lobes, lobes dichotomously branched or not, tapering to long uniseriate subula, margin bearing long uniseriate subulate cilia, the whole arranged ± 3-dimensionally in stem leaves, 2-dimensionally in branch leaves, insertion ± transverse, leaf base 3–4 cells deep, cells elongate; cells in middle of uniseriate subula (40–)60–95 μm long. Underleaves smaller then lateral leaves, bilobed. Gemmae lacking. Perianth lacking, perigynium enclosing capsule, narrowly ellipsoid, mouth ciliate. Capsules very rare. $n = 8 + m^*$, 9. Damp sites in deciduous woodland, sheltered moist stream banks, flushes, ravines, very rare in E.

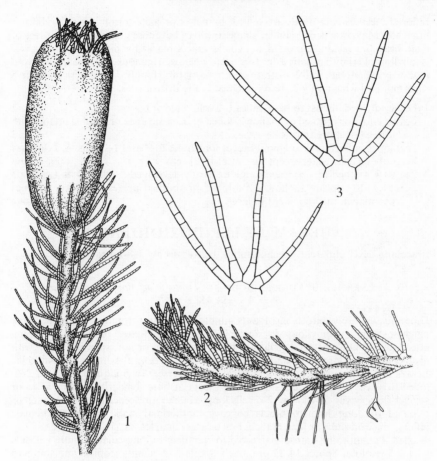

Fig. **5**. *Blepharostoma trichophyllum*: 1 and 2, fertile and sterile shoots (× 15); 3, leaves (× 84).

Anglia and the Midlands, occasional to frequent elsewhere, from Cornwall east to Kent and north to Orkney, in Ireland common in the west, occasional elsewhere. 84, H25. N. Spain and Portugal, S. Italy and Greece north to Fennoscandia and Russia, Bhutan, Shensi, Japan, Azores, Tunisia.

6. LEPIDOZIACEAE

Dioecious, autoecious or heteroecious. Plants minute to robust, whitish-green to brown but never red. Shoots erect and isophyllous to dorsiventral and anisophyllous, stems 1–2-pinnately branched or apparently dichotomously branched, producing from ventral side flagelliform microphyllous branches which may become rhizoidal. Leaves incubous or transverse, very rarely succubous, 3–4-lobed or 2–4-toothed at apex. Underleaves as large as and similar to leaves, to smaller and dissimilar. Vegetative propagules very rare but it is likely that flagelliform branches may act as propagules.

Inflorescences usually on dwarf ventral branches or male terminal on flagelliform branches. Male bracts of ± similar shape to leaves but more concave, underleaves of male branches lacking antheridia. Female bracts markedly different from leaves, isophyllous. Perianth usually ± fusiform, terete below, trigonous above. Capsule ovoid to shortly cylindrical, wall 2–4-stratose. A large, natural family of 27 genera placed in 5 subfamilies of which only 2 are represented in the British Isles.

1 Leaves divided almost to base, lobes 1–2 cells wide at base   2
  Leaves entire or toothed or shallowly lobed at truncate apex or divided up to ⅔ with
      lobes 4 or more cells wide at base   3
2 Leaf lobes uniseriate throughout except sometimes for basal 1–2 cells **6. Telaranea**
  Leaf lobes 2 cells wide, except for apical 1(–2) cells   **7. Kurzia**
3 Leaves 3–4-lobed to ¼–⅓, trigones not or hardly developed   **8. Lepidozia**
  Leaves with truncate shallowly 2–3-lobed or toothed apices or entire, trigones
      prominent, medium-sized to large   **9. Bazzania**

SUBFAMILY LEPIDOZIOIDEAE

Branching never apparently dichotomous. Leaves deeply 3–8-lobed.

6. TELARANEA SPRUCE EX SCHIFFN, *Trans. Proc. Bot. Soc. Edinburgh* 15: 358. 1885

Dioecious or rarely autoecious. Plants minute to robust, forming dense mats or lax patches. Stems isophyllous to anisophyllous, 1–2-pinnate, branches becoming attenuated or not, ventral flagelliform branches present or not, stem in section with distinct cortex of large thin-walled cells surrounding a central cylinder of small cells. Leaves distant to imbricate, symmetrical, ± transversely to obliquely inserted, 2–6-lobed to ½-way or more, lobes uniseriate except at base. Underleaves similar to or smaller than leaves, 2–4-lobed, 2–4 cells long. Male inflorescence on short branch or terminal on long branch, bracts concave, antheridial stalk uniseriate. Female inflorescence on short or long branch, bracts larger than leaves, shallowly to deeply 3–4-lobed. Perianth unistratose, narrowed to crenulate to long-ciliate mouth. Capsule wall 3–5-stratose. Spores 14–17 μm, finely areolate. A mainly tropical and southern hemisphere genus.

Leaf base one cell deep, cells only joined for about half their length
                                                              1. T. nematodes
Leaf base 2–4 cells deep   **2. T. murphyae**

1. T. nematodes (Gott. ex Aust.) Howe, *Bull. Torrey Bot. Club* 29: 284. 1902
(Fig. 6)
T. sejuncta auct. non (Angstr.) S. Arn.

Autoecious. Plants minute, forming small, pale green patches, glossy when dry or growing through other bryophytes. Shoots to 1.5 cm, procumbent, stems irregularly pinnate, branches not attenuated. Leaves distant, patent to spreading, ± transversely inserted, stem leaves divided almost to base into 2–3 lobes 160–360 μm long, leaf base one cell deep, cells joined for about half their length, lobes 4–6 cells long, cells in mid-lobe (35–)45–60(–80) μm long, base of lobe usually 2 cells wide. Underleaves much smaller than leaves, to 70 μm long, bilobed, lobes uniseriate, 2–3 cells long. Clavate to cylindrical or fusiform multicellular gemmae, 170–750(–1300) μm long, frequently present. Inflorescences on very short branches. Male bracts smaller than leaves, 3–4

Fig. **6**. 1–3, *Telaranea nematodes*: 1, shoot (× 22); 2, leaves (× 84); 3, female inflorescence (× 38). 4–9, *T. murphyae*: 4, shoot (× 22); 5, stem leaves (× 84); 6, branch leaf (× 84); 7, underleaves (× 84); 8, female inflorescence (× 60); 9, female bract (× 23); 10, male branch (× 30).

pairs, imbricate, concave with inflexed lobes. Female inflorescence strongly ciliate, bracts larger than leaves, 4-lobed to *ca* $\frac{1}{2}$, margin with long cilia. Perianth fusiform, contracted to ciliate mouth. Capsules rare, June. On moist peat and rotting wood in deeply shaded woodland, especially under *Rhododendron ponticum* and amongst rocks and in turf by the coast, very rare in W. Cornwall, occasional in the extreme west of Ireland, Kerry, W. Cork, W. Galway, W. Mayo. 1, H5. Portugal, Basque region of Spain and France, Azores, Madeira, Tenerife, Natal, Tristan da Cunha, south-east U.S.A., Andes, W. Indies.

Likely to be confused with *Blepharostoma trichophyllum* which differs in the underleaves being of similar size to the lateral leaves and the leaf lobes uniseriate throughout with more numerous, shorter cells. *Kurzia* species differ in the leaf lobes being 2 cells wide except for the apical 1–2 cells. European plants seem considerably smaller than those described from N. and S. America. Originally thought to be an introduction to S.W. Ireland, it now seems more likely that *T. nematodes* is a native species. For the discovery of this plant in the British Isles see Buch, *Ann. Bryol.* **9**, 32–3 (1938) and Richards, *Br. Bryol. Soc. Rep.* **4**, 61–2 (1938).

2. T. murphyae Paton, *Trans. Br. bryol. Soc.* 4: 776. 1965   (Fig. 6)

Dioecious. Yellowish-green, intricate patches. Shoots to 3 cm, procumbent, stems 1–2-pinnate, some branches becoming attenuated. Stem leaves distant to approximate, branch leaves approximate to imbricate, erecto-patent, straight to inflexed and hand-like, insertion slightly oblique, stem leaves 240–400 μm long, mostly 3–4-lobed to $\frac{2}{3}$–$\frac{3}{4}$, leaf base 2–4 cells deep, cells rectangular lobes 4–6 cells long, uniseriate or base of lobe 2 cells wide, cells in mid-lobe 35–55 μm long. Underleaves smaller than leaves, to 180 μm long, 2–4-lobed. Vegetative propagules lacking. Inflorescences on short branches. Male bracts smaller than leaves, imbricate, 3–6 pairs, concave with 2–3 incurved lobes. Female bracts larger than leaves, spinosely 3–4-lobed to $\frac{1}{2}$. Perianths and capsules unknown. $n = 9*$. Shaded, peaty or sandy soil and rotten wood in woodland or under *Rhododendron*, very rare, Tresco, Surrey. 2. Portugal (?).

Differs from *T. nematodes* in its generally larger size, and the leaf base 2–4 cells deep and the bud-like female inflorescence (those in *T. nematodes* being strongly ciliate). *Kurzia* species have a leaf base 1–2 cells deep and the lobes 2 cells wide except for the apical 1–2 cells. *T. murphyae* is known only from the Abbey Gardens, Tresco and the Royal Horticultural Society's Garden, Wisley and is clearly an introduction but its origin is unknown. For the occurrence of this species in England see Paton, *Trans. Br. bryol. Soc.* **4**, 775–7 (1965) and **6**, 228–9 (1971).

7. KURZIA v. MARTENS, *Flora* 53: 417. 1870

Dioecious. Plants minute to medium-sized, dull green to brownish: stems isophyllous or anisophyllous, regularly to irregularly 1–2-pinnate, sometimes producing flagelli-form branches, stem in section with cortical cells thick-walled, large, cells of central cylinder smaller, thinner-walled. Leaves spreading to incurved, transversely inserted or obliquely inserted and incubous (2–)3–4-lobed usually almost to base, lobes symmetrical or not, 2–5 cells wide at base; cell walls thickened, without trigones, oil-bodies usually lacking. Underleaves smaller than or similar to lateral leaves, 2–4-lobed, symmetrical or not. Sexual branches usually short, arising from ventral side of stem. Male bracts larger than leaves, 2–5-lobed to $\frac{1}{2}$–$\frac{2}{3}$, antheridial stalk uniseriate. Female bracts isophyllous, much larger than leaves, simple or lobed, margin entire, denticulate or ciliate. Perianth unistratose, contracted to mouth. Capsule wall 2–3-stratose. Spores 8–14 μm, papillose. About 30 mainly antipodal species.

The three British species of *Kurzia* produce leafless vertical shoots that penetrate the substrate to a depth of up to 20 cm. These may act as organs of perennation or survival after fire (see Pocock & Duckett, *New Phytol.* **99**, 281–304, 1985).

1 Bracts of female inflorescence with uniseriate cilia to 9 cells long; male bracts
   strongly incurved with apical cell of lobes obtuse, 16-32 μm wide
                                                                    1. K. pauciflora
   Female bracts with denticulations or cilia 1–3 cells long; male bracts strongly
   incurved or ± straight, apical cell of lobes acute, 8–16 μm wide   2
2 Female bracts 2–4-lobed to $\frac{1}{3}$–$\frac{1}{2}$ with denticulations and cilia 1–3 cells long; male
   bracts ± straight with lobes ± as long as lamina   **2. K. sylvatica**
   Female bracts 2(–3)-lobed to $\frac{1}{2}$(–$\frac{1}{3}$) with denticulations 1–2 cells long; male bracts
   strongly incurved with lobes ± twice as long as lamina   **3. K. trichoclados**

1. K. pauciflora (Dicks.) Grolle, *Rev. Bryol. Lichén.* 32: 171. 1963   (Fig. 7)
Lepidozia setacea auct., *Microlepidozia setacea* auct., *Telaranea setacea* auct.

Plants minute, dull, dark green, forming small densely interwoven patches or growing
through other bryophytes. Shoots to 2(–3) cm, procumbent to ascending, stems
irregularly 1–2-pinnate, sometimes with small-leaved flagelliform branches, and
vertical leafless branches penetrating substrate. Stem leaves distant except towards
apex, from transversely inserted spreading base erect or incurved and hand-like, 150–
250(–320) μm long, (3–)4-lobed to within 1–2 cells of base, lobes 2(–4) cells wide at
bases, ending in 1(–2) uniseriate cells, branch leaves smaller, approximate to imbricate;
cells thick-walled, in middle of lobes 20–36 μm long, 13–21 μm wide. Underleaves
smaller than leaves. Vegetative propagules lacking. Male inflorescence on short branch
from ventral side of older part of stem or occasionally at end of flagelliform branch,
bracts larger than leaves, strongly incurved and concave, lamina 4–6 cells deep, 2–3-
lobed, lobes ± as long as lamina, entire or toothed, apical cells of lobes obtuse, 32–52
μm long, 16–32 μm wide, 1.6–2.0(–2.8) times as long as wide. Female inflorescence on
short branch from ventral side of older part of stem, bracts much larger than leaves,
outer simple with ciliate margins, inner 3–4-lobed to $\frac{1}{3}$–$\frac{1}{2}$, margins with uniseriate cilia
to 9 cells long. Perianth narrowly ellipsoid, mouth with incurved cilia. Capsules
occasional, late summer to spring. On peat on wet heaths, moorland and in bogs,
growing through *Sphagnum* and *Leucobryum*, rarely on soil, calcifuge, absent from C.
England, rare to frequent elsewhere. 84, H34, C. Europe from Portugal, N. Spain, N.
Italy and Carpathians north to Fennoscandia and Russia, Faroes, Iceland, Azores,
Madeira, N. America.
   Confused with the next two species especially as there has been a tendency to name small
Lepidozias (i.e. *Kurzia* species) *L. setacea* (i.e. *K. pauciflora*) indiscriminately on the assumption
that the other two species are very rare; they are less frequent than *K. pauciflora* but sufficiently
common to make microscopic examination essential. The only reliable characters for discriminating the three British species are those given in the key. Inflorescences occur on the older parts of the
stems and, whilst they are usually present, careful dissection is necessary to locate them. Two
species may occur together.

2. K. sylvatica (Evans) Grolle, *Herzogia* 3: 77. 1973   (Fig. 8)
Lepidozia sylvatica Evans, *Microlepidozia sylvatica* (Evans) Joerg., *Telaranea sylvatica*
(Evans) K. Müll.

Vegetatively similar to *K. pauciflora*. Vegetative propagules unknown. Male bracts
from spreading base ± erect, slightly concave, lamina 4–6 cells deep, 2(–3)-lobed, lobes
1–2 times as long as lamina, toothed, apical cells of lobes acute, 30–44(–52) μm long, 9–
13 μm wide, 3.0–4.0(–5.4) times as long as wide. Female inflorescence with denticulate
outer bracts, inner bracts 2–4-lobed to $\frac{1}{3}$–$\frac{1}{2}$ with denticulations or cilia 1–3 cells long,
occasionally entire. Perianth narrowly ellipsoid, mouth with ± erect cilia 56–80 μm
long. Capsules very rare, autumn. On peat on wet heath and moorland and on

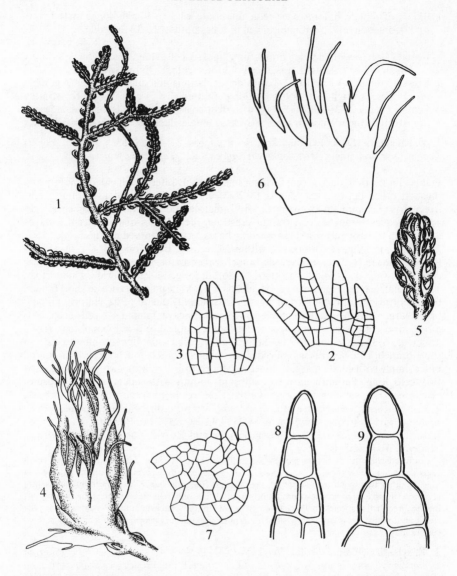

Fig. 7. *Kurzia pauciflora*: 1, shoot (× 15); 2, leaf (× 63); 3, underleaf (× 63); 4, female inflorescence (× 38); 5, male shoot (× 38); 6, female bract (× 63); 7, male bract (× 153); 8, leaf lobe tip (× 450); 9, male bract lobe tip (× 450).

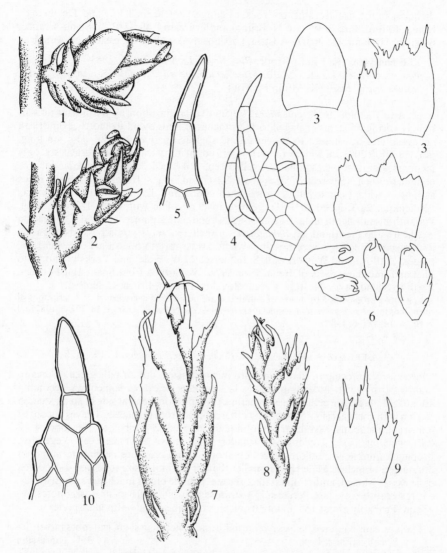

Fig. **8**. 1–5, *Kurzia trichoclados*: 1, female inflorescence (× 63); 2, male branch (× 45); 3, female bracts from different plants (× 63); 4, male bract (× 153); 5, apex of male bract lobe (× 450); 6, bulbils (× 84). 7–10, *K. sylvatica*: 7, female inflorescene (× 63); 8, male branch in side view (× 84); 9, female bract (× 63); 10, male bract lobe tip (× 450).

sheltered, moist, usually sandstone rock, calcifuge, very rare in S. England, Dorset to E. Kent, Stafford, rare in W. and N. Britain and in Ireland. 40, H10, C. Spain, France, Belgium, Holland, Denmark, Austria, Czechoslovakia, Poland, eastern N. America.

3. K. trichoclados (K. Müll.) Grolle, *Rev. Bryol. Lichén.* 32: 171. 1963   (Fig. 8)
Lepidozia trichoclados K. Müll, *Microlepidozia trichoclados* (K. Müll.) Joerg, *Telaranea trichoclados* (K. Müll.) K. Müll.

Vegetatively similar to *K. pauciflora*. Vegetative propagation by frequently present bulbils arising on stems or flagelliform branches, bulbils ovoid to shortly cylindrical, yellowish and translucent to brownish and slightly opaque, 95–450(–600) μm long, (40–)60–170(–200) μm wide, with 3 rows of rudimentary leaves. Male bracts strongly incurved and concave, lamina 2–4 cells deep, 2–4-lobed, lobes to twice as long as lamina, toothed, apical cell of lobes acute, 32–64 μm long, 8–16 μm wide, 3.5–5.0 times as long as wide. Female inflorescence when young bud-like, not ciliate, inner bracts denticulate, 2(–3)-lobed to ⅕(–⅓), ±entire to denticulate with teeth 1–2 cells long. Perianth mouth denticulate. Capsules occasional, late summer, autumn. On peat on wet heath and moorland, on *Sphagnum*, on sheltered, moist, usually sandstone rock and in scree, calcifuge, very rare in lowland areas, rare to occasional elsewhere, W. Sussex, Stafford, and W. Cornwall, S. Somerset, N.W. Wales and Yorkshire north to Shetland, montane parts of Ireland. 40, H21. W. Europe from Spain and Portugal north to Scandinavia, N. Italy, Czechoslovakia, Poland, British Columbia.

Vegetative propagation by means of bulbils is extremely rare in liverworts. For descriptions of the bulbils in *Telaranea nematodes* and *Kurzia trichoclados* see Paton, *J. Bryol.* **14**, 792–3, 1987 and *J. Bryol.* **14**, 181–6, 1986.

8. LEPIDOZIA (DUM.) DUM. *Recueil Observ. Jungerm.*: 19. 1835

Dioecious or autoecious. Plants small to medium-sized, dull or pale green or cream. Shoots prostrate to ascending, stems 1–2-pinnate, branches sometimes becoming attenuated, flagelliform branches sometimes arising from ventral side of stem, stem in section with cortical cells not or hardly differentiated from inner cells. Leaves distant to imbricate, incubous, asymmetrical, obliquely inserted, 2–5-lobed to ¼–¾, lobes 4–12 cells wide at base. Underleaves smaller than leaves, ±symmetrical. Vegetative propagules unknown. Inflorescences on short ventral branches or male terminal on attenuated branches. Male bracts smaller than or of similar size to leaves, 4–10 pairs, concave, 2–3-lobed, antheridial stalk 2-seriate. Female bracts much larger than leaves, entire or toothed at apex. Perianth 2–4-stratose. Capsule wall 3–5-stratose. Spores 11–15 μm. Probably about 150 mainly tropical, subtropical and antipodal species.

1 Plants cream-coloured, leaves overlapping and concealing stem, leaf lobes mostly 7–
    12 cells wide at base   **3. L. cupressina**
   Plants dull or pale green, leaves not concealing stem, leaf lobes 4–7 cells wide at base
                                                                                            2

2 Autoecious, stem leaves 0.9–1.3 times as long as wide, mostly 4-lobed **1. L reptans**
   Dioecious, stem leaves 1.2–1.7 times as long as wide, 3-lobed   **2. L. pearsonii**

1. L. reptans (L.) Dum., *Recueil Observ. Jungerm.*: 19. 1835   (Fig. 9)

Autoecious. Plants small, dull dark green, forming dense patches or growing through other bryophytes. Shoots to 3 cm, procumbent, stems pinnately branched, occasionally slightly bipinnate, branches sometimes becoming attenuated, small-leaved flagelliform branches often arising from ventral sides of older parts of stem. Stem leaves distant to

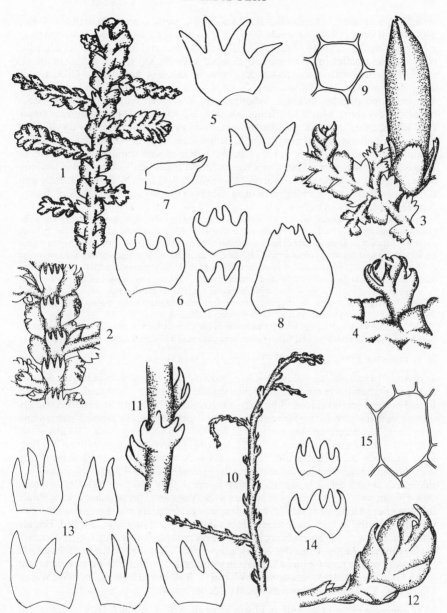

Fig. **9**. 1–9, *Lepidozia reptans*: 1, shoot, dorsal side (× 8); 2, portion of shoot, ventral side (× 27); 3, shoot with perianth and male branch (× 11); 4, male branch (× 27); 5, leaves (× 27); 6, underleaves (× 45); 7, male bract (× 45); 8, female bract (× 27); 9, mid-leaf cell (× 450). 10–15, *L. pearsonii*: 10, shoot, dorsal side (× 8); 11, portion of shoot, ventral side (× 38); 12, male branch (× 38); 13, leaves (× 45); 14, underleaves (× 45); 15, mid-leaf cell (× 450).

imbricate, incubous, ± spreading, not overlapping stem, convex, insertion oblique, 240–560 μm long, 180–480 μm wide, 0.9–1.3(–1.5) times as long as wide, 3–4 lobed to ¼–⅓, lobes triangular, acute to obtuse, 4–7 cells wide at base, ending in 1–2 uniseriate cells, branch leaves smaller, imbricate, mostly 3-lobed; cells 20–32(–40) μm wide in mid-leaf, walls thickened with rounded angles, cells in lamina ± hexagonal. Underleaves somewhat concave, 200–400 μm long, 260–430 μm wide, 0.7–1.0 times as long as wide. Vegetative propagules lacking. Inflorescences on short ventral branches. Male branches very short, bracts imbricate, concave, 2–3-lobed to ⅓. Female bracts larger than leaves, ovate, concave, toothed at apex. Perianths occasional, narrowly ellipsoid or fusiform, plicate above, mouth small, denticulate. Capsules occasional, winter to summer. n = 9, 10. On soil, peat, rocks, banks, tree boles and stumps, rotting wood, in woods, on heaths and moorland, in scree, calcifuge, occasional to frequent in lowland areas, frequent to common elsewhere, 112, H37, C. Portugal, N. Spain, N. Italy and Greece, north to Fennoscandia and Russia, Siberia, Himalayas, China, Taiwan, Japan, Azores, Madeira, N. America.

Only likely to be confused with *L. pearsonii* and the two species may occur together. The pale, yellowish-green colour and very distant, more deeply 2–3-lobed instead of 4-lobed leaves with narrower lobes of *L. pearsonii* are usually distinctive but some attenuated forms of *L. reptans* may be impossible to determine unless fertile. Unlike *L. reptans* the male inflorescences of *L. pearsonii* are borne at the ends of attenuated branches although male inflorescences on dwarf shoots also occur. In *L. reptans* flagelliform shoots frequently arise from the ventral side of older stems but these are rarely present in *L. pearsonii*; in both species the lateral branches may become attenuated and may be confused with the flagelliform ventral branches. In both species attenuated or flagelliform shoots may bear rhizoids with swollen ends.

Although vegetative propagules are unknown in the three British *Lepidozia* species it is likely that deciduous attenuations and flagelliform branches may be reproductive in function.

2. L. pearsonii Spruce, *J. Bot. Br. Foreign* 19: 34. 1881   (Fig. 9)

Dioecious. Plants pale or yellowish-green, growing through other bryophytes. Shoots to 6 cm, procumbent, pinnate, branches usually becoming attenuated, flagelliform branches from ventral side of older parts of stems usually absent. Stem leaves usually remote, occasionally distant, spreading to erecto-patent, not overlapping stem, plane to convex, asymmetrical, insertion oblique, 320–590 μm long, 180–480 μm wide, (1.0–)1.2–1.7 times as long as wide (2–)3-lobed to ½–⅔, rarely an occasional leaf 4-lobed, lobes narrowly triangular, acute, (2–)4–7 cells wide at base, ending in 1–2 uniseriate cells, branch leaves smaller, distant, mostly 3-lobed; cells 20–40 μm wide, walls not thickened, lamina cells ± hexagonal. Underleaves slightly concave, 200–400 μm long, 260–430 μm wide, 0.5–0.9 times as long as wide. Vegetative propagules lacking. Male inflorescence on long attenuated branch, becoming intercalary with age, more rarely on very short ventral branch, male bracts closely imbricate, very concave, 2-lobed. Female inflorescence on short ventral branch, female bracts larger than leaves, ovate, concave, deeply toothed at apex. Capsules very rare. n = 8 + m*. Amongst other bryophytes, on humus and peat in moist, humid habitats in woodland, by waterfalls, on moorland and north-facing rock ledges, occasional in W. and N. Britain and Ireland, from W. Wales, Derby and Yorkshire northwards. 46, H12. S.W. Norway.

3. L. cupressina (Sw.) Lindenb. in Gottsche *et al.*, *Syn. Hep.*: 207. 1845   (Fig. 10)
L. pinnata (Hook.) Dum.

Dioecious. Plants cream, forming mats and cushions to large swelling tufts, rarely scattered shoots growing through other bryophytes. Shoots to 3(–4) cm, procumbent, rarely ascending, stems closely pinnately branched, occasionally ± bipinnate, branches sometimes becoming attenuated, flagelliform branches occasionally rising from ventral

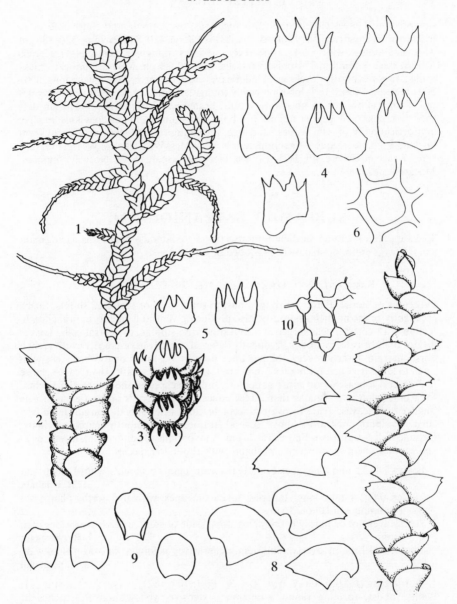

Fig. 10. 1–6, *Lepidozia cupressina*: 1, shoot (× 8); 2, dorsal side of shoot (× 21); 3, ventral side of shoot (× 21); 4, leaves (× 21); 5, underleaves (× 21); 6, mid-leaf cells (× 450). 7–10, *Bazzania pearsonii*: 7, shoot (× 8); 8, leaves (× 23); 9, underleaves (× 23); 10, mid-leaf cell (× 450).

side of stem. Leaves closely imbricate, incubous, overlapping and concealing stem, markedly asymmetrical, very convex, insertion oblique, 480–840 μm long, 520–880 μm wide, ± as long as wide, (2–)3–4-lobed to $\frac{1}{4}$–$\frac{1}{3}$, lobes triangular, acute, (4–)7–12 cells wide at base, ending in 1–4 uniseriate cells; cells 20–28 μm wide in mid-leaf, thick-walled, trigones hardly developed. Underleaves erect, concave, 280–480 μm long, 300–560 μm wide, 4-lobed to $\frac{1}{3}$, lobes narrowly triangular, concave. Vegetative propagules unknown. Inflorescences on dwarf ventral branches. Capsules unknown. On dry, sheltered rocks, soil, peat or humus, tree boles and stumps and logs in woods, in scree and occasionally on cliff ledges, calcifuge, ascending to 300(–400) m. Absent from lowland England, frequent or common in N.W. Wales, W. Scotland, W. Ireland, very rare or rare elsewhere, 40, H21. France, Germany, Spain, S.W. Norway, Yunnan, Macaronesia, Jamaica.

SUBFAMILY BAZZANIOIDEAE

Stems apparently branching dichotomously (vigorous lateral branches displacing main axis). Leaves entire or shallowly 2–3-lobed or toothed.

9. BAZZANIA S.F. GRAY, *Nat. Arr. Br. Pl.* 1: 704, 775. 1821

Dioecious or sterile. Plants small to robust in tufts, mats or scattered shoots. Stems apparently brancing dichotomously, also producing ventral flagelliform branches, in section with cortical cells larger than but otherwise similar, to central cells. Leaves distant to imbricate, incubous, frequently deflexed, usually asymmetrical with arched dorsal margin, insertion very oblique, apex usually 2–3-lobed or toothed; cells with small to large trigones, oil-bodies 2–6 per cell. Underleaves smaller than leaves, entire to 2–several-lobed, margin entire or dentate. Inflorescences on short ventral branches. Male bracts 4–6 pairs, smaller than leaves, concave, 2-lobed or toothed. Female bracts smaller than leaves, ovate to ovate-lanceolate, apex often lobed, margin dentate to ciliate. Perianth unistratose above, several-stratose below, mouth dentate to ciliate. Capsule wall 4–6-stratose. Spores 15–20 μm. A taxonomically difficult, mainly tropical and southern hemisphere genus, probably with about 250 species.

1 Leaves 1.1–2.5 mm long, apex broadly truncate, plants becoming whitish on drying
                                                                1. B. trilobata
  Leaves 0.6–1.4 mm long, tapering to ± acute apex with 0–3 teeth, plants not
    becoming whitish on drying   2
2 Stem leaves not or hardly overlapping dorsal side of stem, underleaves wider than
    long   **2. B. tricrenata**
  Stem leaves broadly overlapping stem, underleaves as long as or longer than wide
                                                                3. B. pearsonii

1. B. trilobata (L.) S.F. Gray, *Nat. Arr. Br. Pl.* 1: 704. 1821   (Fig. 11)
Plants robust, forming tumid, sometimes extensive, yellowish-green patches or growing through other bryophytes, becoming whitish when dry. Shoots flattened or convex, 3–10 cm, procumbent to erect with slightly circinate apex, stems apparently pinnately branched, producing long flagelliform shoots from ventral side. Leaves imbricate, overlapping and concealing stem, spreading ± horizontally, plane or convex, very asymmetrically oblong-ovate, very obliquely inserted, 1.1–2.5 mm long, 1–2 mm wide, apex broadly truncate, shallowly 3-lobed, lobes acute; cells (20–)28–32(–44) μm wide in mid-leaf, trigones large. Underleaves ± erect, usually wider than long,

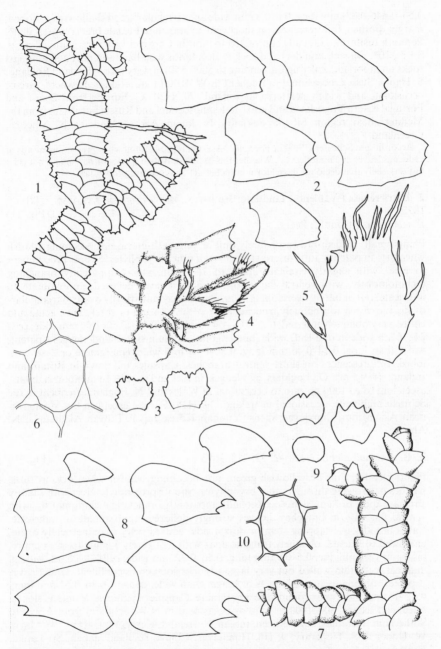

Fig. **11**. 1–6, *Bazzania trilobata*: 1, shoot (× 5); 2, leaves (× 13); 3, underleaves (× 13); 4, female shoot with perianth (× 13); 5, female bract (× 23); 6, mid-leaf cell (× 450). 7–10, *B. tricrenata*: 7, shoot (× 8); 8, leaves (× 23); 9, underleaves (× 23); 10, mid-leaf cell (× 450).

0.50–0.64(–0.88) mm long, 0.6–0.8 mm wide apex irregularly and shallowly 4–5-lobed, margin toothed. Inflorescences on short ventral branches. Female bracts ovate, ciliate. Perianth fusiform, gradually contracted to mouth. Capsules very rare, summer. $n = 9$, $9 + m$, 10^*. On soil, boulders and rock in deciduous woodland, on peat and amongst rocks on moorland, calcifuge, ascending to 500(–850) m, very rare in lowland England, S. Hants, Sussex, occasional to frequent in W. Britain, occasional elsewhere, rare or occasional and widely scattered in Ireland. 70, H20, C. Europe from Spain and Portugal east to Yugoslavia and north to Fennoscandia and Russia but absent from the Mediterranean region, Siberia, Sakhalin, N. Japan, Azores, Madeira, N. America, Greenland.

According to Schuster (1969) in *Bazzania*, 'Asexual reproduction, when present, by means of caducous leaves and underleaves'. Whether this occurs in British species of *Bazzania* is not clear but it is likely that decidous flagelliform branches act as vegetative propagules.

2. B. tricrenata (Wahlenb.) Lindb. in Brotherus, *Musci Fennicae Exs.* fasc. 2: [2]. 1872   (Fig. 11)
B. triangularis (Lindb.) Pears.

Plants small, reddish-brown or rarely dull or yellowish-green, not becoming whitish when dry, in patches, tufts or growing through other bryophytes. Shoots procumbent to erect with slightly circinate apex, to 7(–9) cm, stems apparently branching dichotomously, with ventral flagelliform shoots. Leaves deflexed, approximate to imbricate, not or barely overlapping stem, convex, asymmetrically ovate to triangular-ovate, narrowed to obliquely truncate apex with 2–3 teeth or rarely apex acute and entire, very obliquely inserted, (0.5–)0.6–1.4 mm long, (0.35–)0.50–1.00 mm wide; cells 24–36 μm wide in mid-leaf, walls thin, trigones medium-sized. Underleaves ± patent, wider than long, 0.24–0.40 mm long, 0.36–0.64 mm wide, emarginate or 2–several-lobed. Inflorescences on short ventral branches. Capsules unknown in Britain and Ireland. $n = 9 + m^*$. On boulders, cliff ledges, in turf and in scree in montane habitats, ascending to *ca* 1000 m, rare to occasional in Wales and N. England, frequent in the Scottish highlands, occasional in Ireland. 39, H18. Europe except in the Mediterranean, Asia Minor, Himalayas, Shensi, Yunnan, Korea, Japan, Taiwan, Aleutian Is., N. America.

3. B. pearsonii Steph, *Hedwigia* 32: 212. 1893   (Fig. 10)

Plants slender, dull or yellowish-green, not becoming whitish when dry, in tufts, patches or growing through other bryophytes. Shoots prostrate to erect with slightly circinate apex, to 7 cm, stems occasionally apparently branching dichotomously, with ventral flagelliform branches. Leaves strongly deflexed, approximate to imbricate, leaves broadly overlapping stem or dorsal side, convex, very asymmetrically ovate, narrowed to ± acute, obliquely truncate apex with 2 or rarely 3 acute teeth or entire, insertion very oblique, 0.6–0.9 mm long, 0.50–0.72 mm wide; cells 20–28 μm wide in mid-leaf, cells thin-walled but very large trigones sometimes confluent. Underleaves ± patent, oblong-ovate, as long as or longer than wide, convex, 0.36–0.52 mm long, 0.32–40 mm wide, apex rounded to emarginate. Capsules unknown. Montane slopes, scree, cliff ledges, 300–1000 m, rare or occasional in N.W. Scotland from Argyll to Sutherland, Hebrides, S. Aberdeen, rare in W. Ireland, Kerry, W. Galway, W. Mayo, W. Donegal, S. Tipperary. 9, H6. Himalayas, China, Thailand, Japan, Sri Lanka, Nova Scotia.

May be mistaken for a small form of *B. tricenata* but differing in colour, the stem leaves (but not branch leaves) widely overlapping the stem, the leaf apex usually with only 2 teeth and the underleaves as long as or longer than wide.

7. CALYPOGEIACEAE

Plants small to medium-sized. Shoots prostrate, sparsely branched, branches ventral or apical; stem lacking well defined cortex. Leaves incubous, very obliquely inserted, entire or bidentate; cells large, thin-walled. Underleaves large, entire to deeply bilobed, rhizoids arising from base. Inflorescences on dwarf ventral branches, bracts much smaller than leaves. Perianth lacking. Archegonia becoming situated at bottom of a marsupium. Capsule ovoid or cylindrical, wall bistratose. Two genera, *Calypogeia* and *Metacalypogeia* (N. E. Asia and northern N. America).

10. CALYPOGEIA RADDI CORR. CORDA IN OPIZ, *Naturalientausch*: 653.
1829

Plants small to medium-sized, translucent, pale green, greyish-green or bluish. Shoots prostrate to ascending, sparsely branched. Leaves incubous, distant to imbricate, spreading horizontally, plane to slightly convex, ovate to broadly ovate or triangular, apex rounded to acute or bidentate, very obliquely inserted; cells large, thin-walled, with or without small trigones, oil-bodies compound, colourless or blue, 2–13 per cell. Underleaves large, wider than stem, rounded, entire to deeply bilobed, toothed or not, rhizoids arising from base. Gemmae spherical to ellipsoid, 1–2-celled, very thin-walled, produced in fascicles at apices of usually attenuated shoots. Inflorescences on dwarf ventral branches. Male bracts 3–5 pairs, very small. Female bracts 2–3 pairs, scale-like, persisting or not round mouth of marsupium. Perianth lacking. Archegonia coming to lie at bottom of large, fleshy, rhizoid-covered subterranean marsupium. Capsule cylindrical, wall 2 cells thick. Spores 8–17 μm. A ± cosmopolitan genus of about 90 species.

The species of *Calypogeia* exhibit great phenotypic plasticity, the precise basis of which is unknown. *C. muelleriana*, *C. neesiana*, *C. integristipula* and *C. sphagnicola* form a complex the taxonomy of which is not understood and *C. fissa* integrades with *C. muelleriana*. Specimens may sometimes only be determined by the use of a combination of characters, if then. Although all dimensions given may overlap, opposite extremes may be useful in providing distinctions but juvenile plants or parts of plants often have different shaped leaves with much smaller cells than mature plants. The following key is only a guide and careful comparison with descriptions should be made.

For an account of British *Calypogeia* spp. see Paton, *Trans. Br. bryol. Soc.* **4**, 221–9, 1962.

1 Leaves bilobed with divergent lobes and rounded sinus, underleaves with subulate
   lobes and teeth   **8. C. arguta**
   Leaves entire or if bilobed then lobes not divergent and sinus narrow   2
2 Underleaves bilobed $\frac{1}{2}$–$\frac{3}{4}$, 1–4 cells deep from sinus to rhizoid area   3
   Underleaves entire, emarginate or bilobed to $\frac{1}{3}$, 4–14 cells deep   6
3 Cells in middle of underleaf 40–80 μm long   4
   Cells in middle of underleaf 30–40(–50) μm long   5
4 Oil-bodies colourless, leaf apex bidentate or ± acute, leaves widest *ca* $\frac{1}{3}$ from base
   except on attenuated shoots   **1. C. fissa**
   Oil-bodies deep blue, leaf apex obtuse to rounded, leaves mostly widest near base
                                                                    7. C. azurea
5 Leaves distant to contiguous, mostly widest at base, apex acute to obtuse or bilobed
                                                                    5. C. sphagnicola
   Leaves imbricate, widest $\frac{1}{3}$–$\frac{1}{2}$ from base, apex rounded, truncate or shallowly
   emarginate   **6. C. suecica**
6 Leaves mostly *ca* as wide as long, underleaves 4–6 cells deep along midline, these cells
   36–80 μm long   **2. C. muelleriana**

Leaves 1.0–1.3 times as long as wide, underleaves 7–14 cells deep, cells 20–50 μm long
7

7 At least some leaves with tangentially elongated marginal cells, oil-bodies present
only in 1–3 marginal rows of leaf cells   **3. C. neesiana**
Marginal leaf cells not tangentially elongated, oil-bodies present in ± all leaf cells
                                                                            4. C. integristipula

1. C fissa (L.) Raddi, *Jungerm. Etrusca* 23. 1818   (Fig. 12)

Autoecious or paroecious. Plants slender to medium-sized, pale green, yellowish-green
or brownish-green patches or scattered shoots. Shoots prostrate, to 2(–4) cm long, 1.5–
3.0 mm wide, sparsely branched, gemmiferous shoots ascending, attenuated. Leaves
distant to imbricate, slightly convex, obliquely ovate, decurrent on ventral side, mostly
widest *ca* ⅓ from base, narrowed to bilobed or acute or subacute apex, where apex
bilobed lobes sometimes juxtaposed, 0.8–1.5 mm long, 0.6–1.2(–1.4) mm wide, (1.0–)
1.1–1.4 times as long as wide; marginal cells not tangentially elongated, marginal cells
at apex 20–40 μm wide, mid-leaf cells 40–60(–80) μm long, 30–50(–60) μm wide, usually
thin-walled, trigones lacking or very small, oil-bodies colourless. Underleaves erecto-
patent, bilobed ⅔–¾, 1.5–2.5 times as wide as stem, 0.2–0.3(–0.4) mm long, 0.3–0.5(–0.6)
mm wide, 1.3–1.9 times as wide as long, lobes acute to obtuse, outer margins angular or
with a knob or a tooth, rarely rounded, decurrent, 1–3 cells deep from sinus to rhizoid
area, these cells 40–70(–80) μm long. Gemmae spherical to ellipsoid, 1–2 celled, 16–32
μm. Capsules occasional, late winter, spring. $n = 18$. Damp soil, peat and amongst
Sphagnum, in woods, on banks, heaths, moorland, on sandstone, in scree, by streams,
ditches, calcifuge, ascending to *ca* 700 m, frequent to common except in basic areas,
112, H40, C. Europe north to S. Fennoscandia and S. W. Russia, Turkey, eastern
Himalayas, Morocco, Tunisia, Macaronesia, eastern N. America.

An exceedingly variable species, intergrading with *C. muelleriana*, *C. neesiana*, *C. integristipula*,
C. sphagnicola and *C. azurea*. Differences are indicated under these species but are not absolute
and it may only be possible to determine material using a combination of characters, if then.
Specimens of *C. fissa* may be encountered in which underleaves of *C. fissa* and *C. muelleriana* types
occur on the same stem, in which case the shape of the lateral leaves is the determining character.

2. C. muelleriana (Schiffn.) K. Müll., *Beihefte Bot. Centralbl.* 10: 217. 1901 (Fig. 12)

Autoecious or paroecious. Plants medium-sized, in dark yellowish-green or brownish-
green patches or scattered shoots. Shoots prostrate, to 2(–4) cm long, (1.0–)1.5–3.0(–
4.0) mm wide, sparsely branched, gemmiferous shoots ascending, attenuated. Leaves
usually imbricate, slightly convex, broadly ovate, widest 0–⅓ from base, decurrent on
ventral side, apex rounded, rarely bidentate in immature or etiolated shoots, (0.6–)0.8–
1.5 mm long, (0.7–)0.8–1.4 mm wide, (0.9–)1.0(–1.2) times as long as wide; marginal
cells not tangentially elongated, marginal cells at apex (24–)30–50 μm, wide, mid-leaf
cells 40–70 μm long, 30–50 μm wide, walls usually thin, trigones rarely present, oil-
bodies colourless or rarely pale blue. Underleaves ± patent, emarginate to bilobed to ⅓,
(1.7–)2.0–3.0 times as wide as stem, 0.6–1.3 mm long, 0.8–2.0 mm wide, (1.0–)1.3–1.7
(–1.9) times as wide as long, lobes rounded, outer margins rounded, 4–6 cells deep to
rhizoid area at midline, these cells 36–80 μm long. Gemmae spherical to ellipsoid, 1–2
celled, 20–45 μm. Capsules very rare. $n = 16 + 2m^*$, 18. Moist soil and peat in woods, on
heaths, banks, sandstone, on *Sphagnum*, *Leucobryum*, decaying wood, tussocks in
marshes, calcifuge, rare to occasional in the south and east, frequent to common
elsewhere. 105, H39, C. Europe, mainly N., W. and C., Faroes, Iceland, Macaronesia,
N. America, Greenland.

May be distinguished from typical forms of *C. fissa* by the relatively wider leaves with the larger
apical marginal cells and underleaf shape and number of cells along midline. However,

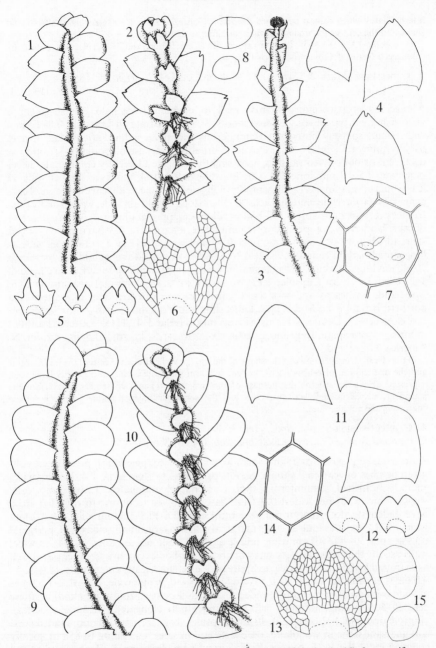

Fig. **12**. 1–8, *Calypogeia fissa*: 1 and 2, dorsal and ventral sides of shoots (× 15); 3, gemmiferous shoot (× 15); 4, leaves (× 21); 5, underleaves (× 21); 6, underleaf (× 60); 7, mid-leaf cell (× 450); 8, gemmae (× 450). 9–15, *C. muelleriana*: 9 and 10, shoots from dorsal and ventral side, (× 15); 11, leaves (× 21); 12, underleaves (× 21); 13, underleaf (× 60), 14, mid-leaf cell (× 330); 15, gemmae (× 450).

intermediates which cannot be named do occur. *C. neesiana* and *C. integristipula* differ in leaf shape and the cells of the underleaf but again intermediates occur.

C. submersa (Arn.) C. Müll. is synonymous with *C. muelleriana* and British plants so named are submerged forms of *C. muelleriana* or *C. fissa*.

3. C. neesiana (Mass. & Carest.) K. Müll. in Loeske, *Hedwigia* 47: 165. 1908.

(Fig. 13)

Autoecious or paroecious. Plants small to medium-sized, pale greyish-green or yellowish-green patches or scattered shoots. Shoots prostrate, to 4 cm long, 0.8–2.5 mm wide, stems sparsely branched, gemmiferous shoots ascending, attenuated. Leaves usually imbricate, slightly convex, ovate to broadly ovate, usually widest ⅓–½ from base, decurrent or not on ventral side, apex slightly truncate to slightly emarginate, never bidentate, 0.5–1.1 mm long, 0.4–1.0 mm wide, (0.9–)1.0–1.3 times as long as wide; at least some of marginal cells tangentially elongated in some or all leaves, marginal cells at apex 34–60 μm wide, mid-leaf cells 30–50 μm long, 30–40 μm wide, walls usually thin, trigones very small or absent, oil-bodies colourless, present only in 1–3 marginal rows of cells. Underleaves appressed, often imbricate, emarginate, 2.3–3.0 times as wide as stem, (0.25–)0.30–0.50 mm long, 0.35–0.70 mm wide, 1.0–1.3(–1.4) times as wide as long, outer margins rounded, 7–12(–17) cells deep to rhizoid area at midline, these cells 20–40 μm long; marginal cells tangentially elongated. Gemmae spherical to ellipsoid, 1–2-celled, 16–26 μm. Capsules very rare. $n = 8 + m* 9, 18$. Damp soil, peat and rocks in woodland, on heaths and moorland, calcifuge, rare to occasional in western and northern Britain from Mid Wales, Lancashire and Yorkshire north to Shetland, very rare elsewhere, S. Devon, S. Hants, Surrey, rare in Ireland. 41, H15. Scattered localities in Europe from Spain, Italy and Yugoslavia northwards, Japan, Sakhalin, Azores, N. America, Greenland.

Differs from *C. integristipula* in the shape of the leaf apex, the elongated cells of the leaf margins and the distribution of oil-bodies in the leaf cells. Intermediates occur and doubts have been expressed about the specific distinctness of the two taxa. These two species differ from the preceding species in underleaf characters and the somewhat smaller mid-leaf cells but these characters are not absolute.

4. C. integristipula Steph., *Spec. Hep.* 3: 394. 1908.   (Fig. 13)
C. meylanii Buch., *C. neesiana* var. *meylanii* (Buch) Schust.

Autoecious or paroecious. Plants small to medium-sized, pale bluish-green to greyish-green patches or scattered shoots. Shoots prostrate, to 3 cm long, 1.4–3.2 mm wide, sparsely branched, gemmiferous shoots ascending, attenuated. Leaves imbricate, slightly convex, ovate, widest 0–⅓ from base, narrowed to obtuse to rounded apex, never bidentate, decurrent or not on ventral side, (0.6–)0.8–1.4 mm long, 0.6–1.2 mm wide, 1.0–1.3 times as long as wide; marginal cells not tangentially elongated, marginal cells at apex 30–40(–50) μm wide, mid-leaf cells 32–50(–60) μm long, 24–44 μm wide, walls usually thin, trigones very small or absent, oil-bodies colourless, in central as well as marginal leaf cells. Underleaves appressed, often imbricate, emarginate, (2.0–)2.5–4.0 times as wide as stem, 0.36–0.54 mm long, 0.44–0.76 mm wide, (1.0–)1.2–1.6 times as wide as long., outer margins rounded, 7–14 cells deep to rhizoid area at midline, these cells 30–50 μm long, marginal cells not tangentially elongated. Gemmae spherical, ellipsoid or ovoid, 1–2-celled, 20–40 μm. Capsules very rare. $n = 9$. Damp, shaded peat, soil and sandstone in woodland, on coasts and in scree, ascending to 850 m, locally common in the Weald (E. Sussex, Kent, Surrey), rare elsewhere, S. Hants, Stafford and Cheshire and Yorkshire north to Ross, very rare in Ireland, N. Kerry, Fermanagh, Antrim. 25, H3. Spain, France and C. Europe north to Fennoscandia and Russia, Japan, Azores, Madeira, N. America, Greenland.

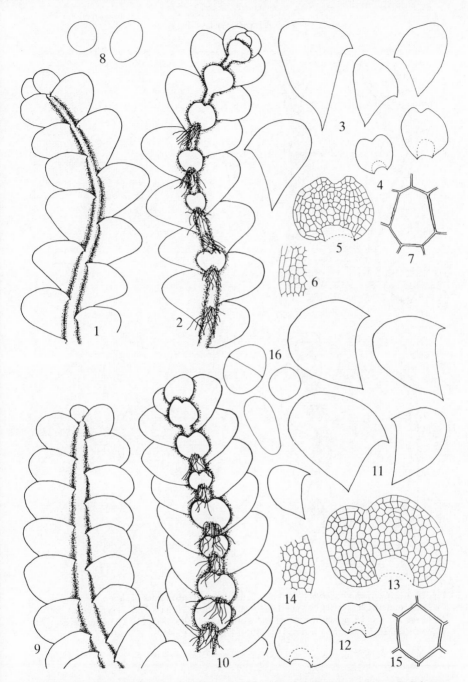

Fig. **13**. 1–8, *Calypogeia neesiana*: 1 and 2, dorsal and ventral sides of shoots (× 15); 3, leaves (× 21); 4, underleaves (× 21); 5, underleaf (× 60); 6, marginal cells of leaf (× 60); 7, mid-leaf cell (× 330); 8, gemmae (× 450). 9–16, *C. integristipula*: 9 and 10, dorsal and ventral sides of shoots; 11, leaves (× 21); 12, underleaves (× 21); 13, underleaf (× 60); 14, marginal cells of lateral leaves (× 60); 15, leaf cell (× 330); 16, gemmae (× 450).

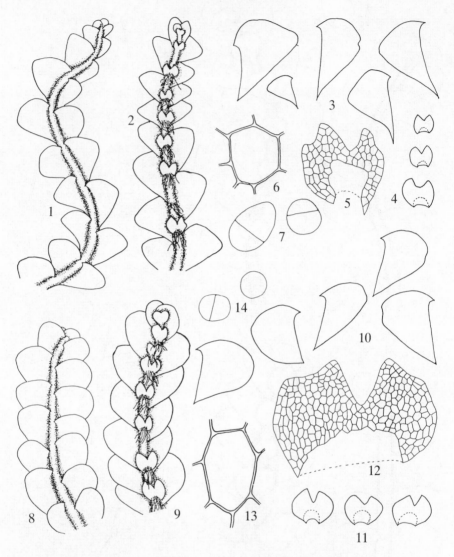

Fig. **14**. 1–7, *Calypogeia sphagnicola*: 1 and 2, dorsal and ventral sides of shoots (× 15); 3, leaves (× 21); 4, underleaves (× 21); 5, underleaf (× 60); 6, mid-leaf cell (× 330); 7, gemmae (× 450). 8–14, *C. suecica*: 8 and 9, dorsal and ventral sides of shoots; 10, leaves (× 21); 11, underleaves (× 21); 12, underleaf (× 60); 13, mid-leaf cell (× 330); 14, gemmae (× 450).

5. C. sphagnicola (H. Arn & J. Perss.) Warnst. & Loeske, *Verh. Bot. Vereins Prov. Brandenburg* 47: 320. 1906   (Fig. 14)

Autoecious. Plants slender, yellowish-green to brownish, small patches or scattered shoots. Shoots prostrate to ascending, to 2 cm long, 0.5–1.6 mm wide, sparsely branched. Leaves distant to subimbricate, plane to slightly convex, obliquely broadly triangular to ovate-triangular, widest at base, decurrent on ventral side, narrowed to acute to obtuse or shallowly bidentate apex, 0.20–0.64(–0.76) mm long, 0.24–0.64 (–0.72) mm wide, (0.8–)1.0–1.4 times as long as wide; marginal cells not tangentially elongated, marginal cells at apex 24–40 μm wide, mid-leaf cells 30–50(–60) μm long, 30–40 μm wide, walls in young leaves thin, in older leaves thickened, trigones lacking or very small, oil-bodies colourless. Underleaves appressed, divided to *ca* $\frac{3}{4}$, 1.5–2.5 (–3.0) times as wide as stem, 0.15–0.40 mm long, 0.20–0.42 mm wide, 1.0–1.7 times as wide as long, lobes acute or subacute, outer margins rounded or with a blunt tooth, 2–4 cells deep from sinus to rhizoid zone, these cells 30–40(–50) μm long. Gemmae on attenuated shoots, ± spherical to ellipsoid, mostly 2-celled, 24–35 μm. Capsules very rare, spring. $n = 18^{*}$. Usually in small quantity on and amongst *Sphagnum* in undisturbed raised, valley and blanket bogs, in suitable habitats throughout Britain and Ireland, ascending to *ca* 300 m. 49, H21. N., W. and C. Europe, Faroes, Iceland, Turkey, Japan, Azores, Madeira, Tenerife, N. America, Greenland, Tierra del Fuego, Tasmania, New Zealand, Oceania.

Other descriptions state that trigones are usually distinct but in British material I have seen they are very small or lacking. Both *C. fissa* and *C. muelleriana* may occur on *Sphagnum*; they differ in underleaf shape and length of the underleaf cells. The taxonomic status of *C. sphagnicola* is open to question as it has been variously stated to be a good species, or an aquatic form of *C. fissa* or of *C. muelleriana*.

6. C. suecica (H. Arn. & J. Perss.) K. Müll., *Beihefte Bot. Centralbl.* 17: 224. 1904
(Fig. 14)

Dioecious or autoecious. Plants dull green to brownish, small patches or scattered shoots. Shoots prostrate, to 1 cm long, 1.0–2.3 mm wide, sparingly branched. Leaves imbricate, slightly convex, ovate to ± orbicular, mostly widest $\frac{1}{3}$–$\frac{1}{2}$ from base, apex rounded, truncate or slightly emarginate, never bidentate, slightly decurrent on ventral side, 0.45–1.00 mm long, 0.4–0.9 mm wide, (0.9–)1.0–1.4 times as long as wide; marginal cells not tangentially elongated, marginal cells at apex 24–40 μm wide, mid-leaf cells 36–40(–60) μm long, 30–50 μm wide, thin-walled, trigones very small but distinct, oil-bodies colourless. Underleaves appressed, bilobed $\frac{1}{2}$–$\frac{2}{3}$, 2–3 times wider than stem, 0.26–0.48 mm long, 0.44–0.80 mm wide, 1.3–1.8 times as wide as long, lobes acute to obtuse, outer margins usually with a small tooth, 2–4 cells deep from sinus to rhizoid area, these cells 30–40 μm long. Gemmae on attenuated shoots, spherical to ellipsoid, 1–2-celled, *ca* 22 μm. Capsules very rare. $n = 9$. On decaying coniferous logs in shaded, humid habitats or on peat, ascending to 210 m, very rare, W. Inverness, Argyll, Kintyre, Ross, W. Sutherland, S. Kerry, Fermanagh, Tyrone. 6, H3. Pyrenees, N. Italy and Yugoslavia northwards, Azores, Tenerife, N. America.

Differs from *C. sphagnicola* in the imbricate leaves, leaf shape and habitat; from *C. fissa* and *C. muelleriana* in size of underleaf cells and from *C. neesiana* and *C. integristipula* in underleaf characters.

7. C. azurea Stotler & Crotz, *Taxon* 32: 74. 1983   (Fig. 15)
C. trichomanis auct.

Autoecious or paroecious. Plants medium-sized, greyish-green with turquoise-blue apices when fresh, patches or scattered shoots. Shoots prostrate, to 5 cm long, 1.5–2.6 mm wide, sparsely branched. Leaves imbricate, slightly convex, ovate to broadly ovate,

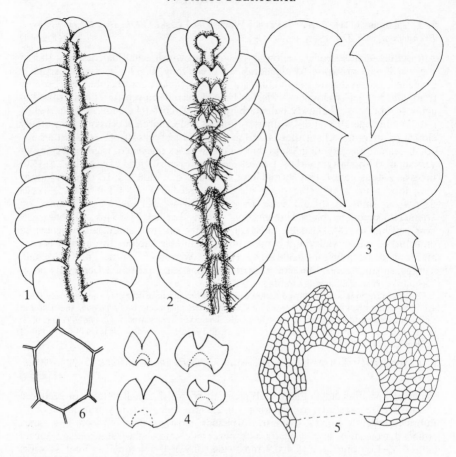

Fig. 15. *Calypogeia azurea*: 1 and 2, dorsal and ventral sides of shoots (× 15); 3, leaves (× 21); 4, underleaves (× 21); 5, underleaf (× 60); 6, mid-leaf cell (× 330).

widest 0–¼(–⅓) from base, apex obtuse to rounded, occasionally bidentate in young shoots, decurrent on ventral side, 0.7–1.4 mm long, 0.7–1.2 mm wide, 0.9–1.4 times as long as wide; marginal cells not tangentially elongated, marginal cells at apex 30–40 μm wide, mid-leaf cells 40–80 μm long, 30–50 μm wide, walls thin or thickened, trigones lacking or very small, oil-bodies deep blue. Underleaves appressed, bilobed ½–⅔, 2.0–3.5 times as wide as stem, 0.32–0.52 mm long, 0.40–0.64 mm wide, 1.2–1.5(–1.7) times as wide as long, lobes acute to obtuse, outer margins entire or with an obscure tooth, 2–4 cells deep from sinus to rhizoid area, these cells 40–80 μm long. Gemmae rare. Capsules unknown in Britain or Ireland. $n = 9$, 16 + 2m, 18. On peat or soil in damp, open habitats on heaths and moorland, calcifuge, ascending to *ca* 1000 m, occasional in western and northern Britain, Brecon, N.W. Wales, Derby, Lancashire and Yorkshire northwards, rare in Ireland. 37, H12. Montane and arctic Europe from Spain, Portugal, Italy and Yugoslavia northwards, Faroes, Iceland, Jenesei, Siberia, Japan, Azores (?), Madeira, Tenerife (?), eastern N. America.

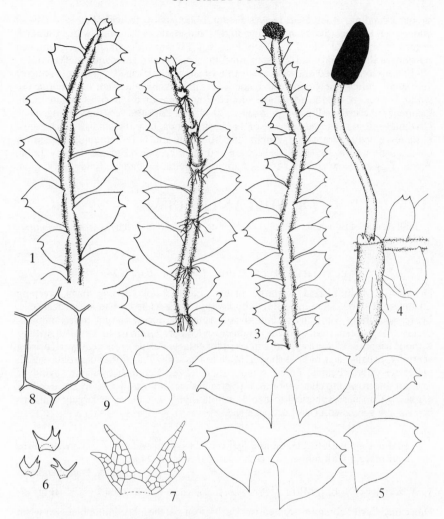

Fig. **16**. *Calypogeia arguta*: 1 and 2, dorsal and ventral sides of shoots, (× 15); 3, gemmiferous shoot (× 15); 4, portion of female shoot with marsupium and capsule (× 15); 5, leaves (× 21); 6, underleaves (× 21); 7, underleaf (× 60); 8, mid-leaf cell (× 330); 9, gemmae (× 450).

When fresh *C. azurea* is readily identified by the turquoise-blue shoot tips and the deep blue oil-bodies. The blue coloration is lost on drying and some forms are then impossible to distinguish from *C. fissa* or *C. muelleriana*. Typical plants of the former differ in the leaves widest about ⅓ from base and the apices frequently bidentate. *C. muelleriana* usually has relatively wider leaves and less deeply lobed underleaves.

8. C. arguta Nees & Mont. in Nees, *Naturg. Europ. Leberm.* 3: 24. 1838   (Fig. 16)

Dioecious. Plants small to medium-sized, pale green, loose patches or scattered shoots. Shoots prostrate, to 2 cm long, 1.0–2.5 mm wide, branching frequently, gemmiferous shoots ascending, attenuated. Leaves distant, obliquely ovate, widest *ca* ⅓ from base, bilobed to ⅙, lobes divergent ending in 1–2 uniseriate cells, sinus wide, rounded, decurrent on ventral side, 0.7–1.2 mm long, 0.4–1.0 mm wide, smaller on attenuated portions of shoots; marginal cells sometimes tangentially elongated, mid-leaf cells (50–) 60–100 μm long, 40–60 μm wide, walls thin or slightly thickened, trigones lacking or very small, cuticle finely papillose. Underleaves not much wider than stem, bilobed to within 1–2 cells of rhiziod area, lobes subulate, outer margin often with subulate tooth. Gemmae spherical to ellipsoid, 1–2-celled, 20–40 μm. Capsules very rare, spring. $n = 9$. On damp soil in sheltered habitats on banks, in woods, by paths and streams, at low altitudes, calcifuge, rare in E. Anglia, the Midlands and N.E. Scotland, frequent or common elsewhere. 105, H38, C. Western and Mediterranean Europe, Turkey, India, Japan, Taiwan, New Guinea, Africa, Azores, Madeira, La Palma, Tenerife, Oceania.

8. ADELANTHACEAE

A small family of two genera, the second genus, *Wettsteinia*, differing from *Adelanthus* in stem anatomy.

1. ADELANTHUS MITT, *J. Proc. Linn. Soc. Bot.* 7: 243. 1864

Dioecious. Medium-sized to large plants. Erect shoots arising from creeping rhizomatous stems; stem in section with layer of thick-walled, opaque, cortical cells. Leaves succubous, ventrally secund, oval to rounded, dorsal margin entire, inflexed, decurrent, ventral margin with one to numerous teeth; cells thin to thick-walled, with or without trigones. Underleaves absent or rudimentary. Fertile branches dwarf, arising ventrally, usually near base of shoots. Male bracts scale-like, antheridia solitary with uniseriate stalks. Female bracts smaller than leaves. Perianth pyriform, ovoid or fusiform, narrowed to ciliate mouth. Calyptra fleshy. Seta massive, many small cells in diameter. Capsule ± spherical to ovoid, wall fleshy, 5–7-stratose. A largely southern hemisphere genus with two European species.

Leaf margin with 0–3 teeth, mid-leaf cells 28–44 μm wide   **1. A. decipiens**
Leaf margin with numerous teeth, mid-leaf cells 16–24 μm wide
                                                           2. A. lindenbergianus

1. A. decipiens (Hook.) Mitt., *J. Proc. Linn. Soc. Bot.* 7: 244. 1864   (Fig. 17)

Dioecious. Tight, sometimes extensive, dark green patches, becoming blackish when dry. Shoots procumbent, often with deflexed tips, arising from mat of rhizomatous stems, 1–2(–3) cm long, producing one or more branches and also flagelliform shoots from lower part. Leaves succubous, fragile, lower small, distant, appressed or not, upper much larger, subimbricate or imbricate, erecto-patent to patent or reflexed, concave with inflexed, longly decurrent dorsal margin, ± orbicular, apex truncate and bispinose or acute to rounded with 0–3 spinose teeth, insertion slightly oblique, mature leaves 0.7–1.7 mm long, 0.7–1.5 mm wide, ± as wide as long; mid-leaf cells 28–44 μm wide, walls reddish-brown, thin to thickened, trigones small to large, marginal row thicker-walled and with larger trigones, forming ill-defined border. Underleaves lacking except small and scale-like on flagelliform branches. Vegetative propagation possibly by deciduous flagelliform branches. Only male plants known in Europe. On rocks and rock outcrops, occasionally on bark, often with *Plagiochila spinulosa*, in humid, deciduous woodland or north-facing slopes, below 300 m except in S.W.

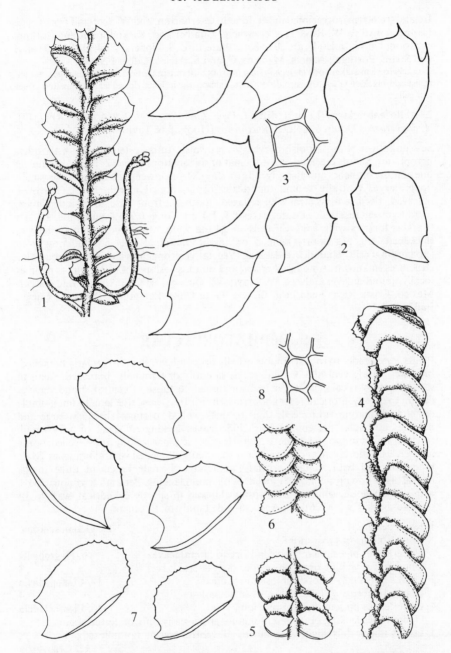

Fig. **17**. 1–3, *Adelanthus decipiens*: 1, shoot (× 12); 2, leaves (× 22); 3, mid-leaf cell (× 330). 4–8, *A. lindenbergianus*: 4, shoot in side view (× 12); 5, part of shoot, ventral side (× 12); 6, part of shoot, dorsal side (× 12); 7, leaves (× 22); 8, mid-leaf cell (× 330).

Ireland (to 600 m), occasional in Merioneth, Caernarfon, and W. Scotland from Clyde Is. and Argyll to W. Ross, rare elsewhere, Cumberland, Kirkudbright and Stirling, occasional in W. Ireland, rare elsewhere, Waterford, S. Tipperary. 14, H14. W. France and Spain, Portugal, Azores, Madeira, C. and S. Africa, Cuba, Ecuador.

Likely to be mistaken for *Plagiochila spinulosa* but differing in the field in being dark green when moist and blackish when dry; also differs in leaf shape, the inflexed dorsal leaf margin and larger leaf cells.

2. A. lindenbergianus (Lehm.) Mitt., *J. Proc. Linn. Soc. Bot.* 7: 244. 1864　(Fig. 17)
A. dugortiensis Douin & Lett, *A. unciformis* (Hook. f. & Tayl.) Mitt.

Sex unknown. Lax, reddish-brown to brownish tufts, often mixed with other bryophytes. Shoots erect, arising from mat of rhizomatous stems, 2–10 cm, producing one or more branches and also flagelliform shoots from lower part. Lower leaves small, upper large, ventrally secund, imbricate, ovate, dorsal margin entire, narrowly incurved, decurrent, ventral side reflexed, toothed ± from base to rounded apex, ± transversely inserted, mature leaves 0.9–1.4 mm long, 0.7–1.0 mm wide, 1.1–1.5 times as long as wide; mid-leaf cells 16–24 μm wide, walls reddish-brown, heavily thickened, trigones poorly defined or absent, marginal cells not differing from submarginal cells. Underleaves lacking. Vegetative propagules lacking. Reproductive structures unknown. Amongst *Calluna* and on mossy slopes, 430–675 m, sometimes locally abundant, W. Galway, W. Mayo, W. Donegal. H3. E. Africa, Madagascar, Mexico, Costa Rica, Venezuela, Bolivia, Peru, Chile, Patagonia, Tierra del Fuego, Falkland Is.

9. CEPHALOZIACEAE

Plants very slender to robust. Shoots usually procumbent, stems irregularly branched, branches arising ventrally or more rarely laterally or terminally branched, stems in section with cortical cells larger than or similar to those of central strand. Leaves distant to closely imbricate, succubous, insertion ± transverse to ± longitudinal, simple or bilobed, margin entire; cells thin- to thick-walled, trigones absent to large and bulging, oil-bodies present or not. Underleaves lacking or small or very small, rarely ± similar in size and shape to lateral leaves. Gemmae lacking or 1–2-celled, borne in fascicles. Inflorescences on dwarf or occasionally elongated ventral branches. Male bracts bilobed with solitary, axillary antheridia. Female bracts in three ranks, increasing in size up stem, 2(–5)-lobed. Perigynium lacking. Perianth trigonous above with one angle on ventral side. Capsule ellipsoid to shortly cylindrical, wall 2(–3)-stratose. Spores 8–20(–26) μm. A world-wide family of 12 genera.

1 Leaves entire　　　　　　　　　　　　　　　　　　　　**17. Odontoschisma**
　Leaves bilobed to $\frac{1}{6}$ or more　　　　　　　　　　　　　　　　　　2
2 Underleaves of ± similar shape and size to lateral leaves　　**16. Hygrobiella**
　Underleaves lacking or much smaller than lateral leaves　　　　　　3
3 Leaves bilobed $\frac{1}{6}$–$\frac{1}{3}$, lobes obtuse to rounded　　　　　　**14. Cladopodiella**
　Leaves bilobed $\frac{1}{4}$–$\frac{2}{3}$, lobes subacute to acuminate　　　　　　　　4
4 Underleaves present, 0.3–0.5 mm long　　　　　　　　**15. Pleurocladula**
　Underleaves lacking or if present minute and usually only on fertile shoots　　5
5 Leaves not or only slightly asymmetrical, ventral margin not inflexed
　　　　　　　　　　　　　　　　　　　　　　　　　　　12. Cephalozia
　Leaves very asymmetrical, ventral margin inflexed and forming a sac
　　　　　　　　　　　　　　　　　　　　　　　　　　　　13. Nowellia

SUBFAMILY CEPHALOZIOIDEAE

Plants usually slender. Branches, and flagelliform branches where present, arising from ventral side of stem, stems with pellucid cortical cells much larger than cells of central strand (except *Cladopodiella*). Leaves ± transversely to ± longitudinally inserted, dorsal margin decurrent or not, bilobed to $\frac{1}{6}$–$\frac{3}{4}$, cells thin- to thick-walled, trigones not well developed, oil-bodies small and numerous or lacking. Underleaves lanceolate or subulate or absent except sometimes on fertile shoots. Gemmae often present, mostly unicellular. Inflorescences on short ventral branches or male on longer branches. Female bracts and bracteoles similar, much larger than leaves, 2–5-lobed, margins toothed or not. Perianth trigonous above. Capsule wall 2(–3)-stratose.

12. CEPHALOZIA (DUM.) DUM., *Recueil Observ. Jungerm.*: 18. 1835

Dioecious or autoecious. Plants slender, pale, yellowish, dull or bright green or sometimes brownish. Shoots procumbent, stems irregularly branched and sometimes with flagelliform branches, both arising from ventral side of stem, stems in section with cortex of a single layer of large pellucid cells and a central strand of small thick-walled cells visible through the cortex with transmitted light. Leaves alternate, succubous, spreading horizontally to dorsally secund, distant to imbricate, orbicular to ovate or ovate-quadrate, usually concave, insertion ± transverse to ± longitudinal, margins entire, dorsal decurrent or not, bilobed to $\frac{1}{3}$–$\frac{3}{4}$, lobes often unequal, acute, straight to strongly connivent, sinus obtuse to rounded; cells thin- to thick-walled. Underleaves absent or small and subulate on fertile stems. Unicellular gemmae sometimes present at apices of ascending stems and branches. Male inflorescence terminal on short or long ventral branch, spicate, sometimes becoming intercalary with age, bracts few to numerous, bilobed, margins usually entire except sometimes for a single tooth on dorsal margin. Female inflorescence terminal on short or more rarely long ventral branch, bracts much larger than leaves, 2–5-lobed to $\frac{1}{3}$–$\frac{3}{4}$, margins entire to sharply toothed, bracteoles similar to or narrower than bracts. Perianth narrowly ovoid, ellipsoid or fusiform-cylindrical, trigonous above, wall 1–3-stratose, mouth crenate-dentate to ciliate. Spores 8–18 μm. A taxonomically difficult genus of 30–40 mainly northern hemisphere species.

Small plants of *Cephalozia* may be mistaken for *Sphenolobopsis* or *Cephaloziella* species but are readily distinguished by the large, pellucid cortical cells through which the central strand is easily visible with transmitted light. Confusion may also arise as different *Cephalozia* species may grow mixed together or mixed with *Cephaloziella* species. With fertile material, in autoecious species, male branches usually arise close to female branches.

Most species of *Cephalozia* are extremely variable morphologically and, except for *C. bicuspidata*, those growing in bogs can only be determined if they are fertile. Although the typical form of such species may appear distinctive, phenocopies of other species make the determination of sterile material from bogs impossible. In order to key out sterile material wherever possible, the following key may in places appear to lack clear alternatives but should provide a means of identifying sterile material from habitats other than *Sphagnum* or Leucobryum.

1 Leaves ± transversely inserted, not decurrent, cells at base of lobes 12–20(–24) μm wide   2

  Leaves transversely, obliquely or ± longitudinally inserted, dorsal margin decurrent or not, cells at base of lobes 20–60 μm wide   3

2 Plants ± brownish, flagelliform shoots often present, leaves distant to imbricate, about as wide as long   **1. C. ambigua**

Plants whitish-green, flagelliform shoots lacking, leaves distant, longer than wide
5. C. leucantha
3 Flagelliform shoots often present, dorsal margin of leaf not or hardly decurrent,
autoecious   4
Flagelliform shoots lacking, leaf margin decurrent or not, autoecious or dioecious
5
4 Leaves ± transversely inserted   **2. C. bicuspidata**
Leaves obliquely inserted   **7. C. pleniceps**
5 Plants not growing on *Sphagnum* or *Leucobryum*   6
Plants growing on *Sphagnum* or *Leucobryum*   10
6 Cells at base of lobes 20–24 μm wide, walls thickened, dorsal margin of leaf hardly
decurrent   **3. C. catenulata**
Cells 30–60 μm wide or if smaller then dorsal margin distinctly decurrent, walls
thickened or not   7
7 Leaf lobes not or slightly connivent, dorsal margin not decurrent **2. C. bicuspidata**
Lobes connivent, dorsal margin decurrent   8
8 Cells at base of leaf lobes 20–28 μm wide   **6. C. lunulifolia**
Cells 32–56 μm wide   9
9 Autoecious, leaf lobes 3–6 cells wide at base, ending in (1–)2 uniseriate cells
9. C. connivens
Dioecious, lobes 2–4 cells wide at base, ending in 2–3 uniseriate cells
10. C. hibernica
10 Plants autoecious   11
Plants dioecious   14
11 Leaf insertion ± transverse or if oblique than at least dorsal part ± transverse, not
decurrent   **2. C. bicuspidata**
Insertion oblique to ± longitudinal, margin decurrent or not   12
12 Perianth mouth crenate-dentate, female bracts bilobed to ⅓   **7. C. pleniceps**
Perianth mouth ciliate, female bracts 3–5-lobed to ½–¾   13
13 Cells at base of leaf lobes 20–32(–36) μm wide, lobes ending in 1–3 uniseriate cells
8. C. loitlesbergeri
Cells 32–56 μm wide, lobes ending in (1–)2 uniseriate cells   **9. C. connivens**
14 Female bracts with dentate margins, perianth mouth ciliate, male bracts with entire
to dentate margins   **4. C. macrostachya**
Female bracts with entire margins, male bracts entire or with a tooth on dorsal
margin near base   **6. C. lunulifolia**

Section *Cephalozia*

Autoecious. Plants often developing reddish or brownish pigmentation. Stems branching terminally and from ventral side, flagelliform shoots usually present. Leaves ± transversely inserted, lobes not connivent. Female inflorescence on long or dwarf ventral branch, bracts bilobed, lobes entire or toothed towards base.

1. C. ambigua Mass., *Malpighia* 21: 310. 1907   (Fig. 19)
C. bicuspidata ssp. *ambigua* (Mass.) Schust., *C. bicuspidata* var. *atra* Arnell

Autoecious. Plants minute, in greenish-brown to brownish-black patches or mixed with other bryophytes. Shoots procumbent, to 5–6 mm long, dorsal cortical cells of stem 16–32 μm long, 10–16 μm wide, stems branching irregularly, flagelliform shoots often present. Leaves distant to approximate or imbricate, somewhat dorsally secund, ovate-quadrate, very concave, ± transversely inserted, not decurrent, 100–210 μm long, 80–200 μm wide, mostly about as long as wide, bilobed to ⅓(–½), lobes acute to obtuse, incurved, 3–6 cells wide at base, ending in one uniseriate cell, sinus obtuse to rounded; cells at base of lobes 16–24 μm long, 14–20 μm wide. Underleaves absent or minute. Reproductive structures as in *C. bicuspidata*. Capsules not known in Britain. *n* = 9. On bare, moist soil on tops of a few of the highest Scottish mountains, very rare, Mid Perth, S. Aberdeen, Banff, E. Inverness. 4. Alpine and arctic Europe from the French Pyrenees, N. Italy and Bulgaria north to Spitzbergen, Siberia, N. America, Greenland.

2. C. bicuspidata (L.) Dum., *Recueil Observ. Jungerm.*: 18. 1835

Autoecious. Plants slender, pale to dark green or brownish, sometimes reddish-tinged, forming patches or creeping through other bryophytes. Shoots procumbent to ascending, to 2.0(–2.5) cm, dorsal cortical cells of stem 40–160 μm long, 24–60 μm wide, stems irregularly branched, flagelliform branches often present. Lower leaves distant, upper approximate to subimbricate or imbricate, erecto-patent to spreading horizontally, broadly ovate to ovate-quadrate, concave, ± transversely inserted or if oblique then at least dorsal part transverse, not decurrent, 200–600 μm long, 160–440(–520) μm wide, bilobed ½–⅔, lobes acute to acuminate, ending in 1–3(–4) uniseriate cells, lobes inflexed, sinus obtuse to rounded; cells at base of lobes 30–70 μm long, 20–50 μm wide, thin- to thick-walled. Small subulate or lanceolate underleaves sometimes present on fertile stems. Gemmae rare, unicellular, at shoot or branch apices. Male inflorescence terminal, becoming intercalary if on long branch, bracts similar to leaves but more concave, imbricate. Female inflorescence on short or long branch, bracts larger than leaves, bilobed to ½, lobes acuminate, entire or toothed towards base, bracteole bilobed. Perianths common, fusiform, trigonous above, mouth ciliate-dentate, teeth 1–3 cells long. Capsules common, autumn to spring.

Key to varieties of *C. bicuspidata*

Dorsal cortical cells of stem 40–80 μm long, cells at base of leaf lobes mostly 20–36 μm wide   var. **bicuspidata**
Cortical cells 70–160 μm long, leaf cells 30–50 μm wide   var. **lammersiana**

Var. **bicuspidata**   (Fig. 18)

Plants pale to dark green, sometimes reddish-tinged. Shoots to 2.0(–2.5) cm, dorsal cortical cells of stem 40–80 μm long, 24–50 μm wide, flagelliform branches often present. Leaves ± transversely inserted, 200–500 μm long, 160–440 μm wide, (0.9–)1.1–1.3 times as long as wide, bilobed to (⅓–)½, lobes straight to incurved, (3–)4–6(–9) cells wide at base, acute to acuminate, ending in 1–3(–4) uniseriate cells; cells at base of lobes 30–50 μm long, (16–)20–36(–42) μm wide, walls thin or thickened. Female inflorescence usually on dwarf branch. *n* = 18. Moist soil, peat, humus, rotting logs, in turf, on paths, by streams, in woods, on heaths, and moorland, in scree and on soil on mountain tops, ascending to 1200 m, common in damp, acidic habitats. 111, H40, C. Throughout Europe north to Spitzbergen, Faroes, Iceland, Turkey, Asia Minor, Caucasus, Siberia, N. Africa, Azores, Madeira, Tenerife, N. America, Mexico, Greenland.

Var. **lammersiana** (Hüb.) Breidl., *Naturw. Ver. Steiermark*: 329. 1893   (Fig. 18)
C. lammersiana (Hüb.) Spruce, *C. bicuspidata* ssp. *lammersiana* (Hüb.) Schust.

Plants pale green sometimes reddish-tinged. Shoots to 2.0(–2.5) cm, dorsal cortical cells of stem 70–160 μm long, (36–)40–60 μm wide, flagelliform shoots usually absent. Leaves distant to approximate, insertion ± transverse or ventral part oblique and

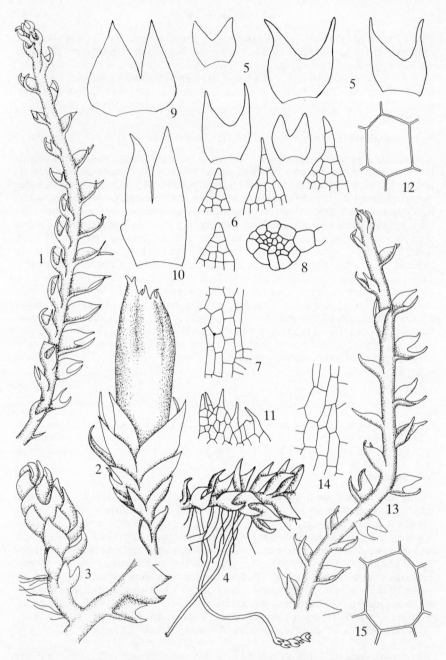

Fig. **18**. 1–12, *Cephalozia bicuspidata* var. *bicuspidata*: 1, shoot (× 33); 2, perianth (× 33); 3, male branch (× 33); 4, part of shoot with flagelliform branches (× 33); 5, leaves (× 63); 6, leaf lobe apices (× 84); 7, dorsal cortical cells of stem (× 84); 8, т.s. stem (× 84); 9, male bract (× 63); 10, female bract (× 21); 11, perianth mouth (× 84); 12, cell at base of leaf lobe (× 450). 13–15, *C. bicuspidata* var *lammersiana*: 13, shoot (× 33); 14, dorsal cortical cells of stem (× 84); 15, mid-leaf cell (× 450).

dorsal part transverse, 360–600(–640) μm long, 240–440(–520) μm wide, 1.0–1.4 times as long as wide, bilobed to ½–⅔, lobes sometimes ± connivent, acute to acuminate, 4–7 cells wide at base, ending in 2–3(–4) uniseriate cells, cells at base of lobes 30–70 μm long, 30–50 μm wide, thin-walled. Female inflorescences on dwarf or long branches. n = 27. Moist soil and peat on banks, in marshes and bogs, dune-slacks, on heaths, occasional to frequent in suitable acidic habitats. 84, H26. Spain, N. Italy and Bulgaria north to Fennoscandia, Sardinia, Faroes, Iceland, Sikkim, Azores, Madeira, N. America.

Biosystematic studies by Kozlicka (in Szweykowski (ed.) *New Perspectives in Bryotaxonomy and Bryogeography*, 15–23, Poznan, 1981) have shown that *C. ambigua* is a good species distinct from *C. bicuspidata*. She treats the two varieties of *C. bicuspidata* recognised here as subspecies but they intergrade morphologically and ecologically and in the absence of geographical differences they cannot be treated as more than varieties. All three taxa show considerable environmentally induced variation which may make determination difficult. Depauperate high alpine forms of *C. bicuspidata* may be mistaken for *C. ambigua*, and the two species may occur together, but *C. ambigua* is a smaller plant with smaller cells.

Var. *lammersiana* tends to be a laxer and softer plant than var. *bicuspidata* and frequently occurs in moister habitats but these differences are not absolute. According to Kozlicka (loc. cit.) var. *lammersiana* may occur in similar habitats to *C. ambigua*.

Section *Catenulatae* Schust., *Hep. Anthoc. N. Amer.* 3: 736. 1974

Plants lacking secondary pigments. Stems usually lacking terminal branching and flagelliform branches. Leaves obliquely inserted, decurrent or not, lobes usually straight. Female bracts dentate (except *C. leucantha*), bilobed. Perianth mouth ciliate-dentate.

3. C. catenulata (Hüb.) Lindb., *Acta Soc. Sci. Fenn.* 10: 262. 1872   (Fig. 19)

Dioecious. Plants very slender, dull green, pale brown or brownish-green, in patches or growing through other bryophytes. Shoots procumbent, stems frequently and irregularly branched, branches procumbent to ascending, flagelliform branches lacking. Leaves distant to subimbricate, often dorsally secund, ovate, concave, insertion oblique, dorsal margin not or scarcely decurrent, 130–220(–270) μm long, 100–170(–220) μm wide, 1.1–1.3(–1.4) times as long as wide, bilobed ⅓–½, lobes straight or slightly connivent, acute, 3–5 cells wide at base, ending in 1–2 uniseriate cells, sinus rounded; cells at base of lobes 16–32 μm long, 20–24(–28) μm wide, walls thickened, pale brownish. Underleaves lacking. Unicellular gemmae sometimes present at stem and branch apices. Male inflorescences terminal on short or long branches, becoming intercalary with age, bracts 4–9 pairs, similar to leaves but imbricate and more concave, sometimes with small tooth on dorsal side. Female inflorescences on short branches, bracts bilobed to ⅓, lobe acute, small third lobe sometimes present, margins toothed, sometimes spinosely so. Perianths common, ellipsoid, trigonous ± to base, narrowed to ciliate-dentate mouth, cilia 1–3 cells long. Capsules common, late winter to summer. On shaded rotting logs, peat or rocks, usually sandstone, ascending to *ca* 350 m, widely distributed but rare, S.W. England, S. Hants, Sussex, Wales, Derby and Cheshire north to Shetland, widely distributed in Ireland. 40, H22. Europe except the far north, Turkey, Caucasus, Siberia, Japan, Madeira, N. America.

Closely allied to *C. macrostachya* but differing in habitat, colour, smaller leaves with thickened cell walls and perianth trigonous almost to base. The colour and perianths, which are usually present, will help to distinguish *C. catenulata* from other *Cephalozia* species in the field.

4. C. macrostachya Kaal., *Rev. Bryol.* 29: 8. 1902

Plants slender, yellowish-green to dark green, in patches on or growing through *Sphagnum*. Shoots procumbent to ascending, to 1.0(–1.5) cm, stems branching

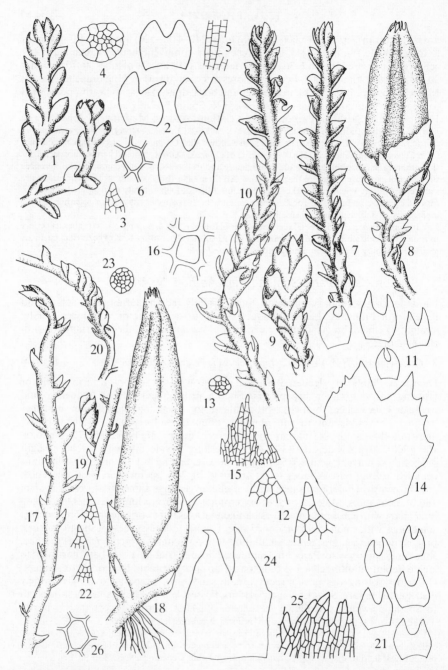

Fig. 19. 1–6, *Cephalozia ambigua*: 1, shoots (×38); 2, leaves (×84); 3, leaf lobe apex (×84); 4, т.s. stem (×84); 5, dorsal cortical stem cells (×84); 6, cell from base of leaf lobe (×450). 7–16, *C. catenulata*: 17, shoot (×33); 8, perianth (×33); 9 and 10, young and old male shoots (×33); 11, leaves (×63); 12, leaf lobe apices (×84); 13, т.s. stem (×84); 14, female bract (×63); 15, perianth mouth (×84); 16, cell at base of leaf lobe (×450). 17–26, *C. leucantha*: 17, shoot (×33); 18 perianth (×33); 19 and 20, young and old male shoots (×33); 21, leaves (×63); 22, leaf lobe apices (×84); 23, т.s. stem (×84); 24, female bract (×63); 25, perianth mouth (×84); 26, cell at base of leaf lobe (×450).

irregularly, flagelliform shoots lacking. Leaves distant to approximate, ± dorsally secund, obliquely orbicular to obliquely ovate-orbicular, in vigorous shoots with ventral margin bulging, ± concave, insertion oblique, dorsal margin decurrent, 240–600 μm long, (180–)240–560 μm wide, 1.0–1.3(–1.5) times as long as wide, bilobed $\frac{1}{3}$–$\frac{1}{2}$ (–$\frac{2}{3}$), lobes straight to connivent and occasionally crossing, (3–)4–9 cells wide at base, subacute to acute, ending in (1–)2(–3) uniseriate cells, sinus rounded to obtuse; cells at base of lobes 24–44 μm long, 20–28(–32) μm wide, in vigorous forms, cells at middle of leaf base 32–64 μm long (24–)32–64 μm wide and conspicuously larger and more translucent than cells at base of lobes, cell walls thin to ± thickened. Underleaves lacking. Spherical to irregular, unicellular gemmae sometimes present at shoot tips. Male inflorescences spicate, terminal on short or long branches, becoming intercalary with age, bracts 4–20 pairs, imbricate, larger and more concave than leaves, ± transversely inserted, bilobed, margins obscurely to sharply serrate but in depauperate forms entire. Female inflorescences on short branches, bracts bilobed, acuminate, margins toothed or spinosely toothed. Perianths occasional, narrowly ovoid to fusiform, mouth ciliate-dentate. Capsules unknown in Britain or Ireland.

Key to varieties of *C. macrostachya*

Female bracts bilobed, bracteole 2–3-lobed, perianth not incised
var. macrostachya
Female bracts 2–4-lobed, perianth incised almost to base into 3 lobes
var. spiniflora

Var. macrostachya (Fig. 20)

Dioecious. Female bracts bilobed to *ca* $\frac{1}{2}$, bracteole 2–3 lobed. Perianth not incised, mouth ciliate-dentate. On *Sphagnum* in bogs, rarely on wet peat or wet decaying vegetation, widely distributed and sometimes locally frequent, ascending to 300 m, from E. Cornwall east to E. Kent and north to E. Inverness and Skye, very rare in Ireland, Wicklow, Donegal. 40, H3. N.W. France, Alps and Hungary north to W. Norway, S. Sweden and Baltic Russia, eastern N. America.

Var. spiniflora (Schiffn.) K. Müll., *Rabenh. Krypt.-Fl. Deutschl.* 2: 28 1916 (Fig. 20)

Dioecious, paroecious or synoecious. Female bracts bilobed from halfway almost to base, bracteole similar. Perianth divided to base into 3 lobes, the margins of which interlock so that the perianth appears tubular, mouth spinose-dentate. In wet turf, rarely on *Sphagnum* or rotting wood. W. Sussex. 1. Germany, Sweden.

I have not seen British material of var. *spiniflora* and the single isotype I examined contained only male plants. The above description and the illustrations are taken from Schiffner's original description.

C. macrostachya is a variable plant likely to be confused with other *Cephalozia* spp. growing in the same habitat and it can only be identified with certainty when fertile. The male plants are smaller than the female but in vigorous individuals the spicate male inflorescences with numerous dentate bracts are characteristic; smaller plants may have inflorescences with as few as 4 pairs of bracts which may be entire and such plants cannot be named. The dentate female bracts are also distinctive. Very vigorous sterile plants have relatively large leaves with a strongly arched ventral margin and the basal cells conspicuously larger and more translucent than those at the base of the leaf lobes.

5. C. leucantha Spruce, *On Cephalozia*: 68. 1882 (Fig. 19)

Dioecious. Plants very slender, pale whitish-green, in patches or creeping through other bryophytes. Shoots procumbent, to 1 cm, stems branching irregularly,

Fig. **20**. 1–10, *Cephalozia macrostachya* var. *macrostachya*: 1, shoot (× 33); 2, young perianth (× 21); 3, male shoot (× 21); 4, leaves (× 38); 5, leaf (× 84); 6, leaf lobe apices (× 84); 7, т.s. stem (× 84); 8, male bract (× 38); 9, female bract (× 38); 10, cell at base of leaf lobe (× 450). 11–12, *C. macrostachya* var. *spiniflora* (redrawn from Schiffner's original figures): 11, paroecious inflorescence with young divisions of perianth (× 33); 12, perianth with bract and bracteole (× 33).

flagelliform shoots lacking. Leaves patent to spreading, usually hardly longer than the width of stem, distant, ovate, concave, insertion ± transverse, dorsal margin not decurrent, 120–210 μm long, 80–180 μm wide, 1.2–1.5 times as long as wide, bilobed to ½ or more, lobes straight or connivent, acute, 2–4(–5) cells wide at base, ending in 1–2 uniseriate cells; cells at base of lobes 16–28(–32) μm long, 12–20(–24) μm wide, walls thickened. Underleaves lacking. Unicellular gemmae sometimes present at shoot tips. Male inflorescence, terminal on branch, becoming intercalary with age, bracts imbricate, larger and more concave than leaves. Female inflorescence on dwarf branch, bracts 2–3-lobed to *ca* ⅓, lobes acute, margins entire to bluntly toothed. Perianths occasional, fusiform-cylindrical, trigonous towards apex, gradually tapered to ciliate-dentate mouth, cilia 1–2 cells long. Capsules occasional, spring, summer. On sheltered peat or peaty soil on banks, growing through *Sphagnum* on wet heaths and moorland, on rotting wood, ascending to *ca* 750 m, very rare in England, Cornwall, Surrey, N. Northumberland, Cumberland, occasional throughout Scotland, in Ireland occasional in the west and very rare elsewhere. 33, H14. From N. Spain, N. Italy and Yugoslavia north to Fennoscandia and Russia, Siberia, Sakhalin, Japan, N. America, Greenland.

Section *Lunulifoliae* Schust., *Hep. Anthoc. N. Amer.* 3: 768. 1974

Plants usually lacking secondary pigments. Stems usually lacking terminal branching, flagelliform branches present or not. Leaves obliquely or very obliquely inserted, decurrent on dorsal side, lobes connivent. Female bracts and bracteoles 2-lobed, margins entire. Perianth mouth crenate-dentate, dentations rarely more than one cell long.

6. C. lunulifolia (Dum.) Dum., *Recueil Observ. Jungerm.*: 18. 1835   (Fig. 21)
C. affinis Lindb. ex Steph., *C. media* Lindb.

Dioecious, very rarely autoecious. Plants slender, in yellowish-green to dull green patches or growing through other bryophytes. Shoots procumbent, to 1.5(–2.0) cm, stems branching irregularly, flagelliform branches lacking. Leaves distant to subimbricate, dorsally secund, ± orbicular to ovate–orbicular, concave, insertion oblique, dorsal margin decurrent, 140–340 μm long, 140–340 μm wide, 1.0–1.2 times as long as wide, bilobed ¼–⅓, lobes connivent, often inflexed, subacute to acute, 3–6 cells wide at base, ending in 1–2 uniseriate cells, sinus rounded to obtuse; cells at base of lobes 20–32(–40) μm long, 20–28(–32) μm wide, walls ± thickened. Underleaves lacking. Ovoid, unicellular gemmae sometimes present at shoot tips. Male inflorescence, terminal on branch, becoming intercalary with age, bracts imbricate, similar to or larger than leaves, sometimes with tooth on dorsal margin. Female infloresence on short branch, bracts larger than leaves, bilobed to *ca* ⅓, lobes acute, entire or with tooth on one side. Perianths occasional, narrowly ovoid to fusiform, plicate above, bistratose in lower half, narrowed to mouth, mouth crenate-dentate, teeth 1 cell long, very rarely 2–3 cells long. Capsules occasional, winter to summer. Rotting wood, peaty banks and tree bases, rarely on sandstone, in woodland, on peat and *Sphagnum* on wet heaths and in bogs, in scree, rarely ascending to more than 500 m, occasional in lowland Britain, frequent or common elsewhere. 92, H28. Throughout Europe north to Fennoscandia, Faroes, Iceland, Siberia, Sakhalin, Japan, S.W. China, Azores, Madeira, N. America, Greenland.

Most likely to be confused with *C. connivens* which is autoecious, differs in its larger leaf cells, 3–5-lobed female bracts and ciliate perianth mouth. *C. bicuspidata* often has flagelliform branches, ± transversely inserted non-decurrent leaves, differs in leaf shape and is also autoecious.

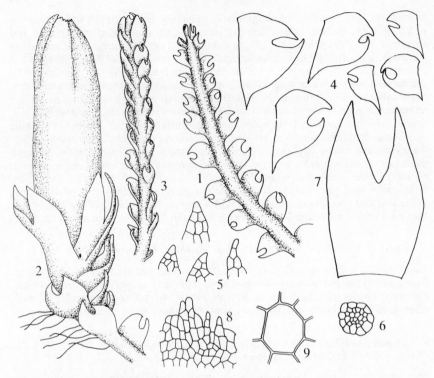

Fig. 21. *Cephalozia lunulifolia*: 1, shoot (× 33); 2, perianth (× 33); 3, male shoot (× 33); 4, leaves (× 63); 5, leaf lobe apices (× 84); 6, T.S. stem (× 84); 7, female bract (× 63); 8, perianth mouth (× 84); 9, cell at base of leaf lobe (× 450).

7. C. pleniceps (Aust.) Lindb., *Bot. Notis.* 1883: 18. 1883 (Fig. 22)
C. pleniceps var. *macrantha* (Kaal. & Nichols.) K. Müll.

Autoecious. Plants slender, in pale green or rarely reddish-brown patches or growing through *Sphagnum*. Shoots procumbent, sometimes fleshy, 0.5–1.5(–2.0), cm, stems branching irregularly, flagelliform branches usually present. Leaves distant to imbricate, usually dorsally secund, ± orbicular, slightly to strongly concave, insertion oblique to very oblique, dorsal margin not or slightly decurrent, (240–)320–600(–720) μm long, (240–)320–600(–680) μm wide, ± as long as wide, bilobed $\frac{1}{3}$–$\frac{1}{2}$, lobes straight to connivent, inflexed, 4–10 cells wide at base, subacute to acuminate, ending in 1–2(–3) uniseriate cells, sinus rounded to obtuse; cells at base of lobes 24–56(–72) μm long, 20–60 μm wide, walls ± thickened. Subulate underleaves occasionally present. Ovoid, unicellular gemmae sometimes present at shoot tips. Male inflorescences on short branches, bracts imbricate, similar to leaves but with tooth on dorsal margin. Female inflorescence on short or long branch, bracts larger than leaves, bilobed to $\frac{1}{3}$, lobes acute, with single tooth on outer margins, bracteole similar. Perianths frequent, narrowly ovoid to narrowly ellipsoid, trigonous above, bistratose on lower half, narrowed to crenate-dentate mouth. Capsules frequent, summer. $n = 18$. On *Sphagnum* at various altitudes, on soil at high altitudes, very rare and at low altitudes in S. Britain,

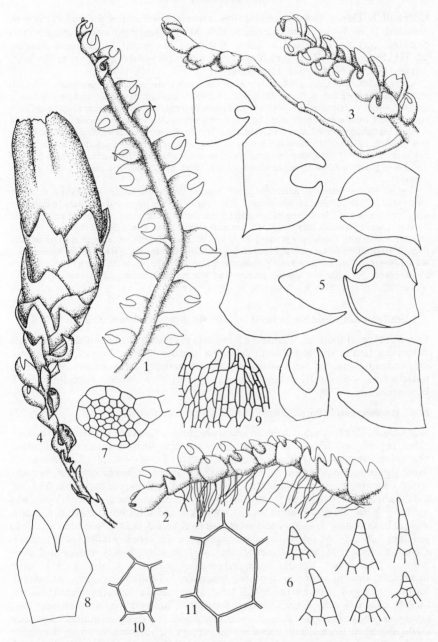

Fig. **22**. *Cephalozia pleniceps*: 1, shoot from wet *Sphagnum* (× 24); 2, shoot from top of *Sphagnum* tussock (× 24); 3, shoot from alpine soil (× 24); 4, perianth (× 24); 5, leaves (× 63); 6, leaf lobe apices (× 84); 7, т.s. stem (× 84); 8, female bract (× 24); 9, perianth mouth (× 84); 10, cell at leaf lobe base from alpine plant (× 450); 11, cell at leaf lobe base from bog plant (× 450).

Cornwall, N. Devon, E. Sussex, E. Norfolk, rare elsewhere and ascending to 1160 m in Scotland, N.W. Wales, Cumberland, Selkirk, Midlothian, Perth and Argyll north to Shetland, very rare in Ireland, S. Kerry, Wicklow, Westmeath, W. Donegal. Antrim. 24, H5. Pyrenees, N. Italy and Yugoslavia north to Spitzbergen, Faroes, Iceland, Turkey, Siberia, N. America, Greenland.

A very variable plant, morphology being related to micro-environmental conditions. Plants in very wet habitats are not succulent, the leaves are distant, only slightly concave and very obliquely inserted, slightly decurrent, the lobes are acuminate and may not be connivent and the cells are large. Plants from soil at high altitudes appear succulent, the leaves imbricate, very strongly concave, obliquely inserted, non-decurrent, the lobes are very connivent and the cells small. Plants from intermediate habitats such as the tops of *Sphagnum* tufts are intermediate in form. Whilst plants from very wet *Sphagnum* may be difficult to determine, those from more elevated parts of *Sphagnum* plants tend to have the characteristic facies of *C. pleniceps* plants found at high altitudes.

C. pleniceps differs from *C. lunulifolia* in the frequent presence of flagelliform shoots, the often larger cells and the autoecious inflorescence. Lax forms from wet micro-habitats may be mistaken for *C. connivens* or *C. loitlesbergeri* but differ when sterile in the non-decurrent or only slightly decurrent leaves; usually however, there are also shoots from drier micro-habitats with more concave leaves with less longly pointed lobes and in which flagelliform shoots are also usually present. High alpine forms from soil may closely resemble *C. bicuspidata* from similar habitats but the very strongly concave leaves and succulent appearance of the plants are characteristic. In such alpine plants perianths are usually present and the perianth mouth differs from that of *C. bicuspidata* and the perianth is bistratose to about halfway.

Section *Lacinulatae* Schust., *Hep. Anthoc. N. Amer.* 3: 795. 1974

Dioecious or autoecious. Plants not becoming reddish or brownish. Stems not branching terminally, flagelliform branches usually absent. Leaves obliquely to longitudinally inserted with decurrent dorsal margin, lobes connivent. Female inflorescence on dwarf ventral branch, bracts 3–5 lobed. Perianth mouth laciniate-ciliate.

8. C. loitlesbergeri Schiffn., *Österr. Bot. Zeitschr.* 62: 10. 1912   (Fig. 23)

Autoecious. Plants slender, pale or yellowish-green, in patches or creeping through other bryophytes. Shoots procumbent, to 1.5 cm, stems sometimes geniculate, branching irregular, flagelliform branches lacking. Leaves distant to approximate, rarely subimbricate or imbricate, ± spreading, rarely erecto-patent, obliquely broadly ovate, concave, insertion oblique to very oblique, dorsal margin decurrent, 220–360 (–440) μm long, 190–220(–385) μm wide, 1.0–1.3 times as long as wide, bilobed $\frac{1}{3}$ to almost $\frac{1}{2}$, lobes straight or more usually connivent and sometimes crossing, 4–6 cells wide at base, ending in 1–3 uniseriate cells, sinus rounded; cells at base of lobes 20–40 μm long, 20–32(–36) μm wide, walls somewhat thickened. Underleaves lacking. Gemmae lacking. Male inflorescences on dwarf branches, bracts smaller and more concave than leaves. Female bracts larger than leaves, 3–5 lobed to $\frac{2}{3}$–$\frac{3}{4}$, lobes acuminate, margin entire or toothed, bracteole 2–3-lobed. Perianths occasional, narrowly ellipsoid, plicate in upper half, unistratose except at base, contracted to ciliate mouth, cilia 4–6 cells long. Capsules occasional, autumn, winter. Growing over *Sphagnum* and *Leucobryum* in bogs and in moorland, very rarely on humus in wet rock clefts, ascending to *ca* 500 m, occasional in parts of N. Scotland, very rare elsewhere. Merioneth, W. Lancashire, M.W. Yorkshire and Durham north to W. Ross, W. Sutherland, Caithness and Shetland, Clare, W. Galway, Meath, Roscommon, W. Donegal, Armagh. 24, H6. N. Spain, Austria and Bulgaria north to Fennoscandia, north-east N. America, Greenland.

Fig. **23**. *Cephalozia loitlesbergeri*: 1, shoot (× 24); 2, perianth (× 24); 3, male branch (× 33); 4, leaves (× 63); 5, leaf lobe apices (× 84); 6, т.s. stem (× 84); 7, female bract (× 45); 8, perianth mouth (× 38); 9, cell from base of leaf lobe (× 450).

Distinguished from *C. connivens* by the smaller leaf cells, from *C. bicuspidata* by the oblique leaf insertion and decurrent dorsal margin. Some states of *C. loitlesbergeri* may only be distinguished with certainty from bog forms of *C. lunulifolia* by the autoecious inflorescence but well developed plants have more attenuated leaf lobes, sometimes ending in 3 uniseriate cells, which may cross, features not found in *C. lunulifolia*.

9. C. connivens (Dicks.) Lindb., *J. Linn. Soc. Bot.* 13: 190. 1872   (Fig. 24)

Autoecious. Plants slender, pale or yellowish-green, in patches or creeping through other bryophytes. Shoots procumbent, to 1.0(–1.5) cm, stems branching irregularly, flagelliform shoots lacking. Leaves distant to approximate, spreading horizontally to dorsally secund, obliquely orbicular, insertion very oblique to ± longitudinal, dorsal margin very decurrent, 240–480 μm long, 240–400 μm wide, mostly about as long as wide, bilobed ⅓–½, lobes connivent, (2–)3–6 cells wide at base, acute, ending in (1–)2 uniseriate cells, sinus rounded; cells at base of lobes (24–)32–56(–80) μm long, (20–)32–56 μm wide, walls ± thickened. Underleaves lacking. Gemmae rare, unicellular, at apices of shoots and branches. Male inflorescences on dwarf branches, bracts

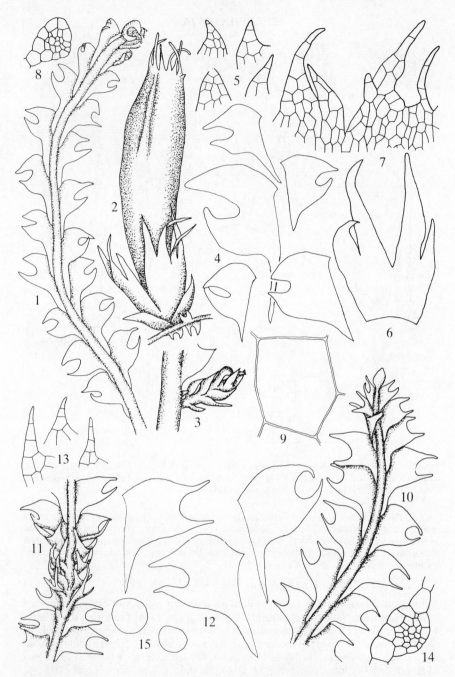

Fig. **24**. 1–9, *Cephalozia connivens*: 1, shoot (× 33); 2, perianth (× 21); 3, male branch (× 33); 4, leaves (× 63); 5, leaf lobe apices (× 84); 6, female bract (× 33); 7, perianth mouth (× 84); 8, т.s. stem (× 84); 9, cell from base of leaf lobe (× 450). 10–15, *C. hibernica*: 10, shoot (× 33); 11, male shoot (× 33); 12, leaves (× 63); 13, leaf lobe apices (× 84); 14, т.s. stem (× 84); 15, gemmae (× 450).

imbricate, concave, bilobed. Female inflorescences on short branches, bracts larger than leaves, 3–5-lobed to $\frac{1}{2}$, lobes acuminate, margin entire or with 1–2 teeth, bracteole bilobed, often with a tooth on one or both margins. Perianths frequent, cylindrical-fusiform, trigonous above, unistratose, contracted to ciliate mouth, cilia 3–5 cells long. Capsules frequent, autumn to spring. Moist peat, soil, sandstone, rotting logs and growing over *Sphagnum* and *Leucobryum* in woods, bogs and on wet heaths, usually at low altitudes but ascending to 650 m, occasional to frequent. 103, H35, C. Europe from the Iberian Peninsula, Corsica and Crete northwards, N. Asia, Japan, Azores, Madeira, N. America.

May be distinguished from *C. bicuspidata* by the very oblique leaf insertion, the longly decurrent dorsal margin and the absence of flagelliform shoots. For differences from *C. lunulifolia* see under that species.

10. C. hibernica Spruce ex Pears., *Irish Natural.* 3: 245. 1894   (Fig. 24)

Dioecious. Plants slender, whitish-green, in small patches or scattered shoots. Shoots procumbent, 3–12 mm, stems branching irregularly, flagelliform shoots lacking. Leaves distant to approximate or subimbricate, spreading horizontally, obliquely ovate, insertion ± longitudinal, dorsal margin very decurrent, 200–480(–560) μm long, 160–400(–440) μm wide, 1.2–1.3 (–1.5) times as long as wide, bilobed to *ca* $\frac{1}{3}$, lobes straight or connivent, 2–4 cells wide at base, ending in 2–3 uniseriate cells, sinus rounded, cells at base of lobes (32–)40–64(–80) μm long, 32–56 μm wide, walls ± thickened. Underleaves lacking. Gemmae spherical to ellipsoid, 20–24 μm, at tips of attenuated shoots. Male inflorescence terminal on stems and branches, becoming intercalary, bracts imbricate, concave, bilobed. Female bracts larger than leaves, bilobed to $\frac{1}{2}$, lobes acuminate, bracteole bilobed. Perianths rare, mouth ciliate, cilia 3–4 cells long (Macvicar, 1926). Heavily shaded, moist peat, sandstone and *Rhododendron ponticum* bark, very rare, Kerry, W. Galway, W. Mayo. H4. Azores, Madeira.

Closely related to *C. connivens* but dioecious, the leaves relatively narrower with the lobes only 2–4 cells wide at the base and frequently ending in 3 uniseriate cells. Slender stems have more shortly pointed leaves and are difficult to distinguish from *C. connivens* and mature shoots, which often have a wispy appearance under the microscope because of spreading leaf tips, must be sought for.

13. NOWELLIA MITT. IN GOODMAN, *Nat. Hist. Azores West Isl.*: 321. 1870

Usually dioecious. Small or medium-sized, frequently reddish or purplish plants. Shoots procumbent, stems branching ventrally, cortex hyaline, central cylinder opaque. Leaves with narrow base, insertion ± transverse, bilobed with ventral margin inflexed to form an inflated sac, margin entire or with tooth on ventral side, margins not decurrent, lobes longly acuminate; cell walls thickened, oil-bodies lacking. Underleaves lacking. Unicellular gemmae sometimes present. Male inflorescence on dwarf ventral branch or at tips of longer branches and becoming intercalary with age, bracts bilobed, subtending solitary antheridia. Female inflorescence on dwarf ventral branch, bracts larger than leaves, bilobed, dentate to ciliate. Perianth trigonous above, wide-mouthed, mouth ciliate. Capsule ovoid, wall 2-stratose. Spores 7–11 μm. A natural, mainly tropical or subtropical genus of 8 species with only a single north temperate representative.

1. N. curvifolia (Dicks.) Mitt. in Goodman, *Nat. Hist. Azores West Isl.*: 321. 1870
      (Fig. 25)

Dioecious or autoecious (?). Plants small, forming dense, thin, green to deep reddish-purple patches or creeping through other bryophytes. Shoots prostrate, to 1(–2) cm,

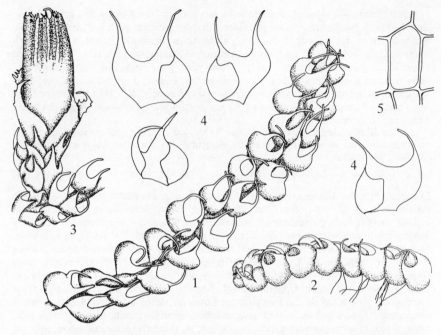

Fig. 25. *Nowellia curvifolia*: 1 and 2, shoots in dorsal and side view (× 21); 3, perianth (× 21); 4, leaves (× 38); 5, mid-leaf cell (× 450).

branching occasionally. Leaves loosely imbricate, dorsally secund, very concave, asymmetrical, base very narrow, insertion ± transverse, 0.5–1.0 mm long, ventral margin strongly inflexed and forming a swollen sac, bilobed to *ca* $\frac{1}{2}$, sinus rounded, lobes decurved, longly acuminate, composed of 4–7 elongate uniseriate cells; cells below sinus, 24–28(–32) μm wide, thick-walled. Underleaves lacking. Gemmae rare, unicellular, at stem tips. Female bracts 2-lobed, dentate. Perianths occasional, narrowly ovoid, plicate above, wide-mouthed, mouth ciliate. Capsules occasional, ovoid, spring, summer. $n = 9$. Rotting logs in humid woodland and on sheltered peaty banks, ascending to 400(–750) m, rare or occasional in S. England but possibly increasing, very rare or absent in the Midlands and E. Anglia, frequent from Mid and N. Wales and N.W. England northwards, frequent in the west and occasional elsewhere in Ireland. 90, H25. S., W. and C. Europe, Caucasus, Turkey, Japan, Taiwan, Java, Ceylon, Azores, Madeira, eastern N. America, Mexico, Guatamala and Costa Rica.

14. CLADOPODIELLA BUCH, *Mem. Soc. Fauna Fl. Fenn.* 1: 89. 1927

Dioecious. Plants slender to medium-sized, green to reddish-brown or brown. Stems irregularly branched, branches and flagelliform shoots arising ventrally, cortical cells not pellucid, stem in section with cortical cells not differentiated from inner cells. Leaves alternate, succubous, obliquely inserted, not decurrent, bilobed $\frac{1}{6}$–$\frac{1}{3}$, lobes obtuse to rounded; cells ± thick-walled. Underleaves small, lanceolate, apex acute or bilobed. Gemmae present or not. Male inflorescence terminal on short or long ventral branch, sometimes becoming intercalary with age, bracts closely imbricate, of similar shape to leaves but more concave. Female inflorescence on short ventral branch, bracts and bracteole larger than leaves, bilobed, margins entire or with a tooth on one or both sides. Perianth mouth crenulate. Spores 12–18 µm. Two species.

Leaves 0.55–1.30 mm long, mid-leaf cells (28–)32–44 µm wide   **1. C. fluitans**
Leaves 0.16–0.48 mm long, cells 20–32 µm wide   **2. C. francisci**

1. C. fluitans (Nees) Buch in Kalliola, *Ann. Bot. Soc. Zool.-Bot. Fenn. Vanamo* 2: 109. 1932   (Fig. 26)
Cephalozia fluitans (Nees) Spruce

Dioecious. Plants medium-sized, green to brown, in swollen patches or growing through other bryophytes. Shoots procumbent, to 5(–20) cm, stems not or hardly branched but frequently producing numerous flagelliform branches from ventral side and also vertical leafless branches penetrating substrate. Leaves distant, rarely approximate or subimbricate, spreading, sometimes slightly recurved, occasionally slightly dorsally secund, ovate-oblong, obliquely inserted, not decurrent, 0.55–1.30 mm long, 0.48–1.10 mm wide, 1.2–1.5(–1.7) times as long as wide, bilobed $\frac{1}{3}$–$\frac{1}{2}$, lobes often unequal, obtuse to rounded, sinus usually narrow, obtuse or rounded; mid-leaf cells (28–)32–44 µm wide, walls thin to thickened, colourless to pale brown. Underleaves appressed, lanceolate, bilobed or not, 150–350 µm long. Gemmae lacking. Male inflorescence spicate, terminal on short or long branch, becoming intercalary with age, bracts numerous, similar to leaves but more concave and imbricate. Female bracts larger than leaves, bilobed to *ca* $\frac{1}{4}$, lobes subacute, margins entire or with a single tooth on one or both sides. Perianths rare, cylindrical, trigonous towards mouth, mouth lobed and crenulate. Capsules rare, spring. $n = 9$. Growing over *Sphagnum* and sometimes submerged, in bogs and on wet heaths, ascending to *ca* 260(–600) m, occasional but sometimes locally common in suitable habitats in England and Wales, rare in Scotland and Ireland. 69, H15. Pyrenees, Alps and Yugoslavia north to Fennoscandia and Russia, Turkey, Siberia, Japan, Azores, Madeira, Morocco, N. America.

Very likely to be confused with *Gymnocolea inflata* which may also occur in similar habitats. *G. inflata* differs in its symmetrical leaves with smaller cells, the usually abundant, differently shaped perianths and lack of underleaves. The underleaves in *C. fluitans* may be difficult to detect as they are small and appressed to the stem; as in *C. francisci* they are most easily seen in side view. Although flagelliform shoots are usually lacking in *G. inflata* they are sometimes abundant and this also causes confusion.

The vertically growing flagelliform shoots which may penetrate the substrate to a depth of up to 30 cm (J.G. Duckett, pers. comm.) may act as organs of perennation and survival of burning.

2. C. francisci (Hook.) Joerg., *Bergens Mus. Skrift*. 16: 274. 1934   (Fig. 26)
Cephalozia francisci (Hook.) Dum.

Dioecious. Plants slender, in green to reddish-brown patches. Shoots to 8 mm, stems leafless and prostrate below, leafy and ascending above, producing branches and numerous flagelliform shoots from ventral side. Leaves subimbricate to closely imbricate, sometimes dorsally secund, rounded-quadrate to ovate-quadrate, concave to strongly concave, obliquely inserted, not decurrent, 160–480 µm long, 140–440 µm wide, 1.0–1.2 times as long as wide, bilobed $\frac{1}{6}$–$\frac{1}{3}$, lobes obtuse to rounded, rarely subacute in very small leaves, sinus acute; mid-leaf cells 20–32 µm wide, walls thickened, colourless or more usually reddish-brown. Underleaves usually appressed, lanceolate, acute or apex asymmetrical or unequally bilobed, 70–200(–240) µm long. Gemmae sometimes present at stem tips, 1–2-celled, ovoid or angular. Male inflorescence on short or long branch becoming intercalary with age, bracts closely

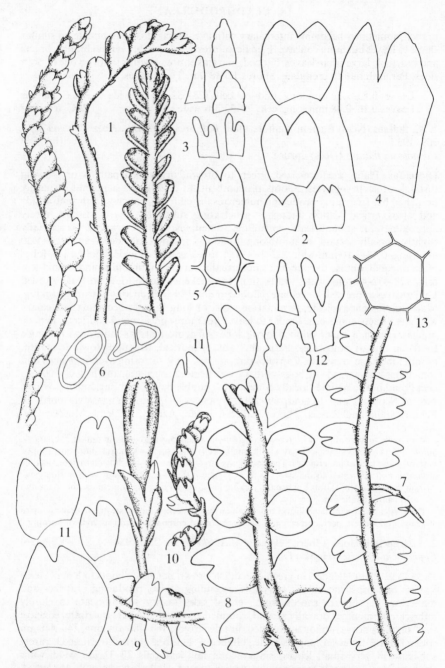

Fig. **26**. 1–6, *Cladopodiella francisci*: 1, shoots (× 38); 2, leaves (× 63); 3, underleaves (× 153); 4, female bract (× 63); 5, mid-leaf cell (× 450); 6, gemmae (× 450). 7–13, *C. fluitans*: 7 and 8, dorsal and ventral sides of shoots (× 15); 9, perianth (× 15); 10, male branch (× 15); 11, leaves (× 21); 12, underleaves (× 153); 13, mid-leaf cell (× 450).

imbricate, strongly concave. Female bracts larger than leaves, bilobed to *ca* ⅓, lobes rounded or obtuse. Perianths occasional, shortly cylindrical, plicate in upper half, hyaline near mouth, mouth lobed, crenulate. Capsules occasional, summer. On peat or peaty soil on wet heaths and in bogs, ascending to 350 m, occasional but sometimes locally frequent in suitable habitats in Britain, rare in Ireland. 41, H7. Spain and N. Italy north to subarctic Fennoscandia, Spitzbergen, Azores, Madeira, north-east N. America.

May be confused with small forms of *Gymnocolea inflata* but differs in the more closely imbricate and concave leaves, the presence of underleaves and in perianth shape.

15. PLEUROCLADULA GROLLE, *J. Bryol.* 10: 269. 1979

A monotypic genus with the characters of the species. Bracteole smaller than bracts. Perianth narrowly fusiform, fleshy towards base.

1. P. albescens (Hook.) Grolle, *J. Bryol.* 10: 269. 1979   (Fig. 27)
Pleuroclada albescens (Hook.) Spruce, *Pleuroclada islandica* (Nees) Pears., *Pleurocladula islandica* (Nees) Grolle

Dioecious. Plants whitish-green, becoming white when dry, in patches, tufts, or growing through other bryophytes. Shoots procumbent to erect, to 2.5(–3.0) cm, stems fleshy, in section with large, hyaline cortical cells, branching occasionally, branches lateral. Leaves distant to approximate or subimbricate, ± orbicular, very concave, insertion ± transverse, not decurrent, (0.4–)0.5–0.7 mm long, (0.44–)0.52–0.85 mm wide, 0.85–1.00 times as long as wide, bilobed to ⅓(–½), lobes incurved, broadly triangular, acute; cells pellucid, in mid-leaf 24–40 μm wide, walls thickened, oil-bodies lacking (Schuster, 1974). Underleaves ovate-lanceolate to ovate, 0.3–0.5 mm long, entire or with a single tooth on one or both sides. Gemmae rare, at shoot apices, spherical to ellipsoid, unicellular, 14–26 μm. Perianth and capsules not known in Britain. $n = 9$. Boulder scree in areas of late snow-lie, at altitudes of (600–)900–1200 m rare, Perth, Angus, Banff, Inverness, Argyll, Arran, Ross. 10. Montane and arctic Europe from N. Italy north to Spitzbergen, Faroes, Iceland, Siberia, Japan, N. America, Greenland.

Although a number of authorities (e.g. Grolle, 1983a), on the basis of isozyme studies by Krzakowa & Szweykowski (*Acta Soc. Bot. Polonicae* 50, 465–79, 1981), treat *P. islandica* as a distinct species, Paton (*J. Bryol.* 9, 1–8, 1976) presents convincing evidence for the reduction of *P. islandica* to synonymy with *P. albescens* even though British material allocated to the former taxon appears distinct.

SUBFAMILY HYGROBIELLOIDEAE

With only one genus.

16. HYGROBIELLA SPRUCE, *On Cephalozia*: 73. 1882

A monotypic genus with the characters of the species.

1. H. laxifolia (Hook.) Spruce, *On Cephalozia*: 74. 1882   (Fig. 27)

Dioecious. Plants very slender, in dull green to greenish-brown or reddish-brown patches or tufts or mixed with other bryophytes. Shoots procumbent to erect, to 2.5 cm, stem in section with large hyaline cortical cells and central strand of small cells, branches occasional, arising laterally, leafless or microphyllous flagelliform branches often present and sometimes constituting the whole plant. Leaves and underleaves ± similar so that leaves appear 3-ranked, lower, very small, distant, upper larger, subimbricate, sometimes successional, ovate, erecto-patent, conduplicate-concave

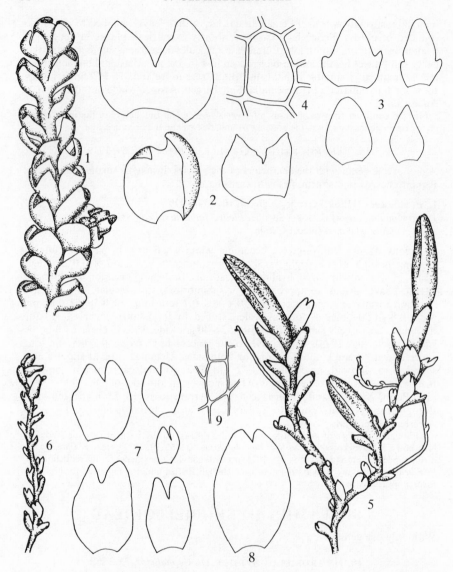

Fig. 27. 1–4, *Pleurocladula albescens*: 1, shoot (× 21); 2, leaves (× 38); 3, underleaves (× 38); 4, mid-leaf cell (× 380). 5–9, *Hygrobiella laxifolia*: 5, female shoot with perianth (× 12); 6, male shoot (× 12); 7, leaves (× 33); 8, female bract (× 33); 9, leaf cell (× 450).

with inflexed tips, insertion ± transverse, not decurrent, upper leaves 0.30–0.65 mm long, *ca* 1.5 times as long as wide, bilobed to *ca* ⅓, lobes obtuse, often unequal in size, sinus acute; mid-leaf cells 20–24 μm wide, walls thickened, yellowish-brown, without trigones, oil-bodies lacking. Gemmae lacking. Male bracts terminal, becoming intercalary, similar to upper leaves but with more concave base and a small lateral tooth, subtending solitary antheridia. Female bracts larger than leaves, uppermost 0.7–1.0 mm long, bilobed ₁₀–⅕, margins entire. Perianth developing whether or not fertilisation occurs, fusiform, trigonous above, mouth ciliate-dentate. Capsules rare, summer. Wet rocks and rocks in streams, on lakeside alluvia, to *ca* 1050 m, occasional to frequent in montane habitats, very rare elsewhere, E. Sussex, Brecon, Cardigan, Merioneth, Caernarfon, Derby, Yorkshire and Westmorland north to Shetland. 44, H15. Alps, Tatra, Norway, Sweden, Finland, Russia, Faroes, Iceland, Japan, Rockies and north-east N. America, Greenland.

Only likely to be mistaken for *Eremonotus myriocarpus* or *Sphenolobopsis pearsonii* but distinguished by the presence of underleaves of similar size to the leaves.

SUBFAMILY ODONTOSCHISMATOIDEAE

Plants medium-sized. Branches arising ventrally and, in some species, laterally, ventral flagelliform shoots often present, stem in section with cortical cells hardly differentiated from those of central strand. Leaves succubous, not lobed, margins entire, insertion oblique, dorsal margin not or only slightly decurrent; cells with large or very large trigones or uniformly thickened, oil-bodies mostly 2–4 per cell. Underleaves usually present, simple or bilobed, often with marginal slime papillae. Gemmae 1–2-celled, often present. Fertile branches dwarf, arising ventrally, male spicate with bilobed bracts. Female bracts and bracteole similar, larger than leaves, bilobed, margins toothed. Perianth trigonous above. Capsule wall 2–3-stratose.

17. ODONTOSCHISMA (DUM.) DUM., *Recueil Observ. Jungerm.*: 19. 1835

Dioecious. Plants medium-sized, green, reddish or reddish-brown to purplish-black. Stem prostrate with prostrate to ascending branches usually arising ventrally and sometimes attenuated and/or gemmiferous, ventral flagelliform shoots often present, stem in section with or without differentiated cortical cells. Leaves succubous, ± orbicular, entire, obliquely inserted; cells with large, often bulging trigones. Underleaves small except on gemmiferous shoots, often with slime papillae. Gemmae frequently present on margins or upper sometimes malformed leaves and underleaves, 1–2-celled, gemmiferous shoots attenuated or not. Inflorescences on dwarf ventral branches, male spicate, bracts much smaller than leaves, bilobed. Female bracts increasing in size up stem, uppermost larger than leaves, bilobed, bracteoles similar. Perianth trigonous above, contracted to lobed, ± entire to ciliate mouth. Capsule ovoid, wall 2–3-stratose. Spores 8–16 μm. A world-wide genus of 20–25 species.

1 Marginal leaf cells often radially elongated with walls ± uniformly thickened, forming ill-defined border, other cells with ± rounded lumens, gemmae lacking
1. O. sphagni
Marginal leaf cells ± similar to intramarginal cells, cell lumens often irregular because of bulging trigones, gemmae frequently present at shoot tips　2
2 Plants green, cell walls colourless, attenuated gemmiferous shoots lacking
2. O. macounii
Plants usually deeply pigmented, rarely green, cell walls usually at least faintly pigmented, attenuated gemmiferous shoots often present　3

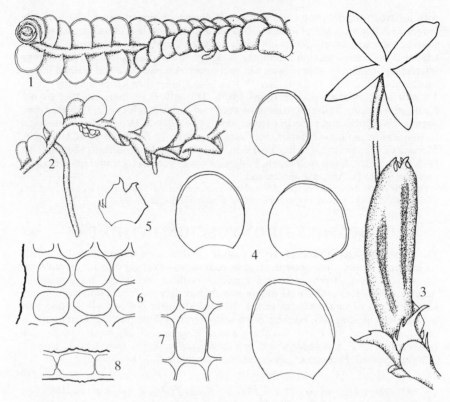

Fig. **28**. *Odontoschisma sphagni*: 1, shoot, dorsal side (× 11); 2, lateral view of shoot with male and flagelliform branches (× 11); 3, perianth and dehisced capsule (× 15); 4, leaves (× 23); 5, male bract (× 84); 6, marginal cells of leaf (× 450); 7, mid-leaf cell (× 450); 8, т.s. mid-leaf cell (× 450).

3 Leaf cell walls usually pale purplish-brown, papillose, middle lamella not visible,
      gemmae almost constantly present, thin-walled   **3. O. denudatum**
   Cell walls brown, smooth, middle lamella distinct, gemmae only occasionally
      present, thick-walled   **4. O. elongatum**

Section *Odontoschisma*

Leaves not strongly concave, 1–2 rows marginal cells radially elongated and with uniformly thickened walls, differing from intramarginal cells, walls ± pigmented; trigones large but not bulging. Capsule wall bistratose.

1. O. sphagni (Dicks.) Dum., *Recueil Observ. Jungerm.*: 19. 1835   (Fig. 28)

Dioecious. Plants medium-sized, green to reddish-brown, usually growing over *Sphagnum*, more rarely forming patches or tufts. Shoots procumbent, rarely ascending, to 5(–8) cm, stems simple or sparsely branched, branches arising ventrally, stout ventral flagelliform shoots common. Leaves approximate to imbricate, somewhat dorsally secund, ± orbicular to ovate-orbicular, concave, margin narrowly incurved, insertion oblique, dorsal margin not or scarcely decurrent, 0.68–1.28 mm long, 0.64–1.20(–1.32) mm wide. (0.9–)1.0–1.1(–1.25) times as long as wide; cells translucent, finely papillose, in mid-leaf 16–28(–32) µm wide, walls thickened, trigones large but not or hardly bulging, cell lumens ± rounded, 1–2 rows marginal cells often radially elongated with walls more uniformly thickened than submarginal cells and forming a poorly defined border. Underleaves rarely present, to 160 µm long larger on young shoots. Gemmae lacking. Male branches ventral, spicate, bracts imbricate, bilobed, with a single or double tooth on each side below. Female bracts bilobed, lobes acute, margins toothed. Perianths very rare, fusiform-cylindrical, plicate above. Capsules very rare, late spring, summer. $n = 8 + m^*$. On *Sphagnum*, wet peat, decaying vegetation and moist humus, ascending to 370(–550) m, frequent or common in suitable habitats, and sometimes locally abundant, widely distributed. 107, H39. Europe except extreme north, Faroes, Iceland, Azores, Madeira, north-east N. America, Greenland.

The inflexed leaf margin with somewhat differentiated cells gives the leaves the appearance of being distinctly bordered and thus distinguishing *O. sphagni* from other entire-leaved bog species as well as from other *Odontoschisma* species. The absence of underleaves is also used to separate *O. sphagni* from the next three species but as the underleaves in these species may be small and very difficult to detect and as underleaves do rarely occur in *O. sphagni* this is not a particularly useful character.

Section *Macouniae* Schust., *Hep. Anthoc. N. Amer.* 3: 848. 1974

Leaves very strongly concave; marginal cells ± similar to intramarginal cells, walls colourless, trigones large, bulging. Capsule wall bistratose.

2. O. macounii (Aust.) Underw., *Bull. Illinois State Lab. Nat. Hist.* 2: 92. 1884

(Fig. 30)

Dioecious. Plants small, forming small, pale green patches. Shoots procumbent, brittle, to 10 mm, stems branching occasionally, branches lateral, flagelliform branches arising ventrally. Leaves distant to imbricate, erecto-patent, somewhat dorsally secund, ± orbicular, very concave to ± hemispherical, dorsal margin inflexed, insertion oblique, not decurrent, 0.28–0.80 mm long, 0.28–0.80 mm wide, (0.8–)1.0(–1.2) times as long as wide; cells smooth, in mid-leaf 24–32 µm wide, walls colourless, trigones very large, bulging, making lumen irregular in outline, marginal cells ± similar to intramarginal cells. Underleaves triangular, 90–140 µm long, younger with marginal slime papillae. Gemmae produced on prostrate to ascending but not attenuated shoot tips, pale green, spherical to ovoid, 1–2-celled, thick-walled, 30–35 µm (Schuster, 1974). Fertile plants unknown in Britain. $n = 9$. Damp, base-rich soil on ledges, 700–1000 m, very rare, Mid Perth, Argyll, Ross. 4. Arctic-alpine habitats from the Italian Alps, Austria, Switzerland and Germany north to N. Russia and Spitzbergen, Iceland, Siberia, northern N. America, Greenland.

O. macounii differs from other British species of *Odontoschisma* in the main branches usually arising laterally rather than from the ventral side of the stem. The pale green colour and the often almost hemispherical leaves are characteristic though some shoots may have less concave leaves approaching those of *O. elongatum* which, however, differs in the brown or reddish-brown cell walls with obvious middle lamellae.

Section *Denudatae* Schust., *Hep. Anthoc. N. Amer.* 3: 833. 1974

Leaves concave but not very strongly so, marginal cells ± similar to intramarginal cells, walls pigmented, trigones large, bulging. Capsule wall bistratose.

3. O. denudatum (Mart.) Dum., *Recueil Observ. Jungerm.*: 19. 1835   (Fig. 29)

Dioecious. Plants in reddish-brown to purplish-brown, rarely green patches. Shoots procumbent, brittle, usually with erect or ascending gemmiferous branches arising

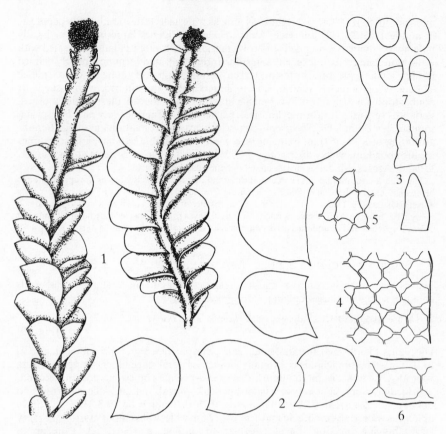

Fig. **29**. *Odontoschisma denudatum* 1, shoots (× 21); 2, leaves (× 38); 3, underleaf (× 84); 4, marginal cells of leaf (× 450); 5, mid-leaf cell (× 450); 6, T.s. leaf cell (× 450) 7, gemmae (× 450).

ventrally, flagelliform branches rare. Leaves of branches small or lacking below, often small and malformed towards gemmiferous apices, mature branch leaves and stem leaves imbricate, erecto-patent, rarely dorsally secund, ± orbicular to ovate-orbicular, very concave, margin incurved and dorsal margin frequently inflexed, insertion oblique, dorsal margin slightly decurrent, 0.48–1.12 mm long, 0.48–1.12(–1.35) mm wide, (0.8–)1.0–1.1(–1.2) times as long as wide; cells opaque, papillose, mid-leaf cells (20–)24–28(–32) μm wide, walls usually pale purplish-brown, middle lamella not visible, trigones very large, frequently bulging so that lumen outline is irregular, marginal 1–2 rows of cells ± similar to submarginal. Underleaves usually only present towards stem apices, very small lanceolate, 120–230 μm long, bilobed or not, sometimes larger and gemmiferous on gemmiferous branches. Gemmae in fascicles, almost always present and especially on attenuated shoots, pale green, ovoid to ellipsoid, thin-walled, 1–2-celled, 20–28 μm long. Perianths and capsules very rare. $n = 9$. Damp peat on banks, heaths and bogs, on tree stumps, rotting logs, shaded sandstone and thin soil over boulders in scree, usually at low altitudes, rare in lowland Britain, occasional elsewhere. 73, H31. Europe, Faroes, Siberia, Nepal, Japan, Formosa, Azores, Madeira, N., C., and S. America south to Brazil, Caribbean.

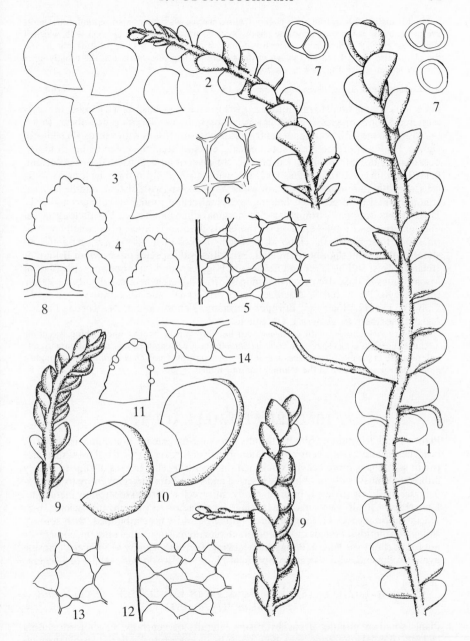

Fig. **30**. 1–7. *Odontoschisma elongatum*: 1, shoot (× 15); 2, gemmiferous shoot (× 15); 3, leaves (× 23); 4, underleaves (× 153); 5, marginal cells of leaf (× 450); 6, mid-leaf cell (× 450); 7, gemmae (× 450); 8, т.s. leaf cells (× 450). 9–14, *O. macounii*: 9, shoots (× 15); 10, leaves (× 38); 11, underleaf (× 153); 12, marginal cells of leaf (× 450); 13, mid-leaf cell (× 450); 14, т.s. leaf cell (× 450).

Readily distinguished from other *Odontoschisma* species when attenuated gemmiferous shoots
are present but patches of prostrate shoots may be mistaken for *O. sphagni* with which it
sometimes grows. Gemmiferous apices may usually be found even in some prostrate shoots and
the plants also differ from *O. sphagni* in the paucity of flagelliform shoots, the usually more
concave leaves and leaf margin areolation.

4. O. elongatum (Lindb.) Evans, *Rhodora* 14: 13. 1912   (Fig. 30)

Dioecious. Plants small to medium-sized, in brown to brownish-black patches or rarely
scattered shoots or growing through other bryophytes. Shoots procumbent, to 1.5
(–3.0) cm, stems with occasional ventral branches, sometimes with ventral flagelliform
shoots. Leaves approximate to subimbricate, erecto-patent to spreading horizontally,
occasionally dorsally secund, ± orbicular, concave, obliquely inserted, not decurrent,
0.44–0.88 mm long, 0.52–0.96(–1.12) mm wide, (0.7–)0.8–1.0 times as long as wide;
cells, smooth, in mid-leaf, 20–28 μm wide, walls brown, with distinct middle lamella,
trigones small to large, often bulging, marginal cells ± similar to submarginal cells.
Underleaves irregularly triangular with marginal slime papillae, *ca* 120 μm long, larger
and sometimes with bifid apex on gemmiferous shoots. Gemmae occasionally present
on shoot tips, brownish, ellipsoid, 2-celled, thick-walled, 24–34 μm, gemmiferous
shoots not or only slightly attenuated. Perianths and capsules unknown in Britain or
Ireland. Moist soil by lakes and flushes, on boggy ground on moorland, calcifuge, at
low altitudes in the extreme north and in W. Ireland, elsewhere usually above 300 m,
rare, Argyll, Perth and S. Aberdeen north to Orkney and Shetland, W. Cork, W.
Galway. 11, H2. Spain, C. Europe, Germany, Fennoscandia, N. Russia, Faroes,
Iceland, northern N. America, Greenland.

Likely to be confused with *O. denudatum* but differing in habitat and lacking ascending
attenuated gemmiferous shoots; in *O. denudatum* the cells are papillose and the walls are usually
pale purplish or purplish-brown, never reddish-brown or brown as in *O. elongatum* and the middle
lamella is not apparent and the gemmae are thin-walled.

10. CEPHALOZIELLACEAE

Plants small to minute. Shoots usually procumbent, stems branching irregularly,
branches usually arising ventrally or laterally, never furcate, cortical cells of similar size
to but sometimes more heavily thickened than central cells. Leaves distant to closely
imbricate, transversely or obliquely inserted and succubous, margins entire to spinose-
dentate, shallowly to deeply bilobed, rarely entire; cell walls thin to strongly incrassate.
Underleaves small, very small or lacking. Gemmae often present, in fascicles, 1–2-
celled. Inflorescences on long or short branches. Male bracts bilobed, with solitary
axillary antheridia. Female bracts and bracteoles increasing in size up stem, connate or
not for part of their length. Perigynium lacking. Perianth usually 4–5-plicate. Capsule
ovoid to ellipsoid, wall 2-stratose. Spores 6–12 μm. A ± world-wide family of 7 genera.

18. CEPHALOZIELLA (SPRUCE) SCHIFFN. IN ENGLER & PRANTL, *Natürl.*
Pflanzenfam. 1(2): 98. 1893

Plants small or minute. Branches arising laterally or ventrally, flagelliform shoots
lacking, stems strongly anisophyllous, not translucent, central cells not visible through
cortex by transmitted light. Leaves usually transversely inserted and orientated,
margins entire to spinose-dentate, bilobed to 0.5 or almost to base; cell walls thin to
strongly thickened, oil-bodies small, 3–9 per cell. Gemmae borne on leaf lobe tips at
stem apices, usually 2-celled, smooth, angular or coarsely papillose. Female inflores-

cence with bracts in 1–several pairs, bilobed, margins entire to spinose-dentate, bracteole not differing markedly in size, often connate with bracts for part or whole of its length. A ± cosmopolitan genus of up to about 90 species.

Many species are phenotypically variable and can often only be determined with certainty by the presence or absence of underleaves in **sterile** stems and the sex of plants. Underleaves are best observed in side view of the stem. Presence of underleaves on gemmiferous or fertile shoots varies within a species and cannot be used in determinations. Male inflorescences may be terminal on long branches and because of the sometimes fragile nature of such branches great care may be needed in deciding whether a plant is autoecious or dioecious. The relative length of the cells of the perianth mouth may help in differientiating some species. The nature of the gemmae, whether smooth, angular or coarsely papillose is useful for separating the subgenera but they are easily lost from dried specimens or when washing soil from plants.

1 Leaf margins of both sterile and fertile stems entire   2
   Leaves toothed or at least some on sterile stems with 1–2 spinose teeth near base 8
2 Sterile non-gemmiferous stems lacking underleaves   3
   Sterile stems with underleaves   7
3 Leaves on sterile stems obliquely inserted, cells at leaf lobe bases mostly 16–20 μm
     wide, innermost female bracts deeply bilobed, calcicole   **5. C. baumgartneri**
   Leaves ± transversely inserted, cells mostly 8–15 μm wide or if wider then innermost
     female bracts not or only shallowly lobed, calcifuge   4
4 Innermost female bracts deeply lobed, margins entire to dentate, leaf lobe cells
     mostly 8–15 μm wide, gemmae smooth   5
   Innermost female bracts fused into unlobed or shallowly lobed entire sheath round
     perianth, leaf cells 10–20(–24) μm wide, gemmae angular   6
5 Usually paroecious, lobes of leaves of sterile stems mostly 3–5 cells wide, cells 8–12
     μm wide   **3. C. rubella**
   Autoecious, leaf lobes mostly 6–9 cells wide, cells mostly 12–15 μm wide
                                                                                4. C. hampeana
6 Sheath round perianth shallowly lobed, leaves of sterile stems bilobed to *ca* 0.5, cells
     of leaves of fertile stems thin-walled   **12. C. integerrima**
   Sheath not lobed, leaves bilobed to 0.6–0.7, cell walls of leaves of fertile stems
     thickened   **13. C. calyculata**
7 Paroecious, cuticle smooth   **7. C. stellulifera**
   Cuticle papillose or if smooth then plants dioecious   13
8 Underleaves absent from sterile stems   9
   Underleaves present on sterile stems   10
9 Lobes of leaves on sterile stems 6–12 cells wide, gemmae angular   **11. C. turneri**
   Lobes mostly 2–4 cells wide, gemmae smooth   11
10 Lobes of leaves of sterile stems mostly 2–4 cells wide, lobe cells mostly about twice
     as long as wide   11
   Lobes mostly 4–9 cells wide, some or all lobe cells 1.5 times as long as wide or less
                                                                                              12
11 Cells at base of leaf lobes of sterile stems mostly 8–13 μm wide, teeth of female
     bracts neither spinose nor recurved   **1. C. spinigera**
   Cells 13–20 μm wide, bract teeth often spinose and recurved   **2. C. elachista**
12 Dioecious, cells at base of leaf lobes of sterile stems 16–24 μm wide, cuticle smooth,
     gemmae coarsely papillose   **10. C. dentata**
   Cells 10–16 μm or if more then papillose and plants paroecious, gemmae smooth
                                                                                              13
13 Dioecious, leaves on sterile stems bilobed to 0.5–0.6, cells at lobe bases 8–14 μm
     wide, cuticle smooth to papillose, cells at perianth mouth mostly 3–4 times as
     long as wide, habitat various   **6. C. divaricata**

Dioecious or paroecious, leaves bilobed to 0.6–0.9, cells 10–24 μm wide, mouth cells mostly 5–11 times as long as wide, cuticle papillose, on cupriferous substrates

14

14 Cortical cells and cells of leaf lobe bases of sterile stems 10–16 μm wide

8. C. massalongi

Cortical and leaf lobe cells 16–20(–24) μm wide   **9. C. nicholsonii**

Subgenus *Cephaloziella*

Leaves of sterile stems entire or dentate, some with 1–2 spinose teeth near base. Underleaves absent or present and small to large. Gemmae green to reddish-brown or purple, ± ellipsoid, never angular or papillose. Bracts bilobed, innermost fused to lobed bracteole for 0–½ their length.

Section *Schizophyllum* (K. Müll.) Joerg., *Bergens Mus. Skrift.* 16: 188. 1934

Stem cortex somewhat translucent on dorsal side. Leaves on sterile stems bilobed to 0.6–0.9, lobes 2–4(–5) cells wide at base, margins entire to spinose-dentate, some with 1–2 spinose teeth near leaf base. Underleaves very small or absent. Plants autoecious, male inflorescences on short branches adjacent to female. Female bracts dentate or spinose-dentate, innermost fused to bracteole at bases.

1. C. spinigera (Lindb.) Joerg., *Bergens Mus. Skrift.* 16: 189. 1934   (Fig. 31)
C. striatula (C. Jens.) Douin, *C. subdentata* Warnst.

Autoecious. Plants very slender, pale green to purplish red, creeping over other bryophytes. Shoots procumbent, stems branching sparsely and irregularly, cortex on dorsal side somewhat translucent, cortical cells 26–40(–52) μm long, 12–20(–32) μm wide. Leaves on sterile stems distant, erecto-patent, transversely inserted, concave, 100–190 μm long, 64–140 μm wide, margins entire or sinuose-dentate, some with a spinose tooth near base, bilobed to 0.6–0.9, sinus acute, lobes lanceolate or narrowly lanceolate, acuminate, often incurved but not connivent. 2–4(–6) cells wide at base, these cells 8–13(–16) μm wide, often thick-walled, cuticle papillose. Leaves on fertile stems dentate to spinose-dentate. Underleaves on sterile stems absent or sparse, subulate or bilobed, 34–80 μm long. Gemmae rare, ellipsoid, 2-celled, 20–25 μm long, 9–10 μm wide (Schuster, 1980). Male inflorescence terminal on short ventral branch near female or successively and becoming intercalary on long branch, spicate, often purplish-red, bracts larger than leaves, imbricate, concave, often subsecund, margins toothed but not spinosely so. Female inflorescence terminal on long shoot, bracts bilobed, margins dentate but not spinosely so and teeth not recurved, cell walls strongly thickened, bracteoles similar. Perianth ellipsoid, plicate above, mouth irregularly crenulate, cells at mouth very heavily thickened, 25–125 μm long, 5–10 μm wide, 3.5–7.5 times as long as wide. Capsules occasional, late spring, early summer. Growing over *Sphagnum* and *Leucobryum* and on peat in bogs and on heaths, calcifuge, ascending to *ca* 600 m, rare, scattered localities from Cornwall east to Sussex and Surrey and north to Shetland, very rare in Ireland, S. Kerry, Clare, Wicklow, Westmeath, W. Mayo, W. Donegal. 25, H6. Rare in Europe from Spain and Hungary north to Fennoscandia and N. Russia, Iceland, Siberia, northern N. America, Greenland.

  C. spinigera and *C. elachista* are very closely related and it is likely that, as Müller (1957) suggests, the latter is a diploid derivative of the former. The cells of *C. spinigera* are generally smaller and more heavily thickened than those of *C. elachista*. In *C. elachista* some of the teeth of

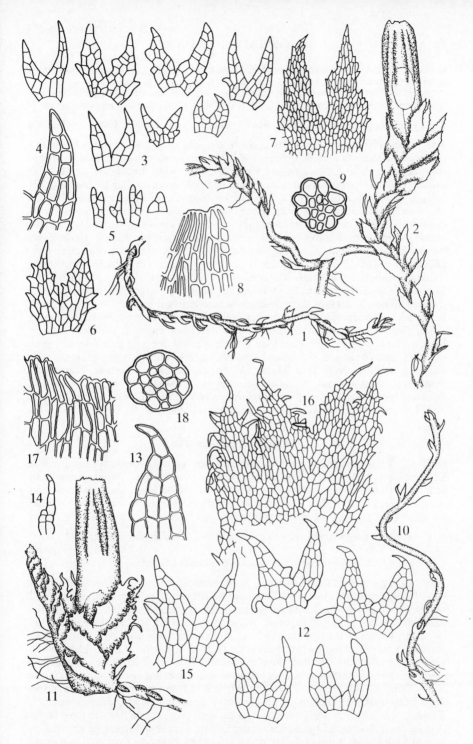

Fig. **31**. 1–9, *Cephaloziella spinigera*: 1, sterile shoot (× 38); 2, autoecious shoot with perianth (× 38); 3, leaves from sterile shoots (× 150); 4, lobe of leaf from sterile shoot (× 270); 5, underleaves from sterile stem (× 150); 6, male bract (× 150); 7, female bract (× 84); 8, portion of perianth mouth (× 270); 9, т.s. sterile stem (× 210). 10–18, *C. elachista*: 10, sterile shoot (× 38); 11, autoecious shoot with perianth (× 38); 12, leaves from sterile stem (× 150); 13, lobe of leaf from sterile stem (× 270); 14, underleaf from sterile stem (× 150); 15, male bract (× 150); 16, female bract and bracteole (× 84); 17, portion of perianth mouth (× 270); 18, т.s. sterile stem (× 210).

the female bracts are spinose and recurved whilst those of *C. spinigera* are neither; the cells at the mouth of the perianth are shorter and wider in the former than in *C. elachista*.

The spinose teeth at the base of some of the leaves of sterile stems will distinguish these two species from other *Cephaloziella* species. The cells of the leaf lobes are also more elongated.

2. C. elachista (Jack ex Gott. & Rabenh.) Schiffn., *Lotos* 18: 338. 1900   (Fig. 31)

Autoecious. Plants very slender, pale green, creeping over other bryophytes. Shoots procumbent, stems branching sparsely and irregularly, cortex on dorsal side somewhat translucent, cortical cells 32–64 μm long, 16–24 μm wide. Leaves on sterile stems distant, patent or erecto-patent, transversely inserted, concave, 100–220 μm long, 50–140 μm wide, margin often with 1–2 spinose teeth near base, otherwise entire, bilobed to 0.5–0.7, sinus acute, lobes lanceolate, acute to acuminate, often incurved and connivent, 2–4(–5) cells wide at base, these cells 13–20 μm wide, thick-walled or not, cuticle smooth. Upper leaves on fertile stems dentate to spinose-dentate. Underleaves on sterile stems absent or sparse, ±subulate, to 80–90 μm long. Gemmae rare, ellipsoid, 2-celled, 25–27 μm long, 10–11 μm wide, (Müller, 1957). Male inflorescence terminal on short ventral branch near female or successively on long branches and becoming intercalary, spicate, bracts imbricate, subsecund, larger than leaves, bilobed, margins spinosely dentate. Female inflorescence terminal on short ventral branch, bracts bilobed, margins spinosely toothed, teeth frequently recurved; bracteoles similar but smaller. Perianth mouth irregularly crenulate, cells at mouth thickened, 32–48 μm long, 8–12 μm wide, 3–5 times as long as wide. Capsules very rare, spring. Creeping over *Sphagnum* or mixed with other bryophytes in bogs and on boggy ground at low altitudes, calcifuge, very rare, Hants., E. Sussex, Berks, E. Norfolk, Shropshire, Pembroke, Westmorland, Cumberland, Berwick, Dublin, Westmeath. 10, H2. Scattered localities in Europe from Spain, Italy and Yugoslavia north to Fennoscandia and N. Russia (but possibly only to Denmark), N. America.

Section *Cephaloziella*

Stem cortex not translucent on dorsal side. Leaves on sterile stems bilobed 0.5–0.7, lobes mostly 4–9 cells wide at base, margins entire or dentate but without spinose teeth near base. Underleaves present or not. Sex various, autoecious plants with male and female inflorescences widely separated. Innermost female bracts and bracteole fused for up to ⅔ their length, lobes with margins entire or dentate.

3. C. rubella (Nees) Warnst., *Kryptogamenfl. Mark Brandenburg* 1: 231. 1902

(Fig. 32)

C. hampeana var. *pulchella* Jens., *C. myriantha* (Lindb.) Schiffn.

Paroecious, paroecious and autoecious, or occasionally completely autoecious. Plants very slender, green or more usually reddish or reddish-brown, patches or scattered shoots. Shoots procumbent, stems branching sparsely and irregularly. Leaves on sterile stems distant, erecto-patent to patent, transversely inserted, 80–160 μm long, 60–100 μm wide, margins entire, bilobed to 0.5–0.7, sinus acute, lobes divergent, lanceolate to narrowly triangular, acute to acuminate, (2–)3–5(–6) cells wide at base, these cells 8–12(–14) μm wide, thick-walled, cuticle ± smooth. Leaves on fertile stems ± similar, margins entire. Underleaves lacking on sterile stems, lacking or very small on fertile stems. Gemmae occasional, irregularly ellipsoid, 14–26 μm long, 12–14 μm wide. Antheridia in axils of bracts below female inflorescence or male inflorescence terminal, spicate, becoming intercalary on long branch, male bracts of similar size to or larger than leaves, imbricate, margins entire or denticulate. Female inflorescence terminal on stem or long branch, bracts bilobed, margins toothed, bracteole narrower, innermost

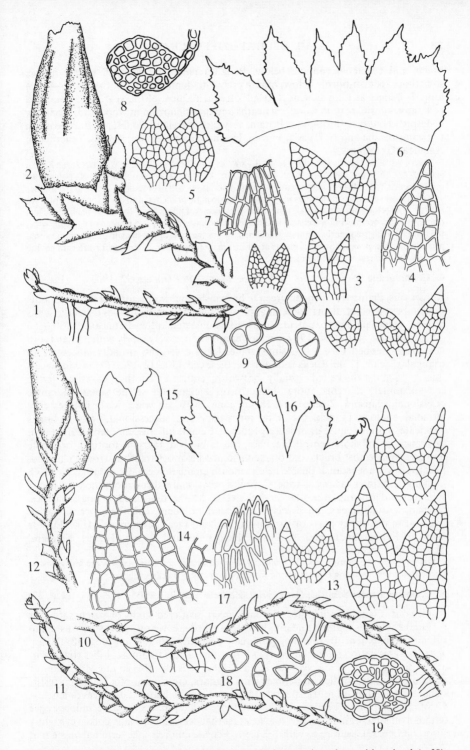

Fig. **32**. 1–9, *Cephaloziella rubella*: 1, sterile shoot (× 38); 2, paroecious shoot with perianth (× 38); 3, leaves from sterile shoots (× 150); 4, lobe of leaf from sterile shoot (× 270); 5, male bract (× 150); 6, female bracts and bracteole (× 84); 7, portion of perianth mouth (× 270); 8, т.s. sterile stem (× 210), 9, gemmae (× 450). 10–19, *C. hampeana*: 10, sterile shoot (× 38); 11, male shoot (× 38); 12, female shoot with perianth (× 38); 13, leaves from sterile stem (× 150); 14, lobe of leaf from sterile stem (× 270); 15, male bract (× 84); 16, female bracts and bracteole (× 63); 17, portion of perianth mouth (× 270); 18, gemmae (× 450); 19, т.s. sterile stem (× 210).

bracts and bracteole connate below. Perianth narrowly ellipsoid, plicate above, sometimes reddish-purple below, mouth crenulate-denticulate, mouth cells 16–42 μm long, 3–7 times as long as wide. Capsules frequent, most of the year. Moist soil and peat, mine-waste, rotting wood, on heaths and moorland and in woodland, calcifuge, widespread but rare throughout Britain, very rare in Ireland. 74, H6, C. Europe north to Fennoscandia and N. Russia, Faroes, Iceland, Siberia, Japan, Azores, Tenerife, N. America, Greenland.

A very variable species for which several varieties have been described. Whether these have a genetic or only an environmental basis is not known. The combination of paroecious inflorescence, entire leaves, cell size and lack of underleaves will distinguish *C. rubella* from other species. However, some gatherings may have separate male branches, and occasional samples are entirely autoecious but can only be referred to this species. Differs from *C. hampeana* in the usual presence of paroecious inflorescences and the leaf-lobe bases mostly 3–5 instead of 6–9 cells wide, the leaf cell walls frequently strongly thickened and in the lack of the brownish coloration often present in *C. hampeana*; however, plants of uncertain identity occur. *C. divaricata* differs in having a dioecious inflorescence and underleaves.

4. C. hampeana (Nees) Schiffn. in Loeske, *Moosfl. des Harzes*: 92. 1903   (Fig. 32)

Autoecious. Plants very slender, green to brownish patches or scattered shoots. Shoots procumbent to erect, stems branching sparsely and irregularly. Leaves on sterile stems distant, erecto-patent to patent, transversely inserted, often ± concave, (130–)160 (–220) μm long, (110–)130–190 μm wide, margins entire, very rarely with a small tooth near base, bilobed to 0.5–0.7, sinus acute, lobes divergent, broadly lanceolate to triangular, acute, (5–)6–9 cells wide at base, these cells (10–)12–15(–16) μm wide, walls thin or slightly thickened, usually brownish, cuticle ± smooth. Leaves on fertile stems ± similar, margins entire. Underleaves lacking on sterile stems. Gemmae occasional, ellipsoid, 2-celled, 14–18 μm long, 8–10 μm wide. Male inflorescence terminal on long branch, becoming intercalary and successional, spicate, bracts imbricate, margins entire or obscurely toothed. Female inflorescence terminal on stem or branch or on short ventral branches, bracts bilobed, margins toothed, bracteole narrower, innermost bracts and bracteole joined below. Perianth narrowly ellipsoid, plicate above, not reddish-purple below, mouth crenulate-dentate, mouth cells 18–48 μm long, 2–5 times as long as wide. Capsules occasional ± throughout the year. Moist soil, peat and growing over other bryophytes, on heaths, moorland, in bogs, marshes, on banks, dune-slacks, mine-waste, cliffs and boulders, calcifuge, occasional to frequent in the wetter parts of Britain and Ireland, rare elsewhere. 92, H32. Europe north to Fennoscandia and N. Russia, Faroes, Iceland, Japan, Turkey, Madeira, Tenerife, N. America, Greenland.

Differs from other entire-leaved *Cephaloziella* species lacking underleaves in the leaf lobes on sterile stems mostly being 6–9 cells wide instead of mostly 3–5 cells wide.

5. C. baumgartneri Schiffn., *Österr. Bot. Zeitschr.* 55: 200. 1905   (Fig. 33)

Autoecious. Plants minute, green to dark green, scattered stems or patches. Shoots procumbent, stems sparsely and irregularly branched. Leaves on sterile stems distant to subimbricate, patent to spreading, obliquely inserted, 130–190 μm long, 80–180 μm wide, margins entire, bilobed to (0.4–)0.5–0.6, sinus acute to obtuse, lobes triangular, acute, divergent, (3–)4–5 cells wide at base, these cells (14–)16–20 μm wide, thin- or more usually thick-walled sometimes with trigones, cuticle smooth; leaves on fertile stems ± similar but more transversely inserted, margins entire. Underleaves lacking. Gemmae ± ellipsoid, 2-celled, 13–20 μm long, 8–12 μm wide. Male inflorescence terminal on long branch, spicate, bracts larger than leaves, imbricate, concave, slightly subsecund, margins entire or with 1–2 teeth. Female inflorescence terminal on stem or

Fig. **33**. *Cephaloziella baumgartneri*: 1, sterile shoot (× 38); 2, male shoot (× 38); 3, female shoot with perianth (× 38); 4, leaves from sterile shoots (× 150); 5, lobe of leaf from sterile shoot (× 270); 6, male bract (× 84); 7, female bracts and bracteole (× 63); 8, т.s. sterile stem (× 210); 9, gemmae (× 450); 10, portion of perianth mouth (× 270).

branch, innermost bracts and bracteole fused to form irregularly lobed sheath round perianth, lobes acute, margins entire or with scattered teeth. Perianth ellipsoid, plicate above, mouth lobed, crenulate-denticulate, mouth cells 16–32 μm long, 1.5–5.0 times as long as wide. Capsules frequent, spring. On chalk and chalky soil in exposed situations, calcicole, very rare, Dorset, I. of Wight, Sussex, E. Kent, Surrey. 6. Mediterranean islands and region of Europe, Portugal, France, Belgium, Cyprus, Turkey, Azores, Madeira, Gran Canaria.

The only markedly calcicole species of *Cephaloziella* in the British Isles. May be confused with *C. rubella* which differs in the smaller leaf cells, and usually paroecious inflorescence.

6. C. divaricata (Sm.) Schiffn. in Engler & Prantl., *Natürl. Pflanzenfam*. 1, 3(1): 99. 1893 (Fig. 34)
C. byssacea sensu Schuster non (Roth) Warnst., *C. starkei* auct.

Dioecious. Plants very slender, reddish-brown, purplish-brown or blackish, rarely green, usually in dense, sometimes extensive patches or growing through other bryophytes. Shoots procumbent to ascending or erect, stems sparingly and irregularly

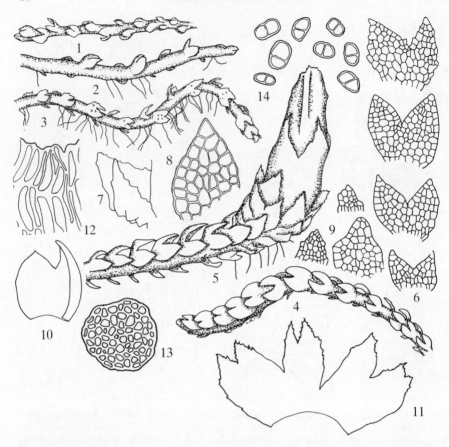

Fig. **34**. *Cephaloziella divaricata*: 1, and 2, sterile shoots in dorsal and lateral view (× 63); 3, shoot of form with dentate, papillose leaves (× 63); 4, male shoot (× 63); 5, female shoot with perianth (× 63); 6, leaves from sterile shoots (× 150); 7, leaf from shoot 3 in side view (× 150); 8, lobe from leaf from sterile stem (× 270); 9, underleaves (× 150); 10, male bract (× 63); 11, female bract and bracteole (× 33); 12, portion of perianth mouth (× 270); 13, т.s. sterile stem (× 210); 14, gemmae (× 450).

branched. Leaves on sterile stems distant to imbricate, erect to erecto-patent from ± spreading base, transversely inserted, channelled or concave, 100–200 μm long, 90–180 μm wide, margins entire or dentate, occasionally spinose-dentate below, bilobed to 0.5–0.6, sinus acute, lobes triangular, acute to obtuse, tips sometimes inflexed, 4–9 cells wide at base, these cells (8–)10–14 μm wide, walls usually thickened, brown, cuticle smooth or with small papillae, sometimes with large conical projections at back below sinus. Leaves on fertile stems similar, entire to dentate. Underleaves present on sterile stems, lanceolate to triangular, sometimes bilobed, 60–112 μm long. Gemmae frequent, yellowish-green to purplish-brown, irregularly ellipsoid, 2-celled, 12–20 μm long, 8–14 μm wide. Male inflorescence terminal, becoming intercalary, longly spicate, bracts numerous, slightly dorsally secund, imbricate, concave. Female inflorescence terminal, bracts and bracteole dentate to spinose-dentate, innermost

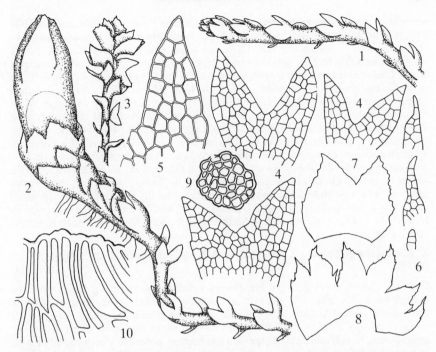

Fig. 35. *Cephaloziella stellulifera*: 1, sterile shoot (× 38); 2, paroecious shoot with perianth (× 38); 3, young fertile shoot tip (× 38); 4, leaves from sterile stem (× 150); 5, lobe of leaf from sterile stem (× 270); 6, underleaves (× 150); 7, male bract (× 84); 8, female bracts and bracteole (× 33); 9, T.S. sterile stem (× 210); 10, portion of perianth mouth (× 270).

connate below, upper margins and apices sometimes hyaline. Perianth strongly plicate above, mouth crenulate to spinosely denticulate, mouth cells 38–48 μm long, 3–4 times as long as wide. Capsules occasional, ± throughout the year. On damp to more usually dryish peat, soil, soil-covered rocks, sandy and gravelly soil, on banks, heaths, moorland, cliffs, walls, mine-waste, dune-slacks, in bogs and scree, rarely on rotting wood, calcifuge, ascending to *ca* 1000 m, frequent to common. 107, H28, C. Europe north to Spitzbergen, Faroes, Iceland, Turkey, N. Africa, Azores, Madeira, Canaries, N. America, western S. America, Greenland.

The most frequent species of the genus in Britain. It is extremely variable and forms with papillose cuticle, large conical outgrowths on the back of the leaf below the sinus and leaf margin dentate, sometimes spinosely so below, have been named var. *asperifolia* (Jens.) Macv. but these intergrade completely with more typical forms and cannot be maintained at varietal rank.

Differs from entire-leaved species except *C. stellulifera*, and forms of *C. massalongi* and *C. nicholsonii* by the sterile stems having underleaves. *C. stellulifera* differs in the usually paroecious inflorescence and relatively longer perianth mouth cells. For the differences from the other two species see dichotomy 13 of key to species.

7. C. stellulifera Schiffn., *Österr. Bot. Zeitschr.* 55: 289. 1905 (Fig. 35)

Paroecious, occasionally autoecious or both. Plants very slender, in yellowish-green to reddish-brown patches. Shoots procumbent to ascending, stems sometimes geniculate, sparsely and irregularly branched. Leaves on sterile stems distant, erecto-patent to spreading, ± transversely inserted, somewhat concave or not, 100–255 μm long, 90–210 μm wide, margins entire, bilobed to 0.6–0.7, sinus acute, lobes lanceloate, acute, 4–6(–8) cells wide at base, these cells 12–16 μm wide, walls ± thickened or not, cuticle smooth. Leaves on fertile stems not much larger, erecto-patent to spreading, margins entire. Underleaves present on sterile stems, lanceolate, sometimes bilobed, 40–110 μm long. Gemmae ± ellipsoid, 2-celled, 17–24 μm long (Müller, 1957). Young inflorescences often with spreading bracts which become erecto-patent with age. Male inflorescence below female or at apex of long branch and becoming intercalary with age, bracts larger than leaves, imbricate, concave with entire to denticulate margins. Female bracts and bracteoles larger than male or subtending leaves if autoecious, innermost fused below, margins entire to dentate. Perianth plicate above, mouth crenulate, mouth cells 30–65 μm long, 4–8 times as long as wide. Capsules frequent, late winter to spring. On soil in turf, amongst rocks, on banks, heaths and mine-waste, calcifuge, at low altitudes, occasional but sometimes locally abundant, from Cornwall and Kent north to Yorkshire and Durham, I. of Man, Lanark, W. Perth, rare in Ireland. 31, H9, C. Europe north to Spitzbergen, Turkey, Madeira, Canaries, northeast N. America.

8. C. massalongi (Spruce) K. Müll., *Rabenh. Krypt.-Fl. Deutschl.* ed. 2. 6(2): 191. 1913
(Fig. 36)

Probably dioecious (in Britain). Plants very slender, in yellowish-green or green to reddish-brown or blackish, sometimes extensive patches, mats or turfs. Shoots prostrate to erect, 120–280(–320) μm wide, stems branching irregularly, 40–80(–100) μm wide, dorsal cortical cells 10–16 μm wide. Leaves on sterile stems distant to approximate or subimbricate, ± transversely inserted, patent to erecto-patent or with basal part patent and lobes ± erect, concave-channelled, 80–180 μm long, 80–160 μm wide, margins irregularly dentate to sinuose-dentate or ± entire, bilobed to 0.6–0.9, sinus acute, lobes lanceolate to narrowly triangular, (2–)4–8 cells wide at base, these cells 10–16 μm wide, walls thin and colourless or more usually thickened, sometimes heavily so and reddish-brown cuticle usually densely papillose, sometimes with conical protrusions on back of leaf below sinus. Underleaves present on sterile stems, subulate to lanceolate and bilobed, 40–90(–120) μm, entire or toothed. Gemmae common, ± ellipsoid, reddish-brown to purplish, (1–)2–celled, (12–)16–26 μm long, 10–16 μm wide. Fertile plants very rare and only female plants with immature perianths known in Britain. n = 18*. On moist to wet mine-spoil, banks, rocks and wall crevices in the vicinity of old copper mines, ascending to *ca* 900 m, calcifuge, rare but sometimes locally abundant, Cornwall, S. Devon, Merioneth, Caernarfon, Anglesey, W. Cork, Waterford. 6, H2. Spain, Portugal, Italy, Corsica, France, Germany, Austria, Switzerland, Czechoslovakia, Norway, Sweden, Vermont, Tennessee.

C. nicholsonii appears to be a polyploid derivative of *C. massalongi* and differs in the wider cells of the stem cortex and leaf lobe bases. The gemmae of *C. nicholsonii* are also frequently asymmetrical. Depauperate forms of *C. nicholsonii* from drier habitats may be difficult to separate from *C. massalongi* but wider cortical cells are usually present. Confusion may also arise where the plants occur mixed. Both species are variable but sterile plants are distinct from other *Cephaloziella* species in the combination of presence of underleaves on sterile stems, toothed leaf margins, papillose leaf cuticle, non-angular or non-papillose gemmae and habitat.

There has been some doubt about the specific status of *C. nicholsonii* but this has arisen as a result of misunderstanding. See Paton, *J. Bryol.* **13**, 1–18, 1984 for discussion of the differences between the two species.

Some forms of *C. massalongi* are, however, completely entire-leaved and may be difficult to distinguish from other entire-leaved species with obvious underleaves (e.g. *C. divaricata* and *C. stellulifera*). *C. stellulifera* has a smooth cuticle. For the differences from *C. divaricata* see dichotomy 13 of key to species.

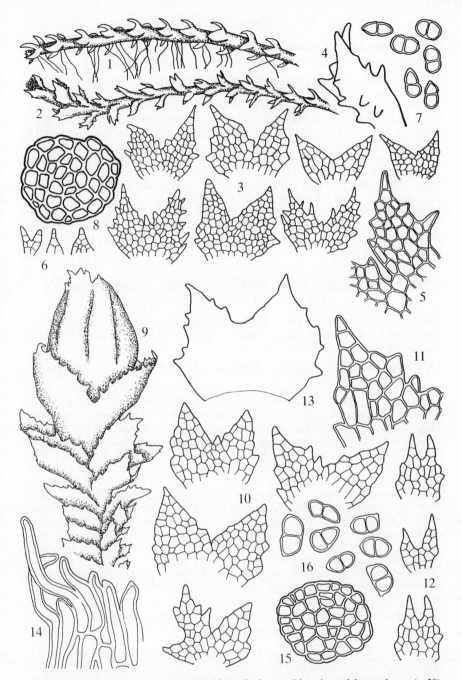

Fig. **36**. 1–8, *Cephaloziella massalongi*: 1 and 2, sterile shoots with entire and dentate leaves (× 38); 3, leaves from sterile stems (× 150); 4, leaf from shoot 2 in side view (× 150); 5, lobe of leaf from sterile stem (× 270); 6, underleaves from sterile stems (× 150); 7, gemmae (× 450); 8, т.s. sterile stem (× 210). 9–16, *C. nicholsonii*: 9, paroecious shoot with immature perianth (× 84); 10, leaves from sterile stems (× 150); 11, lobe of leaf from sterile stem (× 270); 12, underleaves from sterile stem (× 150); 13, male bract (× 84); 14, portion of mouth of immature perianth (× 270); 15, т.s. sterile stem (× 210); 16, gemmae (× 450).

9. C. nicholsonii Douin & Schiffn. in Douin, *Rev. Bryol.* 40: 81. 1913　　　(Fig. 36)
C. massalongi var. *nicholsonii* (Douin & Schiffn.) E.W. Jones

Paroecious and sometimes partly autoecious. Plants similar to *C. massalongi* but often larger. Shoots 200–440 μm wide, sterile stems 80–160 μm wide, dorsal cortical cells 16–20(–28) μm wide. Leaves on sterile stems 110–230 μm long, 130–240 μm wide, cells at base of leaf lobes 16–20(–24) μm wide, cuticle smooth or papillose. Underleaves present on sterile stems, 60–120 μm long. Gemmae ± ellipsoid or somewhat asymmetrical, 17–30 μm long, 12–16 μm wide. Plants sometimes fertile, male inflorescence below female or on separate branch, bracts imbricate, larger, more concave and less deeply bilobed than leaves. Female bracts bilobed, innermost connate with bilobed bracteole for about ½ their length, margins spinose-dentate. Immature perianth plicate above, mouth composed of thick-walled vermiform cells, 70–140 μm long, 7–10 times as long as wide. Capsules unknown. *n* = 36*. On moist wet substrates which may be subject to desiccation, mine-spoil, banks, rocks and crevices in the vicinity of old copper mines at low altitudes, very rare but sometimes locally abundant, Cornwall, Devon, Merioneth. 5. Apparently endemic.

Subgenus *Evansia* (Douin & Schiffn.) K. Müll., *Rabenh. Krypt.-Fl. Deutschl.* ed. 2, 6(2): 787. 1916

Leaves of sterile stems dentate. Underleaves present. Gemmae green, coarsely papillose. Female bracts bilobed, innermost connate with bracteole at base, margins toothed. Perianth ovoid, deeply 5-plicate.

10. C. dentata (Raddi) Steph., *Bull. Herb. Boiss.* 5: 78. 1897　　　(Fig. 37)

Dioecious. Plants very small, green to yellowish-brown scattered shoots or small patches. Shoots prostrate, stems irregularly branched. Leaves on sterile stems distant, ± transversely inserted, spreading, concave or very concave, 160–300 μm long, 120–240(–320) μm wide, margin irregularly dentate to spinose-dentate, bilobed to (0.5–) 0.6–0.7(–0.8), sinus obtuse or rounded, lobes triangular, acute, one often larger than the other, 4–7 cells wide at base, these cells 16–24 μm wide, ± thick-walled, cuticle smooth. Fertile shoots with ascending apices, leaves larger, more coarsely toothed. Underleaves present on sterile stems, lingulate to lanceolate, toothed or not, 80–140 μm long. Gemmae frequent, reddish-brown, ± spherical, coarsely papillose, 1–2 celled, 16–19 μm long. Only female plants known in Britain. Female bracts 2–3 lobed, spinosely dentate, connate with bracteole below. Perianth obloid, deeply plicate above, mouth crenulate-denticulate (Macvicar, 1926). Damp soil on tracks and in hollows, very rare, W. Cornwall. 1. Spain, Italy, Sardinia, Yugoslavia, France, Denmark, Sweden, N. Russia, Madeira.

The concave, spinose-dentate leaves, presence of underleaves, large cells with smooth cuticle and coarsely papillose gemmae will distinguish *C. dentata* from other *Cephaloziella* species.

Subgenus *Prionolobus* (Spruce) K. Müll., *Rabenh. Krypt.-Fl. Deutschl.* ed. 2, 6(2): 115. 1913

Leaves of sterile stems dentate. Underleaves absent. Gemmae angular. Female bracts bilobed, innermost connate with bracteole below, margins dentate. Perianth subcylindrical, deeply 5-plicate.

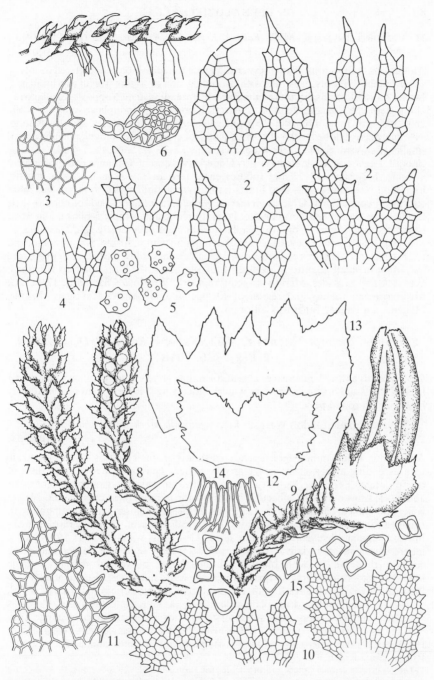

Fig. **37**. 1–6, *Cephaloziella dentata*: 1, sterile shoot (× 38); 2, leaves from sterile shoots (× 150); 3, lobe of leaf from sterile stem (× 270); 4, underleaves from sterile shoots (× 150); 5, gemmae (× 450); 6, т.s. sterile stem (× 210). 7–15, *C. turneri*: 7, sterile shoot (× 33); 8, male shoot with antheridia (× 33); 9, female shoot with perianth (× 33); 10, leaves from sterile stems (× 150); 11, lobe of leaf from sterile stem (× 270); 12, male bract (× 84); 13, female bracts and bracteole (× 63); 14, portion of perianth mouth (× 270); 15, gemmae (× 450).

11. C. turneri (Hook.) K. Müll., *Rabenh. Krypt.-Fl. Deutschl.* ed. 2, 6(2): 202. 1913
(Fig. 37)

Autoecious or dioecious. Plants very small, green to reddish-brown scattered shoots or small patches. Shoots prostrate with ascending apices, stems irregularly branched. Leaves on sterile stems distant to subimbricate, spreading, in two opposite regular rows along stem (pectinate-distichous) transversely inserted, conduplicate-concave, 150–260(–360) μm long, 140–200(–320) μm wide, margins irregularly spinose-dentate, asymmetrically bilobed to 0.6–0.8, sinus acute, lobes narrowly triangular, acute, dorsal smaller than ventral, 6–12 cells wide, these cells 12–20 μm wide, thick-walled, cuticle smooth. Leaves on fertile stems larger. Underleaves absent. Gemmae common, green, angular, 1–2-celled, 14–22 μm. Inflorescences terminal on branches, male spicate, becoming intercalary with age, bracts imbricate, of similar size to leaves but more concave. Female bracts larger than leaves, bilobed, innermost fused for about ¾ their length to bracteole, margins dentate. Perianth elongated, deeply 5-plicate almost to base, narrowed at mouth, mouth crenulate-denticulate, cells mostly 32–52 μm long, 6–8 times as long as wide, ± hyaline. Capsules frequent, winter, spring. Soil on stream and river banks, rock crevices on cliffs, in sheltered habitats at low altitude, calcifuge, rare and often in small quantity, Cornwall, I. of Wight and E. Kent north to Hereford, Warwick and S. Wales, Merioneth, Kintyre, Cork, Wicklow, Channel Is. 17, H3, C. Mediterranean islands and region of Europe and N. Africa, Caucasus, Azores, Madeira, La Palma, Tenerife, California.

Subgenus *Dichiton* (Mont.) K. Müll., *Rabenh. Krypt.-Fl. Deutschl.* ed. 2, 6(2): 785. 1916

Leaf margins entire. Underleaves absent. Gemmae angular, not papillose. Innermost female bracts and bracteole fused for whole of their length to form ± entire or shallowly lobed sheath round perianth.

12. C. integerrima (Lindb.) Warnst., *Kryptogamenfl. Mark Brandenburg* 1: 232. 1902   (Fig. 38)

Autoecious. Plants minute, green, scattered shoots or small patches. Shoots prostrate with ascending apices, stems irregularly branched. Leaves on sterile stems distant, erecto-patent, ± concave, 88–160 μm long, 60–160 μm wide, margins entire, bilobed to about 0.5, sinus obtuse to subacute, 2–5 cells wide at base, these cells 12–20(–24) μm wide, thin-walled, cuticle smooth. Leaves on fertile stems closer, larger, with obtuse to rounded lobe apices and entire margins. Underleaves absent. Gemmae rare, reddish-brown, angular, 2-celled, 14–18 μm long, 12–14 μm wide (Müller, 1957). Male inflorescence on short lateral branch, spicate, bracts imbricate, larger and more concave than leaves, with blunt tooth near base of ventral margin. Female inflorescence terminal, bracts with thickened cell walls, innermost fused to bracteole to form shallowly lobed sheath with entire margin round perianth. Perianth ovoid, plicate, mouth crenulate denticulate, cells 16–26 μm long, walls thickened. Capsules frequent, summer. Damp sandy soil in open habitats at low altitudes, calcifuge, very rare and usually in small quantity, E. Cornwall, N. Wilts, Sussex, Berks, Waterford. 5, H1. Spain, France, Germany, Hungary, Italy, Poland, Fennoscandia, N. and Baltic Russia, Spitzbergen.

The innermost female bracts and bracteole fused into a shallowly lobed sheath with entire margin will distinguish *C. integerrima* from all *Cephaloziella* species except *C. calyculata*. This latter differs in the unlobed sheath round the lower part of the perianth and more heavily thickened cell walls especially in the female bracts. In *C. integerrima* the mouth of the perianth is crenulate-denticulate whilst in *C. calyculata* it is almost entire and the cells are shorter and more heavily thickened.

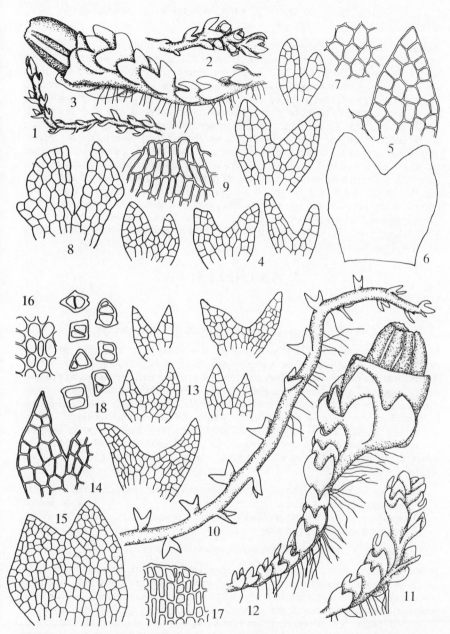

Fig. **38**. 1–9, *Cephaloziella integerrima*: 1–3, sterile, male (antheridia present) and female shoots
(× 63); 4, leaves from sterile stem (× 150); 5, leaf lobe (× 385), 6, upper leaf from female stem
(× 150); 7, cells from leaf on female stem (× 385); 8, male bract (× 150); 9, part of perianth mouth
(× 275). 10–18, *C. calyculata*: 10–12, sterile, male and female shoots (× 38), 13, leaves from sterile
stem (× 150); 14, leaf lobe (× 385); 15, leaf from female stem (× 150), 16, cells from upper leaf on
female stem (× 385), 17, part of perianth mouth (× 275); 18, gemmae (× 450).

13. C. calyculata (Durieu & Mont.) K. Müll., *Rabenh. Krypt.-Fl. Deutschl.* ed. 2
6(2): 787. 1916   (Fig. 38)

Autoecious. Plants minute, green to brownish-green scattered shoots. Shoots prostrate
with ascending apices, stems sparingly branched. Leaves on sterile stems distant,
patent, 96–160 μm long, 48–120 μm wide, margins entire, bilobed to 0.6–0.7, sinus
acute to obtuse, lobes narrowly triangular, acute, (2–)3–6 cells wide at base, these cells
10–18 μm wide, walls thin to thickened, cuticle smooth. Leaves on fertile stems closer,
larger, with obtuse to rounded lobe apices and entire margins, cell walls thickened.
Underleaves absent. Gemmae occasional, angular and frequently irregularly quadrate,
2(–3)-celled, 14–20 μm. Male inflorescence on short lateral branch, spicate, bracts
numerous, imbricate, larger and more concave than leaves. Female inflorescence
terminal, bracts with heavily thickened cell walls, innermost fused to bracteole to form
an unlobed sheath with entire margin round perianth. Perianth ovoid, plicate above,
mouth almost entire, cells 10–16 μm long, walls strongly incrassate. Capsules not
known in Britain. Usually in small quantity amongst other bryophytes, especially
Trichostomum brachydontium, on cliff tops, stream banks and mine waste, at low
altitudes, very rare, W. Cornwall, Glamorgan. 2. Mediterranean islands and region of
Europe, Tunisia, Azores, La Palma.

11. ANTHELIACEAE

With the characters of the genus.

19. ANTHELIA (DUM.) DUM., *Recueil Observ. Jungerm.*: 18. 1835

Dioecious or paroecious. Plants very slender, gregarious or forming thin patches to
extensive tufts or cushions. Shoots procumbent to erect, branching irregular, lateral,
flagelliform shoots absent. Leaves and underleaves similar, imbricate, appressed,
ovate, concave, ± transversely inserted, deeply bilobed, lobes acute, sinus narrow,
margins entire to crenulate-denticulate; cell walls thick or very thick, oil-bodies
lacking. Gemmae lacking. Male bracts similar to leaves with solitary axillary
antheridia. Female bracts appressed, bilobed, becoming hyaline above, margins
denticulate. Perianth ovoid or narrowly ovoid, deeply plicate, slightly narrowed to
lobed, denticulate mouth. Capsule ± globose, wall bistratose. Spores 12–24 μm. Two
species.

> Dioecious, fertile shoots hardly clavate, perianth exserted for about half its length,
> spirals of elaters 4 μm wide   **1. A. julacea**
> Paroecious, fertile shoots markedly clavate, perianth not or hardly exserted,
> spirals of elaters 1.6 μm wide   **2. A. juratzkana**

1. A. julacea (L.) Dum., *Recueil Observ. Jungerm.*: 21. 1835   (Fig. 39)

Dioecious. Plants slender, forming small thin patches to extensive swollen tufts or
cushions 1 m or more across, dark green, dark yellowish-green or brownish, becoming
glaucous when dry. Shoots prostrate to erect, 0.5–0.6 cm long, branching frequent and
irregular. Leaves and underleaves ± similar and so appearing 3-ranked, imbricate,
appressed, ovate, ± concave, insertion transverse, 0.3–0.7 mm long, bilobed to ⅗–⅞,
lobes lanceolate, acute, margins entire to crenulate-denticulate above; mid-lobe cells
12–20 μm wide, walls brownish, very thick, apical cell often hyaline. Male inflorescence
terminal, becoming intercalary with age, bracts similar to leaves but more concave.
Female bracts larger than leaves, bilobed to *ca* ¼, margins crenulate-denticulate.

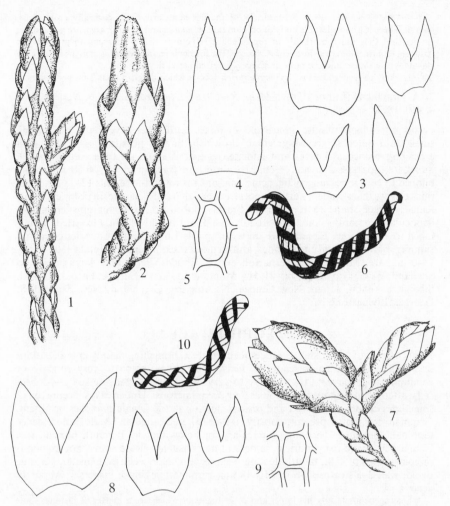

Fig. **39**. 1–6, *Anthelia julacea*: 1, sterile shoot (× 21); 2, shoot with perianth (× 21); 3, leaves (× 38); 4, female bract (× 38); 5, mid-leaf cell (× 450); 6, elater (× 450). 7–10, *A. juratzkana*: 7, shoot with perianth (× 38); 8, leaves (× 63), 9, mid-leaf cell (× 450); 10, elater (× 450).

Perianths rare, narrowly ovoid, extending $\frac{1}{2}$ their length beyond bracts, plicate and hyaline above, wide-mouthed, mouth dentate, sometimes shallowly lobed. Elaters 7–8 μm wide, bispiral, spirals 4 μm wide. Capsules rare, summer. $n = 9$. Moist rocks and soil, flushes, springs, rock crevices and cliff ledges in alpine habitats, to 1340 m but descending to sea-level in N.W. Scotland, occasional to frequent and sometimes locally abundant in montane habitats, Brecon, Cardigan, Merioneth, Caernarfon, Yorkshire and Westmorland northwards. 38, H14. Pyrenees, N. Italy and Roumania north to Spitzbergen, Turkey, Siberia, Himalayas, Yunnan, Japan, N. America, Greenland.

Small sterile forms of *A. julacea* cannot be distinguished from sterile plants of *A. juratzkana*. It is possible, though as yet not proven, that large sterile patches at high altitudes may also be

A. juratkana (Macvicar, 1926; Schuster, 1974) and not necessarily, as is generally assumed, *A. julacea*. There is a considerable body of opinion that *A. julacea* and *A. juratzkana* are not distinct species, but they differ markedly in reproductive structures and there seem no grounds for reducing *A. juratzkana* to a variety of *A. julacea*. The former is readily recognised in the field by the clavate fertile shoots and the non- or scarcely exserted perianth.

Only likely to be confused with *Gymnomitrion* species which differ in their 2-ranked leaves.

2. A. juratzkana (Limpr.) Trev., *Mem. Real. Istit. Lombardo Sci. Mat. Nat.* ser. 3, 4: 410. 1877   (Fig. 39)

Paroecious. Plants minute, gregarious or mixed with other bryophytes or forming small, thin patches, greyish-green to brownish. Shoots prostrate, 3–7 mm, with ascending branches. Leaves and underleaves ± similar and so appearing 3-ranked, imbricate, appressed, ovate, ± concave, insertion transverse, 0.25–0.50 mm long, bilobed to *ca* ⅔, lobes lanceolate, acute, margins entire; mid-lobe cells 12–18 μm wide, thick or very thick-walled. Fertile shoots frequent, markedly clavate, male bracts below female, larger than leaves. Female bracts bilobed to *ca* ½, margins crenulate-denticulate. Perianths common, broadly ovoid, not or only slightly exserted, mouth lobed to *ca* halfway. Elaters *ca* 8 μm wide, spirals 1.6 μm wide. Capsules common, summer. Bare moist soil and amongst and on moist rocks in alpine habitats, (500–)750–1340 m, rare, Merioneth, Caernarfon and S. Ebudes, Perth and Angus north to Shetland, W. Galway, Donegal. 21, H3. Montane and arctic Europe, Faroes, Iceland, Siberia, Sakhalin, Japan, New Guinea, N. America, Greenland, New Zealand, S. Georgia, Livingstone Is.

12. LOPHOZIACEAE

Plants minute to large. Shoots procumbent to erect, branching mainly lateral, stolons or flagella rare. Rhizoids scattered. Leaves succubous, plane to very concave or conduplicate-concave, ± transversely to very obliquely inserted, mostly 2–4-lobed; cells often with trigones, oil-bodies few to numerous. Underleaves deeply bifid, sometimes ciliate, or reduced and present only on fertile female branches or absent. Gemmae commony present, mostly 1–2-celled and angular. Male inflorescence terminal, sometimes becoming intercalary, or below female. Perianth present, free from bracts, ovoid to cylindrical, smooth or more usually plicate above, not trigonous or compressed in T.S., mouth dentate. Seta usually many cells in diameter. Capsule ovoid, wall 2–5-stratose. Spores 10–18 μm, germinating before dispersal. About 19 genera.

A heterogeneous family the limits and constituents of which are a matter of dispute; some authorities (e.g. Grolle, 1983b) treat it as a subfamily of the Jungermanniaceae. Schuster (1969) discusses both matters but comes to no very firm conclusions. I have followed Schuster (1969) and Grolle (1983a) in treating the taxon as a family; it is more practical to recognise two medium-sized heterogeneous units than a single large, highly heterogeneous unit.

The species placed by Schuster (1969) in the single genus *Lophozia* have been placed by other authors (e.g. Arnell, 1956) in as many as six genera. At least in Europe the species fall into three reasonably distinct groups and I have therefore treated them as three separate genera, *Barbilophozia, Lophozia* and *Leiocolea* but it is evident from the literature that such a treatment is merely a matter of opinion.

1 Leaves mostly bilobed   2
  At least upper leaves 3–5-lobed or if 2-lobed then very asymmetrical with one lobe
      tooth-like   12
2 Underleaves present   3
  Underleaves lacking or sparse and subulate   5

3 Leaves ± transversely inserted, perianth plicate in upper half
                                                21. Barbilophozia (p. 92)
   Leaves obliquely inserted, perianth smooth or plicate in upper half   4
4 Gemmae abundant, mostly sharply angular, perianth plicate above, not beaked
                                                23. Lophozia (p. 99)
   Gemmae lacking or if present smooth, perianth smooth, beaked
                                                24. Leiocolea (p. 112)
5 Leaves to *ca* 260 μm long, bilobed to ½ or more   6
   Leaves more than 250 μm long except sometimes at base of stem, rarely bilobed to
      more than ⅓   7
6 Leaves bilobed to *ca* ½, stems not branching dichotomously
                                                26. Eremontus (p. 122)
   Leaves bilobed to ⅔–¾, stems branching dichotomously
                                                27. Sphenolobopsis (p. 122)
7 Leaves ± symmetrical, ± as long as wide   8
   At least some leaves asymmetrical with one lobe or side larger than the other, or
      1.4–2.0 times as long as wide   9
8 Leaf lobes acute to obtuse, female bracts larger than leaves, perianth plicate in
      upper half   **23. Lophozia (p. 99)**
   Leaf lobes rounded, very rarely acute to obtuse, bracts smaller than leaves, perianth
      smooth except at mouth   **25. Gymnocolea (p. 119)**
9 Leaves obliquely ovate-cordate with reflexed ventral margin **22. Anastrepta (p. 98)**
   Leaves not obliquely ovate-cordate, ventral margin plane or inflexed   10
10 Leaves very concave with inflexed margins   **28. Anastrophyllum (p. 123)**
   Leaves not or only slightly concave, margins not inflexed   11
11 Stems not translucent, perianth plicate in upper half   **23. Lophozia (p. 99)**
   Stems translucent, perianth smooth   **24. Leiocolea (p. 112)**
12 Underleaves present   13
   Underleaves absent   14
13 Leaves 3–4-bilobed to ½–⅞, underleaves bilobed but otherwise similar to leaves
                                                20. Tetralophozia (p. 91)
   Leaves mostly bilobed to ⅛–½, underleaves differing in shape from leaves
                                                21. Barbilophozia (p. 92)
14 Lower leaves bilobed, symmetrical or slightly asymmetrical, upper irregularly
      3–4-lobed   **23. Lophozia (p. 99)**
   Lower and upper leaves mostly similarly and regularly 3–4-lobed or if bilobed then
      very asymmetrical   15
15 Leaves symmetrical or slightly asymmetrical, lobed to ¼–½(–⅔)
                                                21. Barbilophozia (p. 92)
   Leaves markedly asymmetrical or if symmetrical then lobed only to *ca* ⅛
                                                29. Tritomaria (p. 127)

20. TETRALOPHOZIA (SCHUST.) SCHLJAK., *Novit. Syst. Pl. Non Vasc.* 13: 227.
                           1976

Dioecious. Slender, cushion- or mat-forming plants. Stems brittle, simple or branched.
Leaves imbricate, ± transversely inserted, from spreading base deeply 3–4-lobed with
erect lobes, lobes channelled, with reflexed margins often toothed or ciliate at base; cells
thick-walled, trigones not pronounced. Underleaves large, deeply bilobed, otherwise
± similar to lateral leaves. Female bracts similar to leaves. Perianth narrowly ovoid,
plicate to base, contracted at ciliate mouth. Three species.

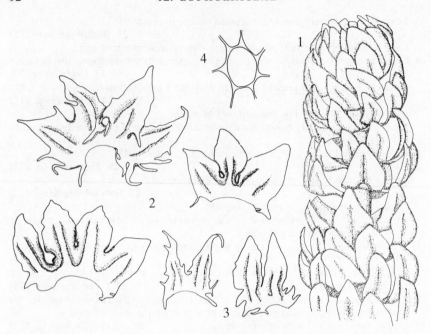

Fig. **40**. *Tetralophozia setiformis*: 1, shoot (× 27); 2, leaves (× 38); 3, underleaves (× 38); 4, cell from middle of leaf lobe (× 450).

1. T. setiformis (Ehrh.) Schljak., *Novit. Syst. Pl. Non Vasc.*13: 228. 1976   (Fig. 40) *Chandonanthus setiformis* (Ehrh.) Lindb.

Yellowish-brown to yellowish-red dense cushions or mats. Shoots slender or very slender, (1–)2–5(–7) cm, stems brittle, procumbent with procumbent or ascending branches, rhizoids near base only, small-leaved innovations with rhizoids near apex sometimes present. Leaves imbricate, 0.5–1.0 mm wide, wider than long, ± symmetrical, insertion ± transverse, base spreading, deeply 3–4-lobed to $\frac{1}{2}$–$\frac{7}{8}$, margins reflexed, entire to ciliate at base; lobes erect, ovate-oblong, channelled, obtuse, cells thick-walled, 16–20 μm wide in middle of lobes. Underleaves deeply bilobed, otherwise similar in shape to leaves. Gemmae lacking. Perianths rare. Capsules unknown in Britain. $n = 9$. On dry acidic rocks, soil, rarely tree roots, 300–1200 m, frequent and sometimes abundant in the eastern Scottish highlands, very rare elsewhere; N. Northumberland, Ross, W. Sutherland, Caithness. 15. From the Ardennes, Harz and Alps northwards, Carpathians, Iceland, Siberia, Ontario, Maine, Ellesmere Is., western N. America, Greenland.

21. BARBILOPHOZIA LOESKE, *Verh. Bot. Vereins Prov. Brandenburg* 49; 37. 1907

Dioecious. Tufts, patches or scattered shoots, stems procumbent to erect, simple or branched. Rhizoids colourless. Leaves slightly to very obliquely inserted, (2–)3–4-lobed, dorsal margin ± decurrent, ventral margin with or without cilia at base, cells with trigones, oil-bodies 4–9 per cell, 4–9 μm long. Underleaves on sterile stems large, deeply bifid and usually ciliate, occasionally small and subulate or lacking. Gemmae 1–2-celled, angular, produced on margins of younger leaves. Perianth ovoid to pyriform, upper part deeply plicate, gradually contracted to ciliate mouth. Capsules generally rare. About 11 species.

1 Lobes of at least some leaves apiculate, cells of cilia of ventral margin of leaves
    elongated   2
  Lobes without apiculus, cells of cilia 1–2 times as long as wide or cilia lacking   3
2 Leaves 0.7–1.5(–1.8) mm wide, usually only some lobes with apiculus, gemmae
    usually present on young leaves   **6. B. hatcheri**
  Leaves (1.6–)1.8–3.2 mm wide, most lobes with apiculus, gemmae very rare
                                                 7. B. lycopodioides
3 Leaves very obliquely inserted, ± horizontal and plane, cilia lacking **8. B. barbata**
  Leaves ± transversely to obliquely inserted, erect to spreading, concave, cilia present
    or not at base of ventral margin   4
4 Leaves mostly 3–4-lobed, lobes longer than wide   **1. B. quadriloba**
  Leaves mostly 2–3-lobed, lobes as wide as or wider than long   5
5 Shoots with attenuated, appressed-leaved, gemmiferous apical innovations
                                                 5. B. attenuata
  Stems lacking apical innovations   6
6 Leaves mostly 2-lobed, gemmae yellowish-green   **2. B. kunzeana**
  Leaves 2–3-lobed, gemmae deep red to reddish-brown   7
7 Leaves mostly 3-lobed, 0.8–1.6 mm wide, cells 16–24 μm wide, cilia present at base of
    ventral margin, underleaves deeply bifid   **3. B. floerkei**
  Leaves 2–3-lobed, 0.6–1.0 mm wide, cells 20–32 μm wide, cilia absent, underleaves
    absent or if present very small, simple   **4. B. atlantica**

Subgenus *Orthocaulis* (Buch) Buch, *Mem. Soc. Fauna Fl. Fenn.* 17: 289. 1942

Leaves nearly transversely to obliquely inserted, ± concave, mostly 3-lobed, basal cilia when present with cells 1–2 times as long as wide.

1. B. quadriloba (Lindb.) Loeske, *Hedwigia* 49: 13. 1909   (Fig. 41)
Lophozia quadriloba (Lindb.) Evans, *Orthocaulis quadrilobus* (Lindb.) Buch

Plants slender, dull green to blackish-brown, forming tufts or mixed with other bryophytes. Shoots 2.5–5.0 cm; stems erect or ascending, simple or branched. Leaves distant to imbricate, nearly transversely inserted, concave, 0.5–1.7 mm wide, wider than long, ± asymmetrical, (2–)3–4-lobed to $\frac{1}{3}$–$\frac{1}{2}$(–$\frac{2}{3}$), lobes longer than wide, obtuse to acute, inflexed, margins reflexed, sinus narrow, rounded to acute, gibbous; ventral margin in younger leaves with cilia at base, cells of cilia 1–2 times as long as wide, cilia lost in older leaves; cells verrucose, (16–)20–28 μm wide, trigones large. Underleaves nearly as long as leaves, divided ± to base into 2 subulate lobes, lobes ciliate at base, cells of cilia 1–2 times as long as wide. Gemmae very rare, not seen on British material, yellowish, rhomboid, 2-celled, 27 × 18 μm (Schuster, 1969). Capsules not known. $n = 9$. On moist ledges, flushes and in moist turf in base-rich montane habitats, very rare; Perth, S. Aberdeen, Argyll. 4. Germany, Alps, Tatra, Norway, Sweden, Finland, Spitzbergen, Iceland, U.S.S.R., Alaska, Greenland.

2. B. kunzeana (Hüb.) Gams, *Kleine Kryptogamenfl.* ed. 2, 1: 44. 1948   (Fig. 41)
Lophozia kunzeana (Hüb.) Evans, *Orthocaulis kunzeanus* (Hüb.) Buch

Plants slender, yellowish-green to light brown, forming dense or lax tufts or mixed with other bryophytes. Shoots (0.5–)1.5–5.0 cm; stems erect or ascending, simple or branched. Leaves distant to imbricate, nearly transversely inserted, patent to

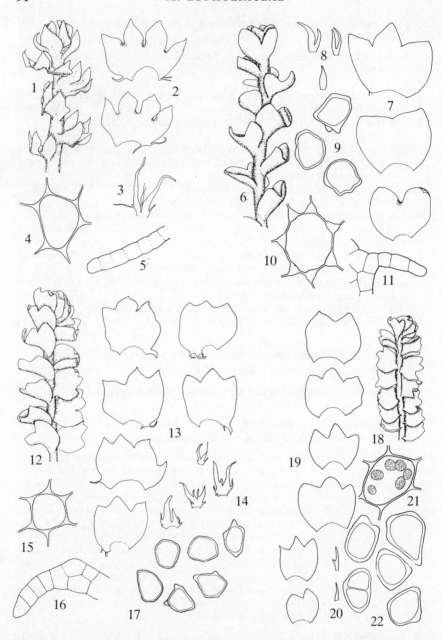

Fig. **41**. 1–5, *Barbilophozia quadriloba*: 1, shoot (× 23); 2, leaves (× 45); 3, underleaf (× 45); 4, leaf cell (× 507); 5, leaf cilium (× 227). 6–11, *B. kunzeana*: 6, shoot (× 15); 7, leaves (× 23); 8, underleaves (× 23); 9, gemmae (× 507); 10, leaf cell (× 507); 11, leaf cilium (× 227). 12–17, *B. floerkei*: 12, shoot (× 9); 13, leaves (× 15); 14, underleaves (× 15); 15, leaf cell (× 507); 16, leaf cilium (× 227); 17, gemmae (× 507). 18–22, *B. atlantica*: 18, shoot (× 9); 19, leaves (× 15); 20, underleaves (× 15); 21, leaf cells (× 507); 22, gemmae (× 507).

spreading, conduplicate-concave with inflexed margins, (0.4–)0.6–1.0(–1.2) mm wide, as wide as or wider than long, ± asymmetrical, bilobed to $\frac{1}{3}$–$\frac{1}{2}$, occasionally 3-lobed, rarely simple, lobes as wide as or wider than long, rounded-obtuse or obtuse, occasionally acute, sinus wide, rounded or obtuse, gibbous or not, ventral basal marginal cilia usually lacking, occasionally a small cilium with cells to twice as long as wide present; cells 16–24(–27) μm wide, trigones small to large. Underleaves small, often concealed by rhizoids, divided almost to base into 2 lanceolate-subulate lobes, occasionally simple, lanceolate, with or without cilia. Gemmae common, pale yellowish-green, obtusely angular, 1–2-celled, (16–)20–24 μm long. Capsules not known in Britain. On boggy ground, in moist turf, on flushed rocks and moist soil in montane habitats, descending to 150 m, rare; Flint and from M.W. Yorkshire and Northumberland north to Inverness and W. Sutherland, Shetland, Sligo. 12, H1. France, Germany, Holland, Belgium, Alps, Tatra, Fennoscandia, Spitzbergen, Iceland, U.S.S.R., Alaska, N. America.

3. B. floerkei (Web. & Mohr) Loeske, *Verh. Bot. Vereins Prov. Brandenburg* 49: 37. 1907 (Fig. 41)
Lophozia floerkei (Web. & Mohr) Schiffn., *Orthocaulis floerkei* (Web. & Mohr) Buch

Dull green to greenish-brown patches or creeping through other bryophytes. Shoots (0.5–)1.5–5.0 cm; stems procumbent or ascending, simple or branched. Leaves distant to imbricate, insertion ± transverse to ± oblique, concave, 0.8–1.6 mm wide, symmetrical or asymmetrical, as wide as or wider than long, (2–)3(–4)-lobed to $\frac{1}{3}$–$\frac{1}{3}$, lobes wider than long, acute to rounded-obtuse, inflexed, sinus very wide, rounded or obtuse, gibbous or not, ventral margin with 1–4 basal cilia with cells 1(–2) times as long as wide; cells 16–24(–26) μm wide, trigones small or large. Underleaves deeply divided into 2 equal or unequal, lanceolate or subulate lobes, ciliate at base. Gemmae occasional, reddish-brown to deep red, obtusely angular, unicellular, 28–40(–48) μm long. Capsules not known in Britain. $n = 8 + m$. On peat, soil, damp rocks, in damp turf, scree and rock crevices, frequent to common in acidic montane habitats, rare elsewhere; S.W. England, E. Sussex, Monmouth and Leicester north to Shetland, rare in Ireland. 80, H21. W., C. and S.E. Europe, Faroes, Iceland, Japan, western N. America, Aleutians, Peru.

Forms of *B. floerkei* and the next three species may be confused. There is an overlap in leaf and cell size between *B. floerkei* and *B. atlantica* and small plants of *B. floerkei* in which the leaves may lack basal cilia may be mistaken for large *B. atlantica*. In such forms of *B. floerkei* the leaf cells are usually at the lower limits of their size range, whilst those of *B. atlantica* are at their upper limits; careful search in small forms of *B. floerkei* will usually reveal some leaves with cilia. However, some plants cannot be named. See Fitzgerald & Fitzgerald, *Trans. Br. bryol. Soc.* 4, 214–20, 1962, for a discussion of the differences between the two species.

Occasional plants of *B. attenuata* lack attenuated gemmiferous shoots but may be distinguished by the frequent occurrence of bilobed leaves, the smaller size of the leaves and the absence of basal marginal cilia.

Although *B. hatcheri* has sometimes been confused with *B. floerkei* it is readily distinguished by the elongated cells of the basal marginal cilia of the leaves, the common occurrence of 4-lobed leaves and the presence of a long apiculus at the apex of one or more lobes of at least some leaves although these may have to be searched for.

4. B. atlantica (Kaal.) K. Müll., *Feddes Repert.* 54: 212. 1951 (Fig. 41)
Lophozia atlantica (Kaal.) Schiffn., *Orthocaulis atlanticus* (Kaal.) Buch

Plants slender, dark green to brown, usually forming patches. Shoots (0.5–)1.0–2.0 cm, procumbent or ascending; stems simple or branched. Leaves imbricate, concave, (0.5–)0.6–1.0(–1.4) mm wide, as wide as or wider than long, symmetrical or asymmetrical, 2–3-lobed to $\frac{1}{4}$, lobes wider than long, acute, unflexed, sinus very wide, acute to rounded, not gibbous, basal ventral margins without cilia but sometimes with a tooth; cells 20–32 μm wide, trigones small or very small. Underleaves where present small, simple, subulate, larger on gemmiferous shoots, not ciliate. Gemmae deep red, unicellular, obtusely angular, 20–40 μm. Capsules not known in Britain. On peat and soil in rock crevices and amongst rocks and on thin humus on damp rocks, ascending to 900 m, occasional in W. and N. Britain, from N.W. Wales, S. Lancashire, and S.W. Yorkshire north to W. Sutherland, Caithness and Orkney, E. Donegal. 33, H1. Norway, Sweden, Finland, Iceland, France, Switzerland, Czechoslovakia, north-east N. America, Greenland.

5. B. attenuata (Mart.) Loeske, *Verh. Bot. Vereins Prov. Brandenburg* 49: 37. 1907
(Fig. 42)
B. gracilis (Schleich.) K. Müll., *Lophozia attenuata* (Mart.) Dum., *Orthocaulis attenuatus* (Mart.) Evans

Plants slender, green to brownish, forming tufts or patches, sometimes extensive, or creeping through other bryophytes. Shoots (0.5–)1.5–4.0 cm; stems ascending, rarely branched, usually developing slender ± cylindrical small-leaved innovations from shoot apices. Leaves distant to imbricate, insertion slightly oblique, patent, concave, 0.4–0.8(–1.1) mm wide, as wide as or wider than long, ± asymmetrical, (2–)3-lobed to ¼–⅓, rarely 4-lobed, lobes about as wide as long, acute or subacute, sinuses wide, acute to rounded, sometimes gibbous; basal marginal cilia lacking; leaves of attenuated shoots appressed, ± transversely inserted, 2–4-lobed, usually gemmiferous and eroded; cells 16–20(–24) μm wide, trigones large. Underleaves rarely present, very small, subulate. Gemmae produced abundantly, light green, obtusely angular, 1–2-celled, 20–34 μm long. Capsules not known in Britain. $n = 8 + m, 9^*$. On peat, soil, thin humus on rocks, rotting logs, usually in sheltered acidic situations, rare in lowland areas, occasional to frequent elsewhere. 92, H22, C. W. and C. Europe, Norway, Sweden, Finland, Japan, Taiwan, Azores, N. America, Greenland.

Subgenus *Barbilophozia*

Leaves very obliquely inserted, nearly flat, rarely concave, mostly 4-lobed, marginal cilia where present with elongated cells.

6. B. hatcheri (Evans.) Loeske, *Verh. Bot. Vereins Prov. Brandenburg* 49: 37. 1907
(Fig. 42)
Lophozia hatcheri (Evans.) Steph.

Green to brownish tufts or patches or mixed with other bryophytes. Shoots 1–3(–5) cm; stems procumbent to erect, simple or sparingly branched. Leaves distant to imbricate, obliquely inserted, patent, concave, variable in shape, 0.7–1.5(–1.8) mm wide, usually wider than long, ± asymmetrical, 3–4-lobed to ¼–⅓, lobes as wide as or wider than long, acute to obtuse, sometimes obscure in gemmiferous leaves, at least some lobes of some lower leaves with long apiculus, sinuses wide, obtuse or acute, gibbous or not, base of ventral margin with 1–several cilia, cells of cilia 4–7 times as long as wide; cells (20–)24–28 μm wide, trigones small. Underleaves large, deeply bifid into lanceolate-subulate lobes, ciliate at base, cilia with elongated cells. Gemmae very common, obtusely angular, reddish, 1–2-celled, 18–28 μm long. Male plants frequent, bracts similar to leaves, saccate. Female bracts several-lobed, lobes with long apiculi. Capsules very rare. $n = 8 + m, 9, 10, 18$. On peat, soil, amongst rocks and on rocks and walls in sheltered habitats, rare to occasional in the Scottish mountains, very rare elsewhere, W. Suffolk, W. Norfolk, Shropshire, N.W. Wales, Lancashire and N.W. Yorkshire north to Caithness. 42. N. and C. Europe, Spain, Portugal, Faroes, Iceland, Caucasus, Siberia, Japan, Taiwan, Himalayas, N. America, Greenland, Antarctica.

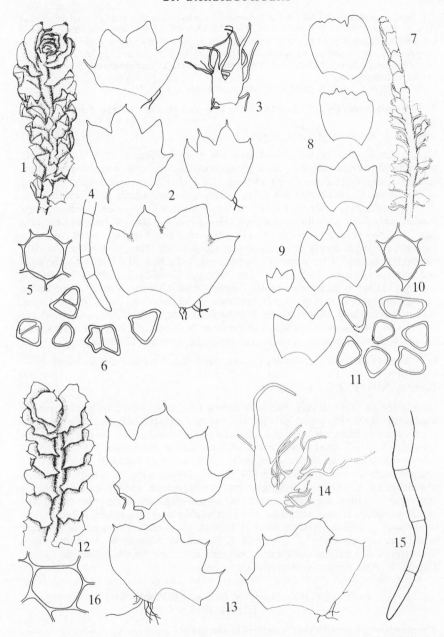

Fig. **42**. 1–6, *Barbilophozia hatcheri*: 1, shoot (× 11); 2, leaves (× 23); 3, underleaf (× 23); 4, leaf cilium (× 227); 5, leaf cell (× 507); 6, gemmae (× 507). 7–11, *B. attenuata*: 7, shoot (× 15); 8, leaves from attenuated part of shoot (× 33); 9, leaves (× 33); 10, leaf cell (× 507); 11, gemmae (× 507). 12–16, *B. lycopodioides*: 12, shoot (× 6); 13, leaves (× 15); 14, underleaf (× 27); 15, leaf cilium (× 227); 16, leaf cell (× 507).

B. hatcheri is usually distinct from *B. lycopodioides* in its smaller size and the usual presence of gemmae. Fewer of the leaves tend to be apiculate. The underleaves of *B. hatcheri* are less ciliate and the species frequently produces male inflorescences which are very rare in *B. lycopodioides*. *B. lycopodioides* only occurs at high altitudes whereas *B. hatcheri* descends to lowland habitats. Intermediates do, however, occur and there is some doubt as to whether the two taxa should be treated as distinct species.

For the differences from *B. floerkei* see under that species.

7. B. lycopodioides (Wallr.) Loeske, *Verh. Bot. Vereins Prov. Brandenburg* 49: 39.
1907   (Fig. 42)
Lophozia lycopodioides (Wallr.) Cogn.

Light green patches or mixed with other bryophytes. Shoots to 5(–8) cm, stems procumbent to erect, simple or sparingly branched. Leaves approximate to imbricate, very obliquely inserted, somewhat crisped, (1.6–)1.8–3.2 mm wide, wider than long, asymmetrical, (3–)4-lobed to $\frac{1}{8}$–$\frac{1}{4}$, lobes wider than long, usually acute with long or short apiculus, sometimes obtuse, sinuses very wide, acute to rounded, often gibbous, base of ventral margin with several cilia, cells of cilia 4–7 times as long as wide; cells thin- to thick-walled, (18–)20–24(–29) μm wide, trigones very small to large. Underleaves large, deeply bifid into lanceolate-subulate lobes, with numerous cilia, cilia with elongated cells. Gemmae very rare, red, 1–2 celled, 20–25 μm long. Capsules not known in Britain. $n = 8 + m$, $8 + 2m$. Amongst boulders, on humus, in turf and amongst heather in sheltered or humid, frequently basic habitats, usually above 600 m, occasional in the Scottish highlands, very rare elsewhere; M.W. and N.W. Yorkshire, N. Northumberland, Westmorland, Cumberland, Argyll, Perth and Angus north to Sutherland. 18. Alpine and arctic Europe from N. Spain and Bulgaria northwards, Faroes, Iceland, Caucasus, Kamchatka, Japan, N. America, Greenland.

8. B. barbata (Schmid. ex Schreb.) Loeske. *Verh. Bot. Vereins Prov. Brandenburg*
49: 37. 1907   (Fig. 43)
Lophozia barbata (Schmid. ex Schreb.) Dum.

Dark green or more usually yellowish-brown patches or mixed with other plants. Shoots to 4(–5) cm, stems procumbent, simple or sparingly branched. Leaves approximate to imbricate, very obliquely inserted, spreading horizontally, plane, 1.0–1.6 mm wide, ± as long as wide, symmetrical or slightly asymmetrical, 2–4-lobed to $\frac{1}{8}$–$\frac{1}{4}$, lobes wider than long, obtuse to rounded, sinus acute to rounded, ventral margin straight, without cilia or teeth; cells (20–)24–32(–36) μm wide, thin- or thick-walled, trigones small. Underleaves lacking or very rarely present, small, subulate. Gemmae very rare, 1–2-celled, angular. Perianths occasional, capsules very rare. $n = 9$. On peat or soil in woodlands and on heaths, amongst boulders, on rock ledges, in turf and on sand-dunes, calcifuge, from sea-level to 1000 m, occasional to frequent in montane areas, very rare elsewhere, E. Cornwall, E. Sussex, W. Suffolk, W. Norfolk, Wales, Shropshire and Leicester northwards, scattered in Ireland. 69, H10. Europe, Faroes, Iceland, N. Asia, Japan, N. America, Greenland.

22. ANASTREPTA (LINDB.) SCHIFFN. in Engler & Prantl, *Natürl. Pflanzenfam.*
1.3(1): 85. 1893

A monotypic genus with the characters of the species.

1. A. orcadensis (Hook.) Schiffn. in Engler & Prantl, *Natürl. Pflanzenfam.* 1.3(1):
85. 1893   (Fig. 43)

Dioecious. Reddish-brown to brown or green, lax tufts or scattered shoots amongst other plants. Shoots 3–9 cm; stems erect, usually simple. Rhizoids colourless, short,

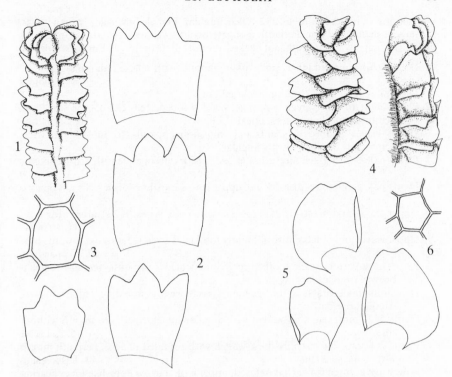

Fig. **43**. 1–3, *Barbilophozia barbata*: 1, shoot (× 6); 2, leaves (× 15); 3, leaf cell (× 507). 4–6, *Anastrepta orcadensis*: 4, shoots (× 6); 5, leaves (× 18); 6, leaf cell (× 507).

numerous. Leaves imbricate, spreading, obliquely inserted, base clasping stem with dorsal margins interdigitating, obliquely ovate-cordate, 0.6–1.1 mm wide, concave at base, becoming ± flat to convex, ventral margin reflexed, apex emarginate with broad shallow sinus forming two short wide acute to obtuse lobes, cells 16–26 μm, thin-walled, trigones present. Underleaves absent or very small, subulate. Gemmae rare, reddish-brown, 1–2-celled, at the apex of leaf lobes. Male bracts leaf-like, saccate, imbricate, subtending 1–2 antheridia. Female bracts shallowly several-lobed, often toothed. Perianths very rare, obovoid, plicate above, contracted to dentate mouth. Capsules unknown in Britain, wall 5-stratose. $n = 8 + m$. On heathery and wooded slopes, sheltered rock ledges, in scree and damp turf, from sea-level to *ca* 1000 m, occasional to frequent in N.W. Wales, the Lake District, W. Scotland, and the W. Highlands, W. Ireland, very rare elsewhere. 42, H10. Pyrenees and N. Italy north to Norway and Sweden, Faroes, Himalayas, S.W. China, Taiwan, Japan, north-west N. America.

23. LOPHOZIA (DUM.) DUM., *Recueil Observ. Jungerm.*: 17. 1835

Tufts, patches or scattered plants; stems procumbent to erect, usually branched. Rhizoids colourless. Leaves ± transversely to obliquely inserted, bilobed $\frac{1}{10}–\frac{1}{3}$, or upper leaves 3–4-lobed, lobes acute to rounded, ventral margin without cilia; cells usually with trigones, oil-bodies 6–50 per cell. Underleaves usually lacking. Gemmae on apices and margins of upper leaf lobes, 1–2-celled, angular. Female bracts larger than leaves, 2–5-lobed, dentate or not. Perianth elongate-ovoid to shortly cylindrical, plicate in upper half, narrowed to mouth, not beaked, mouth ciliate or not. 25 European species.

1 Leaves with rounded lobes, sinus often gibbous (with reflexed margins)
                                                                    9. L. obtusa
   Leaf lobes acute to obtuse, sinus not gibbous   2
2 Plants light to dark green, upper leaves 3–4-lobed, cells (28–)36–56 μm wide, gemmae unicellular, ± spherical   **10. L. capitata**
   Upper leaves 2-lobed or plants pale bluish-green, cells 16–36(–40) μm wide, gemmae 1–2-celled, mostly angular   3
3 Plants pale bluish-green, upper leaves 3–5-lobed, cells opaque with dense contents
                                                                             4
   Plants pale green to reddish-brown, upper leaves mostly 2-lobed, cells not opaque
                                                                             5
4 Upper leaves and bracts dentate to spinose-denate, leaves 1–2-stratose near base
                                                                    11. L. incisa
   Upper leaves and bracts entire or bluntly toothed, leaves 3–5-stratose towards base
                                                                   12. L. opacifolia
5 Paroecious, with concave toothed male bracts below perianth, gemmae reddish-brown, trigones lacking   6
   Dioecious, male bracts entire, gemmae green, yellowish-red or reddish-brown, trigones present   7
6 Plants 5–20 mm, cells thin-walled with angular lumens, perianth mouth without cilia   **6. L. excisa**
   Plants 2–6 mm, cells usually thick-walled with rounded lumens, perianth mouth with 1–4-celled cilia   **13. L. bicrenata**
7 Lower leaves often somewhat reflexed, upper with narrow horn-like lobes bearing globular clusters of reddish-brown gemmae   **1. L. longidens**
   Plants not as above   8
8 Underleaves present, large, on gemmiferous stems to as long as lateral leaves
                                                                 8. L. herzogiana
   Underleaves lacking or very small   9
9 Plants to 5 mm, shoots with apical clusters of yellowish-red gemmae **7. L. perssonii**
   Plants mostly more than 5 mm, gemmae green or reddish-brown   10
10 Leaves plane to slightly concave, mostly lobed ¼–⅓   11
   Leaves distinctly concave, lobed to ½   12
11 Plants usually green or yellowish-green, perianths occasional, green, mouth with 1–2-celled cilia, gemmae usually abundant   **2. L. ventricosa**
   Plants often reddish-tinged, perianths abundant, reddish-tinged at least in lower half, mouth with some cilia 3–4 cells long, gemmae rare **3. L. longiflora**
12 Gemmae light-green, leaf cell walls colourless   **4. L. wenzelii**
   Gemmae reddish-brown, leaf cell walls brownish   **5. L. sudetica**

Subgenus *Lophozia*

Stems rarely more than 240 μm diameter, in section with cells on ventral side distinctly smaller than those on dorsal side. Leaves usually bilobed, margins entire; oil-bodies 5–20 per cell, granular, 6–9 μm. Female bracts mostly entire. Capsule wall 3–5-stratose, epidermal cells of similar size to inner cells, inner layer with semi-annular thickenings.

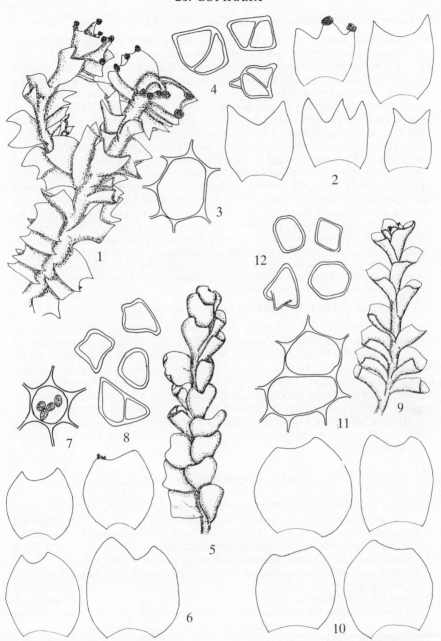

Fig. **44**. 1–4, *Lophozia longidens*: 1, shoot (× 15); 2, leaves (× 23); 3, leaf cell (× 507); 4, gemmae (× 507). 5–8, *L. sudetica*: 5, shoot (× 15); 6, leaves (× 23); 7, leaf cells (× 507); 8, gemmae (× 507). 9–12, *L. wenzelii*: 9, shoot (× 15); 10, leaves (× 38); 11, leaf cells (× 507); 12, gemmae (× 507).

1. L. longidens (Lindb.) Macoun, *Catal. Canad. Pl.* 7: 18.   (Fig. 44)

Dioecious. Dark green, sometimes brownish-tinged patches or growing through other bryophytes. Shoots 0.5–1.0(–3.0) cm, stems procumbent to erect, branched or not, 160–240 μm diameter. Leaves in female plants distant to imbricate, lower spreading to reflexed from slightly concave erecto-patent basal part, upper erecto-patent, slightly obliquely inserted, symmetrical, 0.7–1.1 mm long, 0.6–1.1 mm wide, as long as or longer than wide, bilobed ¼–⅓, occasionally 3-lobed, lobes of lower leaves obtuse, of upper leaves acute, narrow and ± horn-like, sinus crescentic to angular; cells 27–36 μm wide, thin-walled, trigones small. Leaves in male plants imbricate to closely imbricate, erecto-patent, 0.45–0.75 mm long, 0.35–0.75 mm wide, similar in shape to leaves on female plants. Underleaves lacking. Gemmae abundant in reddish-brown clusters at the apices of upper leaves, mostly 2-celled, ellipsoid to angular, 24–32 μm long. Female bracts irregularly 3–4-lobed, toothed. Perianths rare, mouth lobed, with 1–3-celled cilia. Capsules unknown in Britain. *n* = 9. Tree trunks and branches, logs, rocks and walls in shaded humid situations, ascending to *ca* 700 m, rare; Merioneth, N.W, Yorkshire, Durham, Cumberland, Perth and Angus north to Ross and W. Sutherland. 17. N. Italy, Carpathians, C. Europe, France, Germany, Fennoscandia, Spitzbergen, Iceland, Siberia, Asia Minor, China (Shensi), Himalayas, N. America, Greenland.

Readily recognised by the dark green colour contrasting with the clusters of reddish-brown gemmae on the apices of the ± horn-like lobes of the upper leaves. The somewhat reflexed lower leaves of female stems are also characteristic.

2. L. ventricosa (Dicks.) Dum., *Recueil Observ. Jungerm.*: 17. 1835

Dioecious. Green or yellowish-green, occasionally brownish or rarely vinous red patches or scattered shoots. Shoots mostly 1–2(–3) cm; stems procumbent to ascending, simple or branched, 120–280(–400) μm diameter. Leaves distant to closely imbricate, often spreading horizontally, ± symmetrical, plane to slightly concave, insertion slightly oblique, bilobed ⅛–⅓, rarely 3-lobed, lobes acute to obtuse, sinus angular to shallowly crescentic; cells (20–)24–30(–40) μm wide, thin-walled, trigones small to large but not bulging. Underleaves lacking. Gemmae very common on tips of apical leaves, pale green, 1–2-celled, angular, 20–30 μm long. Male bracts 3(–5)-lobed. Perianths occasional to frequent, mouth slightly lobed, with 1–2-celled cilia. Capsules rare, winter, spring.

Key to varieties of L. ventricosa

1 Shoots (1.6–)2.0–2.6 mm wide, stems deep red to reddish-black, at least leaf bases
     reddish-tinged, cells thick-walled with large trigones   var. **longiflora**
   Shoots (0.8–)1.0–2.0 mm wide, stems usually brownish, leaves usually green, rarely
     red-tinged, cells thin-walled, trigones small to large   2
2 Leaves distant to imbricate, plane or slightly concave, lobes not inflexed   3
   Leaves closely imbricate, channelled with dorsal lobe and ventral margin frequently
     inflexed   var. **confertifolia**
3 Oil-bodies homogeneous   var. **ventricosa**
   Oil-bodies biconcentric (with a central translucent body)   var. **silvicola**

Var. ventricosa   (Fig. 45)

Plants bright green or yellowish-green, occasionally red- or brown-tinged. Shoots (0.8–) 1.0–2.0 mm wide, stems green to brown. Leaves distant to imbricate, often spreading horizontally, plane or slightly concave, usually green throughout but sometimes reddish at base, 0.5–1.3 mm long, 0.45–1.30 mm wide, 0.9–1.2(–1.3) times as long as

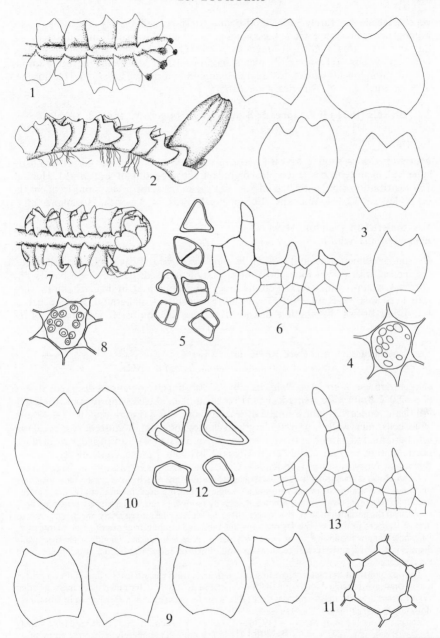

Fig. **45**. 1–6, *Lophozia ventricosa* var. *ventricosa*: 1 and 2, sterile and fertile shoots (× 11); 3, leaves (× 23); 4, leaf cell with oil-bodies (× 450); 5, gemmae (× 507); 6, perianth mouth (× 227). 7, *L. ventricosa* var. *confertifolia*: shoot (× 11). 8, *L. ventricosa* var. *silvicola*: leaf cell with oil-bodies (× 450). 9–13, *L. longiflora*: 9, leaves (× 23); 10, female bract (× 23); 11, leaf cell (× 507); 13, gemmae (× 507); 13, perianth mouth (× 227).

wide, bilobed, very rarely 3-lobed; oil-bodies 10–16 per cell, mostly longer than wide, homogeneous in texture. Gemmae common, green, angular, usually unicellular, 22–32 μm. Perianth usually green throughout. $n = 9^*$. On moist peat and soil on heaths, banks, in woods, soil amongst boulders and in walls, on *Sphagnum*, on rotten wood, occasional in lowland areas, frequent to common in montane areas. 46, H4. Europe, Siberia, north-east N. America, Greenland.

Var. **silvicola** (Buch) E.W. Jones ex Schust., *Hep. Anthoc. N. Amer.* 2: 570. 1969
(Fig. 45)
L. silvicola Buch

Morphologically similar to var. *ventricosa* but oil-bodies 15–25 per cell, mostly spherical, biconcentric with a central highly refractive portion or 'eye'. $n = 9^*$. Habitat and distribution similar to that of var. *ventricosa*, in some areas more frequent, in others less so. 34, H4. W. and C. Europe northwards, N. America, Greenland.

Var. **confertifolia** (Schiffn.) Husn.   (Fig. 45)
L. confertifolia Schiffn.

Of similar dimensions and colour to var. *ventricosa*. Leaves closely imbricate, channelled with dorsal lobe and ventral margin inflexed; nature of oil-bodies not known. On damp soil, or on soil-covered rocks and damp peat, in shaded habitats, very rare; E. Sussex, W. Kent, Herts, E. Suffolk, Shropshire, Cumberland, W. Donegal. 6, H1. Extra-British distribution unknown because of confusion with forms of *L. wenzelii*.

Var. **longiflora** auct. non (Nees) Macoun
L. longiflora sensu Macvicar non *Jungermannia longiflora* Nees

Larger than var. *ventricosa*. Plants usually reddish-tinged, sometimes deep red. Shoots (1.6–)2.0–2.6 mm wide; stems deep red to reddish-black. Leaves approximate to closely imbricate, concave, reddish-tinged at least at base, 1.0–1.5 mm long, 0.85–1.45 mm wide; cells thick-walled, trigones large, oil-bodies spherical, homogeneous, *ca* 20 per cell. Perianth red, mouth as in var. *ventricosa*. Amongst mosses and *Sphagnum* in moist open habitats, very rare, Mid Perth, Angus. 2. Belgium, France, Germany, C. Europe, Romania, Norway, European Russia, Caucasus, Wisconsin, Manitoba, Greenland.

   L. ventricosa is an extremely variable species, some forms of which closely resemble *l. longiflora* or *L. wenzelii*, the three species forming a complex requiring experimental study. *L. longiflora* usually differs in its reddish tint, abundant, usually reddish perianths (sparingly produced in *L. ventricosa*) the perianth mouth with longer, often 3–4-celled cilia and sparsely produced gemmae. Size of trigones in the two species overlaps but the large bulging trigones of some plants of *L. longiflora* are not found in *L. ventricosa*. Whether *L. ventricosa* and *L. longiflora* are specifically distinct is doubtful and urgently requires investigation. For differences from *L. wenzelii* see under that species.

   The only constant character separating var. *ventricosa* from var. *silvicola* is that of the oil-bodies and as these are destroyed on drying the two varieties can only be determined when fresh; whether var. *silvicola* is worth maintaining is debatable. For the occurence of var. *silvicola* in Britain see Banwell, *Trans. Br. bryol. Soc.* **1**, 194–8, 1949.

   The plant referred to by Macvicar (1926) as *L. confertifolia* Schiffn. and later treated as a variety (*L. ventricosa* var. *confertifolia* (Schiffn.) Husn.) is considered merely a habitat form of *L. ventricosa* by Schuster (1969), with a parellel form in *L. wenzelii*, and the taxon is of very doubtful value.

   The large size of var. *longiflora* is distinctive and it could well be a polyploid derivative of var. *ventricosa*. Red forms of *L. ventricosa* (presumably var. *ventricosa*) have been incorrectly identified as either var. *longiflora* or *L. longiflora* (i.e. *L. guttulata*). The type specimens of *Jungermannia longiflora* Nees belong to what has previously been referred to as *L. guttulata* or *L. porphyroleuca*, therefore the plant referred to as *L. ventricosa* var. longiflora requires a new name.

3. L. longiflora (Nees) Schiffn., *Lotos* 7: 145. 1903   (Fig. 45)
L. guttulata (Lindb.) Evans, *L. porphyroleuca* (Nees) Schiffn.

Dioecious. Pale green, often reddish-tinged, to vinous red, usually dense patches. Shoots to 1.0(–1.5) cm; stems procumbent with ascending apices, reddish to reddish-black, 160–280(–360) μm diameter. Leaves imbricate, often reddish at least at base, slightly concave, channelled, ± symmetrical, insertion slightly oblique, 0.8–1.4(–1.9) mm long, 0.6–1.2(–1.4) mm wide, 1.1–1.4 times as long as wide, bilobed $\frac{1}{8}$–$\frac{1}{4}$, lobes acute to obtuse, sinus shallowly crescentic to angular; cells (20–) 24–36 μm wide, thin-walled, trigones medium-sized to large and bulging, oil-bodies 6–12 per cell, homogeneous in texture. Underleaves lacking. Gemmae occasional, angular, mostly unicellular, 20–32 μm long. Female bracts asymmetrical, 2–3-lobed. Perianths very common, at least lower half usually reddish, shortly cylindrical, mouth acutely lobed, ciliate with at least some cilia 3–4 cells long. Capsules very rare. On rotting logs and moist peat, widely distributed but rare in western and northern Britain and Ireland, elsewhere very rare; Cornwall, W. Kent, E. Norfolk, Monmouth, Hereford, Derby and Cheshire northwards. 42, H10. Pyrenees and Alps north to Spitzbergen, Faroes, Azores, Siberia, Sakhalin, Korea, Japan, N. America, Greenland.

Plants of *L. ventricosa* that occur on rotting logs, or that are reddish tinged or have large trigones, may be misidentifed as *L. longiflora*. The latter species usually produces abundant perianths but the acutely lobed and ciliate perianth mouth with some cilia 3–4 cells long provides the only reliable character of distinction from *L. ventricosa*. Cells of *L. ventricosa* frequently have larger trigones than those of some plants of *L. longiflora* but I have seen bulging trigones (as in Fig. **45**) only in *L. longiflora*. *L. ventricosa* produces abundant gemmae but only about 20% of plants of *L. longiflora* are gemmiferous.

4. L. wenzelii (Nees) Steph., *Spec. Hep.* 2: 135. 1902   (Fig. 44)

Dioecious. Pale yellowish-green to yellowish-brown patches or lax tufts. Shoots 1.5–5.0 cm, slender and robust shoots often mixed together, stems procumbent to ascending, mostly simple, green when young, becoming reddish below then reddish-black throughout, 160–400 μm diameter. Leaves distant, approximate or imbricate, upper parts frequently reddish, concave with inflexed dorsal margin and lobes, symmetrical, insertion oblique, 0.7–1.3 mm long, 0.7–1.3 mm wide, 1.0–1.1 times as long as wide, bilobed to $\frac{1}{5}$, lobes obtuse to ± acute, sinus shallowly crescentic; cells 16–28 μm wide, thin-walled, trigones small, oil-bodies 5–10 per cell, homogeneous in texture. Underleaves lacking. Gemmae light green, angular, 1–2-celled, 20–34 μm long. Female bracts more deeply lobed than leaves, sinus angular, sometimes gibbous. Perianths occasional, mouth scarcely lobed, with 1-celled teeth. Capsules not known in Britain. Exposed boggy ground at high altitudes, very rare; Cumberland, Perth, Angus, Banff, Inverness, W. Sutherland. 10. Belgium, France and Alps north to Spitzbergen, Faroes, Iceland, Siberia, Sakhalin, Japan, Taiwan, Sikkim, Maine, Alaska, Greenland.

Distinct from typical forms of *L. ventricosa* in the concave leaves with very shallow sinuses. Slender shoots with distant leaves may have scarcely concave leaves but normal vigorous shoots may usually be found in the same tuft.

5. L. sudetica (Nees ex Hüb.) Grolle, *Trans. Br. bryol. Soc.* 6: 262. 1971   (Fig. 44)
L. alpestris auct. non (Schleich. ex Web.) Evans

Dioecious. Brownish, dense patches or growing through other bryophytes. Shoots 0.5–2.0(–2.5) cm; stems procumbent to ascending, occasionally branched, brown to blackish, 120–240(–320) μm diameter. Leaves distant to imbricate, upper erecto-patent, concave, symmetrical, insertion slightly oblique, 0.5–1.1 mm long, 0.45–1.00 mm wide, mostly about as long as wide, bilobed $\frac{1}{10}$–$\frac{1}{3}$, rarely 3-lobed, lobes inflexed, acute to obtuse, sinus shallowly crescentic, rarely deeper and angular; cells 14–24 μm wide with brown walls, lumens ± rounded, oil-bodies 6–9 per cell. Underleaves rarely present, small, deeply bifid. Gemmae reddish-brown, frequently abundant, angular, 1–2-celled, 20–24 μm long. Perianths rare, mouth with 1–2-celled cilia. Capsules not known in Britain. $n = 9$. On shallow soil and rocks, in scree, on stone walls and peat in acidic habitats from near sea-level to ca 1200 m, occasional to frequent in montane habitats, very rare elsewhere; E. Cornwall, S. Wales, Shropshire, Derby and S.W. Yorkshire northwards. 54, H17. Portugal, Pyrenees, N. Italy and Carpathians north to Spitzbergen, Faroes, Iceland, Siberia, Sakhalin, Japan, Himalayas, Madeira, N. America, Greenland.

Distinguished from the three preceeding species by the reddish-brown gemmae and smaller leaf cells with brownish walls and ± rounded lumens.

6. L. excisa (Dicks.) Dum., *Recueil Observ. Jungerm.*: 17. 1835   (Fig. 46)

Paroecious. Dull green or green, sometimes reddish-tinged patches or scattered shoots amongst other bryophytes. Shoots to 0.5–2.0 cm; stems brownish, simple or branched, procumbent to ascending, 80–280 μm diameter. Lower leaves distant, horizontal, obliquely inserted, upper leaves and bracts imbricate, erecto-patent, usually undulate or crisped, concave, asymmetrical, insertion ± transverse, 0.4–0.8 mm long, 0.40–0.85 mm wide, ± as long as wide, bilobed to $\frac{1}{4}$, lobes usually acute, sinus crescentic to angular, cells 20–32 μm wide, thin-walled, trigones small or absent. Underleaves lacking. Gemmae frequent, reddish, angular, 1–2-celled, 24–37 μm long. Male bracts larger than leaves, undulate, 2–3-lobed, margins bluntly toothed. Female bracts larger than male, crisped, 3–5-lobed, toothed. Perianths common, mouth dentate but cilia lacking. Capsules frequent, spring, late summer. $n = 27$. On peat and soil on heaths, paths, banks, rocks, rotting logs, usually in open and, but not restricted to, acidic habitats, rare in lowland areas, occasional to frequent elsewhere; extending north to Ross, E. Sutherland and Outer Hebrides. 98, H19. C. Faroes, Iceland, Pyrenees, N. Italy and Yugoslavia north to Fennoscandia and east to Siberia, Japan, Tenerife, N. America, Greenland, New Zealand, Antarctica, S. Georgia, southern S. America.

A variable but usually readily recognised species. Var. *cylindricea* (Dum.) Müll. is merely a wet-ground form and although Müller (1954) gives it as a synonym of *L. elongata* (Lindb.) Steph. this is incorrect, the latter being a distinct species (Schuster, 1969) which does not occur in Britain. *L. excisa* differs from *L. ventricosa* in its smaller size, the paroecious inflorescence, reddish gemmae and the somewhat crisped upper leaves and bracts. For the differences from *L. bicrenata* see under that species.

7. L. perssonii Buch & S. Arn. in Buch, *Bot. Notis.* 1944: 384. 1944   (Fig. 46)

Dioecious. Green patches or mixed with other bryophytes. Shoots to ca 5 mm with terminal clusters of yellowish-red gemmae; stems simple or branched, procumbent to erect, green to reddish-brown, 160–240 μm diameter. Leaves distant to imbricate, slightly concave, insertion oblique in lower leaves, ± transverse in upper, 0.45–0.90 mm long, 0.35–0.80 mm wide, bilobed $\frac{1}{4}$–$\frac{1}{3}$ or in upper leaves 3-lobed with third lobe much smaller and sometimes represented by a tooth, leaves often asymmetrical with ventral lobe smaller than dorsal, lobes obtuse to acuminate, sinus rounded to angular, upper gemmiferous leaves often with irregularly toothed margin; cells 20–32 μm wide, thin-walled, trigones lacking or small, oil-bodies 4–10 per cell. Underleaves absent. Gemmae yellowish-red, obtusely angular, 1–2-celled with one cell larger than the other, 18–28 μm long, one or both cells with 1–2 large or several smaller oil bodies, the large

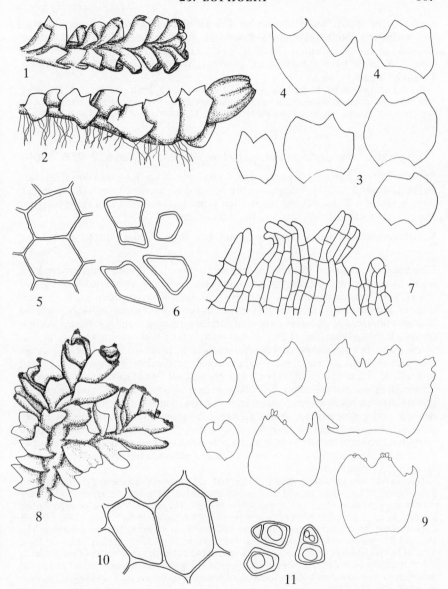

Fig. **46**. 1–7, *Lophozia excisa*: 1 and 2, sterile and fertile shoots (× 15); 3, leaves (× 21); 4, male bracts (× 21); 5, leaf cells (× 415); 6, gemmae (× 415); 7, perianth mouth (× 230); 8–11, *L. perssonii*: 8, shoot (× 13); 9, leaves (× 27); 10, leaf cells (× 507); 11, gemmae (× 507).

persisting for about 2 years after drying. Female bracts larger than leaves, 3–4-lobed and irregularly toothed. Perianths occasional, mouth ciliate, cilia 1–10 cells long. Capsules unknown in Britain. On chalk, limestone or basic soil in open habitats, very rare; E. Kent, S. Essex, Middlesex, Cambridge, W. Yorkshire, Durham. 7. France, Finland, Sweden.

Readily recognised by the terminal clusters of yellowish-red gemmae. Close to *L. excisa* but differing in the gemmae, some cells of which contain one or two large persistent oil bodies. For the occurrence of the plant in Britain see Paton & Birks, *Trans. Br. bryol. Soc.* **5**, 439–42, 1968.

Subgenus *Protolophozia* Schust., *Hep. Anthoc. N. Amer.* 2: 255. 1969

Stems to 250 μm diameter, in section with ventral cells of similar size to dorsal. Mature leaves 2–4-lobed to $\frac{1}{4}-\frac{3}{4}$, margins entire; oil-bodies to *ca* 20 per cell, spherical, homogeneous, 3–5 μm. Female bracts often ± crisped, 2–4-lobed with entire margins. Inner layer of capsule wall with annular thickenings.

8. L. herzogiana Hodgs. & Grolle in Grolle, *Rev. Bryol. Lichén.* 31: 152. 1962

(Fig. 47)

Dioecious. Plants green, scattered or gregarious. Shoots 3–8 mm, stems procumbent with ± ascending gemmiferous tips which may or may not be attenuated, stems green, (96–)128–192 μm diameter, very sparingly branched. Rhizoids often present, sometimes in quantity, on margins of older leaves. Leaves distant to subimbricate, variable in shape and size, mature leaves somewhat flaccid, spreading to subsquarrose, insertion oblique, 0.35–0.64 mm long, 0.21–0.47 mm wide, 1.0–1.6 times as long as wide, bilobed to *ca* $\frac{1}{4}-\frac{1}{3}$, lobes lanceolate to broadly lanceolate, acute to acuminate; leaves on gemmiferous stems smaller, erect, often concave-canaliculate, youngest with eroded tips; cells 25–38 μm wide, thin-walled, trigones small, oil-bodies spherical, to 20 per cell. Underleaves on mature stems 220–380 μm long, bilobed to $\frac{1}{3}-\frac{2}{3}$, underleaves on gemmiferous stems as large as lateral leaves or not, entire and broadly lanceolate to bilobed and ± orbicular to ovate. Gemmae abundant, green, mostly elongate-triangular but sometimes not angular, (1–)2-celled, (22–)28–42(–51) μm long. Fertile plants not known in Britain. Female bracts 2–3-lobed, connate at base, bracteole ovate. Perianth ellipsoid, mouth dentate. *n* = 9*. Under *Calluna* on heathy ground, very rare, N. Hants. 1. New Zealand.

This plant is very variable in stem diameter, leaf and underleaf shape and size, especially on gemmiferous shoots which may or may not be attenuated. Although the gemmae are mostly sharply angular, some are less so or completely smooth. The species is unique amongst British hepatics in producing rhizoids on the margins of older leaves. It differs from other British Lophoziaceae with bilobed leaves in the large underleaves which are not ciliate at the base. The status of this species in Britain is problematical; it occurs in a natural habitat well away from gardens but is elsewhere known only from New Zealand suggesting that it is an introduction. It differs from the New Zealand plant in the gemmae being greenish rather than red and in the smaller cells. For an account of the discovery of this plant in Britain see Crundwell & Smith, *J. Bryol.* **15**, 653–657, 1989.

Subgenus *Schistochilopsis* Kitag., *J. Hattori bot. Lab.* 28: 289. 1965

Stems thick, 200–640 μm diameter, in section with ventral cells of ± similar size to dorsal cells. Upper leaves 2–4-lobed, often toothed, oil-bodies homogeneous, smooth, 2–4 μm long, to 50 per cell. Female bracts often toothed. Capsule wall 3–5-stratose, cells of outer layer larger than inner cells; inner layer with semi-angular thickenings.

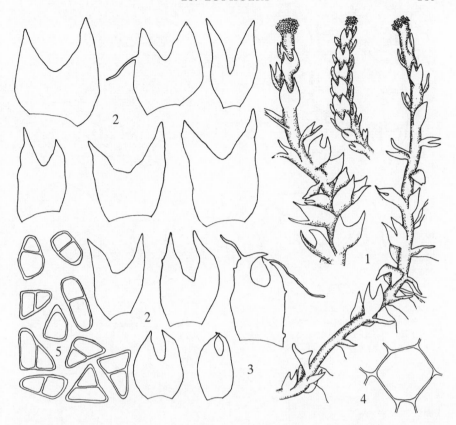

Fig. 47. *Lophozia herzogiana*: 1, shoots (× 21); 2, leaves, two with marginal rhizoids; (× 63); 3, underleaves (× 63); 4, mid-leaf cell (× 450); 5, gemmae (× 450).

9. L. obtusa (Lindb.) Evans, *Proc. Washington Acad. Sci.* 2: 303. 1900   (Fig. 48)
Obtusifolium obtusum (Lindb.) S. Arn.

Dioecious. Light, green scattered plants, rarely forming loose patches. Shoots 1.5–5.0 cm; stems procumbent, occasionally branched, 160–480 μm diameter. Leaves flaccid, distant to approximate, lower spreading horizontally, uppermost erecto-patent, often at least slightly asymmetrical, insertion very oblique, 0.25–0.55 mm long, 0.25–0.60 mm wide, as wide as or wider than long, bilobed ¼–⅓, lobes rounded, similar or one rounded and one obtuse, sinus obtuse to acute, gibbous, or in small leaves lobes sometimes acute and sinus rounded, margins entire or rarely ventral margin with a single tooth; cells (16–)20–32(–36) μm wide, thin-walled, trigones small. Underleaves present where branches arise, absent or rudimentary elsewhere. Gemmae rare, pale green, very thin-walled, mostly unicellular, 18–24 μm long. Perianths very rare. Capsules not known in Britain. $n = 8 + m$. Alluvial detritus, moist turf, basic flushes and scree in open or sheltered habitats from 60–1000 m, rare; Merioneth, Caernarfon, N.W. Yorkshire, S. Northumberland, I. of Man, Dumbarton and Perth north to W. Sutherland. 17. Pyrenees, N. Italy, C. Europe, France, Belgium, Germany, Norway, Sweden, Denmark, U.S.S.R., Faroes, Iceland, Japan, north-east N. America.

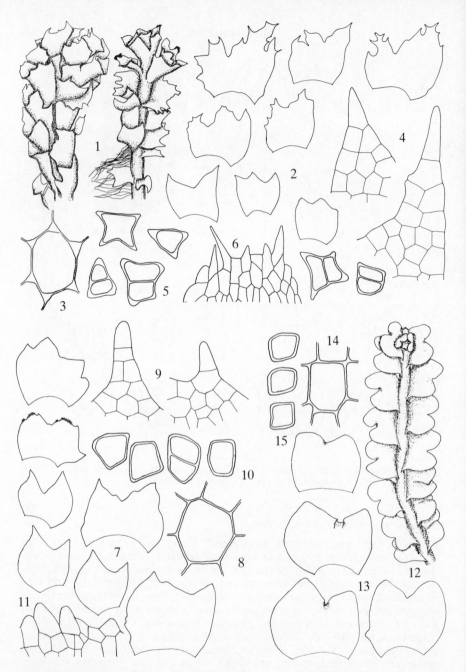

Fig. **48**. 1–6, *Lophozia incisa*: 1, shoots (× 11); 2, leaves (× 15); 3, leaf cell (× 507); 4, tip of upper leaf (× 178); 5, gemmae (× 507); 6, perianth mouth (× 178); 7–11, *L. opacifolia*: 7, leaves (× 15); 8, leaf cell (× 507); 9, upper leaf tips (× 178); 10, gemmae (× 507); 11, perianth mouth (× 178). 12–15, *L. obtusa*: 12, shoot (× 8); 13, leaves (× 23); 14, leaf cell (× 507); 15, gemmae (× 507).

10. L. capitata (Hook.) Macoun, *Catal. Canad. Pl.* 7: 18. 1902   (Fig. 49)

Dioecious. Bright or dull green, sometimes violet-tinted patches or amongst other bryophytes. Shoots to 2 cm; stems erect or ascending, succulent, brittle, green to yellowish brown, 240–600 μm diameter. Leaves on vigorous female plants imbricate below, closely imbricate and forming a capitulum above, 1.4–1.8 mm long, 1.4–2.6 mm wide, wider than long, lower bilobed, upper asymmetrically 3–4-lobed to $\frac{1}{3}$ with obscurely dentate margins, strongly crisped, leaves on depauperate plants smaller, distant, spreading horizontally, about as wide as long, plane, symmetrically bilobed to $\frac{1}{3}$, lobes acute to obtuse, sinus rounded to angular, often gibbous; cells (28–)36–56 μm wide, thin-walled, trigones lacking or very small. Underleaves absent. Male plants smaller than female. Gemmae abundant on margins of upper leaves of small-leaved attenuated shoots, pale green, ± spherical, unicellular, 20–26 μm. Perianths occasional, mouth with mostly 2–3-celled cilia. Capsules not known in Britain. On damp sandy or peaty soil in open, acidic habitats, usually heaths, at low altitudes, rare; S. Hants (1813), N. Hants, E. Sussex, Surrey, N. Essex, Berks, Cheshire. 7. France, Germany, Finland, Norway, Denmark, eastern N. America.

A variable plant distinguished from other species of *Lophozia* and *Barbilophozia* with which it might be confused by the large leaf cells and, when present, the ± spherical gemmae. Vigorous plants most closely resemble *L. incisa* but in that species the plants are pale bluish-green, the upper leaves spinose dentate and cells opaque when fresh. For the occurrence of this plant in Britain see Jones, *Trans. Br. bryol. Soc.* 1, 353–6, 1950. There is an old, unlocalised record from Ireland (see Jones, *ibid.*).

11. L. incisa (Schrad.) Dum., *Recueil Observ. Jungerm.*: 17. 1835   (Fig. 48)

Dioecious. Pale bluish-green patches or scattered shoots amongst other bryophytes. Shoots 3–15 mm; stems brittle, procumbent to erect, simple or branched, bluish-green to blackish, (160–)200–480 μm diameter. Lower leaves and those on slender shoots distant to approximate, erecto-patent, ± concave, symmetrical or not, insertion ± transverse, 0.7–1.0 mm long, 0.7–1.1 mm wide, bistratose near base, 2–3-lobed to $\frac{1}{3}$, upper leaves closely imbricate, crowded towards apex, plicate and crisped, asymmetrical, to 1.2–1.4 mm long and 1.2–1.4 mm wide, irregularly 3–5-lobed to $\frac{1}{3}$, lobes often bristle-pointed, margin dentate to spinose-dentate or with eroded gemmiferous margins; cells opaque with dense contents, 20–36 μm, thin-walled, trigones small. Underleaves lacking. Gemmae usually abundant, pale bluish-green, 1–2-celled, 16–32 μm long. Female bracts similar to but larger than upper leaves. Perianths rare, mouth lobed, with 1–3-celled cilia to 160 μm long. Spores 12–15 μm (Müller, 1954). Capsules very rare, summer. $n = 8 + m^*$, 9. Moist places on soil or peat on heaths and moorland, in rock crevices, moist turf and on rotten logs, very rare in lowland areas but occasional in S.E. England, occasional to common in W. and N. Britain, frequent in W. Ireland, from sea-level to 1200 m. 68, H35. Pyrenees, N. Italy and Bulgaria north to Spitzbergen, Faroes, Caucasus, Siberia, Sakhalin, Japan, China, Taiwan, Himalayas, Azores, N. America, Greenland.

12. L. opacifolia Culm. ex Meylan, *Hepat. Suisse*, 174. 1924   (Fig. 48)

Dioecious. Pale bluish-green patches or scattered shoots amongst other bryophytes. Shoots 5–15 mm; stems brittle, procumbent to erect, simple or branched, bluish-green to blackish, 280–640 μm diameter. Lower leaves approximate, ± symmetrical, upper closely imbricate, undulate to crisped, asymmetrical, insertion ± oblique, 0.7–1.4 mm long, 0.6–1.4 mm wide, 3–5-stratose towards base, unequally 2–4-lobed to $\frac{1}{3}$–$\frac{1}{2}$, lobes acute to obtuse, not bristle-pointed but sometimes ending in a single long cell, margin sometimes bluntly toothed but not spinose-dentate, sometimes eroded if gemmiferous; cells opaque with dense contents, 20–40 μm wide, thin-walled, trigones lacking. Underleaves lacking. Gemmae usually abundant, pale bluish-green, 1–2-celled, 16–32 μm long. Perianths rare, mouth with unicellular cilia to 80 μm long. Spores 18–19 μm (Müller, 1954). Capsules rare, summer. n = 18. On peat, soil, rocks and in scree at altitudes above 500 m; occasional, Westmorland, Perth and Angus north to W. Sutherland, W. Mayo, W. Donegal. 14, H2. Central Europe, Sweden, Finland, Iceland, Alaska, Greenland.

In general appearance very similar to *L. incisa* but differing in the upper leaves lacking spinose teeth and the cells near the leaf base being 3–5-stratose (making the lower part of the leaf darker than the rest when viewed under the microscope). The leaves are often wider than long, the reverse of the situation in *L. incisa* but leaf shape in the two species is too variable to make this a useful character. For the occurrence of *L. opacifolia* in Britain see Jones, *Trans. Br. bryol. Soc.* 3, 180, 1957.

Subgenus *Isopaches* (Buch) Schust., *Amer. Midl. Nat.* 45; 56. 1951

Stems to *ca* 250 μm diameter, in section with cells on ventral side of similar size to those on dorsal side. Leaves usually bilobed with entire margins; oil-bodies 6–12 per cell, granular, 9–12 μm long. Capsule wall 2(–3)-stratose, cells with incomplete semi-annular thickenings.

13. L. bicrenata (Schmid. ex Hoffm.) Dum., *Recueil Observ. Jungerm.*: 17. 1835

(Fig. 49)

Isopaches bicrenatus (Schmid. ex Hoffm.) Buch

Paroecious. Plants very small, usually forming small yellowish-green to reddish-brown patches. Shoots 2–6 mm; stems prostrate with ascending apices or ascending if crowded, simple or branched, 120–240 μm diameter. Leaves imbricate, increasing in size up stem to bracts, concave, ± symmetrical, insertion ± transverse, 0.3–0.7 mm long, 0.2–0.7 mm wide, about as wide as long, bilobed ¼–⅓, lobes acute to obtuse, sinus angular; cells 20–32 μm, usually thick-walled with rounded lumens, occasionally thin-walled with small trigones, oil-bodies 6–12 per cell. Underleaves lacking. Gemmae common on sterile shoots, reddish-brown, 1–2-celled, angular, 18–30 μm long. Male bracts larger than leaves, bilobed, often with a large tooth. Female bracts irregularly 2–4-lobed, lobes dentate or spinose-dentate, rarely entire. Perianths very common, mouth with 1–4-celled cilia. Capsules frequent, at various times of the year. On sandy or peaty acidic soil, especially on heaths, soil crevices amongst stones, in quarries, occasional, mainly at low altitudes, 100, H19. C. Algarve, Pyrenees, N. Italy, Yugoslavia north to Fennoscandia, Iceland, east to Siberia, N. America.

Only likely to be confused with *L. excisa* from which it differs in its smaller size and the ciliate mouth of the perianth; the thick-walled leaf cells with rounded lumens and the dentate or spinose-dentate female bracts are distinctive but some plants of *L. bicrenata* have thin-walled cells and/or entire bracts.

24. LEIOCOLEA BUCH, *Mem Soc. Fauna Fl. Fenn.* 8: 288. 1932

Plants small to robust, often brownish-tinged, calcicole. Shoots procumbent to erect, usually sparsely branched. Rhizoids usually numerous ± to stem apex. Leaves obliquely inserted, dorsal margin decurrent or not, ventral margin lacking cilia (except sometimes *L. rutheana*), bilobed to ⅙–⅓, lobes rounded to acute or obtuse and apiculate, ventral frequently larger than dorsal; cells mostly thin-walled, with or without trigones; oil-bodies mostly 2–5 per cell, opaque. Underleaves present (except *L. badensis* and

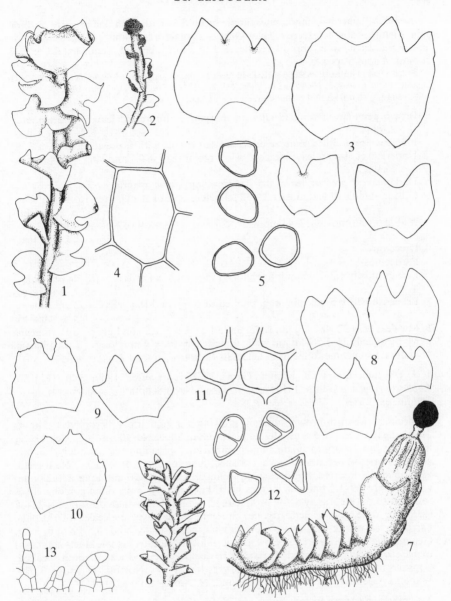

Fig. **49**. 1–5, *Lophozia capitata*: 1 and 2, non-gemmiferous and gemmiferous shoots (× 6); 3, leaves (× 15); 4, leaf cell (× 507); 5, gemmae (× 507). 6–13, *L. bicrenata*: 6 and 7, sterile and fertile shoots (× 15); 8, leaves (× 21); 9, female bracts (× 21); 10, male bract (× 21); 11, leaf cells (× 507); 12, gemmae (× 507); 13, perianth mouth (× 148).

L. turbinata), narrowly lanceolate, simple or bifid, usually with one or more teeth or cilia. Gemmae absent (except *L. heterocolpos*). Female bracts similar to leaves, larger or not. Perianth smooth, abruptly contracted to shortly beaked, denticulate or ciliate mouth. About 8 species.

Several recent authorities suggest that, on morphological grounds, *Leiocolea* is not generically distinct from *Lophozia*. However,' biochemical data (Mues, *Ber. Deutsch Bot. Ges.* **95**, 115–25, 1982) suggest that it should be maintained as a distinct genus.

1 Brown gemmae present, usually on margins of appressed leaves on attenuated shoots   **5. L. heterocolpos**
  Plants without either gemmae or attenuated shoots with appressed leaves   2
2 Underleaves absent, leaves (except below perianth) 0.3–0.7 mm long, stems ± translucent   3
  Underleaves* present, leaves 0.5–2.0 mm long, stems opaque   4
3 Cells in middle of leaf mostly 24–32 μm wide, cells of stem cortex 16–32 μm wide
                                                                   6. L. badensis
  Mid-leaf cells mostly 32–44 μm wide, cortical cells mostly 32–52 μm wide
                                                                   7. L. turbinata
4 Dioecious   5
  Paroecious   6
5 Leaves mostly 1.2–2.0 mm long, bilobed to ⅙, mid-leaf cells (24–)28–44 μm wide
                                                                   3. L. bantriensis
  Leaves mostly 0.5–1.3 mm long, bilobed to ⅕(–⅓), mid-leaf cells 20–32 μm wide
                                                                   4. L. alpestris
6 Mid-leaf cells 32–48(–56) μm wide, underleaves 0.5–1.2 mm long   **1. L. rutheana**
  Mid-leaf cells 24–32 (–40) μm wide, underleaves to 0.64 mm long   **2. L. gillmanii**
*Small and sometimes difficult to detect in *L. bantriensis* and *L. gillmanii*.

1. L. rutheana (Limpr.) K. Müll. in Gams, *Kl. Kryptogamenfl.* 1: 40. 1940   (Fig. 50)
Lophozia rutheana (Limpr.) Howe, *L. schultzii* (Nees) Schiffn., *L. schultzii* var. *laxa* Schiffn. ex Burrell

Paroecious. Reddish-brown tufts or growing through other bryophytes. Shoots ascending to erect, to 6 cm, usually simple; stems opaque, 240–400 μm diameter. Lower leaves distant to approximate, upper imbricate, spreading ± horizontally, slightly convex, upper sometimes somewhat crisped, margins entire or with 1–2 blunt teeth, very obliquely inserted, not decurrent, asymmetrical, 1.4–2.0 mm long, 1.1–2.0 mm wide, mostly 0.9–1.2 times as long as wide, bilobed to ¼, sinus often gibbous, lobes rounded to obtuse, rarely acute or apiculate, ventral larger than dorsal; cells 32–48 (–56) μm wide, thin-walled, trigones present. Underleaves concealed in rhizoids, narrowly lanceolate with two to several cilia, occasionally bifid, 0.5–1.2 mm long. Gemmae lacking. Male bracts below female, bilobed, saccate. Female bracts appressed to base of perianth, bilobed. Perianths frequent, subcylindrical, smooth, beaked. Capsules frequent. Moist, basic habitats, very rare, Berks, Norfolk, W. Mayo. 3, H1. Scandinavia, Finland, N. Germany, Iceland, N. America, Greenland.

2 L. gillmanii (Aust.) Evans, *Bryologist* 38: 83. 1935   (Fig. 50)
Lophozia gillmanii (Aust.) Schust., *Lophozia kauriniii* (Limpr.) Steph.

Paroecious. Green to yellowish-green patches or growing through other bryophytes. Shoots ascending to erect, to *ca* 3 cm, usually unbranched; stems opaque, 160–270 μm diameter. Lower leaves distant, upper imbricate, soft, undulate, hardly contracted at base, obliquely inserted, decurrent on dorsal side, usually asymmetrical, 0.6–1.7 mm long, 0.54–1.70 mm wide, 0.9–1.3 times as long as wide, bilobed to *ca* ⅙, lobes rounded

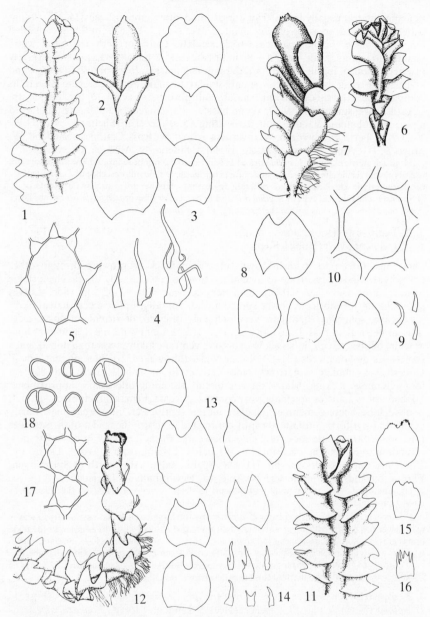

Fig. **50**. 1–5, *Leiocolea rutheana*: 1 and 2, sterile and fertile shoots (× 7); 3, leaves (× 18); 4, underleaves (× 24); 5, leaf cell (× 450). 6–10, *L. gillmanii*: 6 and 7, sterile and fertile shoots (× 13); 8, leaves (× 18); 9, underleaves (× 18); 10, leaf cells (× 450). 11–18, *L. heterocolpos*: 11 and 12, non-gemmiferous and gemmiferous shoots (× 17); 13, leaves (× 46); 14, underleaves (× 46); 15 and 16, leaves and underleaf from attenuated part of gemmiferous shoot (× 46); 17, leaf cells (× 450); 18, gemmae (× 450).

to acute, ventral usually larger than dorsal, sinus often gibbous; cells 24–32(–40) μm wide, thin-walled, trigones present. Underleaves small, 180–640 μm long, concealed amongst coarse rhizoids, subulate to lanceolate, entire or with 1–2 basal teeth. Gemmae lacking. Male bracts 4–8, imbricate below female, saccate, transversely inserted, bilobed, frequently with tooth on dorsal margin, lobes obtuse to obtuse and apiculate. Female bracts erecto-patent, similar to male but not saccate. Perianths common, shortly subcylindrical, smooth, abruptly contracted to beaked mouth. Capsules rare, summer. *n* = 9. On damp basic rocks, cliff ledges and flushes, mainly in alpine or subalpine habitats but descending to sea-level in Shetland, rare, N.W. Yorkshire, Perth, S. Aberdeen, W. Inverness, Argyll, W. Ross, Caithness, Shetland, W. Mayo. 9, H1. Alps, Norway, Sweden, Finland, Siberia, N. America, Greenland.

Although sterile plants of *L. gillmanii* and *L. bantriensis* cannot be separated, careful search will usually reveal fertile shoots and the male bracts below the female inflorescence of *L. gillmanii* are of characteristic appearance quite different from that of either male or female shoots of *L. bantriensis*. The mid-leaf cells in *L. gillmanii* tend to be smaller than in *L. bantriensis* but the latter does sometimes have cells within the size range of *L. gillmanii*.

3. L. bantriensis (Hook.) Joerg. *Bergens Mus. Skrift.* 16: 164. 1934   (Fig. 51)
Lophozia bantriensis (Hook.) Steph.

Dioecious. Green to reddish-brown tufts or patches or growing through other bryophytes. Shoots prostrate to ascending, to 4(–6) cms, usually unbranched; stems opaque, brownish, 200–300 μm diameter, cortical cells 22–32 μm wide. Leaves approximate to imbricate, lower spreading horizontally, upper erecto-patent, soft, ±concave, somewhat undulate, very obliquely inserted, decurrent on dorsal side, ±asymmetrical, (0.8–)1.2–2.0 mm long, (1.0–)1.2–2.5 mm wide, 0.8–1.2 times as long as wide, bilobed to *ca* ⅕, lobes acute to obtuse, ventral usually larger than dorsal, sinus sometimes gibbous; cells (24–)28–44 μm wide, thin-walled, trigones present, small. Underleaves subulate to narrowly lanceolate, entire or with 1–2 teeth, 240–800 μm long. Gemmae lacking. Male bracts terminal, becoming intercalary, approximate, bilobed, lobes acute or apiculate, very concave, saccate. Female bracts erecto-patent, bilobed, lobes acute, sometimes apiculate, margins entire. Perianths occasional, shortly cylindrical-pyriform, smooth, abruptly contracted to shortly-beaked mouth. Capsules rare. *n* = 9. Streamside rocks and silt, moist cliffs, flushes, usually in basic habitats, ascending to *ca* 1000 m, occasional in W. and N. Britain, from S. Wales, Derby and Yorkshire north to Orkney. 56, H17. N. Spain, Italian Alps north to Spitzbergen, Faroes, Iceland, Siberia, western N. America, Newfoundland, Greenland.

The underleaves are often small and difficult to detect being concealed by the rhizoids or overlapping leaf bases.

L. bantriensis is distinguished from *L. alpestris* by the larger, somewhat asymmetrical leaves with larger cells, markedly decurrent dorsal margin and shallower sinus. In *L. alpestris* the male bracts are closely imbricate whereas in *L. bantriensis* they are more distant as in *L. turbinata* and *L. badensis* which are much smaller plants. For differences from *L. gillmanii* see under that species.

According to Grolle (1983a) some authorities suggest that *L. alpestris* may merely be a form of *L. bantriensis* and says the matter requires experimental investigation.

4. L. alpestris (Schleich. ex Web.) Isov., *Ann. bot. Fenn.* 15: 80. 1978.   (Fig. 51)
L. collaris (Nees) Schljak., *L. muelleri* (Nees ex Lindenb.) Joerg., *Lophozia alpestris* (Schleich. ex Web.) Evans, *L. collaris* (Nees) Dum., *L. muelleri* (Nees ex Lindenb.) Dum.

Dioecious. Green to brown patches or growing through other bryophytes. Shoots prostrate to ascending, to 3 cm, simple or occasionally branched; stems opaque, brownish, 130–230 μm diameter. Leaves approximate to imbricate, erecto-patent,

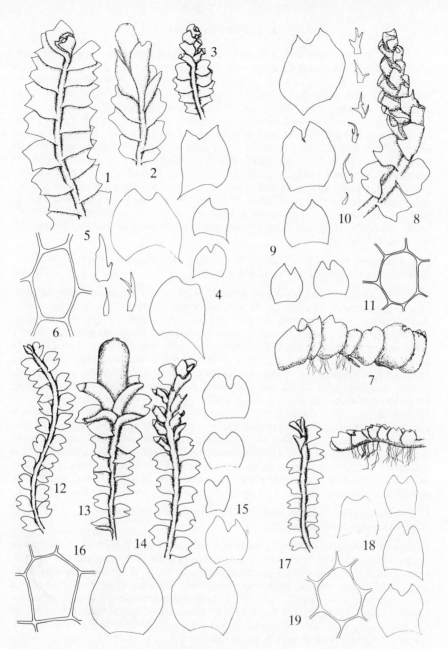

Fig. **51**. 1–6, *Leiocolea bantriensis*: 1, sterile shoot (× 4); 2, shoot with perianth (× 4); 3, male shoot (× 4); 4, leaves (× 7); 5, underleaves (× 11); 6, leaf cell (× 450). 7–11, *L. alpestris*: 7, sterile shoot (× 15); 8, male shoot (× 15); 9, leaves (× 38); 10, underleaves (× 38); 11, leaf cell (× 450). 12–16, *L. turbinata*: 12, sterile shoot (× 11); 13, shoot with perianth (× 11); 14, male shoot (× 11); 15, leaves (× 27); 16, leaf cell (× 450). 17–19, *L. badensis*; 17, shoots (× 11); 18, leaves (× 27); 19, leaf cell (× 450).

concave to very concave with inflexed margins, hardly narrowed at base, obliquely inserted, slightly decurrent on dorsal side, ±symmetrical, 0.5–1.3 mm long, 0.5–1.0 mm wide, 0.8–1.2 times as long as wide, bilobed to $\frac{1}{4}(-\frac{1}{3})$, lobes obtuse to acute, occasionally with a 2-celled apiculus, ±equal or ventral larger than dorsal, sinus sometimes gibbous; cells 20–32 μm wide, thin-walled, trigones medium-sized. Underleaves regularly present, subulate to narrowly lanceolate, margin entire or with tooth on one or both sides at base or spinosely toothed, 80–400 μm long. Gemmae lacking. Male bracts closely imbricate, very concave with inflexed margins. Female bracts erecto-patent, bilobed, lobes apiculate, margins entire to dentate. Perianths occasional, pyriform, abruptly contracted to shortly beaked mouth. Capsules rare. $n = 9$. On damp soil and rocks, ravines, cliffs in basic habitats, ascending to 1050 m, occasional to frequent in basic areas of W. and N. Britain from W. Gloucester, S. Wales, Derby and Yorkshire north to Shetland, occasional in Ireland. 57, H11. N. Spain, Sicily, N. Italy and Carpathians north to Spitzbergen, Faroes, Iceland, N. America, Greenland.

5. L. heterocolpos (Thed. ex Hartm.) Buch, *Mem. Soc. Fauna Fl. Fenn.* 8: 284. 1932
Lophozia heterocolpos (Thed. ex Hartm.) Howe   (Fig. 50)

Dioecious. Green to reddish-brown patches or growing through other bryophytes. Shoots procumbent to erect, to 20 mm branched, appressed-leaved gemmiferous shoots±erect, stems opaque, brown, 200–320 μm diameter. Leaves imbricate, spreading to erecto-patent, ±concave, obliquely inserted, dorsal margin not or slightly decurrent, symmetrical or not, 0.48–0.68 mm long, 0.40–0.64 mm wide, 0.9–1.2 times as long as wide, bilobed to $\frac{1}{3}$, sinus often gibbous, lobes acute to obtuse, subequal or ventral larger than dorsal; appressed leaves on gemmiferous shoots smaller, with eroded margins; cells 20–32(–40) μm wide, thin-walled, trigones large, bulging. Underleaves narrowly lanceolate, entire or with 1–2 cilia or bilobed, 150–300 μm long, those on gemmiferous shoots larger and sometimes resembling leaves. Gemmae very common, on margins of appressed leaves occasionally on other leaves, brown, spherical to ellipsoid, 1–2-celled, 16–24 μm long. Perianths and capsules not known in British or Irish material. $n = 9$. On damp shaded soil and soil-covered rock ledges, often with other bryophytes, in basic habitats, ascending to 500 m, rare, scattered localities from Wales and N.W. Yorkshire north to Ross, Antrim. 24, H1. N. Spain and N. Italy north to Spitzbergen, Siberia, Himalayas, Madeira, eastern N. America, Greenland.

6. L. badensis (Gott.) Joerg., *Bergens Mus. Skrift.* 16: 166. 1934   (Fig. 51)
Lophozia badensis (Gott.) Schiffn.

Dioecious. Closely resembling *L. turbinata*. Pale green to yellowish-green patches or creeping through other bryophytes. Shoots prostrate, to 25 mm, mostly unbranched; stems±translucent, 60–160 μm diameter, cortical cells 16–32 μm wide. Leaves distant and spreading horizontally on slender stems to imbricate and erecto-patent on robust stems, concave, not or hardly narrowed at base, obliquely inserted, shortly decurrent on dorsal side, symmetrical or not, 0.35–0.65 mm long, 0.35–0.70 mm wide, 0.75–1.00 times as long as wide, increasing in size towards perianth, bilobed to $\frac{1}{5}$–$\frac{1}{3}$, lobes acute to obtuse, ±equal or ventral slightly larger than dorsal, very rarely dorsal margin with a tooth: cells 24–32(–36) μm wide, usually thin-walled with small trigones. Underleaves lacking or very rarely present and rudimentary. Gemmae absent. Male bracts terminal, becoming intercalary, approximate, saccate, 2–3-lobed. Female bracts similar to upper leaves, spreading from erect base, bilobed, margin entire. Perianths frequent, pyriform, smooth, abruptly contracted to shortly beaked mouth. Capsules occasional, autumn to spring. $n = 9*$. Damp, basic soil, rocks and dune-slacks, mainly in lowland habitats but ascending to 550 m, sometimes mixed with *L. turbinata*, occasional in lowland Britain, rare elsewhere. 75, H17. N. Spain, N. Italy and Yugoslavia north to Spitzbergen and east to the Urals, Asia Minor, Iran, Siberia, N. America, Greenland.

Less frequent than but possibly overlooked as *L. turbinata*. The two species have been confused and some plants are difficult to name. Typically *L. badensis* differs from *L. turbinata* in the smaller cortical and leaf cells, the presence of small trigones and the slightly decurrent leaf base. Some plants of *L. turbinata* may, however, have slightly decurrent leaf bases, especially if growing in damp shade and can only be separated from *L. badensis* on cell size.

Leiocolea alpestris differs from both *L. turbinata* and *L. badensis* in the constant presence of small underleaves and the distinct trigones though the latter sometimes also occur in *L. badensis*. Male plants of *L. alpestris* have closesly imbricate, very concave male bracts whereas in the other two species the bracts are approximate, somewhat spreading and markedly saccate at the base.

7. L. turbinata (Raddi) Buch, *Ann. Bryol.* 10: 4. 1937   (Fig. 51)
Lophozia turbinata (Raddi) Steph.

Dioecious. Yellowish-green to pale green patches or creeping through other bryophytes. Shoots prostrate, to 10 mm, simple or rarely branched; stems ± translucent, 60–140(–200) μm diameter, cortical cells (28–)32–52 μm wide. Leaves distant, spreading horizontally on slender stems, approximate, patent to erecto-patent on robust stems, ± flat, usually narrowed at base, very obliquely inserted, rarely slightly decurrent, symmetrical or not, 0.28–0.60 mm long, 0.25–0.57 mm wide, 0.8–1.2 times as long as wide, lobes acute to obtuse, ventral lobe sometimes larger than dorsal, leaves towards perianth larger, to 2 mm long; cells (28–)32–44(–56) μm wide, walls not or slightly thickened, trigones lacking. Underleaves lacking or very rarely present and small. Male inflorescence terminal, becoming intercalary with age, bracts approximate, saccate. Female bracts similar to upper leaves, spreading from erect base, bilobed, margins entire. Perianths frequent, narrowly pyriform, smooth, abruptly contracted at mouth, mouth ciliate. Capsules occasional, winter, spring. On damp shaded soil and rocks and in dune-slacks, coastal cliffs, common in basic lowland areas, rare or occasional elsewhere. 99, H35. Mediterranean region of Europe, Spain, Portugal, S. France, Belgium, Germany (Nordrhine-Westphalia).

25. GYMNOCOLEA (DUM.) DUM., *Recueil Observ. Jungerm.*: 17. 1835

Dioecious. Darkish patches. Shoots procumbent to erect; branching various. Leaves obliquely inserted, bilobed; cells usually without trigones. Underleaves lacking or few and subulate. Gemmae usually lacking. Antheridia solitary, male bracts similar to but more concave than leaves. Female bracts similar to leaves, widely spreading so that perianth is naked. Perianth smooth, contracted at dentate mouth, ovoid when unfertilised, ellipsoid to pyriform after fertilisation. Four species.

Leaf lobes obtuse to rounded   **1. G. inflata**
Leaf lobes acute to subacute   **2. G. acutiloba**

1. G. inflata (Huds.) Dum., *Recueil Observ. Jungerm.*: 17. 1835   (Fig. 52)

Dark green to brownish or blackish patches to spongy mats. Shoots flaccid, often interwoven, procumbent to erect, to *ca* 2 cm, innovating from below perianth and sometimes from ventral surface of stem, also sometimes branching laterally. Lower leaves distant, spreading horizontally, ± flat, upper leaves and those on ascending or erect stems approximate to imbricate, patent, concave with inflexed margins, ± symmetrical, obliquely inserted, not decurrent, mostly 0.4–1.0 mm long, 0.5–1.0 mm wide, 0.8–1.2 times as long as wide, margins entire, bilobed to ⅓, lobes obtuse to

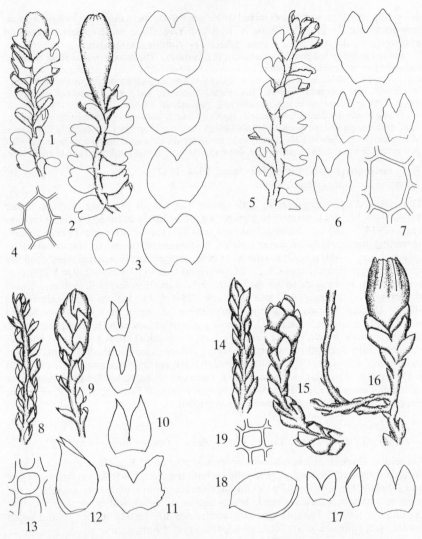

Fig. **52**. 1–4, *Gymnocolea inflata*: 1, shoot with perianth containing unfertilised archegonia (× 9); 2, shoot with perianth with fertilised archegonium (× 9); 3, leaves (× 18); 4, mid-leaf cell (× 450). 5–7, *G. acutiloba*: 5, shoot (× 11); 6, leaves (× 27); 7, mid-leaf cell (× 450). 8–13, *Sphenolobopsis pearsonii*: 8 and 9, sterile and male shoots (× 33); 10, leaves (× 63); 11 and 12, female and male bracts (× 63); 13, leaf cell (× 450). 14–19, *Eremonotus myriocarpus*: 14–16, sterile, male and female shoots (× 33); 17, leaves (× 63); 18, male bract (× 63); 19, leaf cell (× 450).

rounded; cells 20–32(–40) μm wide, thin-walled to moderately thick-walled, trigones usually absent. Underleaves lacking or sparse and subulate. Male bracts similar to but more concave than leaves. Female bracts similar to leaves, spreading. Perianths very common, smooth, contracted at mouth, ovoid when unfertilised, becoming pyriform after fertilisation, unfertilised perianths deciduous, forming vegetative propagules. Capsules rare, winter.

Key to varieties of *G. inflata*

Small-leaved innovations from ventral side of stem rare, lateral branches produced between leaves   var. **inflata**
Small-leaved innovations from ventral side of stem abundant, lateral branches axillary, subtended by unlobed leaves   var. **heterostipa**

Var. inflata

Stems rarely producing small-leaved innovations from ventral surface, lateral branches arising between successive bilobed leaves. $n = 9*$. Wet heaths, sometimes submerged, *Sphagnum* bogs, ditch and stream sides, moist clay, rocks, well drained mineral soils especially where heavy metal-polluted, in acidic habitats, ascending to *ca* 1000 m, frequent or common and sometimes locally abundant. 109, H33, C. Spain and N. Italy north to Scandinavia, Faroes, Iceland, Siberia, Japan, Azores, N. America, Greenland.

Var. heterostipa (Carringt. & Spruce) K. Müll., *Rabenh. Krypt.-Fl. Deutschl.* ed. 2, 6(1): 743. 1910

Abundant innovations produced from ventral side of stem, lateral branches arising in axils of entire, ovate leaves. On rocks, rare, in scattered localities from Caernarfon and Derby north to W. Inverness, Argyll and Angus, Westmeath, W. Donegal. 11, H2. Norway, Washington, Minnesota, Greenland.

    G. inflata occurs in a very wide range of habitats with regard to moisture. Whilst commonly found on wet heath it also occurs, sometimes abundantly, on dry heavy metal-polluted soils. It forms a characteristic community with *Pohlia nutans* on copper spoil heaps on Anglesey, on lead spoil in Caernarfon and industrially derelict sites in the Swansea Valley. It does, however, also occur on non-polluted sandy soils.

    Var. *heterostipa* is probably a habitat modification; the type may produce ventral innovations and occasional leaves associated with lateral branches may be unlobed.

    G. inflata may be confused with *Cladopodiella fluitans* which differs in the form of the perianths which are only rarely produced, and the larger leaf cells. *Cladopodiella fluitans* produces flagelliform shoots but *G. inflata* var. *heterostipa* has been mistaken for it so presence of such shoots cannot be relied upon for separating *Cladopodiella* and *Gymnocolea*. In *C. fluitans* one leaf lobe is often smaller than the other, the sinus is narrower and small underleaves are constantly present. Depauperate saxicolous forms may be confused with small *Marsupella* species which, however, differ in their transverse leaf insertion.

2. G. acutiloba (Schiffn.) K. Müll., *Rabenh. Krypt.-Fl. Deutschl.* ed. 2, 6(1): 745. 1910   (Fig. 52)
G. inflata var. *acutiloba* (Schiffn.) Arnell

Dense reddish-brown to blackish patches. Shoots procumbent, to 1.5 cm, lateral branches arising from axils of entire leaves, occasional innovations arising from ventral side of stem. Leaves distant to approximate, spreading to erecto-patent, ± flat or slightly concave, transversely inserted, margin sometimes with an occasional obscure tooth, to 0.7 mm long, 0.7 mm wide, 1.00–1.25 times as long as wide, bilobed to $\frac{1}{3}(-\frac{1}{2})$, rarely 3-lobed, lobes acute to subacute, sometimes slightly unequal; cells 24–30 μm wide, moderately thick-walled, trigones lacking. Perianths occasional, ovoid, deciduous. Capsules not known in Britain. On and between rocks in boulder scree, altitude 460 m, very rare, Merioneth. 1. Alps, Norway, Sweden, Asia Minor, eastern N. America, Greenland.

According to the key in Macvicar (1926) perianths are absent (presumably in British material) and in his description he says 'Perianth (sterile) rotund or nearly so'. Some gatherings from the only British locality have perianths and these are essentially similar in shape to those of *G. inflata*. There is some doubt as to whether the two taxa are specifically distinct and the similarity in perianths, at least in Britain, supports this view.

G. acutiloba is reputedly associated with copper-bearing rocks but this does not appear to be so in the Welsh locality.

26. EREMONOTUS LINDB. & KAAL. EX PEARS., *Hep. Brit. Isles*: 200. 1900

A monotypic genus with the characters of the species.

1. E. myriocarpus (Carring.) Lindb. & Kaal. ex. Pears., *Hep. Brit. Isles*: 200. 1900
(Fig. 52)

Dioecious. Plants minute, forming reddish-brown patches. Shoots ascending, 3–6 mm, arising from interwoven leafless horizontal stems, simple or branched, shoots and horizontal stems producing small-leaved innovations, female shoots producing innovations from below successive inflorescences, innovations and shoots arising laterally. Lower leaves minute, distant, upper larger, imbricate, erecto-patent, conduplicate-concave, transversely inserted, ± symmetrical, 120–250 μm long, 100–180(–200) μm wide, 1.0–1.4 times as long as wide, bilobed to *ca* ⅔, sinus narrow, acute, lobes acute; cells 10–15 μm wide in middle of lobes, moderately thick-walled, without trigones. Underleaves lacking. Gemmae lacking. Male plants frequent, inflorescences terminal, becoming intercalary, bracts larger than leaves, imbricate, very concave, 200–260 μm long, 180–360 μm wide, bilobed to *ca* ½, lobes obtuse to rounded. Female plants occasional, bracts larger than leaves, imbricate, bilobed to *ca* ½, lobes obtuse. Perianths rare, ellipsoid, deeply plicate above, mouth denticulate. Capsules very rare. On thin soil on damp basic rocks by streams and on shaded cliffs, ascending to *ca* 1000 m, rare, Brecon, Merioneth, Caernarfon, Westmorland, Cumberland, Scottish highlands, Orkney, Kerry, W. Galway, W. Mayo, E. Donegal. 20, H5. Montane parts of Europe from Pyrenees and Alps north to Norway and Finland, Japan.

Clavate shoots may be mistaken for *Marsupella adusta* but that species differs in the larger, less deeply bilobed leaves with marginal row of hyaline cells and in the lack of innovations.

27. SPHENOLOBOPSIS SCHUST. & KITAG. IN SCHUST., *Nova Hedwigia* 22: 152. 1973

With the characters of the species below. There is a second species from Borneo that differs in female reproductive structures and bears sporophytes.

1. S. pearsonii (Spruce) Schust., *Nova Hedwigia* 22: 155. 1973   (Fig. 52)
Cephaloziella pearsonii (Spruce) Douin, *Cephaloziopsis pearsonii* (Spruce) Schiffn., *Sphenolobus pearsonii* (Spruce) Steph.

Dioecious. Plants minute, forming reddish-green to reddish-brown, rarely green, thin patches or rarely amongst other bryophytes. Shoots very slender, 5–15 mm, procumbent, sometimes with erect capitate male inflorescences, stems rigid when dry, simple or dichotomously branched or innovating from below female inflorescences, small-leaved innovations lacking. Leaves distant to imbricate, erecto-patent, conduplicate-concave with inflexed apices, transversely inserted, symmetrical, 120–260 μm long, 100–180 μm wide (0.8–)1.0–1.5 times as long as wide, bilobed to $\frac{2}{3}$–$\frac{3}{4}$, sinus narrow, obtuse to rounded, lobes ± equal, acute to acuminate, a tooth occasionally present near base of dorsal margin; cells thick-walled, without trigones, (10–)12–16 μm wide at base of lobes. Underleaves rare, small and subulate or absent. Gemmae lacking. Male plants common, inflorescences terminal forming capitate heads, becoming intercalary, bracts larger than leaves, imbricate, very concave, 200–300 μm long, 200–280 μm wide, bilobed, lobes acute. Female plants rare, bracts larger than leaves, simple or 2–3-lobed, bluntly toothed, forming small capitate heads. Sporophyte unknown. On sloping to vertical, acidic, exposed or shaded, humid rocks by streams and waterfalls and in block scree, ascending to about 1000 m, rare, N.W. Wales, Lake District, N.W. Yorkshire, W. Scotland from Kirkcudbright north to W. Sutherland, Hebrides, scattered localities in Ireland. 19, H10. S. Norway, Taiwan, eastern N. America, British Columbia, Queen Charlotte Is.

Likely to be confused with *Eremonotus myriocarpus*, *Hygrobiella laxifolia* and *Cephaloziella* spp. from all of which it differs in its dichotomous mode of branching. *Eremonotus* differs in the presence of innovations and *Hygrobiella* in the underleaves being of similar size to the lateral leaves. For the differences from *Marsupella stableri* see under that species.

28. ANASTROPHYLLUM (SPRUCE) STEPH., *Hedwigia* 31: 139. 1893

Dioecious. Plants minute to large. Leaves succubous, notched to bilobed to $\frac{1}{2}$, very concave or canaliculate, often asymmetrical with ventral side larger and more concave than dorsal, insertion ± transverse or dorsal part transverse and ventral part oblique, decurrent or not on dorsal side; cells thick-walled or with large trigones. Underleaves lacking or rudimentary. Gemmae where present reddish, angular. Male bracts similar to leaves but saccate. Female bracts 2–4-lobed, entire or toothed. Perianth plicate above, often slightly dorsiventrally flattened, mouth lobed, denticulate. Capsule wall 2–5-stratose, outer layer with nodulose thickenings, inner with semi-annular thickenings. 25–30 mainly tropical or montane rain-forest species.

1 Leaves to 0.9 mm long, bilobed to $\frac{1}{3}$, trigones small or absent   2
   Leaves 0.9–1.8 mm long, bilobed at most to $\frac{1}{8}$, trigones large, bulging   4
2 Gemmae lacking, leaves mostly 0.6–0.8 times as long as wide, sinus gibbous
                                                                          3. A. saxicola
   Deep red or reddish-brown gemmae usually present, leaves mostly 0.8–2.0 times as
       long as wide, sinus not gibbous   3
3 Gemmiferous shoots not modified, rhizoids few or lacking, leaves mostly 0.8–1.1
       times as long as wide   **1. A minutum**
   Gemmae frequently borne on erect appressed-leaved shoots, rhizoids present almost
       to apex of procumbent shoots, leaves mostly 1.4–2.0 times as long as wide
                                                                       2. A. hellerianum
4 Leaves narrowly triangular, 1.5–2.0 times as long as wide, apex not cucullate
                                                                        4. A. donnianum
   Leaves ovate-rectangular, 1.0–1.4 times as long as wide, apex strongly cucullate
                                                                       5. A. joergensenii

Subgenus *Sphenolobus* (Lindb.) Schust., *Amer. Midl. Nat.* 45: 74. 1951

Plants small. Dorsal side of leaf not or slightly extended across stem; ± transversely inserted, not decurrent; cells ± uniformly thick-walled. Gemmae present, 2-celled, but not produced on attenuated branches. Seta with many cells in section; capsule wall 2–3-stratose.

1. A. minutum (Schreb.) Schust., *Amer. Midl. Nat.* 42: 576. 1949   (Fig. 53)
Sphenolobus minutus (Schreb.) Berggr.

Green to brown patches or growing through other bryophytes. Shoots erect or ascending, to 3 cm long, arising from prostrate basal part, not or sparsely branched, innovating from below perianths; stems brittle. Rhizoids few or lacking. Leaves distant to imbricate, erecto-patent, conduplicate-concave with inflexed apices, transversely inserted, not decurrent, ± asymmetrical, 320–640 μm long, 320–610 μm wide, 0.8–1.1 (–1.3) times as long as wide, bilobed to *ca* ⅓, lobes acute to obtuse, dorsal lobe usually smaller than ventral, dorsal half of leaf flat, ventral half concave; cells (12–)14–18 μm wide, thick-walled, without trigones. Underleaves absent. Gemmae frequent on margins of upper leaves, reddish-brown, angular to knobbly, 1–2-celled, 20–32 μm long. Male bracts similar to leaves but concave at base. Female bracts larger than leaves, 2–4 lobed, lobes obtuse, rounded or apiculate, margin entire. Perianths occasional, shortly cylindrical, contracted to denticulate mouth. Capsules very rare. *n* = 9. On peat, soil and rocks on heaths, in scree and woodland, on rotting logs, ascending to *ca* 1000 m, occasional to frequent in montane habitats, very rare elsewhere, N. Devon, S. Hants, E. Sussex, Hereford, Carmarthen, N.W. Wales, Cheshire, Derby and Yorkshire northwards, generally distributed in Ireland. 46, H18. From the Pyrenees, Italian Alps and the Balkans north to Spitzbergen, Faroes (?), Iceland, east to Siberia, Azores, Japan, S. Africa, Kerguelen Is., N. America, Greenland, Mexico.

Only likely to be confused with some of the smaller *Marsupella* species but differing in the usually asymmetrical leaves.

Subgenus *Crossocalyx* (Meyl.) Schust., *Amer. Midl. Nat.* 45: 74. 1951

Plants minute. Dorsal side of leaf not or slightly extended across stem, ± transversely inserted, base not decurrent; cells ± uniformly thickened. Gemmae unicellular, frequently borne on flagelliform branches. Seta mostly with 12 cells in section; capsule wall 2–3-stratose.

2. A. hellerianum (Nees ex Lindenb.) Schust., *Amer. Midl. Nat.* 42: 575. 1949
                                                                        (Fig. 53)
Crossocalyx hellerianus (Nees ex Lindenb.) Meyl., *Isopaches hellerianus* (Nees ex Lindenb.) Buch, *Sphenolobus hellerianus* (Nees ex Lindenb.) Steph.

Plants minute, in green to brownish patches. Shoots procumbent to ascending, to 4(–6) mm, gemmiferous shoots erect, to 3.5 mm, stems simple or branched. Rhizoids present on decumbent shoots almost to apex. Leaves distant to approximate, erecto-patent to spreading, slightly concave, transversely inserted, not decurrent, symmetrical or slightly asymmetrical, 160–320(–500) μm long, 90–250(–340) μm wide, (1.2–)1.4–2.0 times as long as wide, bilobed to about ⅓, lobes acute, slightly inflexed; cells 16–26(–32) μm wide, walls uniformly thickened or, occasionally, trigones present. Underleaves absent or few, minute, subulate. Leaves of gemmiferous shoots appressed, apices eroded by production of gemmae, which also sometimes occur at apices of normal shoots. Gemmae very common, deep red, ± spherical to bluntly angular or knobbly, unicellular, 12–18 μm long. Fertile plants very rare. Capsules not known in Britain or Ireland. On decorticated logs and tree stumps in shaded humid habitats from sea level to *ca* 300 m, very rare in S. Britain, Carmarthen, Brecon, Montgomery, Merioneth,

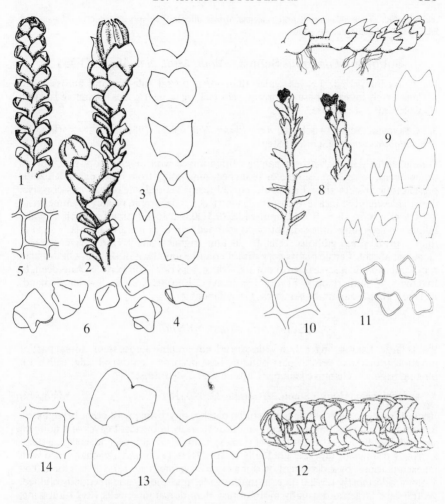

Fig. 53. 1–6, *Anastrophyllum minutum*: 1 and 2, sterile and female shoots (× 13); 3, leaves (× 23); 4, male bract (× 23); 5, mid-leaf cell (× 450); 6, gemmae (× 450). 7–11, *A. hellerianum*: 7 and 8, non-gemmiferous and gemmiferous shoots (× 11); 9, leaves (× 45); 10, mid-leaf cell (× 450); 11, gemmae (× 450). 12–14, *A. saxicola*: 12, shoot (× 11); 13, leaves (× 18); 14, mid-leaf cells (× 450).

Westmorland and Cumberland, rare in the eastern Scottish highlands and W. Scotland from Argyll to W. Ross, very rare in Ireland, Fermanagh, Londonderry. 13, H2. France, Luxemburg, Germany, C. Europe, Norway, Sweden, Finland, Russia eastwards to Siberia, Japan, Bhutan, N. America.

Although the gemmiferous shoots when well developed are very different in appearance from non-gemmiferous shoots, plants may be found which exhibit a complete range between the two types and normal shoots bearing apical gemmae appear to be the first stage in the development of gemmiferous shoots.

The minute size and deep red-tipped gemmiferous shoots and habitat make this plant readily recognisable. *Tritomaria exsecta* and *T. exsectiformis* differ in their very asymmetrical leaves and lack of special gemmiferous shoots.

Subgenus *Eurylobus* Schust., *Amer. Midl. Nat.* 45, 71. 1951

Plants medium-sized. Leaves wider than long, dorsal side arching across stem, ± transversely inserted, base decurrent; cells lacking large trigones. Gemmae lacking. Capsule wall 3–4-stratose.

3. A. saxicola (Schrad.) Schust., *Amer. Midl. Nat.* 45: 71. 1951   (Fig. 53)
Sphenolobus saxicola (Schrad.) Steph.

Dull green to yellowish-brown patches, often mixed with other bryophytes. Shoots procumbent, to *ca* 3 cm, simple or branched, innovating from older parts. Rhizoids present to near stem apex. Leaves closely imbricate, ventral side of leaf very concave, dorsal side slightly concave, transversely inserted, not decurrent, usually asymmetrical, 0.5–0.9 mm long, (0.5–)0.7–1.3 mm wide, (0.5–)0.6–0.8(–1.0) times as long as wide, bilobed to *ca* ⅓, lobes inflexed, obtuse to rounded, rarely acute, dorsal usually smaller than ventral, sinus gibbous; cells 16–30 μm, trigones small. Underleaves absent. Gemmae absent. Fertile plants very rare. Perianths unknown in Britain. On dry acid rocks and in boulder scree from about 300–600 m, very rare, Mid Perth, S. Aberdeen, E. Inverness, W. Sutherland. 4. France, Germany, C. Europe, Norway, Sweden, Finland, Iceland, Siberia, Japan, N. America, Greenland.

Subgenus *Anastrophyllum*

Plants large. Leaves longer than wide, dorsal side arching across stem, dorsal part of insertion transverse, ventral part oblique, base decurrent on dorsal side; cells with bulging trigones. Gemmae lacking. Capsule wall 3–4-stratose.

4. A. donnianum (Hook.) Steph., *Hedwigia* 32: 140. 1893   (Fig. 54)

Reddish-green to deep red or purple tufts, patches or growing with other bryophytes. Shoots erect or ascending, to 10 cm, simple or sparsely branched. Leaves of ± uniform size and distribution except at base of stem, spreading, dorsally secund, clasping stem, narrowly triangular, tapering gradually to apex from ⅛–¼ from base, channelled with inflexed margins, base decurrent on dorsal side, asymmetrical, 1.0–1.8 mm long. 0.6–1.2 mm wide, mostly 1.5–2.0 times as long as wide, apex shortly and unequally bilobed, ventral lobe larger and usually more obtuse than dorsal lobe; cells 13–24 μm wide, trigones very large, bulging. Underleaves lacking. Gemmae lacking. Fertile plants and sporophytes very rare. In well drained habitats in turf on steep slopes, on cliffs and in boulder scree, especially in areas of late snow lie, occasional but sometimes locally abundant from 300–1100 m in the Scottish highlands especially in the west, from Mid Perth, Angus and S. Aberdeen north to Sutherland. 10. S.W. Norway, Faroes, Sikkim, Yunnan.

5. A. joergensenii Schiffn., *Hedwigia* 49: 396. 1910   (Fig. 54)

Light reddish-green to purplish tufts, patches or growing with other bryophytes. Shoots erect or ascending, to 9 cm, simple or sparingly branched. Leaves imbricate, dorsally secund, ovate-rectangular, widest ± at middle, very concave with margins strongly inflexed, base decurrent on dorsal side, slightly asymmetrical, 0.9–1.6 mm long, 0.7–1.3 mm wide, 1.0–1.4(–1.6) times as long as wide, apex bilobed, cucullate, lobes obtuse to rounded; cells 13–20 μm wide, trigones very large, bulging. Underleaves

29. TRITOMARIA 127

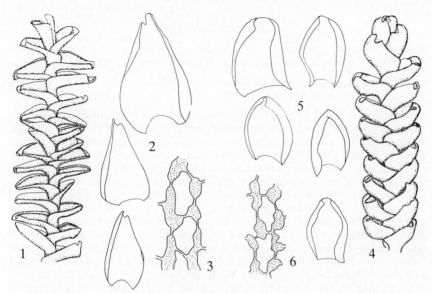

Fig. **54**. 1–3, *Anastrophyllum donnianum*: 1, shoot (× 11); 2, leaves (× 18); 3, mid-leaf cells (× 450). 4–6, *A. joergensenii*: 4, shoot (× 11); 5, leaves (× 18); 6, mid-leaf cells (× 450).

lacking. Gemmae lacking. Fertile plants not known in Britain. In similar habitats to and often with *A. donnianum*, rarely on peat, from 600–900 m, very rare. Inverness, Ross, Sutherland. 6. S.W. Norway.

29. TRITOMARIA SCHIFFN. EX LOESKE, *Hedwigia* 49: 13. 1909

Dioecious. Patches, tufts or scattered shoots. Shoots prostrate to ascending. Rhizoids colourless to pale brown. Leaves usually asymmetrical, insertion transverse or with ventral part oblique, or broadly V-shaped, 3-lobed with ventral lobe larger than dorsal, ventral margin without cilia; cells usually with large, often bulging trigones. Underleaves lacking. Gemmae usually present at apices of upper leaves. Male bracts similar to leaves but saccate. Female bracts 2–5-lobed. Perianth plicate above, contracted or not at mouth, mouth entire or toothed. About 7 species.

1 Leaves ± symmetrical with 3 ± equal lobes **4. T. polita**
  Leaves very asymmetrical, unequally 2–3-lobed, dorsal lobe much smaller than ventral 2
2 Gemmae rare, yellowish-green, leaves undulate when moist, crisped when dry, lobes not tooth-like **3. T. quinquedentata**
  Gemmae very common, deep red, leaves not undulate when moist nor crisped when dry, one lobe lateral and tooth-like 3
3 Gemmae ellipsoid, 12–19 μm long, cells in middle of leaf 12–18 μm wide **1. T. exsecta**
  Gemmae obtusely angled or pyriform, 16–28 μm long, cells in middle of leaf 17–32 μm wide **2. T. exsectiformis**

Subgenus *Tritomaria*

Stems brittle, cortical cells not very thick-walled. Leaves inserted transversely or dorsal side oblique, lobes usually unequal, acute. Perianth mouth toothed.

1. T. exsecta (Schrad.) Loeske, *Hedwigia* 49: 13. 1909   (Fig. 55)
Sphenolobus exsectus (Schrad.) Steph.

Plants minute to small, forming small, yellowish-green to brownish-green patches or mixed with other bryophytes. Shoots procumbent to ascending, to 15(–20) mm, simple or branched. Leaves distant to imbricate, spreading from ± suberect base, canaliculate-concave, ± transversely inserted, not decurrent, asymmetrical, (0.35–)0.65–0.80(–1.10) mm long, very variable, leaves without gemmae unequally 2–3-lobed, lobe on ventral side small, acute and often tooth-like, dorsal lobe larger, frequently deeply notched with two acute lobes, gemma-bearing leaves asymmetrically ovoid with ventral side more rounded than dorsal, apex eroded, the two forms of leaves connected by intermediates; mid-leaf cells 12–18 μm wide, walls thickened, trigones present but not bulging. Underleaves lacking. Gemmae very common, deep red, ellipsoid, mostly 2-celled, 12–19(–22) μm long. Fertile plants very rare. Capsules not known in Britain or Ireland. $n = 9$. Damp decorticated logs and stumps and acidic rocks in shaded, humid habitats, usually at low altitudes, rare. W. Britain from Monmouth and Brecon north to W. Sutherland, very rare in W. Ireland. 20, H6. Pyrenees and N. Italy north to S. Scandinavia and Russia, east to Caucasus, Siberia, Himalayas, China, Taiwan, Korea, Japan, Borneo, Madeira, N. America, Mexico.

2. T. exsectiformis (Breidl.) Loeske, *Hedwigia* 49: 43. 1909   (Fig. 55)
Sphenolobus exsectiformis (Breidl.) Steph.

Superficially very similar to *T. exsecta* but often more vigorous and forming larger patches. Leaves of similar morphology to those of *T. exsecta* but often larger, (0.6–)0.7–1.3(–1.9) mm long; mid-leaf cells 17–32 μm wide, trigones usually present, sometimes bulging. Gemmae very common, deep red, pyriform or obtusely angular, mostly 2-celled, 16–28 μm long. Fertile plants very rare. Capsules very rare, late summer. $n = 18$. In humid habitats on rotting wood, bark, acidic rocks, peaty banks, shallow humus on rocks and sandy soil, usually at low altitudes, rare in lowland areas, occasional elsewhere. 85, H34. Europe from the Pyrenees north to S. Scandinavia and east to the Caucasus and Siberia, N. America.

   T. exsecta and *T. exsectiformis* are unlikely to be confused with any other species. *T. exsectiformis* is usually more vigorous, has larger cells and gemmae, a greater ecological amplitude and is more widespread than *T. exsecta* of which it is in all probability an autodiploid derivative.

3. T. quinquedentata (Huds.) Buch, *Mem. Soc. Fauna Fl. Fenn.* 8: 200. 1932

                                                                    (Fig. 55)

Lophozia quinquedentata (Huds.) Cogn.

Yellowish-green to yellowish-brown patches or mixed with other bryophytes. Shoots procumbent to ascending, to 3(-5) cm, simple or branched, stems fleshy, brittle. Leaves approximate to imbricate, spreading ± horizontally or dorsal side inflexed or reflexed, somewhat undulate, crisped when dry, dorsal side of leaf transversely inserted, ventral side obliquely inserted, not decurrent, strongly asymmetrical, 1.0–1.7 mm long, 1.0–1.8 mm wide, 0.8–1.0(–1.2) times as long as wide, 3–lobed to $\frac{1}{10}-\frac{1}{5}$, rarely 2-lobed, ventral lobe much larger and blunter than dorsal lobe, lobes obtuse to apiculate; mid-leaf cells 16–26(–32) μm wide, thin-walled, trigones usually large. Underleaves absent. Gemmae rare, yellowish-green, ovoid, pyriform or angular, 1–2-celled, 20–32 μm long. Male

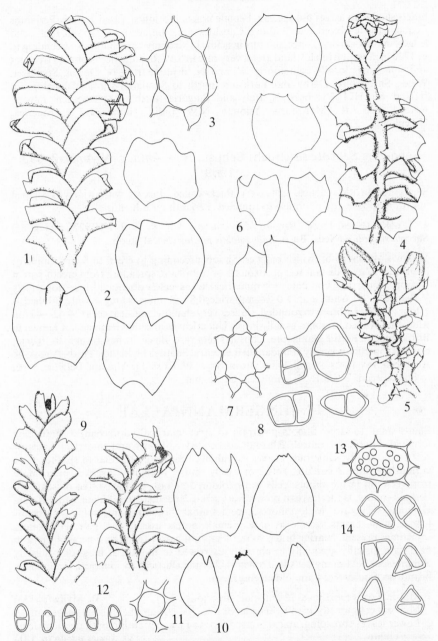

Fig. 55. 1–3, *Tritomaria polita*: 1, shoot (× 11); 2, leaves (× 18); 3, mid-leaf cell (× 450). 4–8, *T. quinquedentata*: 4 and 5, sterile and female shoots (× 6); 6, leaves (× 11); 7, mid-leaf cell (× 450); 8, gemmae (× 450). 9–12, *T. exsecta*: 9, shoots (× 9); 10, leaves (× 23); 11, mid-leaf cell (× 450); 12, gemmae (× 450). 13–14, *T. exsectiformis*: 13, mid-leaf cell (× 450); 14, gemmae (× 450).

bracts similar to leaves but saccate. Female bracts 2–5-lobed, fused at base. Perianths frequent, ovate-obloid, plicate above. Capsules rare, summer. $n = 9$. On soil, rocks, cliff ledges, in rock crevices, scree and turf in acidic to strongly basic habitats, ascending to *ca* 1240 m, absent from lowland areas, very rare in S.W. Britain, occasional to frequent and sometimes locally common in W. and N. Britain, from N. Devon, Somerset, Wales, Shropshire, Derby and Yorkshire north to Shetland, scattered localities in Ireland. 86, H21. Pyrenees, N. Italy and Bulgaria north to Spitzbergen, Faroes, Iceland, U.S.S.R., Manchuria, Yunnan, N.W. India, Himalayas, N. America, Greenland.

Subgenus *Saccobasis* (Buch) Schust., *Hep. Anthoc. N. Amer.* 2: 693. 1969

Stems not brittle, cortical cells very thick-walled. Leaves with widely V-shaped insertion, lobes ± equal, obtuse to rounded. Perianth mouth entire.

4. T. polita (Nees) Joerg., *Bergens Mus. Aarbok. Nat.-R.* 1919: 14. 1922   (Fig. 55)
Saccobasis polita (Nees) Buch, *Sphenolobus politus* (Nees) Steph.

Yellowish-green to brownish patches. Shoots ascending to erect, to 6 cm, simple or branched. Leaves distant to approximate or imbricate, spreading from erecto-patent base or ± squarrose, undulate, ± symmetrical, transversely orientated, insertion widely V-shaped, 0.9–1.6 mm long, 1.0–1.4 mm wide, 0.8–1.3 times as long as wide, 3-lobed to ⅛, lobes ± equal, obtuse to rounded, sinuses very shallow; mid-leaf cells (24–)28–42 μm wide, thin-walled, trigones usually large. Underleaves absent. Gemmae not known in British plants. Fertile plants rare. Capsules very rare. Basic flushes, from 520–1070 m, rare but sometimes locally abundant in the central Scottish highlands, Perth, Angus, N. Aberdeen, Inverness, Argyll, Dumbarton, Skye, W. Ross. 11. Alps and Belgium north to Spitzbergen, U.S.S.R., N. America, Greenland.

13. JUNGERMANNIACEAE

Plants small to large. Shoots prostrate to erect; markedly dorsiventral, branching terminal, lateral or ventral. Rhizoids associated with leaf and underleaf bases or scattered, usually numerous. Leaves succubous, reniform, orbicular to spathulate or variously ovate or cordate, entire or rarely retuse, obliquely inserted, often more transverse on fertile shoots; cells walls coloured or not, trigones often present, oil-bodies 1–many. Underleaves if present very small, lanceolate to subulate, without cilia. Gemmae very rare. Male inflorescence terminal, becoming intercalary, or below female, male bracts saccate at base. Female bracts usually one pair. Perigynium sometimes present, bearing bracts or not. Perianth barely to longly emergent, smooth or plicate, usually gradually to abruptly narrowed to mouth, at most only slightly compressed. Seta many cells in diameter. Capsule ellipsoid to spherical, wall mostly bistratose. Spores 9–24 μm. Nineteen genera.

1 Mid-leaf cells mostly 48–72 μm wide, dioecious   **31. Mylia (p. 133)**
  Mid-leaf cells mostly 16–40 μm wide, or if to 50(–60) μm then plants paroecious 2
2 Lower leaves spreading, apical pressed face to face, perianth mouth denticulate or
  ciliate   **30. Jamesoniella (p. 131)**
  Leaves not as above, perianth mouth at most crenulate   3
3 Perianths exserted from female bracts (except in *J. subelliptica* and *J. obovata*), leaves
  not translucent nor with persistent oil-bodies or retuse or bilobed apices,
  underleaves lacking   **32. Jungermannia (p. 135)**

Perianth not exserted, leaves translucent or oil-bodies persisting on drying or apex retuse or bilobed, underleaves present at least on young parts of stem

33. Nardia (p. 149)

SUBFAMILY JAMESONIELLOIDEAE

Plants usually red- or purple-tinged. Rhizoids in tufts from ventral part of stem and leaf bases. Branches usually arising ventrally. Leaves opposite or alternate, entire, rarely retuse. Underleaves very small or absent. Gemmae lacking. Antheridial stalk 2–4-seriate, jacket cells small and numerous. Female bracts toothed on one or both sides; bracteole well developed, toothed. Perigynium lacking. Perianth emergent. Capsule wall 4–5-stratose.

30. JAMESONIELLA (SPRUCE) CARRING. IN LEES, *London Catal. Brit. Moss. Hepat.* ed. 2: 25. 1881

Dioecious. Usually robust plants with prostrate to erect shoots. Stems simple or branching from ventral side, often innovating from below female inflorescence. Leaves succubous, at least upper often erect and ± pressed face to face, orbicular or quadrate-orbicular, margin entire or retuse, obliquely inserted, ± decurrent on dorsal side; cells with very small to large trigones, oil-bodies 7–20 per cell. Underleaves absent or very small and usually only present on fertile female shoots. Gemmae lacking. Male bracts imbricate, concave; antheridia usually solitary. Female bracts usually toothed near base, bracteole well developed, toothed. Perigynium lacking. Perianth emergent, plicate above, narrowed to usually denticulate mouth. Capsule wall 4–5-stratose. About 40 mainly tropical and southern hemisphere species.

Leaves not undulate, at least some retuse, margin plane, perianth mouth longly ciliate **1. J. autumnalis**
Upper leaves often undulate, not retuse, margin at apex of leaf often inflexed, perianth mouth denticulate **2. J. undulifolia**

1. J. autumnalis (DC) Steph., *Spec. Hep.* 2: 92. 1901 (Fig. 56)

Opaque, dark green, often deep red-tinged patches or growing through other bryophytes. Shoots procumbent, to 2 cm, often innovating from below. Rhizoids numerous, colourless. Lower leaves distant to approximate, ± spreading, horizontally, upper imbricate, ± erect and ± pressed face to face, quadrate-orbicular, sometimes retuse, slightly concave, not undulate, margin plane, obliquely inserted, slightly decurrent on dorsal side, 0.55–1.15 mm long, 0.55–1.15 mm wide, (0.85–)1.00(–1.20) times as long as wide; mid-leaf cells 20–32 μm wide, densely packed with chloroplasts, thin-walled, trigones small or rarely ± absent, marginal row of cells smaller, thicker-walled. Underleaves lacking or very small on upper parts of fertile female stems only. Gemmae lacking. Male inflorescence terminal, becoming intercalary with age, bracts closely imbricate, concave. Female bracts erect, not undulate, apex retuse or not, with ciliate tooth at base on one or both sides. Perianths occasional, shortly cylindrical-pyriform, in upper $\frac{1}{4}$ plicate and tapering to laciniate-ciliate mouth. Capsules rare. $n = 9$. On tree boles, decorticated logs, humus and damp acidic to slightly basic rock in humid birch or oak woodlands and wooded valleys, ascending to 350 m, rare, E. Cornwall, S. Devon, Worcester, Shropshire, Wales, M.W. Yorkshire, Lake District and W. Scotland from Kirkcudbright to W. Ross, formerly also from E. Sussex, W. Gloucester, Angus and Kincardine. 27. N. Italy and Carpathians north to Fennoscandia and Russia, Iceland, Himalayas, Siberia, Korea, China, Japan, Taiwan.

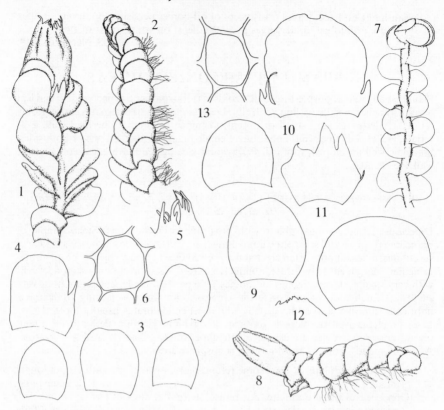

Fig. 56. 1–6, *Jamesoniella autumnalis*: 1 and 2, female and male shoots (× 11); 3, leaves (× 18); 4, female bract (× 18); 5, part of perianth mouth (× 178); 6, mid-leaf cell (× 450). 7–13, *J. undulifolia*: 7 and 8, sterile and female shoots (× 6); 9, leaves (× 18); 10, female bract (× 18); 11, bracteole (× 18); 12, part of perianth mouth (× 178); 13, mid-leaf cell (× 450).

Likely to be confused with *Nardia scalaris*, *Odontoschisma sphagni* and *Mylia* spp. to which it may bear more than a superficial resemblance. *Nardia* differs in the presence of numerous underleaves and large persistent oil-bodies. *Odontoschisma denudatum* has gemmiferous shoots and stolons. *Mylia* species differ in their much larger cells. When perianths are present the ciliate mouth will distinguish *J. autumnalis* from all other entire-leaved native hepatics.

2. J. undulifolia (Nees) K. Müll., *Rabenh. Krypt.-Fl. Deutschl.* ed. 2, 6(2): 758. 1916
(Fig. 56)

J. schraderi (Mart.) Schiffn.

Reddish brown-tinged lax patches or scattered plants. Shoots procumbent, to 5 cm, simple or branched. Lower leaves distant, spreading, upper imbricate, erect and ± pressed face to face, ± rotund, at least upper often undulate, margin entire, not or rarely retuse, often inflexed round apex, obliquely inserted, decurrent on dorsal side, 0.8–1.8 mm long, 0.9–2.0 mm wide, 0.75–1.00 times as long as wide; cells mostly thin-walled, (20–)24–36(–40) μm wide in mid-leaf, trigones very small to medium-sized. Underleaves sparse, lanceolate, *ca* 0.6 mm long. Gemmae lacking. Female bracts erect, undulate, apex retuse or not, with tooth on one or both sides towards base. Perianths occasional, narrowly ellipsoid, plicate in upper half, mouth slightly denticulate, not ciliate. Capsules rare. Amongst *Sphagnum* in bogs, ascending to *ca* 275 m, very rare, E. Cornwall, Westmorland, Argyll, formerly W. Cornwall and W. Gloucester. 5. Fennoscandia, C. Europe, Greenland.

Although confused in the past with *J. autumnalis*, *J. undulifolia* is distinct in habitat, larger, undulate, non-retuse leaves with slightly larger cells and the nature of the perianth mouth. More likely to be confused with *Odontoschisma sphagni* which has stolons and smaller leaf cells and *Mylia* species which have much larger cells.

SUBFAMILY MYLIOIDEAE

With the characters of the genus.

31. MYLIA S.F. GRAY, *Nat. Arr. Br. Pl.* 1: 693. 1821

Dioecious. Medium-sized to large plants with unbranched or sparingly branched stems, branching terminal. Stolons lacking. Rhizoids in tufts associated with leaf bases and underleaves. Leaves succubous, ± orbicular except sometimes when gemmiferous, obliquely inserted, slightly decurrent on dorsal side; cells very large with large bulging trigones. Underleaves small, simple. 1–2-celled gemmae often present at margins of apical leaves. Male bracts similar to leaves but saccate; antheridial stalk uniseriate, jacket cells few and very large. Female bracts similar to leaves; bracteole lacking. Perigynium lacking. Perianth emergent, laterally compressed with broad, closed mouth. Capsule ovoid, wall 3–5-stratose. Four species, the two non-British occurring in eastern Asia.

Leaf cuticle fissured, edges of marginal cells minutely and irregularly crenulate, oil-bodies smooth, leaves and leaf cells not becoming elongated on the production of marginal gemmae.   **1. M. taylorii**
Leaf cuticle and marginal cells smooth, oil-bodies knobbly, gemmiferous leaves and their cells elongated.   **2. M. anomala**

1. M. taylorii (Hook.) S.F. Gray, *Nat. Arr. Br. Pl.* 1: 693. 1821   (Fig. 57)
Leptoscyphus taylorii (Hook.) Mitt.

Tufts, sometimes extensive, yellowish-green or more usually reddish-purple at least above. Shoots erect or ascending, to 7 cm, branching occasionally. Leaves approximate to imbricate, ± spreading and convex from concave, erect base, uppermost erect, pressed face to face, orbicular to ovate, entire or eroded at apex in gemmiferous leaves, obliquely inserted, shortly decurrent on dorsal side, with rhizoids attached at base, 1.4–2.0(–2.7) mm long, 1.2–1.8(–2.3) mm wide, 1.0–1.2 times as long as wide, gemma-bearing leaves not becoming elongated; cells (48–)52–72(–80) μm wide, ± isodiametric in mid-leaf, smaller, isodiametric towards apex, thin-walled, trigones large, bulging, cuticle marbled with reticulate cracks which also result in minutely irregularly crenulate edge to marginal cells; oil-bodies ± smooth, filling lumen. Underleaves concealed by rhizoids, simple, lanceolate-subulate, 0.45–1.00 mm long. Gemmae occasional and sparsely produced from marginal cells of apical leaves, spherical to ovoid, 1(–2)-celled, 45–70 μm long. Bracts similar to leaves. Perianths and capsules very rare. $n = 8 + m^*$, 9. Moist shaded rocks and rock faces and tree boles in sheltered situations, cliff ledges, scree, wet heath, peat and occasionally on *Sphagnum*, in acidic habitats, from ± sea-level to about 1000 m, occasional to frequent in W. and N. Britain, E. Cornwall, Wales, Hereford, Stafford, Shropshire and Yorkshire northwards. 66,

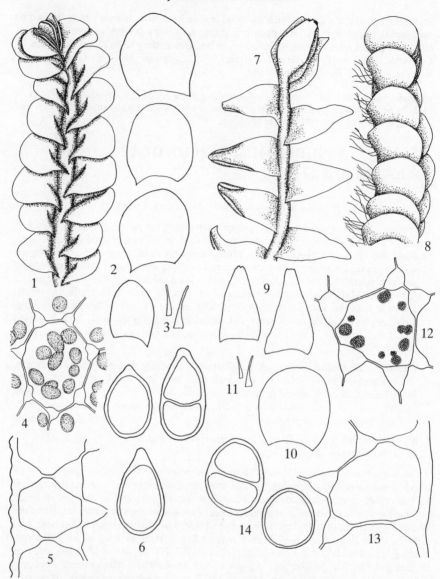

Fig. **57**. 1–6, *Mylia taylorii*: 1, shoot (× 72); 2, leaves (× 11); 3, underleaves (× 11); 4, mid-leaf cell with oil-bodies (× 450); 5, marginal cell (× 450); 6, gemmae (× 450). 7–14, *M. anomala*: 7 and 8, gemmiferous and non-gemmiferous shoots (× 11); 9, leaves from gemmiferous shoot (× 11); 10, leaf from non-gemmiferous shoot (× 11); 11, underleaves (× 11); 12, mid-leaf cell with oil-bodies (× 450); 13, marginal cell (× 450); 14, gemmae (× 450).

H31. N. and W. Europe, Alps, Yugoslavia, Faroes, Japan, China, Sikkim, Nepal, Azores, N. America, S. Greenland.

Although *Mylia* species resemble *Jamesoniella* species in appearance they may be distinguished from these and all other entire-leaved British liverworts by the very large leaf cells (mostly 48–72 μm wide in mid-leaf). Although some authors (e.g. Macvicar, 1926; Schuster, 1969) use habitat as a key character in separating the two *Mylia* species they may occur mixed together in *Sphagnum* bogs, although *M. taylorii* occurs in a number of habitats from which *M. anomala* is absent. Non-gemmiferous shoots of *M. anomala* may only be distinguished from *M. taylorii* by the non-fissured cuticle and smooth edges of the marginal cells; if present, however, the elongated gemmiferous leaves with elongated cells are characteristic.

2. M. anomala (Hook.) S.F. Gray, *Nat. Arr. Br. Pl.* 1: 693. 1821   (Fig. 57)
Leptoscyphus anomalus (Hook.) Mitt.

Yellowish-green to brownish patches or scattered plants. Shoots procumbent, to 4 cm, sparsely branched. Leaves distant to imbricate, lower ± spreading, uppermost erect-appressed, non-gemmiferous leaves orbicular to shortly lingulate, obliquely inserted, slightly decurrent on dorsal side, with rhizoids attached at base, (0.9–)1.2–2.0 (–2.5) mm long, 1.0–2.0(–2.4) mm wide, 0.8–1.2 times as long as wide; cells (40–)48–72 μm wide in mid-leaf, isodiametric or longer than wide, smaller, isodiametric towards apex, trigones large, bulging, cuticle smooth, marginal cells with smooth edges, oil-bodies knobbly. Leaves of gemmiferous shoots ovate-lanceolate to narrowly triangular, often with inflexed margins, apex eroded in apical leaves, 1.5–2.3 mm long, 0.9–1.4 mm wide, 1.2–2.0 times as long as wide; mid-leaf cells 40–64 μm wide, 2–3 times as long as wide, cells towards apex elongated. Underleaves concealed by rhizoids, lanceolate-subulate, to *ca* 0.5 mm. Gemmae very common, lemon yellow to yellowish-green, in fasciculate bunches at tips and margins of upper leaves, spherical to ovoid or ellipsoid, 1–2-celled, 30–56 μm long. Bracts similar to leaves. Perianths very rare. Capsules not known (?) in Britain or Ireland. $n = 8 + m^*$, 9. On moist peat and growing through *Sphagnum*, often in small quantity, ascending to *ca* 700 m, rare to frequent in suitable habitats throughout Britain and Ireland. 93, H32. Throughout Europe from the Pyrenees and Italian Alps north to Spitzbergen, Faroes, Iceland, Siberia, Kamchatka, N. America, Greenland.

SUBFAMILY JUNGERMANNIOIDEAE

Plants very small to large. Rhizoids continuous on ventral surface of stem. Branches arising laterally or terminally. Leaves alternate. Underleaves present or not. Gemmae usually lacking. Antheridial stalk 2–4-seriate, jacket cells small and numerous. Female bracts entire or with marginal slime papillae; bracteoles where present entire. Perigynium present or not. Perianth immersed to longly emergent, rarely lacking. Capsule wall 2–3-stratose.

32. JUNGERMANNIA L., *Sp. Plant.* ed. 1: 1131. 1753

Dioecious or paroecious. Plants small to large, yellowish-green to brown. Where dioecious male plants smaller than female. Shoots prostrate to erect; branches arising laterally, innovations arising from below perianth. Rhizoids usually abundant, colourless to brown or deep red or purple. Leaves alternate, obliquely inserted, succubous, reniform, orbicular, ovate or elliptical, entire; oil-bodies 1–several per cell. Underleaves lacking. Gemmae usually lacking. Male bracts terminal, becoming intercalary, or below female, of similar shape to leaves, saccate at base. Female bracts of similar shape to leaves, entire; bracteoles lacking. Perigynium lacking or very short, or long and bearing 1–2 pairs female bracts. Perianth 0–$\frac{2}{3}$ emergent, shortly cylindrical, obovoid, clavate or fusiform, smooth or more usually plicate above, gradually or abruptly contracted to mouth. Capsule wall bistratose. Spores 10–24 μm. 80–90 species ranging from the Arctic to the Sub-Antarctic.

Key to fertile plants

1 Perigynium present, bearing 1–2 pairs of bracts (see Fig. **62**, 16), cells in upper part of perianth 2–3 times as long as wide, rhizoids purplish or colourless to brownish   2
  Perigynium lacking (see Fig. **58**, 20) or if present not bearing bracts, cells in upper part of perianth similar to leaf cells, rhizoids colourless to brownish   5
2 Perigynium bearing 1 pair of bracts, perianth to about $\frac{1}{2}$ emergent, leaves with rhizoids at base   3
  Perigynium bearing 2 pairs of bracts, perianth hardly longer than upper bracts, leaves without basal rhizoids   4
3 Dioecious   **11. J. hyalina**
  Paroecious   **12. J. paroica**
4 Plants 0.3–0.8 cm, rhizoids colourless to brownish, mid-leaf cells 24–36(–40) μm wide   **13. J. subelliptica**
  Plants 0.5–3.0 cm, rhizoids colourless to purple, mid-leaf cells 20–50 μm wide   **14. J. obovata**
5 Plants 2–7 cm, rhizoids few, confined to lower part of stem, leaves embracing stem   **6. J. exsertifolia** spp. **cordifolia**
  Rhizoids abundant almost to stem apex, plants to 2 cm or if more then leaves not embracing stem   6
6 Marginal row of leaf cells with thick walls, up to twice the size of the submarginal row and forming a distinct border   **9. J. gracillima**
  Marginal row of leaf cells not or scarcely forming a border   7
7 Dioecious   8
  Paroecious   11
8 Perianth ± gradually tapered to mouth, mouth not beaked, leaves ovate to broadly ovate or broadly ovate-cordate*   9
  Perianth abruptly narrowed to beaked mouth, leaves orbicular to wider than long   10
9 Plants 0.5–4.0 cm, yellowish-green to dark green, leaves on fertile stems ± concave, male inflorescence with 5–12 pairs of bracts, bracts saccate at base   **2. J. atrovirens**
  Plants 0.3–1.0 cm, brownish-green to blackish, leaves on fertile stems very concave, male inflorescence with 2–3 pairs of bracts, saccate part constituting $\frac{1}{2}$ or more of each bract   **5. J. borealis**
10 Shoots 0.5–2.5(–4.0) cm, mid-leaf cells mostly 28–32 μm wide, oil-bodies 2–3 per cell, gemmae lacking   **9. J. gracillima**
  Shoots 0.2–0.6 cm, mid-leaf cells 24–40 μm wide, oil-bodies usually solitary, shoots sometimes terminating in a brownish-yellow globular mass of gemmae   **10. J. caespiticia**
11 Perianth smooth, abruptly contracted to sunken, beaked mouth, leaves 1.25–1.75 times as long as wide, mid-leaf cells 32–48 μm wide   **1. J. leiantha**

*It is necessary to dissect leaves off stem to determine their shape accurately.

Perianth plicate above, mouth if beaked not sunken, leaves 0.7–1.3 times as long as wide or if longer then mid-leaf cells 16–28 μm wide   12

12 Perianth tapered to mouth, leaves longer than wide, leaf bases lacking rhizoids 13
Perianth abruptly narrowed to beaked mouth, leaves as wide as or wider than long, with rhizoids at base   14

13 Perianth fusiform, leaves on fertile stems broadly elliptical to broadly ovate, mid-leaf cells 16–28(–32) μm wide   **3. J. pumila**
Perianth obovoid to broadly fusiform, leaves on fertile stems ovate-cordate to rounded-cordate, mid-leaf cells 12–16(–24) μm wide   **4. J. polaris**

14 Leaves ± reniform, mid-leaf cells 28–44 μm, male bracts closely appressed   **7. J. confertissima**
Leaves orbicular, mid-leaf cells 20–32(–40) μm, male bracts with reflexed tips   **8. J. sphaerocarpa**

Key to sterile plants

1 At least some rhizoids purple to reddish-purple   2
Rhizoids colourless to brownish   3

2 Leaves 0.7–1.0 times as long as wide   **J. hyalina** or **J. paroica**
Leaves 1.0–1.3 times as long as wide   **13. J. obovata**

3 Leaves longer than wide, without basal rhizoids, or if wider than long then broadly cordate and embracing stem   4
Leaves as wide as or wider than long with rhizoids at base   7

4 Leaves 1.25–1.75 times as long as wide, mid-leaf cells (24–)32–48 μm wide, trigones distinct, bulging   **1. J. leiantha**
Leaves 0.8–1.3 times as long as wide or if longer then mid-leaf cells 16–28(–32) μm wide and trigones small or absent   5

5 Plants 2–7 cm, leaves embracing stem, widest $\frac{1}{5}$–$\frac{1}{4}$ from base   **6. J. exsertifolia** spp. **cordifolia**
Plants to 4 cm, leaves not or only slightly embracing stem, widest $\frac{1}{3}$–$\frac{1}{2}$ from base 6

6 Plants 0.5–4.0 cm, leaves 0.6–1.8 mm long, 0.6–1.5 mm wide   **2. J. atrovirens**
Plants 0.3–1.0(–2.0) cm, leaves 0.3–1.2 mm long, 0.3–1.0 mm wide   **J. pumila, J. polaris, J. borealis** or **J. subelliptica**

7 Leaves with distinct border of enlarged thick-walled cells   **9. J. gracillima**
Leaves lacking such a border   8

8 Shoots prostrate, scattered or forming patches   9
Shoots usually erect, forming tufts   **J. confertissima** or **J. sphaerocarpa**

9 Leaves orbicular with 2–3 small oil-bodies per cell, gemmiferous shoots lacking   **9. J. gracillima**
Leaves reniform, usually with one large oil-body per cell, shoots sometimes terminating in brownish-yellow mass of gemmae   **10. J. caespiticia**

Subgenus *Liochlaena* (Nees) S. Arn., *Illustr. Moss Fl. Fennosc. Hep.*: 107. 1959

Rhizoids colourless to brown, not attached to leaf bases. Leaves longer than wide, often parallel-sided. Perigynium lacking, perianth longly emergent, cylindrical, smooth, abruptly contracted to beaked mouth. Three European species.

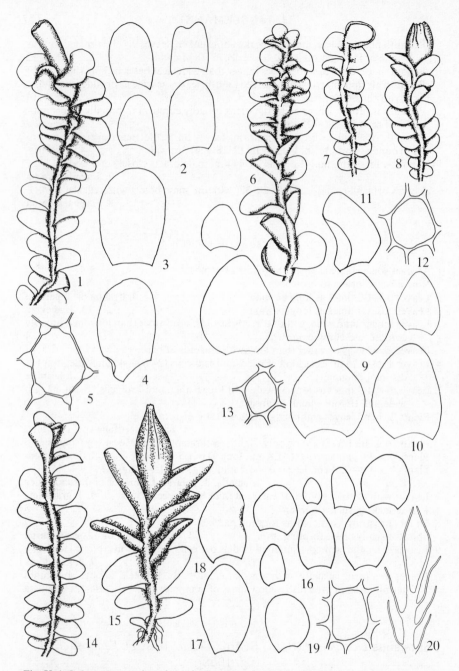

Fig. **58**. 1–5, *Jungermannia leiantha*: 1, fertile (paroecious) shoot (× 6); 2, leaves (× 11); 3, female bract (× 11); 4, male bract (× 11); 5, mid-leaf cell (× 450). 6–13, *J. atrovirens*: 6–8, male, sterile and female shoots (× 8); 9, leaves (× 15); 10, female bract (× 15); 11, male bract (× 15); 12, mid-leaf cell (× 450); 13, cell near perianth mouth (× 450). 14–20, *J. pumila*: 14 and 15, sterile and fertile (paroecious) shoots (× 13); 16, leaves (× 38); 17, male bract (× 38); 18, female bract (× 38); 19, mid-leaf cell (× 450); 20, diagrammatic L.S. of perianth (× 13).

1. J. leiantha Grolle, *Taxon* 15: 187. 1966 (Fig. 58)
Aplozia lanceolata sensu Dum., *J. lanceolata* auct. non L.

Usually paroecious. Bright green to brownish patches. Shoots procumbent, 1.0–2.5 cm, innovating from below perianth. Rhizoids colourless to brownish, not arising from leaf bases. Leaves distant to imbricate, spreading to reflexed and convex from erecto-patent, concave base, ovate-oblong, widest ⅓–⅔ from base to lingulate and parallel-sided, apex rounded or occasionally retuse, obliquely inserted, decurrent on dorsal side, 0.9–2.4(–2.8) mm long, 0.7–1.5(–1.6) mm wide, 1.25–1.75 times as long as wide; cells (24–)32–48 µm wide, thin-walled, trigones distinct, often bulging. Underleaves lacking. Gemmae very rare, on erect, attenuated shoots, 23–33 µm (Schuster, 1969). Male bracts 2–3 pairs below female bracts, similar to leaves but more transversely inserted, saccate. Female bracts similar to or larger than leaves, squarrose. Perigynium lacking. Perianths common, longly emergent, shortly cylindrical, often slightly curved, smooth, abruptly contracted to concave apex, mouth beaked. Capsules frequent, spring. On sandstone, slate and in basic flushes in sheltered habitats, very rare, M.W. Yorkshire, W. Ross, formerly also E. Sussex, S.W. Yorkshire and Westmorland. 5. Spain, N. Italy and Yugoslavia north to Fennoscandia and western U.S.S.R., Madeira, Tenerife, N. America.

Subgenus *Jungermannia*

Rhizoids colourless to brown, not arising from leaf bases. Leaves usually longer than wide; trigones small or absent. Perigynium lacking. Perianth slightly dorsiventrally compressed, plicate, not beaked, wall 2-stratose below, cells in upper third similar to leaf cells. Five European species.

2. J. atrovirens Dum., *Sylloge Jungerm.*: 51. 1831 (Fig. 58)
Aplozia atrovirens (Dum.) Dum., *A. riparia* (Tayl.) Dum., *Solenostoma atrovirens* (Dum.) K. Müll., *S. sphaerocarpoideum* (De Not.) Paton & Warburg, *S. triste* (Nees) K. Müll.

Dioecious. Plants small, yellowish-green to dull green, in patches or growing through other bryophytes. Shoots procumbent, 0.5–4.0 cm, branching occasionally, sometimes producing stolons. Rhizoids numerous, colourless to brownish, not growing from leaf bases. Lower leaves distant, spreading to suberect, obliquely inserted, upper imbricate, ±concave, partly embracing stem, erecto-patent to ± reflexed, more transversely inserted, not decurrent, broadly ovate, widest about ⅓ from base, 0.6–1.8 mm long, 0.6–1.5 mm wide, 1.0–1.3(–1.4) times as long as wide; mid-leaf cells 20–32 µm wide, walls thin or thickened, ±colourless to yellowish, trigones very small or lacking. Gemmae lacking. Male bracts usually 5–12 pairs, concave, base saccate, tips ± reflexed. Female bracts ovate, reflexed. Perigynium lacking. Perianths common, ellipsoid to obovoid, about ⅔ emergent, plicate above, cells in upper ⅓ similar to mid-leaf cells. Capsules occasional, spring, summer. $n = 9^*$. Moist sheltered soil, soil-covered rocks, and rocks, often where basic, in woods, by streams, on cliffs, in ravines, rarely on submerged rocks, ascending to *ca* 1000 m, rare in lowland areas, occasional to frequent in W. and N. Britain and Ireland. 91, H28. Europe north to Spitzbergen, Faroes, Iceland, Caucasus, Middle East, Sakhalin, Japan, Korea, Macaronesia, Morocco, N. America, Greenland.

J. atrovirens varies very much in size and whilst large forms may be identified when sterile, smaller plants cannot reliably be separated from the following three species when sterile. Fortunately all four species are frequently fertile and *J. pumila* and *J. polaris* may be distinguished by their paroecious inflorescences (in *J. pumila* the male bracts may be poorly differentiated or even absent from an occasional fertile shoot but the fusiform perianth is characteristic). For differences from *J. borealis* see under that species.

3. J. pumila With., *Arrang. Br. Pl.* ed. 3, 3: 883. 1796   (Fig. 58)
Aplozia pumila (With.) Dum., *Solenostoma pumilum* (With.) K. Müll.

Paroecious. Plants small, dull green to blackish, in patches or growing through other bryophytes. Fertile shoots erect or ascending from prostrate branched stems, sterile shoots to 0.3–1.0(–2.0) cm, fertile shoots to 1 cm, stolons lacking. Rhizoids numerous, colourless to brownish, not arising from leaf bases. Lower leaves distant, spreading, elliptical to ovate, obliquely inserted, upper imbricate, erecto-patent, broadly elliptical to broadly ovate, widest ⅓–½ from base, concave, ± transversely inserted, not decurrent on dorsal side, 0.35–1.20 mm long, 0.32–0.85 mm wide, (1.0–)1.1–1.7 times as long as wide; mid-leaf cells 16–28(–32) μm wide, walls thickened, colourless, trigones small or absent. Gemmae lacking. Male bracts 2–3 pairs below female, ovate, saccate with spreading tips. Female bracts of similar shape to male, reflexed. Perianths common, fusiform to clavate, about ⅔ emergent, plicate above, cells in upper ⅓ similar to mid-leaf cells. Capsules common, winter, spring. Damp shaded rocks and soil in woods, by streams, on cliffs, etc., often where basic, ascending to *ca* 1000 m, rare in lowland areas, occasional to frequent in W. and N. Britain and Ireland. 81, H27. Iberian peninsula, Sicily, N. Italy and Yugoslavia north to Fennoscandia and W. Russia, Faroes, Iceland, Ukraine, Siberia, Japan, Java, Azores, Tenerife, La Palma, Tanzania, N. America, Greenland.

4. J. polaris Lindb., *Öfv. K. Vet. Akad. Forh.* 23: 560. 1887   (Fig. 59)
Aplozia schiffneri Loitl., *Solenostoma schiffneri* (Loitl.) K. Müll., *Solenostoma pumila* ssp. *polaris* (Lindb.) Schust.

Paroecious. Plants small or very small, dull green to brownish, in patches or growing through other bryophytes. Sterile shoots procumbent, fertile shoots erect or ascending, 0.3–1.0 cm. Rhizoids numerous, colourless, not arising from leaf bases. Lower leaves on sterile stems distant, elliptical to ovate, obliquely inserted, upper leaves and those on fertile stems imbricate, partly embracing stem, ovate-cordate to rounded-cordate, more transversely inserted, widest ⅓ to ½ from base, 0.3–0.8 mm long, 0.25–0.70 mm wide, mostly 1.0–1.3 times as long as wide; mid-leaf cells 16–24(–32) μm wide, thin-walled, trigones small or absent. Gemmae lacking. Male bracts 2–3 pairs below female, ovate-cordate, saccate with spreading tips. Female bracts rounded-cordate. Perianths common, obovoid, to broadly fusiform, about ⅔ emergent, plicate above, cells in upper third similar to mid-leaf cells. Capsules common, summer. On basic soil on cliff ledges and amongst rocks, 500–1000 m, very rare, Perth, S. Aberdeen, W. Inverness. 4. Pyrenees, N. Italy, Balkans, Alps, Norway, Sweden, Finland, Spitzbergen, Iceland, Novaya Zemlya, Siberia, N. America, Greenland.

Although the perianth may be obovoid and hence differing from the fusiform perianth of *J. pumila*, perianths occur which are broadly fusiform and somewhat similar to those of *J. pumila*. The only reliable character that can then be used to separate the two species is the ovate-cordate to rounded-cordate leaves of fertile shoots and upper parts of sterile shoots of *J. polaris*.

5. J. borealis Damsh. & Váňa, *Lindbergia* 4: 5. 1977   (Fig. 59)
Solenostoma oblongifolium auct. non (K. Müll.) K. Müll.

Dioecious. Plants small, brownish green to blackish, in patches or growing through other bryophytes. Sterile shoots procumbent to ascending, sometimes very slender, 0.3–1.0 cm, fertile shoots erect or ascending. Rhizoids numerous, colourless to brownish, not arising from leaf bases. Leaves on sterile stems distant, spreading, slightly concave, obliquely inserted, to closely imbricate, patent, very concave, partly transversely inserted, leaves on fertile shoots larger, embracing stem, ovate-cordate to

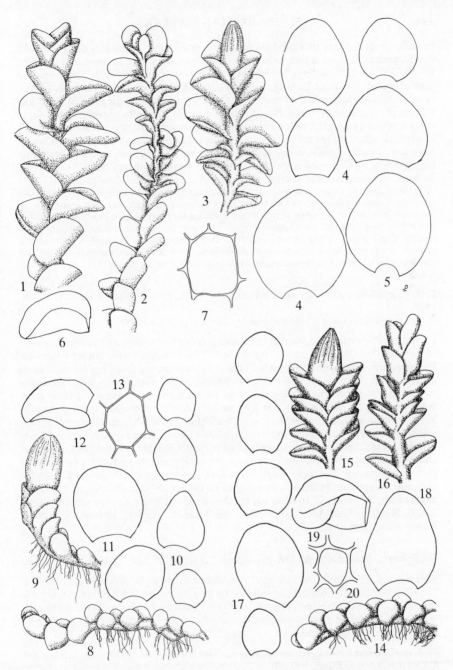

Fig. **59**. 1–7, *Jungermannia exsertifolia* ssp. *cordifolia*: 1–3, sterile, male and female shoots (× 6); 4, leaves (× 11); 5, female bract (× 11); 6, male bract (× 11); 7, mid-leaf cell (× 450). 8–13, *J. polaris*: 8, 9, sterile and fertile (paraecious) shoots (× 11); 10, leaves (× 21); 11, female bract (× 21); 12, male bract (× 21); 13, mid-leaf cell (× 450). 14–20, *J. borealis*: 14–16, sterile, female and male shoots (× 18); 17, leaves (× 21); 18, female bract (× 21); 19, male bract (× 21); 20, mid-leaf cell (× 450).

broadly ovate-cordate or rounded-cordate, very concave, cucullate, not decurrent, 0.3–1.2 mm long, 0.3–1.0 mm wide, 1.1–1.3(–1.5) times as long as wide, mid-leaf cells 16–24(–32) μm wide, walls yellowish-brown to brown, thin to thick, trigones small to medium-sized. Gemmae lacking. Male bracts 2–3 pairs, very concave with saccate basal part constituting half or more of bract, upper part spreading, often cucullate. Female bracts erect-spreading, broadly ovate-cordate. Perianths common, obovoid to clavate, about ⅔ emergent, becoming completely emergent with age, plicate above, cells in upper third similar to mid-leaf cells. Capsules occasional, summer. Damp, sheltered, basic soil or rock on cliffs, in ravines and by streams, 150–1050 m, rare but sometimes locally common, Caernarfon, Perth, Kintyre and Angus north to Ross. 12. C. Europe, Norway, Sweden, Faroes, Iceland, Caucasus, Alaska, Canada, Greenland.

The very concave, ovate-cordate to rounded-cordate leaves with partially transverse insertion are characteristic but forms occur with more distant, flatter leaves which cannot be separated from the three preceding species when sterile. When fertile the very saccate male bracts are distinctive and female plants may be distinguished from *S. atrovirens* by the very concave, broadly ovate-cordate upper leaves. These shoots, and also some sterile shoots, resemble a miniature *J. exsertifolia* spp. *cordifolia* which, however, differs in its larger dimensions, narrower leaf insertion and decurrent leaf base. For the occurrence of this plant in Britain see Paton, *Trans. Br. bryol. Soc.* 5, 435–8, 1968.

6. J. exsertifolia Steph. ssp. **cordifolia** (Dum.) Váňa, *Folia Geobot. Phytotax.* 8: 268. 1973   (Fig. 59)
Aplozia cordifolia Dum., *Solenostoma cordifolium* (Dum.) Steph.

Dioecious. Plants medium-sized to large, dull green to blackish-purple or black, in tufts or patches. Shoots procumbent, mostly 2.0–7.5 cm, occasionally more, branched. Rhizoids few, mostly near stem base, colourless, not arising from leaf bases. Leaves distant to approximate or subimbricate, flaccid, erecto-patent, concave, embracing stem, cordate or broadly ovate-cordate to rounded-cordate, widest ⅛–¼ from base, insertion oblique, very narrow, 1.0–2.6 mm long, 0.9–3.1 mm wide, 0.8–1.2 times as long as wide; mid-leaf cells 20–32 μm wide, walls thin, brownish, trigones absent or very small. Gemmae lacking. Male bracts numerous, reflexed, base saccate. Female bracts orbicular, erecto-patent, embracing base of perianth. Perianths occasional, fusiform, about ½ emergent, plicate above, cells towards mouth similar to mid-leaf cells. Capsules rare, summer. *n* = 9. Wet rocks in and near streams, in springs and flushes, occasional to common in montane habitats, ascending to *ca* 1000 m, Wales, Derby and Yorkshire northwards, W. and N.E. Ireland. 62, H9. Montane and arctic Europe from S. Spain, Sicily, N. Italy and the Carpathians northwards, Faroes, Iceland, Caucasus, Kamchatka, Greenland.

Subgenus *Solenostoma* (Mitt.) Amak., *J. Hattori bot. Lab.* 22: 53. 1960

Rhizoids colourless to brownish, rarely violet, some arising from basal part of leaves. Leaves orbicular; trigones usually small or lacking. Gemmae lacking. Perigynium low or absent. Perianth emergent, slightly laterally compressed, plicate above, ± abruptly contracted to beaked mouth, wall 2-stratose below, cells in upper part similar to mid-leaf cells. Six European species.

7. J. confertissima Nees, *Naturg. Europ. Leberm.* 1: 277, 291. 1833   (Fig. 60)
Solenostoma levieri (Steph.) Steph.

Paroecious. Plants yellowish-green to brownish, in dense tufts or growing with other bryophytes. Shoots procumbent to erect, 0.3–1.0 cm, innovating from below perianth. Rhizoids numerous, colourless to pale brown, occasionally pale purple, some arising

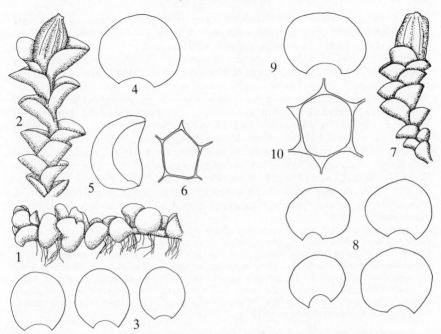

Fig. 60. 1–6, *Jungermannia sphaerocarpa*: 1 and 2, sterile and fertile (paroecious) shoots (× 11); 3, leaves (× 18); 4, female bract (× 18); 5, male bract (× 18); 6, mid-leaf cell (× 450). 7–10, *J. confertissima*: 7, fertile (paroecious) shoot (× 11); 8, leaves (× 18); 9, female bract (× 18); 10, mid-leaf cell (× 450).

from leaf bases. Leaves imbricate or closely imbricate, erecto-patent, slightly concave, obliquely inserted, decurrent on both sides, ± reniform, 0.4–1.1 mm long, 0.5–1.2 mm wide, 0.7–0.9 times as long as wide; marginal cells not forming a border, mid-leaf cells 28–44 μm wide, walls colourless to brownish, thin, trigones medium-sized. Gemmae lacking. Male bracts 2–6 pairs below female, imbricate, saccate, erect or with slightly reflexed tips. Female bracts embracing lower part of perianth, reniform. Short perigynium present. Perianths common, $\frac{1}{2}$–$\frac{2}{3}$ emergent, obovoid, plicate, abruptly contracted to shortly-beaked mouth, cells in upper part ± similar to mid-leaf cells. Capsules common, summer, autumn. $n = 8 + m^{*}$, 9. On damp soil amongst rocks and on ledges, on damp rocks and in flushes, in basic habitats, 90–1000 m, rare, Carmarthen, Yorkshire, Durham, Peebles and Argyll north to W. Ross. 12. Alpine and arctic Europe from the Pyrenees, Alps and Balkans north to Spitzbergen, Iceland, Caucasus, Kashmir, Himalayas, Siberia, Japan, N. America, Greenland.

For the occurrence of this plant in Britain see Paton, *Trans. Br. bryol. Soc.* **6**, 50–5, 1970.

8. J. sphaerocarpa Hook., *Brit. Jungerm.* Tab. 74. 1815   (Fig. 60)
Aplozia sphaerocarpa (Hook.) Dum., *Solenostoma sphaerocarpum* (Hook.) Steph.

Paroecious. Green to reddish-brown, often dense tufts. Shoots procumbent to erect, 0.5–3.0 cm, innovating from below perianth, small-leaved innovations sometimes present. Rhizoids numerous, colourless to pale brown or rarely pale purple, some arising from leaf bases. Leaves on sterile stems distant to approximate, soft, obliquely inserted, on fertile stems distant to subimbricate, erecto-patent, embracing stem, more transversely inserted, orbicular, decurrent on dorsal side, 0.6–1.4 mm long, 0.7–1.5 mm wide, 0.8–1.0 times as long as wide; marginal cells quadrate to rectangular, sometimes forming slight border, 16–24 μm wide, mid-leaf cells 20–32(–40) μm wide, walls usually colourless, thin, trigones usually distinct. Gemmae lacking. Male bracts of similar size to upper leaves, 2–5 pairs below female, saccate, with reflexed tips. Female bracts of similar size but wider than male bracts, wider than long, embracing lower part of perianth, apices reflexed. Perigynium lacking or very short. Perianths common, $\frac{1}{2}$–$\frac{2}{3}$ emergent, obovoid, plicate, narrowed to shortly beaked mouth, cells in upper part ± similar to mid-leaf cells. Capsules occasional, spring, summer. $n = 8 + m^*$, 9, $16 + 2m^*$, 18. On often basic rocks in and by streams and waterfalls, flushed rocks and occasionally on damp soil by streams, ascending to about 1000 m, rare or occasional in W. and N. Britain, from Devon, S. Wales, Hereford and Stafford northwards, Jersey. 53, H16, C. Montane and alpine Europe from the Iberian peninsula and Balkans north to Spitzbergen, Iceland, Turkey, Caucasus, Japan, E. Africa, Réunion, N. America, Greenland.

  J. confertissima and *J. sphaerocarpa* differ from the preceding species of the genus in their frequently erect tufted habit, the leaves with rhizoids arising from the bases and the perianth abruptly narrowed to a shortly beaked mouth (which may be lost in older perianths). In *J. sphaerocarpa* the leaves are laxer, more rounded and with smaller cells than in *J. confertissima* but the two cannot be separated with certainty when sterile; the closely imbricate, usually appressed male bracts of *J. confertissima* are distinctive. When cultured from spores *J. sphaerocarpa* may develop deep purple rhizoids (Dr M.E. Newton, pers. comm.), a coloration usually regarded as being diagnostic of subgenus *Plectocolea*.

9. J. gracillima Sm. in Sowerby, *Engl. Bot.* 32: Tab. 2238. 1811   (Fig. 61)
Aplozia crenulata (Mitt.) Lindb., *Solenostoma crenulatum* (Sm.) Mitt., *S. gracillimum* (Sm.) Schust.

Dioecious. Yellowish-green to green or more usually reddish-brown tinted, lax to dense patches or growing with other bryophytes. Shoots procumbent, 0.5–2.5(–4.0) cm, without or more usually with innovations and sometimes consisting entirely of small-leaved slender shoots. Rhizoids numerous, colourless, some arising from leaf bases. Lower leaves small, distant to approximate, above and on fertile shoots larger, imbricate, erecto-patent, concave, orbicular, widest at middle, obliquely inserted, not decurrent, 0.3–1.0 mm long, 0.3–1.0 mm wide, ± as long as wide; marginal cells thicker-walled and 1.5–3.0 times as wide as submarginal row, forming distinct border, 24–52 μm wide, mid-leaf cells (20–)28–32(–40) μm wide, walls thin or slightly thickened, colourless to brownish, trigones small or lacking, oil-bodies 2–3, small. Innovations and sometimes whole plants with distant to closely imbricate small concave leaves inserted obliquely to transversely, often very variable in size; marginal cells often not differentiated from submarginal row except sometimes in largest leaves and in female bracts. Gemmae lacking. Male bracts 2–6 pairs, imbricate, with reflexed tips, saccate at base. Female bracts orbicular, of similar size to upper leaves, embracing lower part of perianth, apices reflexed, margin bordered. Perigynium lacking. Perianths frequent, $\frac{1}{2}$- emergent, shortly cylindrical to obovoid, 4–5-plicate, ± abruptly narrowed to shortly beaked mouth. Capsules occasional, winter, spring. $n = 9$. Soil, especially where clayey, on ditch and stream banks, woodland paths, on heaths, mine wastes, etc., in acidic habitats, from sea-level to *ca* 1000 m, occasional to frequent in lowland areas, common elsewhere. 109, H31, C. Throughout Europe, Faroes, Iceland, Caucasus, Lebanon, Tunisia, Algeria, Azores, Madeira, Tenerife, Jamaica, N. America, Greenland.

  An extremely variable species for which a number of varieties have been described and, although it is not known whether these variants have a genetic basis or are environmentally

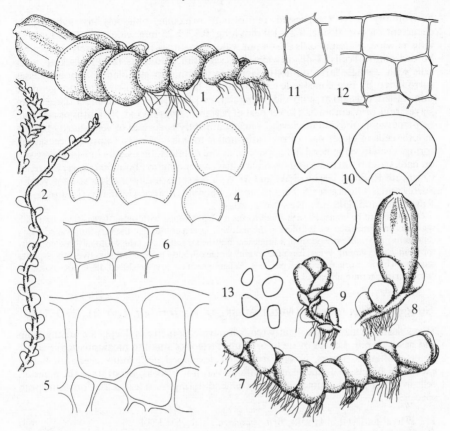

Fig. **61**. 1–7, *Jungermannia gracillima*: 1, female shoot (× 23); 2 and 3, sterile and male shoots from slender form (× 23); 4, leaves (× 23); 5 and 6, marginal cells from leaves of normal and slender forms (× 450). 7–13, *J. caespiticia*: 7 and 8, sterile and female shoots (× 43); 9, gemmiferous shoot (× 43); 10, leaves (× 110); 11, mid-leaf cell (× 450); 12 marginal cells of leaf (× 450); 13, gemmae (× 450).

induced, they integrade to such an extent that they cannot be maintained. Small forms and innovating shoots may have leaves lacking a border but may be recognised by their orbicular, usually concave leaves and acidic habitat. Larger forms lacking a leaf border are rare but confusing and may occur in waterlogged habitats but again the orbicular leaves are characteristic, other *Jungermannia* species that might possibly occur in a similar habitat having elliptical, ovate or cordate leaves. *J. caespiticia* differs in the production of gemmiferous shoots, the leaf cells with solitary oil-bodies, and *Nardia scalaris* in the presence of underleaves and persistent oil-bodies.

10. J. caespiticia Lindenb., *Nova Acta Leop. Carol Suppl.* 14: 67. 1829   (Fig. 61)
Aplozia caespiticia (Lindenb.) Dum., *Solenostoma caespiticium* (Lindenb.) Steph.

Dioecious. Plants small, delicate, in yellowish-green patches, scattered plants or growing through other bryophytes. Shoots procumbent, 0.2–0.6 mm, branching sparingly, sometimes producing small-leaved innovations. Rhizoids numerous, colourless, some arising from leaf bases. Leaves distant to subimbricate, spreading horizontally to erect-appressed, reniform to orbicular, obliquely inserted, slightly decurrent on dorsal side, 0.25–1.0 mm long. 0.32–1.25 mm wide, 0.65–0.95 times as long as wide; marginal cells 28–48 μm wide, sometimes forming a barely discernible border, mid-leaf cells 24–40 μm wide, walls very thin, colourless, trigones lacking, most cells with a single large oil-body. Gemmae terminal in globular gelatinous masses surrounded by two concave leaves forming a brownish gall-like structure, gemmae thin-walled, unicellular, ellipsoid or angular, *ca* 12 μm long. Female bracts larger than leaves, reniform, embracing lower part of perianth, tips reflexed. Perianths occasional, *ca* $\frac{2}{3}$-emergent, obovoid or clavate, plicate, abruptly narrowed to very shortly beaked mouth, cells in upper part of perianth similar to mid-leaf cells. Capsules occasional, spring. Usually ephemeral in temporary, acidic habitats, on usually fine or clayey soil on banks and by streams and on mine waste, ascending to about 300 m, very rare, E. Cornwall, Radnor, Derby, S.W. and M.W. Yorkshire and formerly also I. of Wight, Surrey and E. Gloucester. 8. Fennoscandia, Belgium, France, Germany, C. Europe, W. Russia, Iceland, Ukraine, New York, Alaska.

Only likely to be confused with small forms of *J. gracillima* with unbordered leaves and with which it sometimes occurs. It differs in the yellowish-green colour and the very thin-walled larger cells (cells of variable size occur in a single leaf but the largest exceed the width of those normally found in *J. gracillima*). When fresh the usually single oil-body per cell is distinctive, other species having two or more oil-bodies per cell. When gemmae are produced these are borne in characteristic terminal heads.

Subgenus *Plectocolea* (Mitt.) Amak., *J. Hattori bot. Lab.* 21: 270. 1960

Stems fleshy. Rhizoids frequently reddish-purple, sometimes deeply so, arising from leaf bases or not. Leaves ovate to orbicular; trigones small to medium-sized. Female bracts similar to leaves, arising from well developed perigynium. Perianth barely exceeding bracts to $\frac{1}{2}$-emergent, plicate above, not beaked, unistratose throughout, cells in upper part 2–3 times as long as wide, differing from leaf cells. Four European species.

11. J. hyalina Lyell in Hook., *Brit. Jungerm.*, Tab. 63. 1814   (Fig. 62)
Eucalyx hyalinus (Lyell) Carring., *Plectocolea hyalina* (Lyell) Mitt., *Solenostoma hyalinum* (Lyell) Mitt., *S. ontariensis* Schust.

Dioecious. Yellowish-green to pale green, occasionally reddish-tinged, glistening patches or growing through other bryophytes. Shoots procumbent to erect, 0.5–2.0 cm, sparingly branched; stems fleshy, brittle. Rhizoids numerous, at least some deep reddish-purple, some arising from leaf bases. Lower leaves distant, spreading horizontally, obliquely inserted, upper leaves approximate to imbricate, erect-spreading, more transversely inserted, often undulate, orbicular or obliquely orbicular with wide base, decurrent on dorsal side, 0.7–1.5 mm long, 0.8–1.7 mm wide, 0.7–1.0 times as long as wide; mid-leaf cells 24–40(–50) μm wide, walls usually thin, trigones small to medium-sized and bulging, cuticle smooth. Gemmae lacking. Male bracts 3–8 pairs, saccate with spreading tips. Perigynium present, bearing one pair of reniform bracts. Perianths occasional, apparently about $\frac{1}{2}$-emergent, obovate, plicate, cells in upper part elongated. Capsules rare, spring. $n=9^*$. Damp, often sheltered soil, alluvium and rocks by streams, on banks, tracks, cliffs, in woods, in flushes, in acidic or basic habitats, at low altitudes, widespread but local, extending north to Sutherland, scattered localities in Ireland. 75, H19. Europe, Faroes, Iceland, Crete, Turkey, Caucasus, Siberia, Japan, Korea, N. Africa, Macaronesia, N. America.

In this and the next three species, all belonging to the subgenus *Plectocolea*, and also in *J. confertissima* and sometimes *J. sphaerocarpa*, a tubular structure, the perigynium, bearing the perianth, develops at the apex of fertile female shoots. In the subgenus *Plectocolea* the female bracts and sometimes two or more preceding leaves are also borne on the perigynium and under the microscope these structures appear as if adnate to the lower part of the perianth which, because of this tubular perigynium, appears to be longer than it really is.

J. hyalina and *J. paroica* are very similar and although the latter tends to be a larger plant the two species can only be distinguished with certainty when fertile and it is questionable whether they are specifically distinct. The leaf cell size character given by Macvicar (1926) and later authors is of no use. Care must be taken with plants of *J. paroica* collected in summer as at that time fertile shoots may be male only.

12. J. paroica (Schiffn.) Grolle, *J. Jap. Bot.*, 39: 238: 1965 (Fig. 62)
Eucalyx paroicus (Schiffn.) Macv., *Plectocolea paroica* (Schiffn.) Evans, *Solenostoma paroicum* (Schiffn.) Schust.

Paroecious, otherwise very similar to *J. hyalina*. Shoots 1–3 cm; stems fleshy, brittle. Leaves 0.8–1.6 mm long, 1.0–2.0 mm wide, 0.70–0.95 times as long as wide; mid-leaf cells 30–50(–60) μm wide. Male bracts 2–6 pairs below female. Perianths frequent. Capsules occasional, spring. $n = 9^*$. Damp, sheltered soil and rocks by streams, on banks and cliffs, ascending to about 800 m, rare but widely distributed mainly in W. and N. Britain, N. Hants and from S. Somerset, Wales, Stafford and Yorkshire northwards, rare in Ireland. 55, H10. Germany, Faroes.

13. J. subelliptica (Lindb. ex Kaal.) Lev., *Bull, Soc. Bot. Hal.* 1905: 211. 1905
(Fig. 62)
Eucalyx subellipticus (Lindb. ex Kaal.) Breidl., *Plectocolea subelliptica* (Lindb. ex Kaal.) Evans, *P. obovata* var. *minor* (Carring.) Schljak., *Solenostoma subellipticum* (Lindb. ex Kaal.) Schust.

Paroecious. Small, yellowish-green to brownish tufts, patches or mixed with other bryophytes. Shoots prostrate to erect, 3–8 mm; stems fleshy, sparingly branched. Rhizoids numerous, colourless to brownish, none arising from leaf bases. Leaves on prostrate stems approximate to imbricate, often very concave, ± obliquely inserted, on fertile stems subimbricate, erecto-patent, ovate to broadly ovate, concave, somewhat narrowed at base, ± transversely inserted, 0.5–1.1 mm long, 0.45–1.00 mm wide, 1.0–1.4 times as long as wide; mid-leaf cells 24–36(–40) μm wide, walls thin, colourless, trigones small or lacking, cuticle ± smooth. Gemmae lacking. Male bracts 2–5, below female, larger than leaves, erect to reflexed, saccate at base. Female bracts larger than male bracts, 4, borne on perigynium, erect to spreading. Perianths common, hardly longer than upper pair of bracts, upper part broadly conical, plicate, cells in upper part 2–3 times as long as wide, mouth very shortly ciliate when young. Capsules occasional. $n = 9$. On moist, usually basic, damp soil and rocks by streams, waterfalls and on banks and cliffs, ascending to about 1000 m, rare in W. and N. Britain from E. Cornwall, S. Somerset, Brecon, Merioneth, Caernarfon, I. of Man, S. Northumberland and Dumfries north to Shetland, W. Galway, W. Mayo, Sligo, Donegal, Antrim. 33, H6. Scattered localities in alpine and arctic Europe from N. Italy and the Balkans north to Spitzbergen, Faroes, Iceland, U.S.S.R. (Behring Is.), N. America, Greenland.

J. subelliptica resembles small forms of *J. obovata*, which is probably a polyploid derivative, and the two have been confused. Whether they are specifically distinct is open to question. *J. subelliptica* is usually smaller with smaller mid-leaf cells with a ± smooth cuticle; the female bracts are sometimes but not always erect and the free part of the perianth is sometimes more widely conical; leaves on sterile stems are only slightly obliquely inserted. The rhizoids in *J. obovata* are

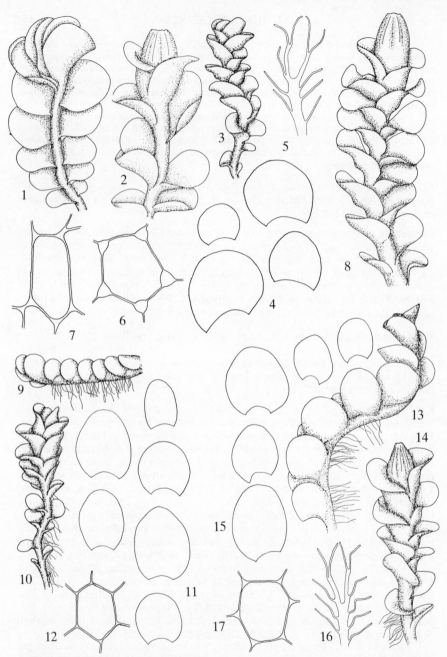

Fig. **62**. 1–7, *Jungermannia hyalina*: 1–3, sterile, female and male shoot (× 11); 4, leaves (× 15); 5, L.S. female shoot showing perigynium and perianth (× 8); 6, mid-leaf cell (× 450); 7, cell near perianth apex (× 450). 8, *J. paroica*: fertile (paroecious) shoot (× 11). 9–12, *J. subelliptica*: 9 and 10, sterile and fertile (paroecious) shoots (× 11); 11, leaves (× 18); 12, mid-leaf cell (× 450). 13–17, *J. obovata*: 13 and 14, sterile and fertile (paroecious) shoots (× 11); 15, leaves (× 15); 16, diagrammatic L.S. of fertile shoot with perigynium and perianth (× 8); 17, mid-leaf cell (× 450).

usually purple. When sterile the plant cannot be separated from sterile plants of other *Jungermannia* species such as *J. pumila* with similar leaf shape.

14. J. obovata Nees, *Naturg. Europ. Leberm.* 1: 332. 1833 (Fig. 62)
Eucalyx obovatus (Nees) Carring., *Plectocolea obovata* (Nees) Lindb., *Solenostoma obovatum* (Nees) Mass.

Paroecious. Dark green to reddish-brown tufts or patches, often mixed with other bryophytes. Shoots prostrate to erect, 0.5–3.0(–6.0) cm; stems fleshy, sparingly branched, producing small-leaved innovations. Rhizoids numerous, at least some pale to deep reddish-purple, none arising from leaf bases. Lower leaves and those on prostrate shoots distant to approximate, obliquely inserted, upper large, imbricate, more transversely inserted, patent, concave, broadly ovate to ± orbicular, somewhat narrowed at base, 0.6–1.5 mm long, 0.6–1.5 mm wide, 1.0–1.3 times as long as wide; mid-leaf cells 20–50 μm wide, walls thin or slightly thickened, colourless to brownish, trigones small to medium-sized and bulging, cuticle verruculose. Gemmae lacking. Male bracts 2–4 pairs below female, saccate with spreading tips. Female bracts 4, spreading to squarrose, borne on perigynium. Perianths common, hardly longer than upper pair of bracts, upper part narrowly conical, plicate, mouth 2–3-lobed, cells towards apex 2–3 times as long as wide. Capsules occasional, spring, summer. $n = 18$. Damp rocks and soil by streams, on cliffs and banks, rarely on submerged rocks, ascending to about 1000 m, rare to frequent in W. and N. Britain, from Wales, Derby and Yorkshire northwards, occasional in Ireland. 59, H18. Spain, Italy and the Balkans north to Norway, Sweden and Finland, Faroes, Iceland, Turkey, Caucasus, N. America, Greenland.

33. NARDIA S.F. GRAY, *Nat. Arr. Br. Pl.* 1: 679. 1821

Dioecious or paroecious. Plants very small to robust. Stems only sparsely branched, branches arising from ventral side. Rhizoids usually numerous. Leaves alternate, obliquely inserted, succubous, reniform to orbicular, entire, retuse or rarely bilobed; oil-bodies large, 1–several per cell. Underleaves lanceolate or subulate, usually restricted to young parts of stems. Gemmae lacking. Male bracts terminal, becoming intercalary, or below female, similar to leaves but saccate at base. Female bracts 2–3 pairs, arising from perigynium, similar to but larger than leaves, sometimes emarginate; toothed bracteoles present. Perianth very short, immersed in bracts. Calyptra derived from perigynium. Capsule ± globose, wall bistratose. Spores 8–22 μm. About 14 species, mainly north temperate and arctic in distribution.

1 Leaves usually simple, oil-bodies glistening, perigynium not bulbous (see Fig. **63**, 11)
2

Leaves bilobed or at least upper emerginate or retuse, oil-bodies opaque, perigynium bulbous (see Fig. **63**, 6) 3

2 Plants 2–25 cm, leaves reniform, translucent, extending ± equally both above and below stem, oil-bodies not persisting on drying **1. N. compressa**

Plants 0.5–3.0(–5.0) cm, leaves ± orbicular, not translucent, hardly extending below ventral side of stem, oil-bodies persisting for several years after drying
2. N. scalaris

3 Plants 0.3–2.0 cm, leaves usually retuse to emarginate, mid-leaf cells 20–48 μm wide
3. N. geoscyphus

Plants 0.15–0.35 cm, leaves usually bilobed, mid-leaf cells 10–20 μm wide
4. N. breidleri

Subgenus *Nardia*

Plants robust. Stem cortex cells larger than inner cells. Rhizoids few, restricted to stem base. Leaves erect, reniform, appressed face to face giving shoot appearance of being laterally compressed, cells decreasing markedly in size from mid-leaf to apex. One species.

1. N. compressa (Hook.) S.F. Gray, *Nat. Arr. Br. Pl.* 1: 694. 1821   (Fig. 63)
Alicularia compressa (Hook.) Nees

Dioecious. Plants usually robust, deep green to reddish-brown or deep purple, forming often large tufts. Shoots procumbent, 2–15(–25) cm, stems tough, sparsely branched, sometimes producing innovations. Leaves imbricate, erect-appressed so that shoots appear laterally compressed, extending ± equally above and below stem, translucent, orbicular-reniform to reniform, entire, ± flat, obliquely inserted, slightly decurrent on dorsal side, 0.8–1.6 mm long, 0.9–2.8 mm wide (0.5–)0.6–0.9 times as long as wide, cells decreasing in size markedly from mid-leaf to apex, mid-leaf cells 30–40(–48) μm wide, walls thin to thick, colourless to purplish, trigones ± lacking to medium-sized, oil-bodies not persisting after drying. Underleaves small and sparse, usually only on younger parts of stem, subulate to lanceolate, rarely bifid, 140–240 μm long. Gemmae lacking. Fertile plants very rare. *n* = 8 + m*, 9. Wet and submerged rocks in and by streams, ascending to about 1000 m, rare to occasional but sometimes locally abundant in W. and N. Britain, Ireland, very rare elsewhere, E. Sussex, E. Cornwall, Devon, Wales, Derby and Yorkshire north to Shetland. 52, H23. Scattered localities in Europe from the Iberian Peninsula, N. Italy and Balkans north to Norway and Sweden, Iceland, Turkey, Caucasus, Kamchatka, Japan, Yunnan, N. America, Greenland.

Subgenus *Geoscypharia* Trev. *Mem Real. Istit. Lombardo Sci. Mat. Nat.* 13: 101. 1877

Plants small to medium-sized. Stem cortex cells smaller than inner cells. Rhizoids numerous. Leaves not appressed face to face, ± orbicular, entire to bilobed, cells not decreasing in size markedly from mid-leaf to apex.

2. N. scalaris S.F. Gray, *Nat. Arr. Br. Pl.* 1: 694. 1821   (Fig. 63)
Alicularia scalaris (S.F. Gray) Corda

Dioecious. Green to deep reddish-brown patches or cushions, sometimes extensive, or growing through other bryophytes. Shoots procumbent, rarely ± erect, 0.5–3.0(–5.0) cm, sparsely branched, stems brittle. Rhizoids numerous, ± to stem apex, not arising from perigynium. Leaves distant to ± imbricate, erect-spreading to spreading, not or scarcely extending beyond ventral side of stem, not translucent, orbicular to broadly orbicular or reniform-orbicular, entire, concave, obliquely inserted, scarcely decurrent, 0.6–1.3 mm long, 0.6– 1.3 mm long, 0.7–0.9(–1.0) times as long as wide; cells scarcely decreasing in size from mid-leaf to apex, mid-leaf cells (24–)28–40 μm wide, walls thin to slightly thickened, trigones small to medium-sized and bulging, oil-bodies 2–4, glistening, persisting several years after drying. Underleaves lanceolate, spreading, 180–320 μm long. Gemmae lacking. Male bracts 2–6 pairs, subimbricate, ± erect from spreading saccate base. Female bracts 2–3 pairs, erect, arising from perigynium, larger than leaves, sometimes emarginate. Perianths occasional, very short, pyramidal, not emergent from involucral bracts. Capsules occasional, winter, spring. *n* = 8 + m, 9*. On usually damp soil on waste ground, stream banks, on and by tracks, in woodland, on moorland, on damp rocks and rocks in and by streams, rarely in basic habitats,

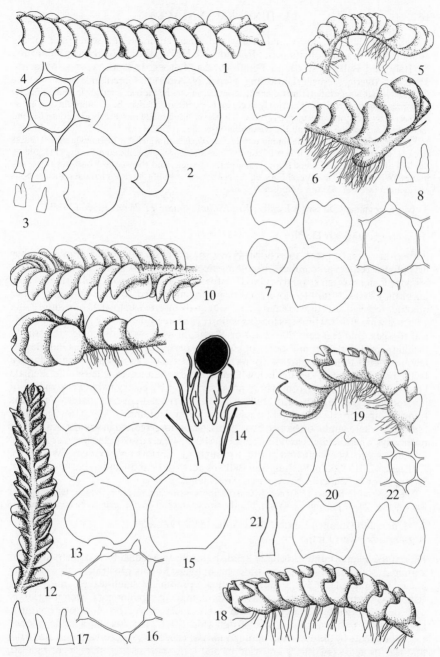

Fig. **63**. 1–4, *Nardia compressa*: 1, shoot (× 7); 2, leaves (× 11); 3, underleaves (× 45); 4, mid-leaf cell (× 450). 5–9, *N. geoscyphus*: 5 and 6, sterile and fertile (paroecious) shoots (× 13); 7, leaves (× 18); 8, underleaves (× 45); 9, mid-leaf cell (× 450). 10–17, *N. scalaris*: 10–12, sterile, female and male shoots (× 118); 13, leaves (× 18); 14, l.s. female shoot with sporophyte (× 11); 15, female bract (× 18); 16, mid-leaf cell (× 450); 17, underleaves (× 45);. 18–22, *N. breidleri*: 18 and 19, sterile and young fertile (paroecious) shoot (× 32); 20, leaves (× 45); 21, underleaf (× 178); 22, mid-leaf cell (× 450).

ascending to 1100 m, occasional in lowland areas, common and sometimes locally abundant in W. and N. Britain and Ireland. 108, H37, C. From the Iberian Peninsula, N. Italy and Balkans north to Fennoscandia, Faroes, Iceland, Azores, Madeira, Tenerife, Algeria, Tunisia, Darjeeling, Nepal, N. America, Greenland.

Usually readily separated from other entire-leaved species by the persistent glistening oil-bodies and spreading underleaves. The rarely occurring aquatic forms may be distinguished from *N. compressa* by the more orbicular leaves much less overlapping, not or scarcely extending beyond the ventral side of the stem and the persistent oil-bodies.

N. geoscyphus is usually a smaller plant than *N. scalaris* with which it sometimes grows. It differs from *N. scalaris* in the upper leaves of sterile stems and leaves on fertile stems being emarginate, the lack of underleaves except from young parts of stems and the opaque, non-persistent oil-bodies. When fertile, mature shoots of *N. geoscyphus* usually have a characteristic bulbous perigynium (see Fig. **63**) not found in *N. scalaris*.

3. N. geoscyphus (De Not.) Lindb., *Not. Sällsk. Fauna Fl. Fenn.* 13: 371. 1874

(Fig. 63)

Alicularia geoscyphus De Not.

Paroecious. Small, green to reddish-brown or purplish patches or amongst other bryophytes. Shoots procumbent, becoming erect if crowded, 0.3–1.2(–2.0) cm, sparsely branched; stems often strongly curved, brittle. Rhizoids numerous, colourless to pale purplish. Leaves distant to closely imbricate and appressed, lower ± orbicular, upper quadrate-orbicular, retuse to emarginate, obliquely inserted, 0.45–0.90 mm long, 0.48–1.05 mm wide, 0.7–1.0 times as long as wide; cells scarcely decreasing in size from mid-leaf to apex, mid-leaf cells 20–32 μm or 32–48 μm wide, walls thin, trigones usually medium-sized, bulging, oil-bodies opaque, not persisting after drying. Underleaves few, restricted to upper part of young stems, lanceolate, *ca* 180 μm long. Gemmae lacking. Male bracts 2–3 pairs below female, larger than leaves, emarginate. Female bracts 2 pairs, arising from perigynium, ± crisped, emarginate. Perigynium in horizontal shoots ± erect with bulbous, fleshy, rhizoid-covered bases. Perianths frequent, immersed. Capsules frequent, spring to autumn. $n = 9^*$, $16 + 2m$. Soil, especially where sandy, on banks, by streams, paths, on scree and in woods, usually but not always where acidic, ascending to about 800 m, widely distributed but rare, from S. Devon east to W. Kent and north to Shetland, Carlow, W. Galway, W. Mayo, Donegal. 42, H5. Spain, W. Italy and Balkans north to Fennoscandia, Faroes, Iceland, Japan, Azores, Madeira, N. America, Greenland.

There appear to be two cytotypes, with $n = 9$ and $n = 16 + 2m$, respectively, and two ranges of leaf cell size, 20–32 μm and 32–48 μm, but the two characters do not seem to be correlated.

4. N. breidleri (Limpr.) Lindb., *Bot. Notis.* 1880: 197. 1880   (Fig. 63)

Alicularia breidleri Limpr.

Dioecious. Plants minute, usually reddish-purple to reddish-brown, in patches or scattered, sometimes mixed with other bryophytes. Shoots prostrate with ascending apices when fertile, 1.5–3.5 mm, sometimes producing small-leaved innovations. Rhizoids numerous, colourless. Leaves approximate to subimbricate, becoming larger and more imbricate at apices of fertile stems, suberect to erect, concave, obliquely inserted, 0.24–0.48 mm long, 0.24–0.40 mm wide, 1.0–1.2 times as long as wide, bilobed $\frac{1}{4}$–$\frac{1}{3}$, lobes acute to obtuse, sinus acute to lunate; cells scarcely decreasing in size from mid-leaf to apex, (8–)10–20 μm wide in mid-leaf, walls usually thickened, reddish, trigones lacking or small. Underleaves subulate, *ca* 80 μm long, on youngest parts of stems only. Gemmae lacking. Perigynium bulbous, fleshy as in *N. geoscyphus*. Male bracts imbricate, bilobed. Female bracts larger and wider than leaves, less deeply bilobed. Perianths occasional, immersed. Capsules not known in Britain (?). On bare soil at high altitudes, particularly on mountain summits, often with *Anthelia juratzkana*, from 900–1240 m, very rare, Perth, Angus, S. Aberdeen, Banff, Inverness, Argyll, Ross. 11. Rare in alpine and arctic Europe from N. Italy, France and C. Europe north to Norway, Sweden, Finland and western U.S.S.R., Iceland, Japan, western N. America, Greenland.

Only likely to be confused with some small or depauperate *Marsupella*, *Gymnomitrion* and Lophoziaceae species. It differs from the first two in the oblique insertion of the leaves and in the presence of underleaves although these are difficult to detect. The lack of gemmae, smaller leaf cells and presence and form of the underleaves distinguish *N. breidleri* from any possibly confusing Lophoziaceae species.

14. GYMNOMITRIACEAE

Minute to large plants. Shoots prostrate to erect, not markedly dorsiventral; branching usually from ventral side of stem bases or as innovations from below female inflorescence. Rhizoids sparse, not arising from leaf bases. Leaves ± transversely inserted, not noticeably succubous unless secund, distant to imbricate, spreading horizontally to appressed, bilobed; cell walls usually thickened, pigmented, trigones distinct or not. Underleaves lacking. Gemmae lacking. Male bracts larger and more concave than leaves. Perigynium present or not, marsupium developing or not. Perianth, where present, immersed in female bracts. Capsule spherical, wall bistratose. Spores 7–18 μm. Eight genera.

> Shoots julaceous or not, leaves distant to imbricate, spreading horizontally to appressed, stem visible or not when leaves are moist, leaf margins entire, not hyaline (except *M. adusta*), marginal cells not more heavily thickened than inner cells **34. Marsupella**
> Shoots julaceous, leaves closely imbricate, appressed, stem not visible, leaf margins crenulate, hyaline or marginal cells strongly thickened
> **35. Gymnomitrion**

34. MARSUPELLA DUM., *Comment. Bot.*: 112. 1822

Dioecious or paroecious. Plants minute to large, in tufts or patches. Shoots julaceous or not. Rhizoids colourless to purplish. Leaves distant to imbricate, spreading to appressed, concave, ± transversely inserted, bilobed $\frac{1}{20}$–$\frac{1}{2}$, margin entire, not hyaline; cells thin- to thick-walled, trigones present or not, marginal cells not differentiated; oil-bodies 2–3 per cell. Underleaves lacking. Gemmae lacking. Male bracts larger than leaves. Female bracts 4–6, larger than preceding male bracts or leaves. Perigynium present or not. Perianth where present, immersed in bracts. Calyptra unistratose, not derived from perigynium. Capsule spherical. Spores 7–18 μm. About 40 mainly northern hemisphere species.

The following key is to typical plants; depauperate or juvenile plants may deviate from the characters given and may be difficult to determine. Leaves must be carefully dissected from the stem to determine the depth of the sinus correctly.

1 Plants 0.5–10 cm, leaf lobes obtuse to rounded   2
   Plants to 1 cm or if up to 3 cm, then lobes of at least upper leaves acute   3
2 Leaves bilobed $\frac{1}{20}$–$\frac{1}{3}$, basal part of leaf not overlapping stem, leaf margin usually
      narrowly revolute on one or both sides below   **1. M. emarginata**
   Leaves bilobed $\frac{1}{3}$–$\frac{1}{2}$, basal part of leaf overlapping stem, leaf margin plane
                                                                                                                          2. M. sphacelata
3 Leaves appressed or if erecto-patent then very distant   4
   Leaves erecto-patent to spreading horizontally, distant to imbricate   6

4 Leaves somewhat secund, sinus lunate **8. M. condensata**
  Leaves not secund, sinus acute or obtuse 5
5 Dioecious, leaf lobes and sinus acute, mid-leaf cells 12–20 μm wide **7. M. boeckii**
  Paroecious, leaf lobes and sinus obtuse, mid-leaf cells 8–12 μm wide **9. M. adusta**
6 Leaves somewhat secund, bilobed $\frac{1}{8}-\frac{1}{4}$ **10. M. brevissima**
  Leaves not secund, bilobed $\frac{1}{4}-\frac{1}{2}$ 7
7 Dioecious, plants only occasionally fertile, leaves of ± equal size along sterile stems 8
  Paroecious, plants abundantly fertile, leaves increasing in size up sterile stems, fertile
  stems markedly clavate 9
8 Leaves subimbricate when moist, $\frac{1}{3}-\frac{1}{2}$ bilobed, mid-leaf cells 12–20 μm wide, trigones
  not distinct **3. M. funckii**
  Leaves distant when moist, $\frac{1}{4}-\frac{1}{3}$ bilobed, mid-leaf cells 12–16 μm wide, trigones
  distinct **11. M alpina**
9 Plants 1–7 mm, leaves erecto-patent, lobes acute, rarely obtuse
      4. M. sprucei and **5. M. profunda**
  Plants 3–10 mm, leaves patent to spreading, lobes obtuse **6. M. sparsifolia**

Subgenus *Marsupella*

Vestigial marsupium not developing with maturation of sporophyte; rhizoids not arising from perigynium. Perianth present.

1. M. emarginata (Ehrh.) Dum., *Comment. Bot.*: 114. 1822

Dioecious. Dull green to brownish or black or deep red tufts or patches. Shoots erect to prostrate, 1–10 cm, branching occasionally and innovating from below female inflorescence. Rhizoids restricted to stem bases. Leaves ± equal in size along stem, distant to imbricate, patent to spreading horizontally, base embracing stem, one or both margins narrowly revolute at least below, 0.7–1.4 mm long, 0.8–2.2 mm wide, 0.6–1.0 times as long as wide, bilobed $\frac{1}{20}-\frac{1}{3}(-\frac{1}{4})$, lobes obtuse, rounded or rounded and apiculate, sinus obtuse to very shallowly lunate; mid-leaf cells 16–24 μm wide, becoming smaller towards apex, walls heavily thickened and trigones large to walls very heavily thickened and trigones not discernible. Male bracts more erect than leaves, saccate at base. Perianths occasional, immersed in bracts. Capsules occasional, late winter, spring.

Key to varieties of *M. emarginata*

1 Leaves bilobed $\frac{1}{6}-\frac{1}{3}(-\frac{1}{4})$, lobes obtuse, not apiculate, margin revolute on dorsal side
  only, cell walls hardly thickened var. **emarginata**
  Leaves bilobed $\frac{1}{20}-\frac{1}{6}$, lobes rounded with or without an apiculus, margin narrowly
  revolute on both sides, cell walls thickened 2
2 Plants green to black, leaves not shiny when dry, cell walls thickened, colourless to
  brownish var. **aquatica**
  Plants deep red, leaves shining when dry, cell walls very heavily thickened, reddish
      var. **pearsonii**

Var. **emarginata**

Dull green or more usually brown or reddish-brown tufts or patches. Shoots 1.0–5.0 cm. Leaves not shiny when dry, patent to ± spreading, dorsal margin and rarely also ventral margin narrowly revolute at least in some leaves, 0.6–1.4 mm long, 0.7–1.6 mm

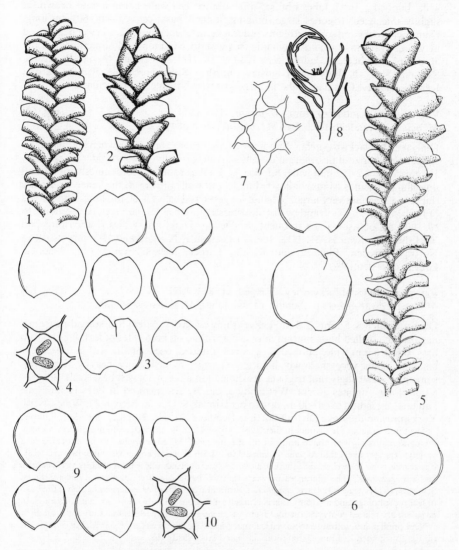

Fig. **64**. 1–4, *Marsupella emarginata* var. *emarginata*: 1 and 2, sterile and female shoots (× 11); 3, leaves (× 15); 4, mid-leaf cell (× 450). 5–8, *M. emarginata* var. *aquatica*: 5, male shoot (× 11); 6, leaves (× 15); 7, mid-leaf cell (× 450); 8, l.s. female inflorescence (× 11). 9–10, *M. emarginata* var. *pearsonii*: 9, leaves (× 15); 10, mid-leaf cell (× 450).

wide, bilobed $\frac{1}{6}-\frac{1}{5}(-\frac{1}{4})$, lobes obtuse, sinus obtuse; cell walls colourless to brownish, slightly thickened, trigones large, bulging, longest basal cells 30–40(–60) μm long. Damp acid rocks and soil by streams and tracks, in ravines, on cliffs and wet heaths and in woods, rare in lowland England, frequent to common and sometimes locally abundant elsewhere, ascending to *ca* 1200 m. 84, H31, C. Corsica, Iberian Peninsula, Sardinia, Sicily, N. Italy and Yugoslavia north to S. Fennoscandia, Russia, Faroes, Iceland, Turkey, Caucasus, Siberia, Macaronesia, N. America, Greenland.

Var. **aquatica** (Lindenb.) Dum., *Hep. Eur.*: 126. 1874   (Fig. 64)
M. aquatica (Lindenb.) Schiffn., *M. robusta* (De Not.) Evans

Dull green to blackish patches. Shoots 2–10 cm. Leaves not shining when dry, patent to spreading ± horizontally, margin usually narrowly revolute on both sides of leaf, 1.0–1.5 mm long, 1.2–2.2 mm wide, bilobed $\frac{1}{20}-\frac{1}{6}$, lobes rounded, sometimes with a small unicellular apiculus, sinus shallowly lunate; cell walls brownish, thickened or heavily thickened, trigones very large, bulging, longest basal cells 50–80 μm long. Submerged or flushed acidic rocks in and by fast-flowing streams in montane habitats, ascending to *ca* 1000 m, occasional to frequent, S. Devon, Hereford, Wales, Lancashire and Yorkshire northwards. 51, H18. Iberian Peninsula, N. Italy and Hungary north to Norway, Sweden, Finland and Russia, Faroes, Iceland, Azores, N. America, Greenland, Aleutians.

Var. **pearsonii** (Schiffn.) Corley, *J. Bryol.* 11, 843. 1981   (Fig. 64)
M. aquatica var. *pearsonii* (Schiffn.) E.W. Jones, *M. pearsonii* Schiffn.

Deep red patches. Shoots 2–8 cm. Leaves shiny when dry, patent to ± spreading from a narrow, channelled base, margin narrowly revolute on both sides of leaf, 0.8–1.3 mm long, 0.8–1.6 mm wide, bilobed $\frac{1}{20}-\frac{1}{8}$, lobes rounded, occasionally with a unicellular apiculus, sinus very shallowly lunate; cell walls reddish, very heavily thickened, sometimes so strongly that trigones are difficult to discern, longest basal cells (40–)50–80 μm long, trigones absent. Wet acidic rocks by streams and in bogs in montane habitats, ascending to *ca* 900 m, rare to occasional, S. Devon, Brecon, N. Wales, N.W. Yorkshire and the Lake District north to Shetland. 21, H8. Norway, Sweden, Russia.

   The three varieties are distinct in their typical forms but the characters do not always occur in the combinations given above and not all specimens of *M. emarginata* can be allocated to a variety. It is very likely that *M. emarginata* exhibits both genetical and environmental variability; it is also possible, however, that there are two species involved, one represented by var. *aquatica* and var. *pearsonii*, the differences between the two being obscured by phenotypic plasticity. Schuster (1974) suggests that var. *aquatica* is an aquatic ecotype of *M. emarginata* but on the basis of leaf morphology and anatomy this is unlikely; it is more probably that var. *aquatica* and var. *pearsonii* are related ecotypes, distinct from var. *emarginata*. Experimental study is required.

   Very small plants without a recurved leaf margin may be mistaken for *M. funckii* but differ in the obtuse to rounded leaf lobes, shallower sinus and larger leaf cells.

2. M. sphacelata (Gieseke ex Lindenb.) Dum., *Recueil Observ. Jungerm.*: 24. 1825
                                                                     (Fig. 65)

M. jörgensenii Schiffn., *M. sullivantii* (De Not.) Evans, *M. sphacelata* var. *media* (Gottsche) E.W. Jones

Dioecious. Dull green to brownish, purplish-black or blackish, small compact to large tumid tufts. Shoots prostrate to erect, 0.5–6.0 cm, sparingly branched, innovating from below female inflorescence; stems brittle. Leaves on sterile stems gradually increasing in size up stems of small plants, of similar size in larger plants, distant below, subimbricate above, erecto-patent to spreading from ± erect basal part which partly

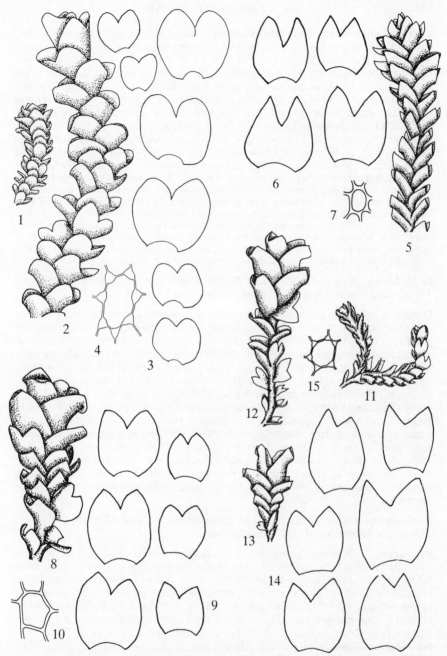

Fig. **65**. 1–4, *Marsupella sphacelata*: 1 and 2, shoots from terrestrial and aquatic habitats (×8); 3, leaves (×15); 4, mid-leaf cell (×450). 5–7, *M. funckii*: 5, shoot (×27); 6, leaves (×63); 7, mid-leaf cell (×450). 8–10, *M. sparsifolia*: 8, fertile (paroecious) shoot (×23); 9, leaves (×38); 10, mid-leaf cell (×450). 11–15, *M. sprucei*: 11, sterile shoot (×18); 12 and 13, fertile (paroecious) shoots (×18); 14, leaves (×23); 15, mid-leaf cell (×450).

embraces and overlaps stem, concave, margin plane, 0.6–1.6 mm long, 0.7–1.7 mm wide, 0.8–1.0 times as long as wide, bilobed ⅓–½, lobes rounded or rarely obtuse, apices sometimes erect, sinus acute; mid-leaf cells 20–28 μm wide, walls not or slightly thickened, brownish, trigones medium-sized. Fertile plants rare. Female bracts larger than leaves. Perianths occasional, immersed. Capsules very rare. On acidic to mildly basic submerged rocks in streams, damp or flushed rocks by streams, lakes, on cliffs, in turf and on damp soil in turf and on heaths, 400–1200 m, very rare in England, Wales and Ireland, occasional in Scotland, N. Devon, Brecon, Caernarfon, N.E. and N.W. Yorkshire, Cumberland, Kirkcudbright, Stirling and Dumbarton north to Shetland, rare in Ireland. 28, H6. Iberian Peninsula, N. Italy and Greece north to Fennoscandia and Russia, Corsica, Iceland, Caucasus, Japan, Azores, N. America, Greenland.

A variable species occurring in a wide range of habitats from completely submerged to ± dry rocks and on soil and may be confused with *M. emarginata* and *Gymnocolea inflata*. It differs from both in the more deeply lobed leaves, from *M. emarginata* in the plane leaf margin and the basal part of the leaf overlapping the stem; from *Gymnocolea inflata* in the transversely inserted leaves and thinner cell walls with much more pronounced trigones. The very fragile nature of the leaves and stems of *M. sphacelata* is also characteristic and is such that it is difficult to dissect off intact leaves. The upper leaves of dry ground forms may be obtuse or even acute and this may lead to confusion with *M. funckii* which, however, has smaller leaf cells and always occurs on soil.

3. M. funckii (Web. & Mohr.) Dum., *Recueil Observ. Jungerm.*: 24. 1835   (Fig. 65)
M. pygmaea (Limpr.) Steph., *M. hungarica* Boros & Vajda

Dioecious. Brownish-green to blackish patches. Shoots procumbent to erect, 4–12 mm, branching sparsely, innovating from below female inflorescence; sterile shoots ± parallel-sided, fertile shoots clavate. Leaves of ± equal size along stem, subimbricate, patent, concave, margins plane, 0.18–0.55(–0.65) mm long, 0.18–0.65(–0.85) mm wide, 0.8–1.1 times as long as wide, bilobed (¼–)⅓–½, lobes acute to obtuse, sinus acute, sometimes rounded at base; mid-leaf cells 12–20 μm wide, thick-walled, trigones hardly developed. Plants only occasionally fertile. Male bracts larger than leaves, saccate. Female bracts and subtending leaves larger and more concave then leaves. Perianths occasional, immersed. Capsules very rare. Sandy or stony soil and dry peat on paths, by tracks, roads and streams, ascending to *ca* 550 m, very rare in S. England, occasional elsewhere, Cornwall, S. Devon, N. Wilts, Dorset, E. Sussex, Worcester, Shropshire, Wales, Lake District and N. Northumberland north to W. Ross and W. Sutherland, Jersey, Guernsey. 48, H12, C. Portugal, Pyrenees, N. Italy, Greece, W. and C. Europe north to S. Scandinavia and Finland, Faroes, Iceland, Turkey, Japan, Macaronesia, Tennessee.

4. M. sprucei (Limpr.) H. Bern., *Catal. Hep. Sud-Ouest Suisse*: 33. 1888   (Fig. 65)
M. ustulata Spruce non (Hüb) Spruce ex Pears.

Paroecious. Plants minute, in purplish-brown to blackish tufts or patches. Shoots erect, sometimes arising from stolons, 1–7 mm, producing one to several innovations from below perianth, sometimes successively, sterile shoots narrowly clavate, fertile broadly clavate. Leaves of sterile stems increasing in size up stem, imbricate, erecto-patent, concave, 0.25–0.66 mm long, 0.18–0.50 mm wide, 1.0–1.3 times as long as wide, bilobed ⅕–½, lobes acute, somewhat inflexed, sinus acute, sometimes rounded at base; mid-leaf cells 12–28 μm wide, walls thin or thickened, brownish, trigones distinct but not bulging. Plants usually fertile. Male bracts larger than leaves, below female, bilobed, lobes acute to obtuse. Female bracts larger than male, bilobed, lobes acute to obtuse. Capsules common, summer, autumn. On moist or shaded rocks and bare soil, ascending to *ca* 1200 m, occasional in montane parts of Britain and Ireland, very rare elsewhere, Cornwall, S. Devon, W. Sussex, Wales, Yorkshire and Lake District north to Ross and W. Sutherland. 36, H8. Portugal, Pyrenees and Bulgaria north to Norway,

Sweden and Finland, Russia, Faroes, Iceland, Madeira, N. America, Greenland, Tierra del Fuego, New Zealand (?).

Differs from *M. funckii* in the usually present paroecious inflorescence and the sometimes larger leaf cells; *M. sprucei* may occur on rocks or soil whilst *M. funckii* is restricted to soil.

5. M. profunda Lindb. – see p. 338.

6. M. sparsifolia (Lindb.) Dum., *Bull. Soc. Roy. Bot. Belgique* 13: 128. 1874

(Fig. 65)

Paroecious. Small, reddish-brown to purplish-brown or blackish patches. Shoots erect, arising from leafless stoloniform shoots, 3–10 mm, branching sparsely and producing single innovation from below perianth, sterile shoots narrowly clavate, fertile shoots broadly clavate. Leaves of sterile stems increasing in size up stem, approximate to imbricate, patent to spreading, concave, margins plane, 0.45–0.80 mm long, 0.4–0.9 mm wide, 0.7–1.1 times as long as wide, bilobed ¼–⅓, lobes obtuse, rarely subacute, somewhat inflexed, sinus acute; mid-leaf cells 16–20 μm wide, walls thickened, reddish-brown, trigones distinct but not bulging. Plants usually fertile. Male bracts below female, larger than leaves, concave, lobes obtuse. Female bracts larger than male bracts, lobes obtuse. Perianths common, immersed. Capsules frequent. Granite rocks on the summit plateau of Lochnagar, S. Aberdeen, *ca* 1075 m. 1. N. Italy, Alps, France, Germany, Norway, Sweden, Finland, Russia, Faroes, Azores, S. Africa, New Zealand.

May occur mixed with or near *M. sphacelata* and *M. sprucei*. *M. sphacelata* differs in its rounded leaf lobes, leaf bases overlapping the stem and the only occasionally present dioecious inflorescence; *M. sprucei* is a smaller plant with usually acute leaf lobes.

7. M. boeckii (Aust.) Kaal., *Nyt. Mag. Naturv.* 33: 409. 1893

Dioecious. Plants minute, loosely to tightly intertwined in green to deep reddish-purple patches. Shoots procumbent to erect, 5–15 mm, sterile shoots ± cylindrical, fertile shoots clavate, branching sparsely and producing small-leaved innovations, also innovating from below female infloresence. Leaves of sterile stems of ± equal size along stem, lower leaves eroded or absent, above distant and erecto-patent or subimbricate and appressed, concave, 0.16–0.40 mm long, 0.11–0.32 μm wide, 1.0–1.5 times as long as wide, bilobed ¼–½, lobes acute to acuminate, sinus acute to very acute; mid-leaf cells 12–20 μm wide, walls thickened, colourless to deep purplish-red, trigones distinct or not. Fertile plants common. Male inflorescences produced successively, bracts larger and more concave than leaves, bilobed, lobes acute. Female bracts much larger than leaves, bilobed, lobes acute. Capsules unknown in Britain.

Key to varieties of *M. boeckii*

Leaves very distant to distant — var. **boeckii**
Leaves approximate to subimbricate — var. **stableri**

Var. boeckii (Fig. 66)

Plants loosely intertwined. Leaves very distant to distant, appressed to erecto-patent, 0.2–0.4 mm long, 0.15–0.32 mm wide; mid-leaf cells 12–20 μm wide. Only female plants known in Britain. Damp or wet rocks above 900 m, very rare, E. Inverness, Argyll. 2. Pyrenees, France, Austria, Germany, Poland, Carpathians, Norway, Sweden, Finland, Russia, Greenland.

Var. stableri (Spruce) Schust., *Hep. Anthoc. N. Amer.* 3: 107. 1974 (Fig. 66)
M. stableri Spruce

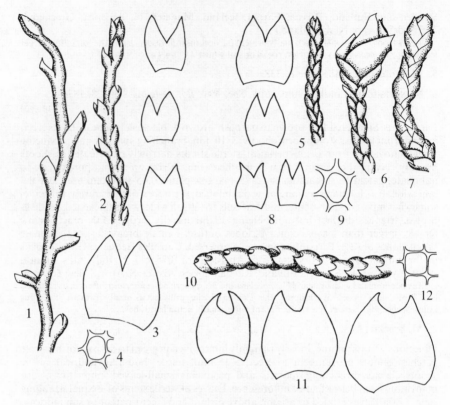

Fig. **66**. 1–4, *Marsupella boeckii* var. *boeckii*: 1 and 2, shoots (× 24); 3, leaves (× 63); 4, mid-leaf cell (× 450). 5–9, *M. boeckii* var. *stableri*: 5–7, sterile, male and female shoots (× 24); 8, leaves (× 63); 9, cells (× 450). 10–12, *M. condensata*: 10, shoot (× 24); 11, leaves (× 63); 12, mid-leaf cell (× 450).

Plants densely intertwined. Leaves approximate to subimbricate, appressed, 0.16–0.35 mm long, 0.11–0.35 mm wide; mid-leaf cells 12–16 μm wide. Male and female plants frequent but sporophyte unknown in Britain. Acidic to basic, damp subalpine and alpine rocks, also rarely on soil in areas of late snow-lie, 140–1220 m, rare, Merioneth, Caernarfon, Westmorland, Cumberland, Argyll, Mid Perth and Angus north to Ross. 14. Norway, British Columbia

Although *M. boeckii* var. *stableri* is treated as a distinct species by recent European authors, Macvicar (1926) expressed doubts about its specific status and Schuster (1974) reports intermediates between *M. boeckii* and *M. stableri* from Greenland. The only difference between the two taxa is the slight difference in size and the positioning of the leaves; varietal status seems the most satisfactory treatment of *M. stableri*.

The usually deep reddish-purple coloration, at least of the upper parts and bracts, will distinguish this species from any other *Marsupella* species except some forms of *M. emarginata* which differ in their much larger size and leaf shape. May be mistaken for *Cephaloziella* species or *Sphenolobopsis pearsonii* but these differ in the oblique insertion of the more deeply divided leaves, and, in the presence in about half the *Cephaloziella* species, of underleaves.

Subgenus *Hyalacme* (Lindb.) Lindb., *Finland (Helsingfors)* 1855 (290):
2. 1885

Protuberance formed by vestigial marsupium developing on ventral side of gynoecium with maturation of sporophyte so that stem apex is bilateral, rhizoids sometimes produced on this structure, perianth present. One species.

8. M. condensata (Ångstr. ex Hartm.) Lindb. ex Kaal., *Nyt. Mag. Naturv.* 33: 420. 1893 (Fig. 66)

Dioecious. Plants filiform, in reddish-brown to brown patches. Shoots ± erect, densely packed, 5–20 mm, branching and innovating from near base, mature sterile shoots ± cylindrical, fertile shoots clavate. Leaves of sterile stems ± equal in size along stem, approximate to subimbricate; ± appressed, somewhat secund, concave, 0.33–0.42 mm long, 0.32–0.42 mm wide, about as long as wide, bilobed $\frac{1}{4}$–$\frac{1}{3}$, lobes acute, inflexed, sinus lunate; mid-leaf cells 12–16(–20) μm wide, walls slightly thickened, yellowish-brown, trigones distinct, ± bulging. Fertile plants occasional. Male bracts larger than leaves. Female bracts much larger than leaves. Capsules unknown in Britain. Bare moist soil or humus, above 900 m, very rare, Mid Perth, S. Aberdeen, Banff, Inverness. 5. Alps, Tatra, Norway, Sweden, Finland, Russia, Spitzbergen, Iceland, Greenland.

Subgenus *Homocraspis* (Lindb. ex Schiffn.) Grolle

Vestigial marsupium lacking, gynoecium radially symmetrical at maturity of sporophyte and without rhizoids, perianth lacking.

9. M. adusta (Nees emend. Limpr.) Spruce, *Rev. Bryol.* 8: 98. 1881 (Fig. 67)
Gymnomitrion adustum Nees emend. Limpr.

Paroecious. Plants minute, in dull green to brownish-black patches. Shoots erect or ascending, 2–4(–6) mm, branching below, sterile shoots ± cylindrical, fertile clavate. Leaves on sterile stems ± similar in size along stem, on fertile stems increasing in size to bracts, subimbricate, appressed, concave, 0.20–0.45 mm long, 0.20–0.35 mm wide, 1.1–1.5 times as long as wide, bilobed $\frac{1}{10}$–$\frac{1}{5}$, lobes obtuse, sinus obtuse; marginal row of cells hyaline, mid-leaf cells 8–12 μm wide, walls thickened, trigones not distinct. Usually abundantly fertile. Male bracts larger and more concave than leaves, erect, sinus rounded, lobes obtuse. Female bracts erect, sinus shallow, rounded, lobes obtuse. Perianth absent. Capsules occasional, spring to autumn. Rocks, stones and soil, 400–1340 m, rare, N.W. Wales, Lake District, Dumfries, Lanark, Dumbarton, Perth and Angus north to Ross and W. Sutherland, Hebrides, S. Kerry, W. Mayo. 22. H2. Bulgaria, Alps, Tatra, Norway, Azores, Madeira.

The clavate, constantly present fertile shoots with appressed leaves are of characteristic appearance. The leaves of *M. brevissima* are of a somewhat similar shape to those of *M. adusta* and may cause confusion; the former is, however, a generally larger plant with somewhat secund leaves and larger leaf cells. Likely also to be confused with other *Marsupella* species with appressed, subimbricate leaves or with *Gymnomitrion* species. The *Marsupella* species differ in their acute leaf lobes and all except *M. condensata* have acute sinuses; the secund leaves with lunate sinuses of the latter are distinctive. All *Gymnomitrium* species have more closely imbricate, tightly appressed leaves, larger leaf cells and, except for *G. concinnatum*, a crenulate leaf margin.

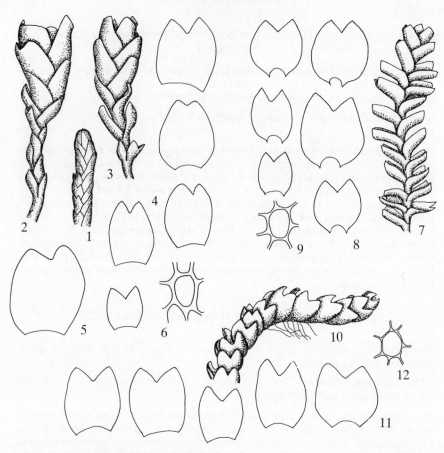

Fig. **67**. 1–6, *Marsupella adusta*: 1–3, sterile and mature and young fertile (paroecious) shoots
(× 27); 4, leaves (× 45); 5, male bract (× 45); 6, mid-leaf cell (× 450). 7–9, *M. alpina*: 7, shoot
(× 18); 8, leaves (× 27); 9, mid-leaf cell (× 45). 10–12, *M. brevissima*: 10, shoot (× 27); 11, leaves
(× 45); 12, mid-leaf cell (× 450).

10. M. brevissima (Dum.) Grolle, *J. Jap. Bot.* 40: 213. 1965   (Fig. 67)
Gymnomitrion varians (Lindb.) Schiffn., *M. varians* (Lindb.) K. Müll.

Paroecious, autoecious or apparently dioecious. Reddish-brown to blackish, dense,
often extensive patches. Shoots erect or ascending, 4–12 mm, branching from below
and innovating from below female inflorescence, fertile shoots clavate. Leaves of sterile
stems of ± equal size along stem, subimbricate, patent to erecto-patent, somewhat
secund, concave, 0.25–0.55mm long, 0.25–0.50 mm wide, 1.0–1.3 times as long as wide,
bilobed ⅛–¼(–⅓), lobes inflexed, acute to subobtuse but becoming rounded with age,
sinus acute to obtuse; mid-leaf cells 12–20 μm wide, walls thickened, yellowish-brown,
trigones ± distinct. Fertile plants common. Male bracts below female, on separate
branches or on apparently separate plants, larger and more concave than leaves, lobes
acute, sinus acute. Female bracts larger and more concave than leaves and male bracts.

Perianth absent. Capsules common, summer, autumn. Acidic soil on cliff ledges and on or near mountain summits especially in areas of late snow-lie, 850–1340 m, rare but sometimes locally abundant, Perth, Angus, S. Aberdeen, Banff, Inverness, Argyll, Ross, E. Sutherland. 14. Alps, Tatra, Bulgaria, Romania, Norway, Sweden, Finland, Faroes, Iceland, E. Himalayas, N. America, Greenland.

11. M. alpina (Gott. ex Limpr.) Bern., *Catal. Hep. Sud-Ouest Suisse*: 29. 1888

(Fig. 67)

Gymnomitrion alpinum (Gott. ex Limpr.) Schiffn.

Dioecious. Plants slender, in dense, dark reddish-brown tufts or patches. Shoots procumbent, (0.3–)1.0–3.0 cm, branching occasionally. Leaves of ± equal size along stem, distant to approximate when moist, patent to spreading from erect, sheathing, decurrent base, concave, 0.35–0.72 mm long, 0.32–0.72 mm wide, about as wide as long, bilobed ¼–⅓, lobes acute to obtuse, becoming rounded with age, somewhat inflexed, sinus acute to obtuse; mid-leaf cells 12-16 μm wide, walls thickened, reddish-brown, trigones distinct but not bulging. Fertile plants rare. Capsules unknown in Britain. On dry, sloping to vertical, usually acidic rocks, occasionally on soil in areas of late snow-lie, sea-level to 1200 m, rare in N.W. Wales, Lake District and Kirkcudbright, occasional from Dumbarton, Perth and Angus north to Ross and W. Sutherland. 19. Alps, Tatra, Pyrenees, Spain, Portugal, Norway, Sweden, Japan, Alaska, British Columbia.

Unlikely to be mistaken for any but small forms of *M. emarginata* or for *M. funckii* which, however, differ in leaf shape and larger leaf cells.

35. GYMNOMITRION CORDA IN OPIZ, *Naturalientausch* (12): 651. 1829

Dioecious. Small plants, usually in dense patches. Shoots julaceous, dorsiventrally compressed or not. Rhizoids colourless. Leaves closely imbricate, appressed, often difficult to discern, concave, ± transversely inserted, as long as or longer than wide, emarginate to bilobed, margin entire or crenulate but not toothed, often hyaline, leaves often eroded with age; cells thick-walled, marginal often hyaline, trigones present, sometimes bulging, marginal cells sometimes thin-walled and hyaline, oil-bodies 2–3 per cell. Underleaves lacking. Gemmae lacking. Male bracts similar to leaves or larger, saccate at base. Female bracts several pairs, larger than leaves, inner sometimes lobed. Perigynium vestigial, not tubular, after fertilisation constituting part of calyptra which bears unfertilised archegonia. Perianth lacking or rarely present and very short. Capsule spherical, wall 2–3-stratose. Spores 10–16 μm. About 15 species, mainly of dry, acidic, montane habitats.

1 Plants usually whitish, leaf lobes rounded **2. G. obtusum**
  Plants whitish or not, lobes acute or acuminate at least in young leaves 2
2 Leaf margin usually entire, not hyaline, cuticle finely papillose, mature sterile shoots
    300–600 μm wide **1. G. concinnatum**
  Leaf margin, at least in young leaves, crenulate, hyaline, cuticle smooth, shoots 160–
    320(–400) μm wide 3
3 Plants whitish to greyish, mature sterile shoots clavate, dorsiventrally compressed
    3. G. corallioides
  Plants green to brownish, mature sterile shoots ± cylindrical 4
4 Leaf lobe apices acute, marginal cells elongated with projecting apices, mature sterile
    shoots mostly 160–280 μm wide **4. G. crenulatum**
  Apices of youngest leaf lobes apiculate, marginal cells not elongated nor with
    projecting ends, shoots mostly 240–400 μm wide **5. G. apiculatum**

Subgenus *Gymnomitrion*

Perianth lacking.

1. G. concinnatum (Lightf.) Corda in Opiz, *Naturalientausch* (12): 651. 1829

(Fig. 68)

Yellowish-brown to dull greenish-white, dense patches. Shoots prostrate to erect, densely packed, 5–12(–25) mm, mature shoots ± cylindrical, 300–600 μm wide, branching occasionally, fertile shoots clavate, to 1000 μm wide. Rhizoids sparse. Leaves closely imbricate, appressed, individually discernible, very concave, margin not hyaline, entire or obscurely crenulate above, 0.30–0.85 μm long, 0.30–0.72 mm wide, bilobed $\frac{1}{8}$–$\frac{1}{3}$, lobes acute or occasionally obtuse, sinus acute; cells 16–24 μm wide in mid-leaf, thick-walled, trigones large, usually bulging, cells towards margin very incrassate, not hyaline, cuticle finely papillose. Fertile plants occasional. Capsules rare, summer. On dry, acidic to basic soil and rocks in sub-alpine and alpine habitats but descending to 75 m by the coast, frequent in Scotland, rare or occasional in suitable habitats elsewhere, E. Cornwall, N. Wales, Yorkshire and Westmorland north to Shetland, W. Galway, Donegal, Antrim. 40, H4. Alpine and arctic Europe from N. Spain, N. Italy and Bulgaria north to Spitzbergen, Faroes, Iceland, Asia Minor, Caucasus, Himalayas, Siberia, Japan, N. America, Greenland.

2. G. obtusum (Lindb.) Pears., *J. Bot. Br. Foreign* 18: 337. 1880   (Fig. 68)

Whitish to pale yellowish-green, rarely brownish, dense patches. Shoots prostrate to erect, densely packed, 4–20 mm, mature shoots ± cylindrical, 240–400 μm wide, branching occasionally, fertile shoots clavate, *ca* 600 μm wide. Leaves imbricate, appressed, individually discernible, very concave, margin and leaf apices hyaline, margin crenulate to base, 0.32–0.64 μm long, 0.24–0.56 μm wide, 1.0–1.3 times as long as wide, bilobed *ca* $\frac{1}{5}$, lobes rounded, sinus acute; cells 16–24 μm wide, moderately thick-walled, trigones medium-sized, bulging or not, cells towards margin not more heavily thickened, not hyaline except in upper part, cuticle finely papillose. Fertile plants occasional. Capsules rare. On dry, acidic to basic, subalpine and alpine rocks, rarely on soil, 400–1200 m, rare to occasional, sometimes locally common, in suitable habitats, E. Cornwall, N. Devon, North Wales and N.W. Yorkshire north to Sutherland and Caithness, rare in Ireland. 39, H9. Scattered localities from Spain, Portugal and C. Europe north to Norway, Sweden, Finland and Russia, Faroes, Caucasus, N. America, Greenland.

In *G. obtusum* the hyaline marginal and submarginal cells have similar walls to the inner cells and only differ in lack of chloroplasts and oil-bodies. In the succeeding species the hyaline marginal cells, in addition to lacking contents are thin-walled, contrasting with the thicker-walled inner cells. *G. obtusum* is distinct from all other species of the genus in that the upper leaves have rounded lobe apices, the lobes often being semicircular (older leaves in other species may appear similar because of erosion of the apices). In the field the whitish colour of *G. obtusum* usually distinguishes it from *G. concinnatum* which is mostly brownish; *G. corallioides* has clavate, dorsiventrally compressed shoots, and *G. crenulatum* is smaller and reddish- or purplish-brown.

3. G. corallioides Nees, *Naturg. Europ. Leberm.* 1: 118. 1833   (Fig. 68)

Patches, whitish to greyish, sometimes becoming black with age. Shoots arcuate to erect, densely packed, 4–12(–15) mm, mature shoots clavate, sometimes successionally so, dorsiventrally flattened, 180–320(–360) μm wide, branching frequently. Leaves closely imbricate, appressed, hardly discernible, very concave, uppermost leaves shortly bilobed, lobes acute, sinus acute, margin shallowly crenulate, marginal row of cells longer than wide, thin-walled, hyaline, in older leaves apex and margins becoming

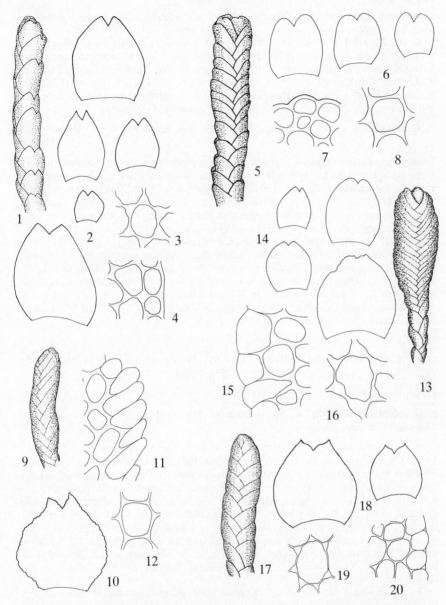

Fig. **68**. 1–4, *Gymnomitrion concinnatum*: 1, shoot (× 15); 2, leaves (× 27); 3, mid-leaf cell (× 450); 4, marginal cells near leaf apex (× 450). 5–8, *G. obtusum*: 5, shoot (× 15); 6, leaves (× 27); 7, marginal cells at leaf apex (× 450); 8, mid-leaf cell (× 450). 9–12, *G. crenulatum*: 9, shoot (× 27); 10, leaf (× 45); 11, marginal cells near leaf apex (× 450); 12, mid-leaf cell (× 450). 13–16, *G. corallioides*: 13, shoot (× 15); 14, leaves (× 27); 15, marginal cells from side of leaf (× 450); 16, mid-leaf cell (× 450). 17–20, *G. apiculatum*: 17, shoot (× 15); 18, leaves (× 27); 19, mid-leaf cell (× 450); 20, marginal cells near leaf apex (× 450).

eroded and upper part and ultimately almost whole leaf hyaline, leaves 0.4–0.9 mm long, 0.35–0.90 mm wide, (0.9–)1.0–1.3 times as long as wide, mid-leaf cells 16–28 μm wide, thick-walled, trigones large, bulging or not, cuticle smooth. Fertile plants rare. Capsules rare. On acidic, usually dry, basic soil and rocks in subalpine and alpine habitats, from *ca* 500 m, rare, Perth, S. Aberdeen, W. Inverness, Skye, Ross, W. Sutherland, Caithness, E. Donegal, formerly also Caernarfon, S. Kerry. 10, H2. Alpine and arctic Europe from the Pyrenees, Alps and Carpathians north to Spitzbergen, Faroes, Iceland, Caucasus, Siberia, Japan, N. America, Greenland.

4. G. crenulatum Gott. ex Carring., *Trans. Bot. Soc. Edinburgh* 7: 444. 1863

(Fig. 68)

Dark green or more usually reddish-brown to purplish-brown patches. Shoots filiform, wiry, arcuate, intricate, 3–7(–10) mm, mature shoots slightly dorsiventrally compressed, ± parallel-sided, (120–)160–280(–400) μm wide, branching frequently, fertile shoots clavate, to 500 μm wide. Rhizoids sparse. Leaves closely imbricate, appressed, barely discernible, very concave, margin crenulate, hyaline, 0.25–0.50 mm long, to 0.5 mm wide, longer than wide, bilobed $\frac{1}{5}$–$\frac{1}{4}$, lobes acute, sinus acute; mid-leaf cells (10–) 12–20(–24) μm wide, marginal cells hyaline, thin-walled, elongated with projecting apices, submarginal and inner cells thick-walled, trigones present, cuticle smooth. Fertile plants occasional. Capsules very rare. On acidic to slightly basic, usually dry subalpine and alpine rocks, from 60 m, very rare in Devon, S.W. Wales, occasional to frequent in N. Wales and the western Scottish highlands, rare in N. England and E. Scotland, scattered localities in Ireland, especially in the west. 32, H21. S.W. Norway, Spain, France, Portugal, Baffin Is.

Subgenus *Nardiocalyx* (Joerg.) S. Arn., *Illustr. Moss F. Fennoscand.* 1 (Hep.): 225. 1956

Perianth present, immersed in female bracts.

5. G. apiculatum (Schiffn.) K. Müll., *Hedwigia* 81: 113. 1942   (Fig. 68)
Marsupella apiculata Schiffn.

Green, often reddish-tinged patches or growing with other bryophytes. Shoots arcuate, 4–12(–15) mm, mature shoots cylindrical, 240–400 μm wide, branching frequently. Rhizoids sparsely present along stem. Leaves closely imbricate, appressed, hardly discernible when moist, very concave, margin entire, hyaline, 0.35–0.55 mm long, 0.32– 0.52 mm wide, about as long as wide, bilobed to $\frac{1}{3}$, lobes apiculate, apiculus composed of 2 cells, sinus acute, older leaves becoming eroded; mid-leaf cells 16–24 μm wide, marginal and apical cells thin-walled, hyaline, submarginal and inner cells thick-walled, trigones present, cuticle smooth. Capsules not known in Britain. Damp soil in areas of late snow-lie, above 1000 m, very rare, S. Aberdeen, Banff, Inverness. 4. Bulgaria, Alps, Tatra, Norway, Sweden, Finland, Iceland, Siberia, Japan, Alaska, Greenland.

  This very rare plant may occur mixed with small *Marsupella* species such as *M. condensata* with which it may be confused. It may be distinguished by the hyaline leaf margin as well as by the overall shape of the leaves.

15. SCAPANIACEAE

Usually dioecious. Plants very small to very robust, often with red, purple or brown pigmentation. Shoots dorsiventral, prostrate to erect, stems simple or branching laterally, rarely terminally, innovating from below female inflorescence, in section usually with a well defined cortex of 1–3 rows of small, thick-walled cells. Rhizoids scattered. Leaves distichous, succubous, bilobed, usually conduplicate, lobes non-decurrent to longly decurrent, margins usually denticulate or dentate, keel 0.2–0.8 length of ventral lobe, usually sharp, often winged, very rarely shorter or absent; cells usually with trigones, cuticle often papillose, oil-bodies (1–)2–10 per cell. Underleaves lacking. Gemmae commonly present in fascicles at tips of upper leaves, 1–2-celled. Male inflorescence terminal, becoming intercalary, bracts ± similar to or smaller than leaves, saccate with (1–)2–4(–7) axillary antheridia with uniseriate stalks. Female inflorescence terminal, bracts ± similar to leaves but larger, bracteoles and perigynium lacking. Perianth dorsiventrally flattened, smooth, frequently deflexed above, mouth wide, truncate, entire to ciliate-dentate, rarely subterete or terete, plicate above and narrowed to mouth. Seta in section of numerous similar cells. Capsule usually ovoid, wall (3–)5–6(–7)-stratose. Spores 8–22 μm. Seven genera mainly distributed in the cooler parts of the Northern Hemisphere.

1 Leaf lobes lanceolate, tapering to acute apex, waxy cuticle present
                                                 36. Douinia (p. 167)
  Leaf lobes narrowly lingulate to ± orbicular, not tapering to apex, waxy cuticle lacking   2
2 Ventral leaf lobe narrowly lingulate to lingulate, ± parallel-sided, gemmae stellate, perianth plicate above, narrowed to mouth   **37. Diplophyllum (p. 168)**
  Ventral lobe various, rarely lingulate or parallel-sided, gemmae ellipsoid, ovoid or pyriform, perianth usually smooth, truncate at mouth   **38. Scapania (p. 172)**

SUBFAMILY DOUINIOIDEAE

Leaves conduplicate-concave, not keeled, lobes lanceolate, tapering to apex; waxy cuticle present. Perianth terete, plicate above, narrowed to mouth. Elaters with annular or unispiral thickening. One species.

36. DOUINIA (C. JENS.) BUCH, *Commentat. Biol. Soc. Sci. Fenn.* 3(1): 13. 1928

A monotypic genus with the characters of the species.

1. D. ovata (Dicks.) Buch, *Commentat. Biol. Soc. Sci. Fenn.* 3(1): 14. 1928 (Fig. 69)

Dioecious. Plants olive-green, bluish-green or brownish, in patches or growing through other bryophytes. Shoots procumbent with ascending apices or ascending, stems sparsely branched. Lower leaves distant, upper larger, imbricate, conduplicate-concave, transversely inserted, bilobed, margins entire or obscurely toothed, dorsal lobe smaller, narrower and more sharply pointed than ventral and ± at right angles to and appressed to it; ventral lobe ovate-lanceolate to lanceolate, tapering to acute apex, (0.3–)0.5–1.3 mm long; cells in middle of ventral lobe 16–24(–32) μm wide, walls thickened, trigones small or lacking, rarely large, cuticle waxy. Underleaves absent. Gemmae lacking. Male inflorescence terminal, spicate, becoming intercalary, bracts 3–6 pairs, imbricate, smaller and more symmetrical than leaves, saccate. Female bracts larger and more symmetrical than leaves, margins entire to toothed. Perianths occasional to frequent, ovoid, deeply plicate almost to base, mouth ciliate. Capsules globose, wall 3-stratose, occasional to frequent, spring to autumn. Spores 12–15 μm. Tree bark (usually *Quercus* or *Betula*), sheltered rocks and rock faces, crevices and scree, ascending to *ca* 900 m, rare in W. Britain from Cornwall to Kirkcudbright, Derby, N.W. Yorkshire, N. Northumberland, rare to occasional from Ayr and Stirling

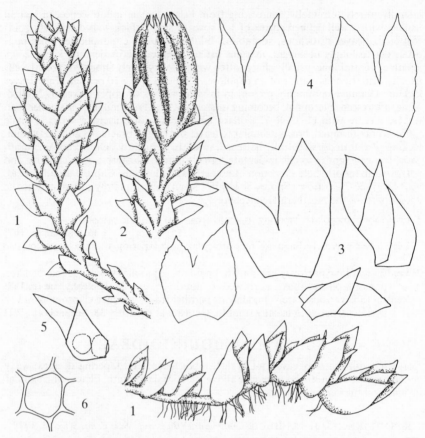

Fig. **69**. *Douinia ovata*: 1, shoots (× 21); 2, perianth (× 21); 3, leaves (× 38); 4, female bracts (× 27); 5, male bract (× 38); 6, mid-leaf cell (× 450).

north to Shetland, rare in Ireland. 32, H13. W. Europe from N. Portugal and N. Spain, north to Norway, S. Sweden and S. Finland, Faroes, Japan, western N. America.

SUBFAMILY SCAPANIOIDEAE

37. DIPLOPHYLLUM (DUM.) DUM., *Recueil Observ. Jungerm.*: 15. 1835

Plants small to medium-sized, in patches, turfs or scattered. Stems prostrate to ascending, in section with distinct small-celled cortex, branches sparse, lateral. Leaves distant to imbricate, succubous, transversely inserted, decurrent or not, bilobed, lobes conduplicate, keel rounded below and sheathing stem, sharp above, dorsal lobe similar to and 0.3–0.7 length of ventral lobe, margins usually denticulate; cell walls thickened, trigones lacking, oil-bodies 2–16 per cell, cuticle often papillose. Underleaves lacking. Gemmae often present, bluntly stellate, unicellular. Male bracts ± similar to leaves or smaller, base saccate. Perianth slightly to strongly dorsiventrally compressed, plicate above, narrowed to mouth. Capsule ellipsoid, wall 4-stratose. A mainly northern hemisphere genus of about 20 species.

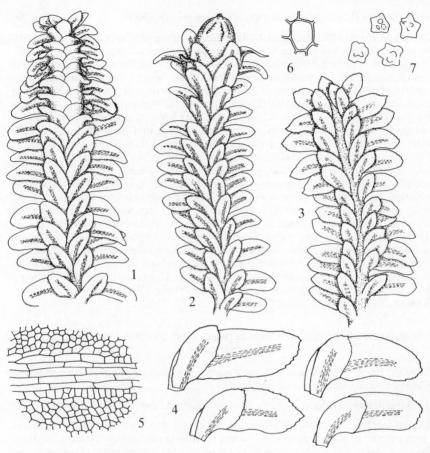

Fig. **70**. *Diplophyllum albicans*: 1–3, male, female and sterile shoots (× 13); 4, leaves (× 18); 5, vitta and surrounding cells in middle of ventral lobe of leaf (× 150); 6, leaf cell (× 450); 7, gemmae (× 450).

1 Leaf lobes with conspicuous vitta of elongated cells (3–8 times as long as wide) in middle of ventral lobe   **1. D. albicans**
  Leaf lobes without a vitta (cells in middle of ventral lobe ± 1–2 times as long as wide)
      2
2 Dioecious, ventral leaf lobes mostly 2.0–2.5 times as long as wide, oil-bodies coarsely papillose   **2. D. taxifolium**
3 Paroecious, ventral lobes mostly 1.6–2.3 times as long as wide, oil-bodies smooth
      3. D. obtusifolium

Subgenus *Diplophyllum*

Stem in section with cortex of very thick-walled cells. Leaf lobes with vitta of very elongated cells.

1. D. albicans (L.) Dum., *Recueil Observ. Jungerm.*: 16. 1835   (Fig. 70)

Dioecious. Plants yellowish-green to dark green or brownish, occasionally tinged red, or whitish, in patches, tufts or turfs. Shoots procumbent to erect, to *ca* 5 cm, stems occasionally branched, innovating from below perianth. Rhizoids very sparse. Leaves distichous, distant to subimbricate, dorsal lobe 0.3–0.7 length of and of similar shape to ventral lobe, appressed to stem and ventral lobe; ventral lobe narrowly lingulate to ovate-oblong, sometimes curved, 0.5–1.7 mm long, 0.2–0.6 mm wide, 2.0–3.3 times as long as wide, margin toothed at least above, apex acute to obtuse and apiculate or obtuse or rounded; keel strongly curved; both lobes with a band of elongated cells forming a distinct vitta, vitta cells (28–)50–90 μm long, (9–)12–16 μm wide, (2.5–)3.0–8.0 times as long as wide in middle of ventral lobe, other cells ± isodiametric, oil-bodies coarsely papillose, trigones lacking, cuticle ± smooth. Gemmae frequent on tips and margins of upper leaves, yellowish-green, bluntly stellate, unicellular, 14–16 μm. Male inflorescence spicate, usually reddish-purple, bracts closely imbricate with erect saccate basal part and strongly recurved ventral lobe. Female bracts with erect basal part and recurved ventral lobe. Perianths frequent, narrowly ovoid to narrowly obovoid, plicate above, narrowed to ciliate mouth. Capsules frequent except in drier parts of Britain, winter, spring. $n = 8, 9$. Peaty, sandy and soil banks in woods and sheltered habitats, on rocks, cliffs and in scree, heaths and moorland, calcifuge, from sea-level to 1340 m, common and sometimes locally abundant. 109, H39, C. Europe from Spain, Portugal and Greece north to Spitzbergen, Faroes, Iceland, Turkey, Kamchatka, Korea, Japan, Taiwan, N. America, Greenland, Hawaii.

Subgenus *Protodiplophyllum* Schust., *Hep. Anthoc. N. Amer.* 3: 192. 1974

Stem in section with cortical cells only slightly thickened and poorly differentiated. Cells in middle of leaf not forming vitta.

2. D. taxifolium (Wahlenb.) Dum., *Recueil Observ. Jungerm.*: 16. 1835   (Fig. 71)

Dioecious. Green to yellowish-brown patches or mixed with other bryophytes. Shoots procumbent to ascending, to 2 cm, stems branching occasionally, innovating from below perianth. Rhizoids very sparse. Leaves distichous, approximate to subimbricate, dorsal lobe of similar shape to and appressed to ventral lobe and *ca* 0.7 its length, ventral lobe narrowly lingulate, frequently curved, (0.4–)0.7–1.2 mm long, (0.2–)0.3–0.5 mm wide, 2.0–2.5 times as long as wide, entire, crenulate-denticulate or toothed especially where gemmiferous, apex rounded or more rarely acute or obtuse; cells in middle of ventral and dorsal lobes somewhat elongated but not forming a vitta, in middle of ventral lobe 16–32 μm long, 12–16 μm wide, 1.0–2.0 times as long as wide, trigones lacking, oil-bodies coarsely papillose, cuticle papillose. Gemmae occasional on margins of upper leaves, bluntly stellate, unicellular, 12–16 μm. Perianths occasional, narrowly ovoid to ellipsoid, plicate above, mouth shallowly lobed, crenulate. Capsules not known in Britain. $n = 8 + m$. On moist soil in rock crevices in sheltered or exposed habitats, calcifuge, usually above 1000 m but descending to 600 m in N. England, rare in the Scottish highlands, very rare elsewhere, Caernarfon, Westmorland, Argyll, Perth and Angus north to Sutherland. 15. Arctic and alpine Europe from Spain, Corsica, Italy and Yugoslavia north to Spitzbergen, Iceland, Turkey, Siberia, Sakhalin, Korea, Japan, N. America, Greenland.

Similar in appearance to, and sometimes growing mixed with *D. albicans* which differs in the presence of a vitta in the leaf lobes.

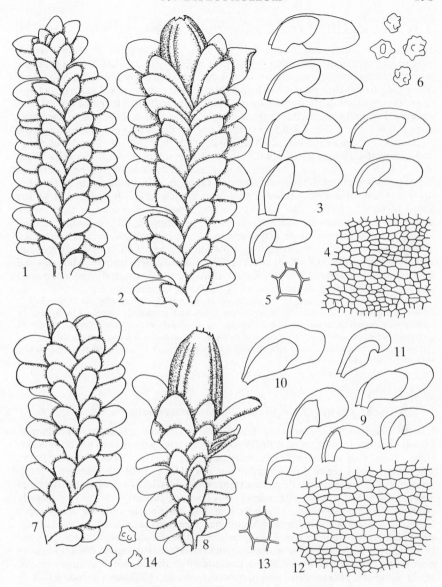

Fig. **71**. 1–6, *Diplophyllum taxifolium*: 1 and 2, sterile and female shoots (× 15); 3, leaves (× 23); 4, cells from middle of ventral leaf lobe (× 150); 5, leaf cell (× 450); 6, gemmae (× 450). 7–14, *D. obtusifolium*: 7 and 8, sterile and female shoots (× 15); 9, leaves (× 23); 10, female bract (× 23); 11, male bract (× 23); 12 cells from middle of ventral lobe (× 150); 13, leaf cell (× 450); 14, gemmae (× 450).

3. D. obtusifolium (Hook.) Dum., *Recueil Observ. Jungerm.*: 16. 1835   (Fig. 71)

Paroecious or autoecious. Small, pale green to yellowish-green or reddish-purple patches or growing through other bryophytes, especially *D. albicans*. Shoots prostrate with ascending apices to ascending, to *ca* 1 cm, stems sparsely branched, only rarely innovating from below perianth. Rhizoids numerous to stem apex. Leaves distichous, approximate to imbricate, dorsal lobe of similar shape to 0.3–0.7 length of ventral lobe, appressed or not to ventral lobe; ventral lobe lingulate or broadly lingulate, sometimes curved, 0.4–1.1 mm long, 0.2–0.5 mm wide, 1.6–2.3 times as long as wide, margin entire to denticulate, apex rounded, rarely apiculate; cells in middle of ventral and dorsal lobes somewhat elongated but not forming a vitta, in middle of ventral lobe 20–32 μm wide, 14–16 μm wide, 1.2–2.1 times as long as wide, cell walls thickened, trigones lacking, oil-bodies smooth, cuticle papillose. Gemmae rare, bluntly stellate, unicellular, 12–14 μm. Male bracts below female or at branch tips, 2–3(–5) pairs, of similar shape to leaves but base saccate. Female bracts similar to but larger than leaves. Perianths very common, narrowly ellipsoid to ellipsoid, plicate above, mouth shallowly lobed, crenate-ciliate. Capsules common, spring. On soil on banks, by tracks, in quarries, very rarely on rotting logs, from sea-level to 350 m, widely scattered but rare from Cornwall and W. Kent north to Shetland and absent from the Midlands and E. Anglia, rare in Ireland. 49, H9. Rare in Europe from Spain, Italy and Yugoslavia north to Fennoscandia, Japan, western N. America, Greenland.

Possibly overlooked because of frequently occurring mixed with *D. albicans* from which it differs in lack of vitta, relatively broader leaf lobes and numerous rhizoids. In Britain *D. obtusifolium* occurs at much lower altitudes than *D. taxifolium* with which it has been confused and from which it differs in the somewhat broader leaf lobes and almost constantly present paroecious inflorescences readily recognised by the saccate male bracts below the perianth and smooth oil-bodies.

38. SCAPANIA (DUM.) DUM., *Recueil Observ. Jungerm.*: 14. 1835

Usually dioecious. Plants small to robust, pale green to deep reddish-brown or purple. Shoots prostrate to erect, often with deflexed apices, arising from creeping rhizomatous stems, stem with or without differentiated cortex. Leaves bilobed, conduplicate, spreading, usually keeled, keel winged or not, dorsal lobe usually smaller than ventral, margins entire to dentate or ciliate, decurrent or not; cells thin- to thick-walled, trigones absent to large and bulging, oil-bodies (1–)2–12 per cell, cuticle smooth to very coarsely papillose. Gemmae frequent, sometimes abundant, on margins and lobe tips of upper leaves, 1–2-celled. Male inflorescence terminal, becoming intercalary, or intercalary, bracts similar to or smaller than leaves, saccate at base. Female inflorescence terminal, bracts usually larger and more strongly toothed or ciliate than leaves. Perianth usually dorsiventrally flattened, often deflexed, truncate with entire to dentate or ciliate mouth, rarely subterete, plicate above and narrowed to mouth. Capsules ± ovoid, wall 3–7-stratose. Spores 8–20(–22) μm. A taxonomically difficult, mainly northern genus of about 60 species found mainly in moist or sheltered habitats but with a few species extending into the tropics.

Many *Scapania* species exhibit great phenotypic plasticity, colour is often influenced by habitat, and juvenile forms may differ from mature plants, these factors sometimes making identification very difficult. The following key is based on typical, mature plants and determinations should be confirmed by comparison with descriptions. Not all specimens can be named with certainty. Where I have not seen fresh material number of oil-bodies per cell has been taken from Buch, *Soc. Sci. Fenn. Comm. Biol.* **3**(1), 1–177, 1928.

1 Leaves with keel 0.2 or more length of ventral lobe*, margins entire or dentate
    (ciliate-dentate in *S. nemorea*)   2
  Leaves divided to or almost to base, margins ciliate-dentate or spinose-dentate 24
2 Plants to 1.5(–2.0) cm, leaves with both lobes acute and serrate, often coarsely so,
    ± from widest part   **18. S. umbrosa**
  Plant size various, leaves not as above   3
3 Cuticle coarsely papillose, dorsal lobes of leaves usually crossing stem   4
  Cuticle smooth to slightly papillose, if more strongly papillose then dorsal lobes not
    crossing stem   5
4 Dorsal lobe of leaf standing away from stem, lobe base not decurrent, gemmae 16–
    27(–30) μm long   **19. S. aequiloba**
  Dorsal lobe at most reflexed in older leaves, bases decurrent, gemmae 28–40 μm
    long   **20. S. aspera**
5 Leaf lobes subequal with dorsal lobe mostly 0.8–1.0 length of ventral lobe   6
  Dorsal lobe markedly smaller than ventral, 0.4–0.8 its length   8
6 Dorsal lobe crossing stem, keel 0.4–0.6 length of ventral lobe   **14. S. subalpina**
  Dorsal lobe not crossing stem, keel 0.5–0.8 length of ventral lobe   7
7 Plants yellowish-brown to reddish-brown, leaf lobes ± rounded-quadrate, keel 0.7–
    0.8 length of ventral lobe   **1. S. compacta**
  Plants whitish-green to brownish-green, leaf lobes ± ovate-rectangular, keel 0.5–0.7
    length of ventral lobe   **4. S. cuspiduligera**
8 Ventral lobe of leaf not decurrent to below insertion of keel   9
  Ventral lobe of leaf decurrent to below insertion of keel   18
9 Dorsal lobes of leaves narrow, mostly 1.5–2.5 times as long as wide (q.v. Fig. 1), oil-
    bodies large, to 20 μm long, filling cell lumens in middle of leaf lobes, plants
    calcicole   10
  Dorsal lobes wider, oil-bodies smaller, not filling cell lumens, calcicole or not 11
10 Oil-bodies 1 per cell, total area of dorsal lobe of leaf $\frac{1}{4}$–$\frac{1}{3}$ total area of ventral lobe
                                                                      2. S. gymnostomophila
  Oil-bodies 2–4(–5) per cell, area of dorsal lobe *ca* $\frac{1}{2}$ area of ventral lobe
                                                                           3. S. calcicola
11 Plants small, 2–8(–10) mm, ventral lobe of leaves mostly 1.5–2.0 times as long as
    wide (q.v. Fig. 1)   12
  Plants larger, usually 1–6 cm, ventral lobe of leaves mostly 1.0–1.5 times as long as
    wide   16
12 Marginal cells towards apex of ventral lobe 20–28 μm wide (measured parallel to
    leaf edge), cells in mid-lobe with 6–10 oil-bodies, perianth mouth dentate
                                                                             9. S. lingulata
  Marginal cells 12–24 μm wide, cells in mid-lobe with 2–6 oil-bodies, perianth mouth
    entire, dentate or ciliate-dentate   13
13 Marginal 1–2 rows of cells of ventral leaf lobe uniformly thickened, differing from
    inner cells with trigones   14
  Marginal cells with trigones, similar to inner cells   15
14 Marginal cells towards apex of ventral leaf lobe mostly 12–16 μm wide, all with oil-
    bodies, perianth mouth entire   **5. S. parvifolia**
  Marginal cells 16–24 μm, some or all without oil-bodies, perianth mouth dentate
                                                                                 6. S. curta
15 Keel 0.25–0.50 length of ventral lobe, gemmiferous leaf lobes sometimes with apical
    finger-like teeth, perianth mouth entire or with sparse teeth   **7. S. scandica**

*See Fig. 1 for measurement of leaf dimensions.

Keel 0.5–0.7 length of ventral lobe, gemmiferous leaves without finger-like teeth, perianth mouth ciliate-dentate   **8. S. praetervisa**

16 Dorsal leaf lobe not or hardly crossing stem, keel ± straight to weakly arched at least in most leaves, gemmae green   **10. S. irrigua**
   Dorsal lobe widely crossing stem, keel strongly arched, gemmae green to purplish-red or yellowish-brown   17

17 Keel 0.25–0.40 length of ventral lobe, lobes not decurrent below keel insertion, calcicole   **11. S. degenii**
   Keel 0.2–0.3 length of ventral lobe, both lobes often decurrent below keel insertion, calcifuge   **12. S. paludicola**

18 Keel strongly arcuate to semicircular, 0.2–0.3 length of ventral lobe (q.v. Fig. 1), ventral lobe shortly decurrent to just below keel insertion   **12. S. paludicola**
   Keel not strongly arcuate or semicircular or if so then dorsal and ventral lobes longly decurrent, keel 0.2–0.6 length of ventral lobe   19

19 Leaf margins entire to dentate or if coarsely dentate then plants green to reddish or purple   20
   Leaf margins coarsely toothed and plants yellowish-brown to olive-green or margins ciliate-dentate   23

20 Leaf lobes often subequal, dorsal lobe 0.7–1.0 length of ventral lobe, keel 0.4–0.6 length of ventral lobe, gemmae 20–40 μm long   **14. S. subalpina**
   Dorsal lobe smaller than ventral, 0.4–0.7 its length, keel 0.20–0.45 length of ventral lobe, gemmae 13–24 μm   21

21 Dorsal lobe rounded-quadrate to ovate-reniform, not decurrent
                                                                                                                                                      13. S. undulata
   Dorsal lobe reniform to cordate, decurrent to below insertion of keel   22

22 Keel curved, plants frequently reddish to purple   **15. S. uliginosa**
   Keel arcuate to semicircular, plants usually pale green   **16. S. paludosa**

23 Leaf margins ciliate-dentate, gemmae in brownish clusters, unicellular, thin-walled
                                                                                                                                                      17. S. nemorea
   Leaf margins coarsely toothed, gemmae in greenish clusters, 1–2-celled, thick-walled   **21. S. gracilis**

24 Dorsal leaf lobes pointing ± to stem apex, apex rounded to acute, ventral lobe not or hardly reflexed   **22. S. ornithopodioides**
   Dorsal leaf lobes not pointing to stem apex, apex acute, with spinose tooth, ventral lobe strongly reflexed   **23. S. nimbosa**

Subgenus *Jensenia* S. Arn., *Ill. Moss Fl. Fenn.* 1: 173. 1956

Dioecious or monoecious. Cortical cells of stem grading into inner cells. Leaves transversely inserted, lobes ± equal, hardly conduplicate, entire or finely and sparsely denticulate, not decurrent, without a sharp keel; marginal cells not thick-walled, trigones small to medium-sized, cuticle smooth or weakly papillose. Gemmae 2-celled. Perianth compressed or subterete, mouth entire or denticulate.

1. S. compacta (A. Roth) Dum., *Recueil Observ. Jungerm.*: 14. 1835   (Fig. 72)

Paroecious, autoecious or dioecious. Plants small to medium-sized, yellowish-brown to reddish-brown, in patches or mixed with other bryophytes. Shoots prostrate with ascending tips to ascending, 0.5–2.0 cm, stems rarely branched. Lower leaves distant, upper approximate to imbricate, dorsal lobe 0.8–1.0 size of and of similar shape to ventral, not or barely crossing stem, reflexed to inflexed, rounded-quadrate, not decurrent, apex obtuse to rounded, margin entire; ventral lobe 0.6–1.6 mm long, not

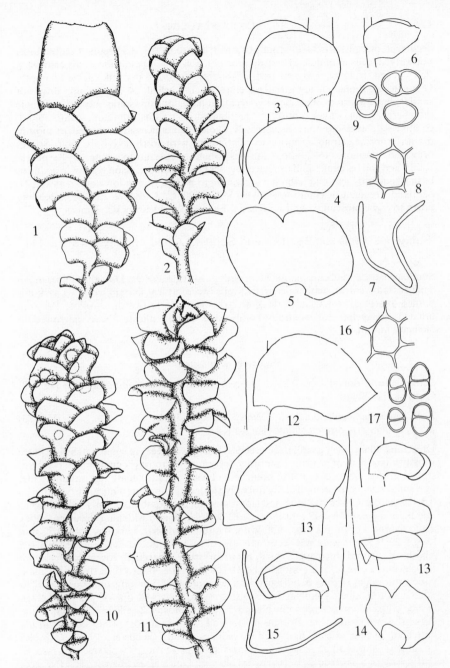

Fig. **72**. 1–9, *Scapania compacta*: 1, fertile shoot (× 21); 2, sterile shoot (× 21); 3, leaves, dorsal side (× 21); 4, leaf, ventral side (× 21); 5, flattened leaf (× 21); 6, male bract (× 21); 7, section of leaf through keel (× 33); 8, mid-leaf cell (× 450); 9, gemmae (× 450). 10–17, *S. cuspiduligera*: 10 and 11, male and sterile shoots (× 15); 12, leaf, ventral side (× 21); 13, leaves, dorsal side (× 21); 14, flattened leaf (× 21); 15, section of leaf through keel (× 33); 16, cell from middle of ventral lobe (× 450); 17, gemmae (× 450).

decurrent, margin entire to obscurely toothed; keel (0.6–)0.7–0.8 length of ventral lobe, not winged; cells in middle of ventral lobe (20–)24–28(–32) μm wide, walls brownish, ± thickened, trigones medium-sized, oil-bodies small, 2–7 per cell, cuticle ± smooth. Gemmae occasional, yellowish-green, ellipsoid, 1–2-celled, 24–30 μm long. Bracts of similar shape to leaves, male slightly saccate at base, lobe tips sometimes acute, female more erect than leaves, entire to denticulate. Perianths common, ovate-obloid, compressed, wide-mouthed to narrowly ovoid, subterete and narrowed to mouth, mouth truncate, sinuose, entire to denticulate with teeth 1–2 cells long. Spores brownish, papillose, 16–20 μm. Capsules common, autumn to spring. $n = 8 + m^*$, 9. Rocks and soil on banks, cliffs, crevices in rockfaces, in quarries, on heaths, in acidic to basic, frequently exposed habitats, ascending to 850 m, rare in lowland areas and N. Scotland, frequent or common elsewhere. 95, H20, C. S. and W. Europe, extending north to S. Fennoscandia and N. Russia, Faroes (?), Macaronesia, Tunisia.

Subgenus *Kaalaasia* Buch emend Schust., *Amer. Midl. Nat.* 49(2): 441. 1953

Small, prostrate, calcicole plants. Stem with unistratose cortex. Dorsal leaf lobe much smaller than ventral, neither decurrent; cells except at leaf margin with 1–5 dark oil-bodies, marginal not usually thick-walled. Gemmae brownish, 2-celled. Perianth inflated, somewhat dorsiventrally compressed or not, plicate above, narrowed to laciniate mouth.

Section *Kaalaasia*

Leaves entire, dorsal lobe 0.25–0.35 area of ventral lobe; oil-bodies one per cell. Gemmae 12–32 μm.

2. S. gymnostomophila Kaal., *Bot. Notis.* 1896: 21. 1896　　　　　　　　(Fig. 73)
Diplophyllum gymnostomophilum (Kaal.) Kaal.

Dioecious. Plants very small, dark green to brownish-green, in small patches or more usually mixed with or growing over other bryophytes. Shoots procumbent, 0.3–1.0 cm, stems simple or occasionally branched. Leaves approximate to subimbricate, of ± equal size along stem, usually dorsally subsecund, margins entire, dorsal lobe *ca* 0.5–0.7 length of ventral lobe and appressed to it, rhomboidal, only extending to middle of stem, 0.24–0.48 mm long, (0.12–)0.16–0.28 mm wide, not decurrent, apex acute to obtuse; ventral lobe subfalcate, ovate-rectangular, concave, 0.44–0.72 mm long, 0.20–0.52 mm wide, not decurrent, apex obtuse to rounded; keel ± arched, 0.3–0.5 length of ventral lobe; cells opaque when fresh, marginal row smaller, more translucent, in middle of ventral lobe 16–24 μm wide, thin-walled, trigones small, cuticle smooth or slightly papillose, oil-bodies large, to 20 μm long, only one per cell in middle of leaf lobes. Gemmae in reddish-brown clusters at tips of younger leaves, ovoid, ellipsoid or occasionally bluntly angular, mostly 2-celled, 12–24 μm long. Perianths not known in Britain or Ireland, obovoid, somewhat dorsiventrally flattened, plicate above, narrowed to ciliate mouth (Schuster, 1974). Capsules unknown. Damp soil on ledges and in crevices of basic rock, 75–600 m, very rare, E. Perth, Angus, S. Aberdeen, E. Inverness, Skye, E. Donegal, E. Mayo. 4, H2. Pyrenees and Alps north to Spitzbergen, Iceland, Siberia, N. America, Greenland.

　　May occur mixed with *S. calcicola* or *S. cuspiduligera* but differing in the small narrow dorsal leaf lobes which give the plants the appearance of a *Diplophyllum*. The large, usually single oil-

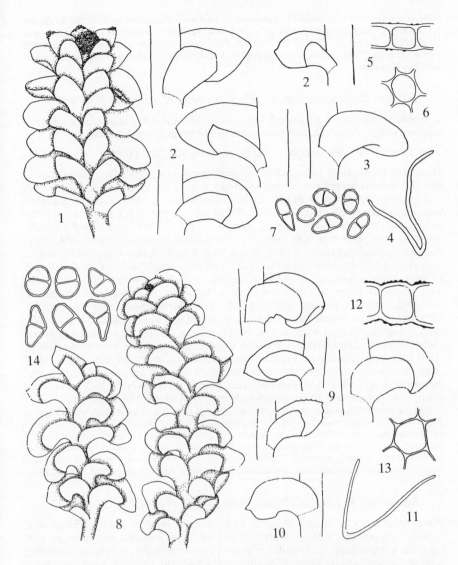

Fig. **73**. 1–7, *Scapania gymnostomophila*: 1, shoot (× 21); 2, leaves, dorsal side (× 38); 3, leaf, ventral side (× 38); 4, section of leaf through keel (× 63); 5, leaf cells in T.S. (× 450); 6, cell from middle of ventral lobe (× 450); 7, gemmae (× 450). 8–14, *S. calcicola*: 8, shoots (× 18); 9, leaves, dorsal side (× 30); 10, leaf, ventral side (× 30); 11, section of leaf through keel (× 45); 12, leaf cells in T.S. (× 450); 13 cell from middle of ventral lobe (× 450); 14, gemmae (× 450).

body in the leaf cells is an absolute distinguishing character. According to Buch (1928) the oil-bodies of *S. gymnostomophila* persist for as long as 10 years after collection. However, Paton & Duckett (*Bot. J. Linn. Soc.* **37**, 177–87, 1970) say that very few oil-bodies persisted after 12 months drying. The use of the persistence or otherwise of oil-bodies as a taxonomic character must be treated with caution.

Section *Calcicolae* Schust., *Hep. Anthoc. N. Amer.* 3: 305. 1974

Upper leaves often with a few small teeth, dorsal lobe 0.4–0.6 area of ventral lobe; oil-bodies mostly 2–5 per cell. Gemmae green to brownish, 25–40 μm long.

3. S. calcicola (H. Arn. & J. Perss.) Ingham, *Naturalist* 561: 11. 1904   (Fig. 73)

Dioecious. Plants small, pale green to yellowish-green, in small patches or more usually mixed with other bryophytes. Shoots ascending, 0.5–1.5 cm, simple or rarely branched. Leaves approximate to imbricate, of ± equal size along stem, margins entire or uppermost leaves with unicellular teeth, dorsal lobe 0.5–0.8 length of ventral lobe, flat or convex, rhomboidal, not crossing stem, 0.4–0.6(–0.8) mm long, 0.24–0.40(–0.48) mm wide, not decurrent, apex obtuse or obtuse and apiculate; ventral lobe ovate, subfalcate, usually concave, 0.32–1.16 mm long, 0.32–0.68 mm wide, not decurrent, apex rounded to obtuse and apiculate; keel weakly to strongly arched, 0.35–0.55 length of ventral lobe, sometimes winged near insertion; cells opaque when fresh, in middle of ventral lobe 20–28 μm wide with 2–4(–5) oil-bodies filling cell lumens, walls thickened, brownish, trigones present, marginal cells not differentiated, cuticle ± papillose. Gemmae in green to pale brown clusters, pyriform, ovoid or ellipsoid, occasionally bluntly angular, 2-celled, thick-walled, 20–40 μm long, 16–20(–24) μm wide. Only male plants known in Britain. Perianth ovoid, inflated, plicate above, narrowed to lobed, denticulate mouth (Schuster, 1974). Capsules unknown. $n = 18$. In rock clefts, on cliff ledges and damp soil in basic, montane habitats, 50–1000 m, very rare, Brecon, Perth, Inverness, Argyll, Skye. 8. Alpine and arctic Europe from Spain, N. Italy and Greece north to Fennoscandia, Russia and Spitzbergen (?), Iceland, Newfoundland.

Usually in small quantity growing through mosses and sometimes mixed with *Scapania aequiloba*, *S. cuspiduligera* and *S. gymnostomophila* but only likely to be confused with small forms of *S. aequiloba*. It differs from these in its neat habit, the dorsal lobe of the leaf not crossing the stem and in colour and size of gemmae. *S. gymnostomophila* differs in the relatively smaller dorsal leaf lobe. For a detailed account of *S. calcicola* see Perry, *Trans. Br. bryol. Soc.* **5**, 137–244, 1967, and for a comparison with *S. aequiloba* see Paton & Duckett, *Bot. J. Linn. Soc.* **63**, 177–87, 1970.

Subgenus *Buchiella* Schust., *Amer. Midl. Nat.* 43(3): 592. 1949

Plants small, calcicole. Stem with cortex of 1–2 rows of small, markedly differentiated cells. Leaves subequally bilobed, lobes spreading, entire, ventral lobe decurrent, keel rounded; marginal cells thick-walled, trigones present, oil-bodies 2–5 per cell. Gemmae reddish-brown, 2-celled. Perianth dorsiventrally flattened, truncate, mouth entire.

4. S. cuspiduligera (Nees) K. Müll., *Rabenh. Krypt.-Fl. Deutschl.* ed. 2, 6(2): 427. 1915   (Fig. 72)
S. bartlingii (Hampe) Nees

Dioecious. Plants small, whitish-green to brownish-green, in patches or growing through other bryophytes. Shoots prostrate to ascending, 0.3–2.0 cm, stems simple, rarely branched. Leaves distant to subimbricate or imbricate, undulate in more vigorous forms, margins entire, dorsal lobe subequal to or slightly smaller than ventral lobe, 0.8–1.0 its length, often reflexed, not crossing stem, ovate to ovate-rectangular, 0.4–1.7 mm long, 0.2–1.0 mm wide, not decurrent, apex rounded or obtuse, sometimes apiculate in upper leaves, scarcely decurrent; ventral lobe ± similar in shape to dorsal, sometimes reflexed, 0.3–1.4 mm long, 0.16–0.90 mm long, decurrent, apex rounded; keel 0.5–0.7 length of dorsal lobe, ± straight, not sharp or winged; cells in middle of ventral lobe 16–24 μm wide, with 2–6 oil-bodies per cell, walls ± thickened, trigones distinct, marginal cells thick-walled, apical cells of upper leaves sometimes elongated, cuticle ± smooth. Gemmae brownish, rarely greenish-white, ellipsoid, ovoid or pyriform, (1–)2-celled, thick-walled, (16–)20–40 μm long, 12–24 μm wide. Male bracts closely imbricate, strongly saccate. Perianths very rare, dorsiventrally flattened, truncate, mouth entire (Schuster, 1974). Capsules unknown in Britain or Ireland. $n = 9$. On soil on rock ledges, in rock crevices and in short turf in basic habitats, 100–600 m, rare, scattered localities from S. Wales, Shropshire and Derby north to S. Harris, W. Sutherland and Caithness, Sligo. 25, H1. Arctic and alpine Europe from the Pyrenees and Italian and Transylvanian Alps north to Spitzbergen, Iceland, east to Novaya Zemlya and Siberia, Japan, Africa, N. America, Greenland.

Subgenus *Scapania*

Plants minute to robust. Stem with 1–5-stratose cortex. Leaves with keel 0.2–0.8 length of ventral lobe, dorsal lobe usually smaller than ventral, margins usually toothed. Gemmae 1–2-celled. Perianth strongly dorsiventrally compressed, smooth, mouth truncate, entire to ciliate-dentate.

Section *Curtae* (K. Müll.) Buch, *Soc. Sci. Fenn. Comm. Biol.* 3(1): 55. 1928

Plants small or very small. Stem with 1–2-stratose cortex. Leaf lobes narrowed to base, neither decurrent, margins entire or with small teeth, dorsal lobe of similar shape to ventral but wider, ventral lobe ovate, obovate or lingulate 1.2–2.2 times as long as wide; marginal cells differentiated or not, oil-bodies 4–12 per cell, cuticle faintly papillose.

5. S. parvifolia Warnst., *Hedwigia* 63: 78. 1921   (Fig. 74)

Dioecious. Plants very small, greenish, gregarious or mixed with other bryophytes, especially *Scapania* sect. *Curtae* species. Shoots prostrate to ascending, 3–6 mm, stems very rarely branched. Leaves approximate to imbricate, dorsal lobe ± appressed to ventral lobe and ± 0.7 its length, not crossing stem, ovate, 0.44–0.70 mm long, 0.36–0.44 mm wide, not decurrent, margin entire, apex pointed, sometimes apiculate; ventral lobe ovate-lanceolate, slightly curved, 0.6–1.0 mm long, 0.44–0.66 mm wide, not decurrent, margin entire or sometimes with a few small, scattered teeth, in gemmiferous shoots apical finger-like teeth sometimes present, apex rounded, sometimes apiculate; keel slightly winged, 0.35–0.45 length of ventral lobe; walls of 1–2 rows marginal cells uniformly and frequently heavily thickened, 12–16(–20) μm wide (parallel to leaf edge) towards apex, all containing small oil-bodies, other cells with distinct trigones, 12–20(–24) μm wide with 2–5 oil-bodies per cell in middle of ventral lobe. Gemmae bright green, ellipsoid, 2-celled, 12–15 μm long, 12–15 μm wide (Buch, 1928), seasonal. Perianths unknown in Britain, mouth entire, marginal cells with uniformly thickened walls (Buch, 1928). Capsules unknown. In small quantity on acidic soil in montane turf and on cliff ledges, ascending to 1000 m, very rare, Perth, W. Inverness, Argyll. 4. Rare in arctic and alpine Europe from the Alps north to Spitzbergen, Iceland, Siberia, Hokkaido, Quebec, Greenland.

Differing from other British species of section *Curtae* in the smaller leaf cells, from all except *S. curta* in the marginal 1–2 rows of leaf cells lacking trigones. The mouth of the perianth is entire but it may also be entire in *S. scandica*; in the other species the mouth is dentate or ciliate-dentate. First reported from Scotland by Buch (1928) and discussed by Jones, *Trans. Br. bryol. Soc.* 4, 463–4, 1963.

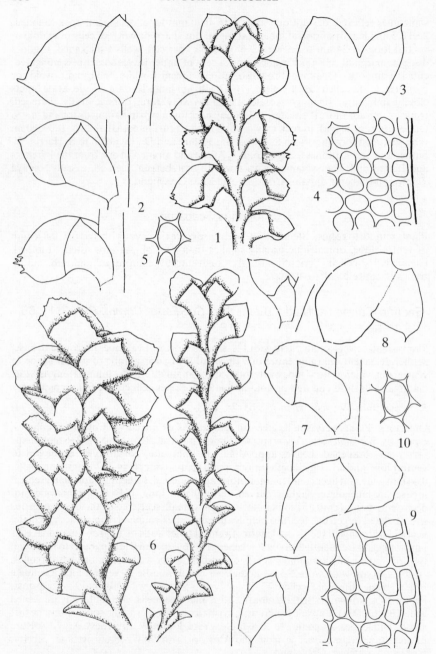

Fig. **74**. 1–5, *Scapania parvifolia*: 1, shoot (× 21); 2, leaves, dorsal side (× 38); 3, flattened leaf (× 38); 4, leaf margin cells (× 450); 5, cell from middle of ventral lobe (× 450). 6–10, *S. curta*: 6, shoots (× 21); 7, leaves, dorsal side (× 38); 8, flattened leaf (× 38); 9, leaf margin cells (× 450); 10, cells from middle of ventral lobe (× 450).

6. S. curta (Mart.) Dum., *Recueil Observ. Jungerm.*: 14. 1835 (Fig. 74)

Dioecious. Plants very small, yellowish-green, gregarious or growing with other bryophytes. Shoots prostrate, 2–6 mm, stems very rarely branched. Leaves distant to imbricate, dorsal lobe appressed to ventral and 0.6–0.8 its length, not crossing stem, ovate, 0.28–0.56 mm long, 0.16–0.44 mm wide, not decurrent, margin entire, apex pointed; ventral lobe plane or slightly concave, obovate, 0.40–0.84 mm long, 0.24–0.56 mm wide, not decurrent, margin entire or with sparse, mostly 1-celled teeth, apex rounded and frequently apiculate; keel slightly arched, 0.5–0.6 length of ventral lobe; walls of 1–2 rows marginal cells uniformly and frequently heavily thickened, (12–)16–24 μm wide, some or all without oil-bodies, other cells with small but distinct trigones, 16–24 μm wide with 2–4(–5) oil-bodies in middle of ventral lobe. Gemmae colourless, ellipsoid, 2-celled, 20–24 μm long, 7–12 μm wide (Buch, 1928). Perianths very rare in Britain, dorsiventrally compressed or not, mouth dentate, teeth 1–3 cells long, marginal cells thickened (Buch, 1928). $n = 16 + 2m*$. In small quantity on soil in woodland or turf at low altitudes, very rare but probably under-recorded, E. Cornwall, N. Hants, Sussex, W. Kent, Yorkshire, N. Lancashire, E. Inverness, W. Mayo. 10, H1. Europe north to Spitzbergen, Iceland, Siberia, Japan, Azores, Madeira, Tenerife, Gomera, N. America, Greenland.

In the absence of perianths, which are not known in Britain in *S. parvifolia*, that species may only be distinguished from *S. curta* by the smaller leaf cells and relatively shorter keel. In *S. parvifolia* and other species of sect. *Curtae* all the marginal and submarginal cells contain oil-bodies whilst they are absent from most or all of the marginal cells in *S. curta*. However, this cannot be regarded as a totally reliable character as oil-bodies tend to disintegrate rapidly and may only be absent from marginal cells as the result of drying whether before or after collection. Further, the marginal cells may not be uniformly thickened, making the separation of *S. curta* from the next two species difficult if not impossible in the absence of perianths. Most early British and Irish records of *S. curta* refer to *S. scandica*.

7. S. scandica (H. Arn. & Buch) Macv. *Stud. Handb. Br. Hep.* ed. 2: 394. 1926
(Fig. 75)

Dioecious. Plants small, green, sometimes tinted purplish-red, in patches or mixed with other bryophytes. Shoots prostrate to ascending, 2–10 mm, stems unbranched. Leaves distant to imbricate, dorsal lobe appressed to ventral lobe and 0.5–0.7(–0.8) its length, not crossing stem, ovoid to rhomboidal, (0.16–)0.40–0.90 mm long, (0.12–)0.24–0.56 mm wide, not decurrent, margin entire, apex rounded or more usually pointed; ventral lobe plane to concave, sometimes with narrowly erect margins, obovate or obovate-lingulate, (0.32–)0.64–1.40 mm long, (0.16–)0.40–0.72 mm wide, not decurrent, margin entire or in gemmiferous shoots with finger-like teeth at apex, apex rounded, rarely mucronate; keel slightly arched or more usually ± straight, 0.3–0.5 length of ventral lobe; marginal cells with trigones, similar to inner cells, 16–24 μm wide towards apex, oil-bodies present, cells in middle of ventral lobe 16–24 μm wide with 2–5(–6) oil-bodies, trigones small, distinct. Gemmae green to reddish-purple, ellipsoid to ovoid, 2-celled, 14–24 μm long, 8–12 μm wide. Male bracts 1–6 pairs, strongly saccate. Perianths strongly compressed, mouth shallowly lobed or sinuose, entire or occasionally with sparse 1(–3)-celled teeth, marginal cells similar to lower cells. Capsules unknown. On peaty or mineral soil, rock ledges, on banks, on moorland, in exposed or sheltered habitats, ascending to *ca* 1100 m, occasional in W. and N. Britain, E. Cornwall, S. Somerset, S. Wales, Derby and Yorkshire north to Shetland, rare in Ireland. 52, H7. Montane and arctic Europe from N. Spain, N. Italy and Romania north to Fennoscandia and N. Russia, Faroes, Iceland, N. America, E. Greenland.

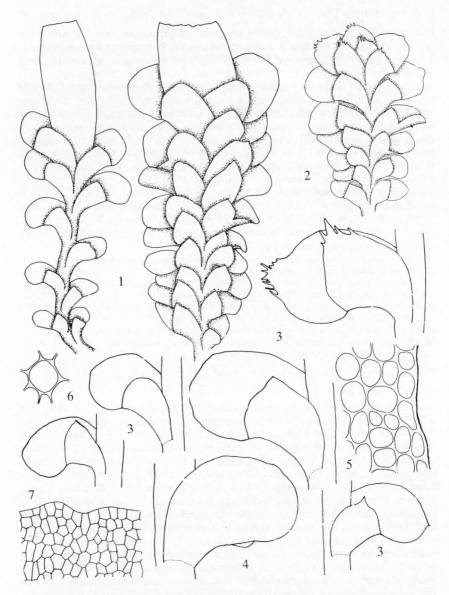

Fig. **75**. *Scapania scandica*: 1 and 2, fertile and gemmiferous shoots (× 21); 3, leaves, dorsal side (× 38); 4, leaf, ventral side (× 38); 5, leaf margin cells (× 450); 6, cell from middle of ventral lobe (× 450); 7, perianth mouth (× 230).

The most frequent species of section *Curtae* and because of morphological variability likely to be confused with the other species. *S. scandica* and *S. parvifolia* which are regarded by some authorities as synonymous differ from other species of the section in the ± entire perianth mouth, the relatively short keel and the gemmiferous leaves sometimes producing apical finger-like teeth. *S. parvifolia* differs from *S. scandica* in smaller leaf cells and uniformly thickened marginal cells. For differences from *S. curta*, *S. praetervisa* and *S. lingulata* see under those species. The presence or absence of small marginal teeth in the leaves, used by some authors to differentiate between species of section *Curtae*, is not a satisfactory character. They may at times be present in any of the five British species and although reputed to be present in some species only on gemma-producing leaves, it is clear from observations on *S. parvifolia* and *S. scandica* gemmae are not always present on leaves with marginal teeth.

8. S. praetervisa Meyl., *Jahresber. Naturf. Ges. Graubundens N.F.* 61: 364. 1926

(Fig. 76)

S. mucronata auct. angl., *S. mucronata* ssp. *praetervisa* (Meyl.) Schust., *S. mucronata* var. *arvernica* (Culm.) K. Müll., *S. mucronata* var. *praetervisa* (Meyl.) Buch

Dioecious. Plants very small, yellowish-green to brown, gregarious or growing with other bryophytes. Shoots prostrate, 4–8 mm, stems occasionally branched. Leaves imbricate, dorsal lobe 0.6–0.8 length of ventral lobe, not crossing stem, ovate, 0.5–1.1 mm long, 0.24–0.64 mm wide, not decurrent, margin entire, apex pointed; ventral lobe concave, standing away from dorsal lobe, ovate, 0.6–1.2 mm long. 0.36–0.80 mm long, not decurrent, margin entire or with a few 1(–2)-celled teeth, apex rounded and usually mucronate; keel curved, 0.5–0.7 length of ventral lobe; marginal cells with trigones, similar to inner cells, 12–20 μm wide towards apex, oil-bodies present, cells in middle of ventral lobe 16–24 μm wide with (2–)3–6(–7) oil-bodies, trigones distinct. Gemmae green to reddish, ovoid to ellipsoid, 2-celled. 18–28(–30) μm long, 10–16 μm wide. Male bracts strongly saccate. Perianth weakly dorsiventrally compressed, often plicate above, mouth lobed, ciliate-dentate, teeth 1–8 cells long, marginal cells not thickened. Capsules unknown in Britain. *n* = 9*. Usually in small quantity on moist shaded soil and rocks by streams and rivers in base-rich habitats below 200 m(?), very rare, M.W. Yorkshire, Mid Perth, Kincardine, Moray. 4. Rare in arctic and alpine Europe from Germany and C. Europe north to Spitzbergen, Iceland, Novaya Zemlya, Alaska, N. Greenland.

The mouth of the perianth distinguished *S. praetervisa* from other British species of section *Curtae*, but in the sterile state it cannot always be distinguished from certainty from *S. curta* or *S. scandica*. For the occurrence of *S. praetervisa* in Britain see Perry, *Trans. Br. bryol. Soc.* **4**, 785–9, 1965 (as *S. mucronata*) and Paton, *J. Bryol.* **11**, 409–10, 1981. *S. praetervisa* is very closely related to and possibly not specifically distinct from *S. mucronata* Buch.

9. S. lingulata Buch, *Meddeland Soc. Faun. Flor. Fenn.* 42: 90, 92, 94. 1916 (Fig. 76)
S. microphylla Warnst.

Dioecious. Plants small, green to brownish, gregarious or growing with other bryophytes. Shoots prostrate to ascending, 2–6 mm, stems unbranched. Leaves distant to imbricate, dorsal lobe 0.5–0.7 length of ventral lobe, not crossing stem, ovate to rhomboidal, 0.24–0.60 mm long, 0.16–0.48 mm wide, not decurrent, margin entire, apex pointed; ventral lobe concave, often with narrowly erect margin, standing away from dorsal lobe, ovate to lingulate, 0.4–1.0 mm long, 0.24–0.60 mm wide, not decurrent, margin entire or toothed towards apex in gemmiferous shoots, apex rounded occasionally mucronate; keel slightly arched, 0.4–0.5 length of ventral lobe; marginal cells with trigones similar to inner cells, (16–)20–28(–32) μm wide towards apex, oil-bodies present, cells in middle of ventral lobe (16–)20–28 μm wide with (4–)6–10(–12) oil-bodies, trigones distinct, bulging. Gemmae green to purplish-red, ovoid or

Fig. **76**. 1–9, *Scapania praetervisa*: 1 and 2, sterile and fertile shoots (× 15); 3, leaves, dorsal side (× 24); 4, leaf, ventral side (× 24); 5, flattened leaf (× 24); 6, perianth mouth (× 170); 7, leaf margin cells (× 450); 8, cell from middle of ventral lobe (× 450); 9, gemmae (× 450). 10–16, *S. lingulata* 10, shoot (× 15): 11, leaves, dorsal side (× 38); 12, leaf, ventral side (× 38); 13, flattened leaf (× 38); 14, leaf margin cells (× 450); 15, cell from middle of ventral lobe (× 450); 16, gemmae (× 450).

ellipsoid, (1–)2-celled, (15–)20–32(–36) μm long, (10–)12–20 μm wide. Male bracts 1–3 pairs slightly saccate. Perianths occasional, strongly compressed, mouth weakly lobed, dentate, teeth 1–3(–4) cells long (Paton, 1981). Capsules unknown. $n = 27*$. Usually in small quantity, on soil on banks and on rotting logs, usually in acidic habitats, ascending to ca 560 m, rare, Brecon, Anglesey, E. Perth, Angus, Moray, E. Inverness, Argyll, Skye, E. Ross, N. Tipperary, W. Mayo. 10, H2. Rare in Europe from France, Germany and C. Europe north to Fennoscandia and N. Russia, Iceland, Novaya Zemlya, N. America, Greenland.

Distinguished from other members of section *Curtae*, with species of which it sometimes grows mixed, by the larger leaf cells, more numerous oil-bodies and larger gemmae. For the occurrence of this plant in Britain and Ireland see Paton, *J. Bryol.* **11**, 399–403, 1981.

Section *Irriguae* (K. Müll.) Buch, *Soc. Sci. Fenn. Comm. Biol.* 3(1): 84. 1928

Plants small to medium-sized, of moist or wet habitats. Stem with 1–2-stratose cortex. Dorsal lobe of leaf 0.5–0.7 length of ventral, not or scarcely decurrent below insertion of keel, margins entire or with small teeth, ventral lobe ovate-oblong, 0.85–1.35 times as long as wide, abruptly narrowed to base, marginal cells not differentiated, cuticle ± papillose. Gemmae 1–2-celled.

10. S. irrigua (Nees) Nees in Gottsche et al., *Syn Hep.*: 67, 1844   (Fig. 77)

Dioecious. Plants small to medium-sized, often flaccid, yellowish-green, rarely brown, in tufts or mixed with other bryophytes. Shoots ascending, (0.5–)1.0–5.0 cm, stems rarely branched. Leaves distant or approximate below, subimbricate or imbricate above, margins entire or uppermost leaves with a few sparse teeth near the apex, dorsal lobe not or scarcely crossing stem, ± convex, 0.5–0.7(–0.8) length of ventral lobe, obliquely rectangular to cordate-triangular, 0.5–1.4 mm long, (0.3–)0.4–1.1 mm wide, not decurrent, apex obtuse to acute, sometimes apiculate, ventral lobe obovate to broadly obovate, sometimes reflexed, 0.8–1.8 mm long, 0.4–1.2 mm wide, 1.2–1.5 times as long as wide, not or rarely decurrent almost to insertion of keel, apex rounded to acute, sometimes apiculate; keel ± straight to weakly arched, strongly arched in occasional leaves, (0.3–)0.4–0.5(–0.7) length of ventral lobe, not winged; marginal cells not differentiated from inner cells, cells in middle of ventral lobe 20–28 μm wide with 2–5 oil-bodies, trigones small but distinct. Gemmae green, ellipsoid to ovoid, (1–)2-celled, 16–22 μm long, 10–14 μm wide. Male bracts slightly smaller than leaves with blunter lobes, slightly saccate. Perianths very rare, mouth sparsely dentate to ciliate, teeth to 4 cells long (Schuster, 1974). Capsules very rare, winter. On damp, especially clayey or sandy soil on tracks and woodland rides, on damp ground on heaths and moorland, by streams and flushes and in damp turf, in acidic habitats, frequent or common in the wetter parts of Britain and Ireland, rare elsewhere. 108, H32. Europe from the Pyrenees, southern Alps and Yugoslavia to Fennoscandia and Russia, Faroes, Iceland, Caucasus, Siberia, Sakhalin, Japan, N. America, Greenland.

11. S. degenii Schiffn. ex. K. Müll in K. Müll., *Rabenh. Krypt.-Fl. Deutschl.* ed. 2, 6(2): 497. 1915   (Fig. 78)

Dioecious. Plants medium-sized, greenish-brown to yellowish-brown, in tufts or mixed with other bryophytes. Shoots 2–5 cm, stems rarely branched. Leaves subimbricate, lower entire, upper usually with sparse 1–2-celled teeth, dorsal lobe convex, widely crossing stem, 0.6–0.7 length of ventral lobe, frequently inflexed, reniform to broadly cordate-triangular, 0.9–1.4(–1.7) mm long, 0.7–1.1(–1.5) mm wide, not decurrent, apex

186   15. SCAPANIACEAE

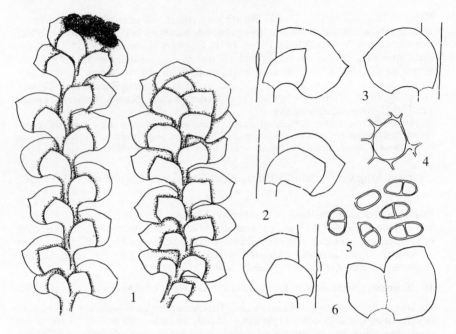

Fig. 77. *Scapania irrigua*: 1, shoots with and without gemmae (× 8); 2, leaves, dorsal side (× 15); 3, leaf, ventral side (× 15); 4, cell from middle of ventral lobe (× 450); 5, gemmae (× 450); 6, flattened leaf (× 15).

obtuse to rounded; ventral lobe convex, often reflexed, broadly ovate to obliquely cordate, 1.3–1.8(–2.3) mm long. 1.0–1.5(–1.9) mm wide, not decurrent, apex obtuse, sometimes minutely apiculate in upper leaves; keel usually strongly arched, (0.20–) 0.25–0.40 length of dorsal lobe; marginal cell walls ± more heavily thickened than in inner cells, cells in middle of ventral lobe 20–32 μm wide, with 2–8 oil-bodies, walls yellowish to yellowish-brown, trigones large and sometimes bulging. Gemmae in dark brown clusters, ovoid, 1(–2)-celled, 12–20 μm long, 10–12 μm wide. Perianth plicate above, mouth truncate, sinuose-lobed, entire or with small sparse teeth. Capsules unknown. In flushes and on damp cliff ledges in basic habitats (30–)500–1000 m, very rare except in the central Scottish highlands, N.W. Yorkshire, Durham, Cumberland, Perth, S. Aberdeen, Moray, W. Inverness, Argyll, Skye, E. Sutherland, Shetland. 13. France, Austria, Czechoslovakia, Switzerland, Poland, Romania, Norway, Sweden, Finland (?), N. Russia, Iceland, Siberia, Alaska, eastern N. America, Greenland.

Likely to be confused with *S. irrigua* and *S. paludicola*. Differs from both in being a calcicole and having more heavily thickened marginal cells, and from *S. irrigua* in the usually strongly arched keel, the dorsal lobe crossing the stem and the brownish, mostly unicellular gemmae. *S. paludicola* differs in habitat and the more strongly arched, relatively shorter keel, and in both leaf lobes often decurrent to below insertion of keel. For an account of *S. degenii* see Paton, *Trans. Br. bryol. Soc.* 5, 83–5, 1966.

12. S. paludicola Loeske & K. Müll. in K. Müll., *Rabenh. Krypt.-Fl. Deutschl.* ed. 2, 6(2): 415. 1915   (Fig. 78)

Dioecious. Plants medium-sized, yellowish-green, in small tufts or scattered amongst *Sphagnum*. Shoots ascending, to 6 cm. Leaves distant to approximate, clasping stem,

Fig. **78**. 1–6, *Scapania degenii*: 1 and 2, sterile and female shoots (× 5); 3, leaves, dorsal side (× 13); 4, leaf, ventral side (× 13); 5, cell from middle of ventral lobe (× 450); 6, gemmae (× 450). 7–10, *S. paludicola*: 7, shoot (× 10); 8, leaves, dorsal side (× 15); 9, leaf, ventral side (× 15); 10, gemmae (× 450).

entire or with a few sparse unicellular teeth, dorsal lobe widely crossing stem, convex, 0.4–0.5(–0.6) length of ventral lobe, reniform to ovate-cordate, 0.5–1.0 mm long, 0.5–0.9 mm wide, often decurrent to below insertion of keel, apex rounded to obtuse and apiculate; ventral lobe rounded-quadrate to ovate-cordate, 1.2–1.8 mm long, 1.0–1.5 mm wide, usually decurrent to below insertion of keel, apex rounded and minutely apiculate to acute; keel strongly arched to ± semicircular, 0.2–0.3 length of ventral lobe; marginal cells not differentiated from inner cells, cells in middle of ventral lobe 20–32 μm wide, with 2–5 oil-bodies; trigones medium-sized. Gemmae sometimes present, in reddish-brown clusters, ellipsoid to ovoid, 1–2-celled, 14–22(–30) μm long, 10–14 μm wide. Only male plants known in Britain, bracts ± saccate. Perianth truncate with ciliate-dentate mouth, teeth 1–4 cells long. Capsules unknown. $n = 8 + m$, 9. In *Sphagnum* in bogs and wet heaths, calcifuge, ascending to *ca* 360 m, very rare, E. Cornwall, Merioneth, Denbigh, M.W. Yorkshire. 4. France, N. Italy, Romania and Bulgaria north to Spitzbergen, Iceland, Siberia, Sakhalin, Japan, northern N. America, Greenland.

Likely to be confused with *S. irrigua* or *S. paludosa*. In *S. irrigua* the keel is longer relative to the ventral lobe, and although some leaves on some plants may have a strongly arched keel other leaves on such plants have a weakly arched or ± straight keel as is usually found in most plants. The gemmae of *S. paludicola* differ from those of *S. irrigua* in frequently being unicellular; en masse they are in reddish-brown clusters whilst those of *S. irrigua* form dark green clusters. The leaf shape and strongly arched keel give *S. paludicola* a strong superficial resemblance to *S. paludosa*. The latter differs, however, in habitat, the longly decurrent leaf bases and lack of gemmae. For the occurrence of *S. paludicola* in Britain see Paton, *J. Bryol.* **11**, 405–7, 1981.

Section *Scapania*

Plants usually large, stem cortex 1–5-stratose. Leaf margins entire to toothed, dorsal lobe 0.8–1.3 times as long as wide, decurrent or not, apex rounded, ventral lobe usually larger, 0.80–1.25 times as long as wide, longly decurrent, apex rounded, keel 0.20–0.65 length of ventral lobe, if winged wings not toothed; marginal cells thick-walled or not, oil-bodies 3–10 per cell. Gemmae (1–)2-celled.

13. S. undulata (L.) Dum., *Recueil Observ. Jungerm.*: 14. 1835   (Fig. 79)
S. dentata Dum., *S. intermedia* (Husn.) Pears.

Dioecious. Plants medium-sized to robust, yellowish-green to dark green, reddish or purple, rarely brownish, in tufts, cushions or patches, sometimes very extensive. Shoots erect or ascending, 1–10 cm, sometimes denuded below, stems sparingly branched. Leaves distant to imbricate, sometimes ventrally secund, flaccid or not, undulate or not, crisped or not when dry, margins entire or more usually dentate, sometimes coarsely so, dorsal lobe not crossing to widely crossing stem, convex, 0.5–0.7 length of ventral lobe, rounded-quadrate to ovate-reniform, 0.5–2.0 mm long, 0.4–1.75 mm wide, not decurrent, apex rounded to obtuse, rarely ± acute; ventral lobe often deflexed, ovate, ovate-rectangular or ovate-orbicular, 0.9–3.2 mm long, 0.5–2.6 mm wide, distinctly decurrent, ventral margin often deflexed, apex rounded to very obtuse; keel ± straight to curved, winged, 0.20–0.45 length of ventral lobe; marginal cells not differentiated from inner cells, cells in middle of ventral lobe (16–)20–28 μm wide with 2–6 oil-bodies, walls thin-walled to uniformly thickened, colourless to deep red or purple, rarely brownish, trigones present or not. Gemmae occasionally present in greenish clusters, ellipsoid to pyriform, very thin-walled, 1–2-celled, 18–24 μm long, 11–14 μm wide. Male bracts similar to leaves but saccate at base. Perianths occasional to frequent, smooth or contracted and plicate towards mouth, truncate, mouth sinuose, entire to dentate. Capsules occasional, winter, spring. $n = 8 + m*$, 9. On rocks and tree roots in and by streams, rivers and lakes, often submerged, on wet rocks on cliffs, in ravines, waterfalls, in bogs and flushes, on damp soil on banks and paths, in open or shaded, acidic habitats, occasional in lowland areas, common to very common elsewhere, ascending to *ca* 1300 m. 103, H37, C. Europe north to Spitzbergen, Faroes, Iceland, Turkey, Ukraine, Korea, Sakhalin, Japan, Tunisia, Macaronesia, N. America, Greenland.

Very variable and treated by Macvicar (1926) as three species and a number of varieties but these all intergrade to such an extent as to render taxonomic recognition impossible. It is clear from field observation and literature reports that much of the variation is environmentally induced.

14. S. subalpina (Nees ex Lindenb.) Dum., *Recueil Observ. Jungerm.*: 14. 1835

(Fig. 80)

Dioecious. Plants small to medium-sized, often silt-encrusted, in green to reddish-brown tufts. Shoots ascending to erect, 1–4 cm, stems rarely branched, innovating from below perianth. Lower leaves distant to approximate, often eroded, upper approximate to imbricate, flaccid or not, sometimes undulate, dorsal lobe smaller than to subequal to ventral lobe, (0.6–)0.7–1.0 its length, widely crossing stem, reflexed or not, ovate-orbicular. 0.7–2.0 mm long, 0.7–1.8 mm wide, not decurrent; ventral lobe ovate-orbicular, 0.9–3.0 mm long, 0.7–2.4 mm wide, decurrent, apex rounded or obtuse; keel 0.4–0.6 length of ventral lobe, ± straight, sharp, sometimes winged; cells in middle of ventral lobe 20–28 μm wide with 4–6(–8) oil-bodies, walls thickened or not, marginal cells usually more thick-walled, trigones small to medium-sized, cuticle slightly papillose. Gemmae green, ellipsoid, ovoid or pyriform, (1–)2-celled, 20–40 μm long, 16–24 μm wide. Male bracts slightly smaller than leaves, markedly saccate. Female bracts larger than leaves, lobes subequal or dorsal smaller. Perianths rare to occasional, dorsiventrally compressed, truncate, mouth denticulate. Capsules rare to occasional, spring, summer. On alluvial detritus and wet rocks by streams and rivers, ascending to *ca* 1000 m rare, in S.W. England and S. Wales, occasional to frequent in N. Wales, N. England, Scotland and Ireland, absent from lowland Britain. 59, H13. Portugal, Spain, Italy and Greece north to Spitzbergen, Faroes, Iceland, Siberia, Kamchatka, Japan, Madeira, N. America, Greenland.

A variable species compact forms of which are likely to be confused with *S. compacta* with which it sometimes grows. It differs in the more deeply divided leaves with a sharp rather than rounded keel, in the more rounded leaf lobes, the dorsal lobe widely crossing the stem and in the marginal leaf cells frequently being heavily thickened. *S. compacta* is paroecious and sometimes has a subterete perianth contracted at the mouth. Large, lax, semi-aquatic forms with the dorsal leaf lobe somewhat smaller than the ventral may be difficult to distinguish from *S. undulata* and although it has a longer keel and larger gemmae some plants cannot be named with certainty.

15. S. uliginosa (Sw. ex Lindenb.) Dum., *Recueil Observ. Jungerm.*: 14. 1835

(Fig. 81)

S. obliqua (Arnell) Schiffn.

Dioecious. Plants robust, green or more usually reddish to deep reddish-purple, in often swelling tufts or mixed with other bryophytes. Shoots erect, to 10 cm, stems usually branched. Leaves distant below, subimbricate to imbricate above, flaccid, crisped when dry, margins entire or occasionally ventral lobe with a few small 1–2-celled teeth, dorsal lobe scarcely to widely crossing stem, convex, 0.3–0.6(–0.7) length of ventral lobe, reniform to cordate, 0.5–1.6(–2.2) mm long, 0.5–1.3(–2.0) mm wide, decurrent, apex rounded to very obtuse, rarely ± pointed; ventral lobe strongly deflexed, ovate to rounded-quadrate, 1.4–2.9(–3.5) mm long, 1.2–2.5(–2.9) mm wide, decurrent, apex rounded or very obtuse; keel curved, not arcuate, 0.2–0.4 length of

Fig. **79**. *Scapania undulata*: 1, sterile shoots (× 9); 2, female shoot (× 6); 3, leaves, dorsal side (× 13); 4, leaf, ventral side (× 13); 5, perianth mouth (× 140); 6, cell from middle of ventral lobe (× 450); 7, gemmae (× 450).

ventral lobe, winged; marginal cells occasionally more heavily thickened than other cells, cells in middle of ventral lobe 24–32(–36) μm wide with 3–4 oil-bodies, walls thin or very thin, coloured or not, trigones absent or very small. Gemmae rare, greenish to reddish, ellipsoid, unicellular, 13–19 μm long, 8–10 μm wide (Buch, 1928). Male plants occasional, bracts smaller than leaves, saccate, Perianths occasional, mouth entire or sparsely denticulate. Capsules very rare. Flushes, springs and on rocks in streams, (275–)550–1550 m, Hereford, very rare in N.W. Wales and N. England, occasional in Scotland, very rare in Ireland. 26, H6. Alpine and subarctic Europe from Spain (?), N. Italy and Romania north to Fennoscandia and N. Russia, Iceland, eastern N. America, Aleutian Is., Greenland.

Zehr, *Bryoph. Biblltheca* **15**, 1–140, 1980, reduces *S. paludosa* to synonymy with *S. uliginosa* on the grounds that the two taxa intergrade in all the characters used to separate them. He provides no statistical or experimental data in support of this but, although the two taxa are sometimes difficult to separate, they are discrete. *S. uliginosa* is often deeply reddish-tinged and the leaf keel is

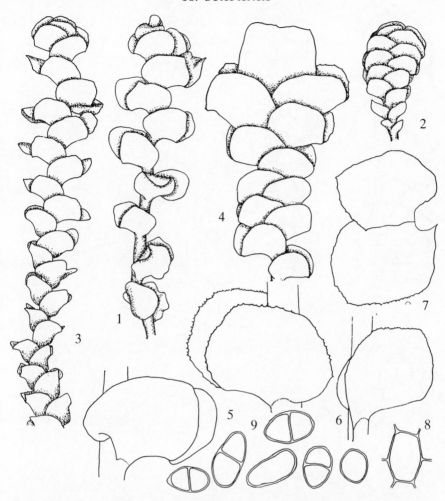

Fig. **80**. *Scapania subalpina*: 1, shoot of lax plant (× 8); 2, shoot of compact plant (× 8); 3 and 4, male and female plants (× 8); 5, leaves, dorsal side (× 21); 6, leaf, ventral side (× 21); 7, flattened leaf (× 21); 8, cell from middle of ventral lobe (× 450); 9, gemmae (× 450).

not arcuate, *S. paludosa* is, at the most, only lightly reddish-tinted and the keels, or at least some, are strongly arcuate to semicircular. Both species may be confused with *S. undulata* with which they sometimes grow, but differ in the longly decurrent dorsal leaf bases.

16. S. paludosa (K. Müll.) K. Müll., *Bull Herb. Boiss.* ser. 2, 3: 40. 1903   (Fig. 81)

Dioecious. Plants usually robust, yellowish-green to pale green or pinkish-tinged, sometimes reddish below, in tufts or mixed with other bryophytes. Shoots erect, to 7(– 8) cm, stems branching occasionally. Leaves distant below, approximate to imbricate above, flaccid, crisped when dry, margins entire or more usually with small 1–2-celled teeth, dorsal lobe widely crossing stem, plane to convex, 0.4–0.7 length of ventral lobe,

Fig. **81**. 1–6, *Scapania uliginosa*: 1 and 2, sterile and male shoots (× 6); 3, leaves, dorsal side (× 13); 4, leaf, ventral side (× 13); 5, leaf section through keel (× 27); 6, cell from middle of ventral lobe (× 450). 7–10, *S. paludosa*: 7, shoot (× 5); 8, leaves, dorsal side (× 8); 9 leaves, ventral side (× 8); ¹0. cell from middle of ventral lobe (× 450).

reniform to cordate-reniform, 0.6–2.0 mm long, 0.7–1.8 mm wide, decurrent, apex rounded; ventral lobe reflexed or not, rounded-cordate to cordate-reniform, 1.3–3.5 mm long, 1.4–2.8 mm wide, longly decurrent, apex rounded; keel 0.2–0.4 length of ventral lobe, winged, strongly arcuate to semicircular at least in some leaves; marginal cells sometimes more heavily thickened than other cells, cells in middle of ventral lobe 24–32 μm wide with 4–10 oil-bodies, cell walls very thin or thin, usually colourless, trigones lacking. Gemmae not known. Male plants occasional, bracts smaller than leaves, saccate. Perianths and capsules unknown in Britain. Flushes, springs and on rocks in streams, 490–1000 m, very rare, Merioneth, Caernarfon, Cumberland, Mid Perth, Angus, S. Aberdeen, E. Inverness, Argyll, Ross. 10. Alpine and subarctic Europe from Spain, N. Italy and Romania north to Fennoscandia and N. Russia, Faroes, Iceland, Siberia, Japan, northern N. America, Greenland.

Section *Nemorosae* (K. Müll.) Buch, *Soc. Sci. Fenn. Comm. Biol.* 3(1): 152. 1928

Dioecious. Plants robust. Stem with 2–4-stratose cortex with tangentially flattened cells. Leaves unequally bilobed, margins ciliate-dentate with apical cell of teeth usually 2–3 times as long as wide, both lobes decurrent, keel short, winged. Gemmae in fasciculate bunches, usually unicellular, very thin-walled.

17. S. nemorea (L.) Grolle, *Rev. Bryol. Lichén.* 32: 160. 1963.   (Fig. 82)
S. nemorosa (L.) Dum.

Dioecious. Plants medium-sized to large, green or dull green, rarely reddish or brownish-tinted, in lax patches. Shoots erect or ascending, apices sometimes deflexed, 1–6 cm, stems rarely branched. Leaves imbricate above, approximate to subimbricate, rarely distant below, margins of both lobes of at least upper leaves usually ciliate-dentate, many teeth with apical cell *ca* twice as long as wide, in stunted forms sometimes only dentate or subentire, dorsal lobe plane or convex, crossing stem, ± appressed to ventral lobe and 0.4–0.6 its length, ovate-cordate, 0.6–1.2 mm long, 0.4–1.1 mm wide, decurrent, apex obtuse to pointed; ventral lobe plane to convex, sometimes with upper margin reflexed, ovate or ovate-orbicular, 1.1–2.2 mm long, 0.8–1.7 mm wide, decurrent, apex rounded to obtuse, rarely apiculate; keel ± straight or occasionally curved, 0.2–0.3 length of ventral lobe, winged; cells in middle of ventral lobe 16–24 μm wide with 3–5 oil-bodies, marginal cells smaller, more thick-walled, trigones very small or small, rarely medium-sized. Gemmae in brownish clusters, individually reddish-brown, ovoid or ellipsoid, unicellular, 14–20 μm long, 8–11 μm wide. Male bracts saccate. Perianths occasional, mouth truncate, ciliate-dentate. Capsules occasional, autumn to spring. $n = 8, 9$. On moist soil, rotting logs, tree stumps and rocks in woodland, sheltered banks, ravines, marshy ground, bogs, moist turf and moorland, rare to occasional in S.E. England and the Midlands, frequent to common and sometimes locally abundant elsewhere. 106, H30, C. Europe north to Fennoscandia and N. Russia, Turkey, Siberia, Azores, Madeira, Tenerife, Gomera, eastern N. America.

A widespread and variable species usually readily recognised by the ciliate-dentate leaf margins and the apical cells of the teeth *ca* twice as long as wide, a feature found elsewhere only in *S. ornithopodioides* and *S. nimbosa*. *S. gracilis* differs in colour, reflexed leaf lobes with coarsely toothed margins and the gemmae 2–celled. Forms of *S. undulata* with dentate leaf margins differ in the coarser teeth, shorter keel and non-decurrent dorsal lobe. *S. crassiretis* Bryhn, also in section *Nemorosae*, has been reported from Caernarfon and Mid Perth but these records are based on misidentifications (see Paton, *J. Bryol.* **8**, 493, 1975).

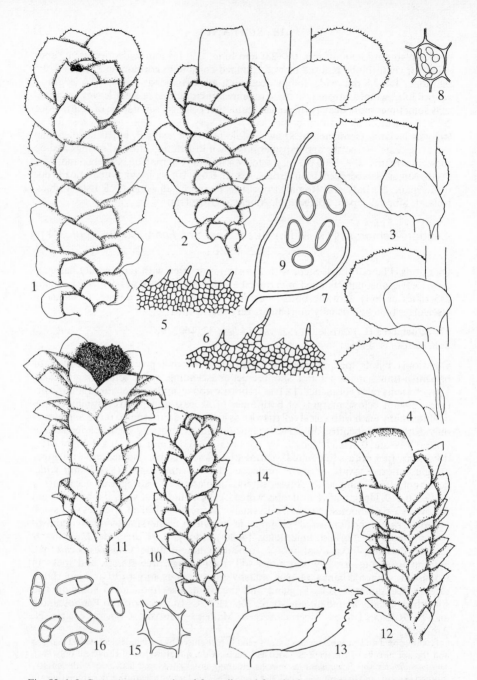

Fig. **82**. 1–9. *Scapania nemorea*: 1 and 2, sterile and female shoots (× 9); 3, leaves, dorsal side (× 15); 4, leaf, ventral side (× 15); 5, leaf margin (× 84); 6, perianth mouth (× 84); 7, т.s. leaf through keel (× 38); 8, cell from middle of ventral lobe (× 450); 9, gemmae (× 450). 10–16, *S. umbrosa*: 10–12, sterile, gemmiferous and female shoots (× 15); 13, leaves, dorsal side (× 38); 14, leaf, ventral side (× 38); 15, cell from middle of ventral lobe (× 450); 16, gemmae (× 450).

Section *Umbrosae* Buch., *Soc. Sci. Fenn. Comm. Biol.* 3(1): 120. 1928

Dioecious. Plants small. Shoot tips deflexed, stem with 3–4-stratose cortex. Leaves ventrally subsecund, lobes acute, dorsal smaller than ventral, ventral decurrent to below insertion of keel, margin ± coarsely toothed, keel short, winged, Gemmae 2-celled.

18. S. umbrosa (Schrad.) Dum., *Recueil. Observ. Jungerm.*: 14. 1835   (Fig. 82)

Dioecious. Plants small, whitish-green, yellowish-green or pale reddish, in patches. Shoots ascending, 0.3–1.5(–2.9) cm, tips deflexed, stems rarely branched. Leaves distant to subimbricate below, imbricate above, usually ventrally subsecund, toothed, sometimes coarsely so, ± from widest part, dorsal lobe ± appressed to and 0.6–0.7 length of ventral lobe, not crossing stem, lanceolate to ovate, 0.4–0.8 mm long, 0.24–0.48 mm wide, not decurrent, apex acute; ventral lobe ovate to lanceolate, 0.64–1.28 mm long, 0.38–0.72 mm wide, decurrent, apex acute; keel ± straight to weakly arched, 0.2–0.4 length of ventral lobe, winged; cells in middle of ventral lobe 16–24 μm wide with 4–12 oil-bodies, walls usually thickened, colourless to yellowish-brown, trigones medium-sized, marginal cells smaller and more heavily thickened. Gemmae in dark brown clusters, ellipsoid to shortly cylindrical or sausage-shaped, 2-celled, (12–)20–24 μm long, 6–10 μm wide. Male bracts smaller than leaves, saccate, lobes ± equal. Perianths occasional, deflexed above, truncate, mouth entire. Capsules rare, winter, spring. $n = 8 + m$. Usually in small quantity on rotting wood, schistose and sandstone rocks and peat, rarely on soil, in moist, shaded habitats, ascending to *ca* 450 m, occasional to frequent from N. Wales, Derby and Yorkshire northwards, very rare elsewhere, Cornwall, Devon, S. Somerset, Sussex, W. Kent, W. Gloucester. Worcester, Stafford, S. Wales, occasional in Ireland. 63, H26. From the Pyrenees, N. Italy and Yugoslavia north to Fennoscandia and N. Russia, Turkey, Azores, N. America.

Section *Aequilobae* Buch, *Soc. Sci. Fenn. Comm. Biol.* 3(1): 110. 1928

Plants medium-sized to robust. Stem cortex 2–4-stratose. Leaf margins usually denticulate, dorsal lobe widely crossing stem, decurrent or not; cells with trigones, cuticle coarsely papillose.

19. S. aequiloba (Schwaegr.) Dum., *Recueil Observ. Jungerm.*: 14. 1835   (Fig. 83)

Dioecious. Plants small to medium-sized, yellowish-green to brownish, in tufts or mixed with other bryophytes. Shoots erect or ascending, 0.5–2.0(–4.0) cm, stems rarely branched, innovating from below perianth. Leaves imbricate to closely imbricate, dorsal lobe divergent from stem, barely to widely crossing stem, smaller than or subequal to ventral lobe and 0.6–0.8(–0.9) its length, ovate-quadrate to ovate-rectangular, 0.55–1.10 mm long, 0.3–0.9 mm wide, entire or with a few unicellular teeth towards apex, base often semi-auriculate, not decurrent, apex acute, sometimes apiculate; ventral lobe of similar shape, 0.8–1.5 mm long, 0.5–1.0 mm wide, not decurrent, apex acute, sometimes apiculate; keel 0.4–0.5 length of ventral lobe, straight or slightly curved, winged, cells in middle of ventral lobe 16–24(–28) μm wide with 2–4 oil-bodies, walls thickened, yellowish-brown to brown, trigones well developed, marginal cells more thick-walled, cuticle coarsely papillose. Gemmae pale green, ellipsoid, 16–27(–30) μm long, 11–14 μm wide, 2-celled. Male bracts smaller than leaves, with subequal, obtuse to rounded, spreading lobes. Female bracts larger than leaves, lobes unequal, more strongly toothed with teeth sometimes 2-celled. Perianths occasional, plicate above, mouth lobed, ciliate-dentate. Capsules not known in Britain or Ireland. $n = 8 + m$. On rocks, rock ledges, cliffs and soil on rocks, rarely in dune-

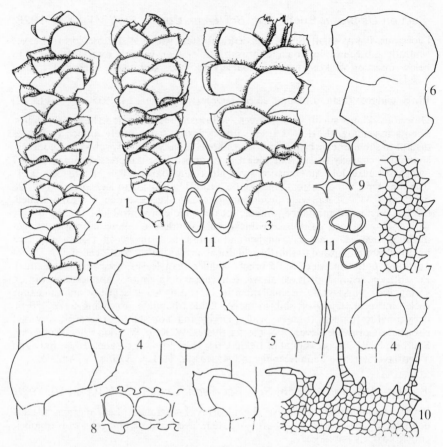

Fig. 83. *Scapania aequiloba*: 1–3, sterile, male and female shoots (× 6); 4, leaves, dorsal side (× 21); 5, leaf, ventral side (× 21); 6, flattened leaf (× 21); 7, margin of ventral leaf lobe (× 130); 8, т.s. leaf cells (× 450); 9, cell from middle of ventral lobe (× 450); 10, portion of perianth mouth (× 130); 11, gemmae (× 450).

slacks, above 300 m except in N.W. and N. Scotland, calcicole, occasional in the Scottish highlands, very rare elsewhere, Carmarthen, Caernarfon, Yorkshire and Westmorland north to W. Sutherland, Outer Hebrides, S. Tipperary, Sligo, Leitrim, Donegal, Antrim. 31, H6. Montane Europe from Spain, Italy, Yugoslavia and Greece north to Fennoscandia and N. Russia, Turkey, Ellesmere Is. (?), Greenland (?).

20. S. aspera M. & H. Bern. in H. Bernet, *Catal. Hep. Sud-Ouest Suisse*: 42. 1888
(Fig. 84)

Dioecious. Plants medium-sized, dull green to brownish, in tufts or mixed with other bryophytes. Shoots erect or ascending, (0.5–)1.0–4.0(–7.0) cm, stems rarely branched, innovating from below perianths. Leaves imbricate to closely imbricate, dorsal lobe convex, appressed to stem or, in older leaves, reflexed, widely crossing stem, 0.50–0.70(–0.75) length of ventral lobe, ovate or ovate-quadrate, (0.3–)0.5–1.3 mm long, 0.4–1.0 mm wide, margin entire towards base, entire or with 1–2-celled teeth above,

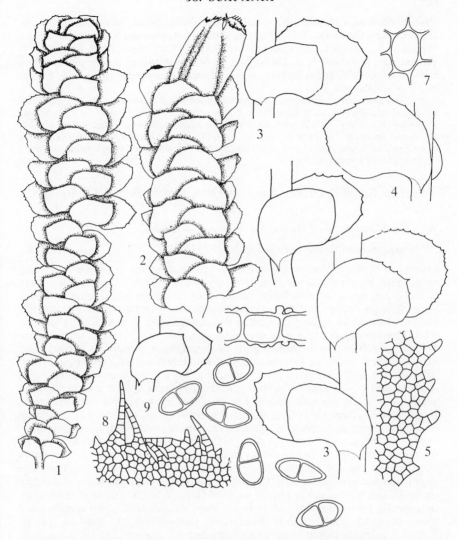

Fig. **84**. *Scapania aspera*: 1 and 2, sterile and female shoots (× 9); 3, leaves, dorsal side (× 23); 4, leaf, ventral side (× 23); 5, margin of ventral leaf lobe (× 130); 6, т.s. leaf cells (× 450); 7, cell from middle of ventral lobe (× 450); 8, portion of perianth mouth (× 130); 9, gemmae (× 450).

decurrent, sometimes weakly so, apex rounded or obtuse; ventral lobe ovate or ovate-lingulate, 0.7–2.2 mm long, 0.4–1.6 mm wide, decurrent, apex rounded to obtuse, occasionally subacute, sometimes apiculate, margin toothed, teeth mostly 2 cells long; keel (0.3–)0.4–0.5 length of ventral lobe, ±straight, winged; cells in middle of ventral lobe (16–)20–24 μm wide with 2–8 oil-bodies, walls thickened, yellowish-brown to brown, trigones small to large and bulging, marginal cells more thick-walled; cuticle coarsely papillose. Gemmae pale green, ellipsoid, 28–40 μm long, 14–18 μm wide, 2-celled. Male bracts with subequal lobes, saccate. Female bracts larger than leaves, more strongly toothed. Perianths occasional, plicate above, mouth lobed, ciliate-dentate. Capsules rare. On rocks, cliff ledges, soil, sand-dunes, amongst boulders and in turf, calcicole, a mainly lowland species but ascending to 800 m, occasional to frequent in S. Britain, rare in the north, S. Devon and N. Somerset east to Surrey and north to Sutherland and Caithness, Orkney, scattered localities in Ireland. 76, H24. Europe north to Spitzbergen, Turkey.

S. aspera, particularly small forms, have been confused with S. aequiloba. S. aequiloba differs in the dorsal lobe standing away from the stem, the non-decurrent leaf bases, the margins with mostly unicellular teeth and the smaller gemmae. For a discussion of the differences between the two species see Long, *Bull. Br. bryol. Soc.* **31**, 26–9, 1978. S. gracilis may also be mistaken for S. aspera but is calcifuge, usually has coarser marginal teeth, the dorsal leaf margin being toothed to the base and the gemmae smaller.

Section *Gracilidae* Buch, *Soc. Sci. Fenn. Comm. Biol.* 3(1): 46, 106. 1928

With the characters of the species. Stem cortex 2–4-stratose.

21. S. gracilis Lindb., *Helsingfors Dagblad* 1873 (353): 2. 1873   (Fig. 85)

Dioecious. Plants medium-sized to robust, yellowish-brown to olive-green, rarely bright green, in tufts or patches, sometimes very extensive. Shoots procumbent to erect, to 6(–12) cm, stems occasionally branched, innovating from below perianth. Lower leaves distant to imbricate, upper subimbricate to closely imbricate, margins coarsely toothed, dorsal lobe crossing the stem or not, convex, erect or more usually reflexed, 0.6–0.7 length of ventral lobe, obliquely ovate-orbicular, ovate-reniform or ovate, 0.5–1.7 mm long, 0.4–1.5 mm wide, slightly decurrent, apex rounded to obtuse, basal margin entire to strongly ciliate-dentate; ventral lobe reflexed, ovate-orbicular, ovate-lingulate or broadly lingulate, 0.8–2.4 mm long, 0.6–2.0 mm wide, decurrent, apex rounded; keel 0.2–0.4 length of ventral lobe, slightly curved, winged; cells in middle of ventral lobe 16–24 μm wide with 2–4 large oil-bodies, walls brownish, thickened, trigones small to large and bulging, marginal cells more thick-walled or not, cuticle not or scarcely papillose. Gemmae in green clusters, ovate to ellipsoid, 1–2-celled, thick-walled, 18–30 μm long, 12–18 μm wide. Male bracts imbricate, smaller than leaves, saccate at base. Female bracts similar to and slightly larger than leaves. Perianths occasional to frequent, reflexed above, truncate, mouth sinuose-lobed, sparsely and slightly toothed to ciliate-dentate. Capsules occasional to frequent, winter to summer. On rocks and tree trunks in woods, scree, ravines, on heaths and moorland, cliffs, occasionally on rotting logs and peat, calcifuge, occasional to very common and sometimes locally abundant in W. Britain from Cornwall north to Shetland, and W. Ireland, less frequent towards the east and very rare in S. England, Dorset, S. Hants, E. Sussex, W. Kent. 82, H31, C. Atlantic coast of Europe from Portugal north to S.W. Norway and Sweden, Corsica, Italy, Romania, Faroes, Morocco, Azores, Madeira, La Palma, Tenerife.

Subgenus *Protoscapania* Amak. & Hatt., *J. Hattori bot. Lab.* 12: 110. 1954

Dioecious. Plants robust, strongly pigmented. Stem rigid, sparingly branched, in section with cortex of several layers of thick-walled cells. Leaves deeply bilobed, keel very short or absent, leaf margin spinose- or ciliate-dentate, apical cell of teeth elongated; cells with large, bulging trigones. Perianth dorsiventrally flattened, not plicate, truncate, sparsely ciliate.

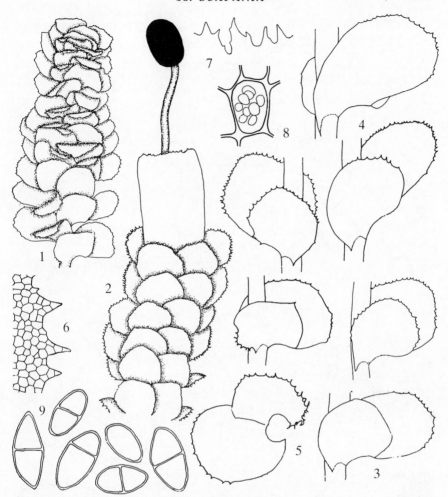

Fig. **85**. *Scapania gracilis*: 1 and 2, sterile and fertile shoots (× 9); 3, leaves, dorsal side (× 15); 4, leaf, ventral side (× 15); 5, flattened leaf (× 15); 6, leaf margin (× 110); 7, perianth mouth (× 60); 8, cell from middle of ventral lobe (× 450); 9, gemmae (× 450).

22. S. ornithopodioides (With.) Waddell, *Moss Exchange Club Catal. Brit. Hep.*: 4. 1897 (Fig. 86)

Dioecious. Plants robust or very robust, reddish-purple, in patches or mixed with other bryophytes. Shoots prostrate to erect, to 13 cm, stems occasionally branched. Leaves approximate to imbricate, bilobed to base, margins ciliate-dentate, spinose-dentate at base, teeth often eroded in older leaves, dorsal lobe appressed to and crossing stem, pointing ± towards stem apex, convex, (0.5–)0.6–0.8 length of ventral lobe, ovate or ovate-orbicular, decurrent, apex rounded, 1.1–1.7 mm long, 1.0–1.7 mm wide; ventral lobe plane or slightly convex, not or only slightly reflexed, ovate, decurrent, apex

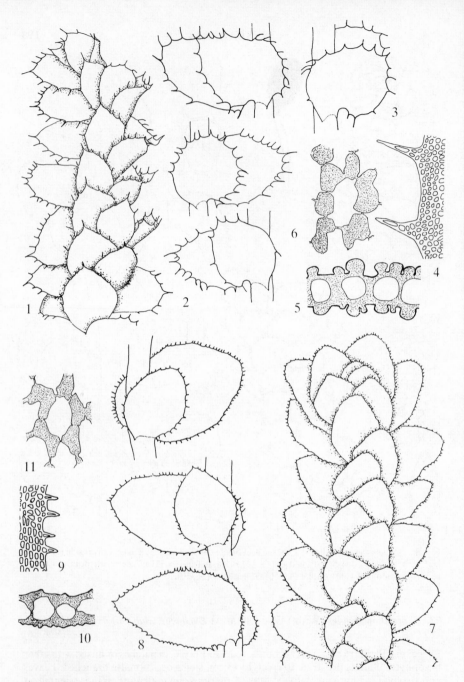

Fig. **86**. 1–6, *Scapania nimbosa*: 1, shoot (× 8); 2, leaves, dorsal side (× 15); 3, leaf, ventral side (× 15); 4, leaf margin (× 84); 5, leaf cells in T.S. (× 450); 6, cell from middle of ventral lobe (× 450). 7–11, *S. ornithopodioides*: 7, shoot (× 8); 8, leaves, dorsal side (× 15); 9, leaf margin (× 84); 10, leaf cells in T.S. (× 450); 11, cell from middle of ventral lobe (× 450).

rounded to acute, 1.6–2.5 mm long. 1.2–2.0 mm wide; keel absent; cells in middle of ventral lobe 20–24 μm wide with 1–4 oil-bodies, walls yellowish-brown to reddish-purple, trigones large or very large, bulging, cuticle papillose or not. Perianths unknown in Europe. $n = 9$. Heathy slopes, boulder scree and cliff ledges, 280–1000 m, rare in N. Wales, the Lake District and the Cairngorms, occasional in W. Scotland from Argyll north to Sutherland and Shetland, very rare in Ireland, S. Kerry, W. Galway, W. Mayo, W. Donegal. 20, H4. S.W. Norway, Faroes, Bhutan, Sikkim, Yunnan, Taiwan, Japan, Hawaii.

Growing in similar habitats to but more frequent than *S. nimbosa. S. ornithopodioides* may be distinguished in the field by the non-reflexed ventral leaf lobe and the smaller, much more numerous marginal teeth. Müller (1957) reports that reddish-black, ovoid gemmae, 20–28 μm long, 17–21 μm wide are produced rarely by *S. ornithopodioides.*

23. S. nimbosa Tayl. ex Lehm., *Novar. Min. Cogn. Stirp. Pugillus* 8: 6. 1844

(Fig. 86)

Sex unknown. Plants robust, reddish-brown, in lax tufts or mixed with other bryophytes. Shoots erect or ascending, to 7(–10) cm, stems occasionally branched. Leaves approximate to imbricate, bilobed almost to base, margins spinose-ciliate with distant teeth, teeth sometimes curved, dorsal lobe appressed to and crossing stem, convex, (0.6–)0.7–0.9 length of ventral lobe, ovate, decurrent, apex ± acute and ending in a spinose tooth, 1.2–1.9 mm long, 0.8–1.4 mm wide; ventral lobe of ± similar shape, deflexed with reflexed upper margin, 1.5–2.2 mm long, 1.0–1.9 mm wide, decurrent, apex acute and ending in a spinose tooth; keel very short, arcuate; cells in middle of ventral lobe 16–24(–28) μm wide, walls reddish-brown, trigones large or very large, bulging, cuticle coarsely papillose. Gemmae unknown. Heathy slopes, boulder scree and cliff ledges, 400–1000 m, Caernarfon (extinct?), rare to occasional in W. Scotland from Argyll and Perth north to W. Sutherland, Hebrides, Cairngorms, very rare in Ireland, S. Kerry, W. Mayo, W. Donegal. 15, H3. S.W. Norway, Yunnan.

16. GEOCALYCACEAE

Plants small to robust, frequently whitish-green, yellowish-green or pale green. Shoots usually prostrate, stems irregularly branched, branches arising laterally, ventrally or terminally. Rhizoids scattered or restricted to bases of underleaves. Leaves succubous, alternate or opposite, obliquely or very obliquely inserted, retuse to deeply bilobed, rarely entire; cell walls thin or thickened, trigones often present, oil-bodies usually *ca* 12 per cell. Underleaves smaller than leaves, usually bilobed. Vegetative propagation, where present, by gemmae or caducous leaves or flagelliform branches. Fertile branches either usually lateral (Lophocoleoideae) or ventral (Geocalycoideae). Male inflorescence spicate, occasionally intercalary, bracts few to numerous. Female infloresence terminal on a leafy shoot or branch or on modified ventral branch. Female bracts large, reduced or vestigial. Perianth present, wide-mouthed and perigynium or marsupium lacking, or reduced and narrowed at mouth or absent and perigynium or marsupium present. Capsule ovoid to cylindrical, wall 2–5-stratose. A mainly tropical and southern hemisphere family of 21 genera.

1 Leaves ± opposite   **44. Saccogyna (p. 218)**
  Leaves alternate 2
2 Leaves wedge-shaped, 0.22 mm long, male bracts and leaves below female inflorescence larger **39. Leptoscyphus (p. 202)**
  Leaves more than 0.22 mm long, not wedge-shaped, male bracts and leaves on female branches similar in size to or smaller than other leaves 3

3 Rhizoids restricted to underleaf bases, underleaves bilobed $\frac{1}{2}$–$\frac{3}{4}$ with 0–2 teeth on one or both sides, perianth wide-mouthed, male inflorescence not concealed by lateral leaves   4
  Rhizoids not restricted to underleaf bases, underleaves simple, or bilobed almost to base with entire margins, perianth narrowed at mouth, or absent and marsupium present, male inflorescence very small and hidden under lateral leaves   5
4 At least lowest leaves bilobed or if all leaves rounded to emarginate then dorsal margin not or hardly decurrent, perianth well developed, male inflorescence spicate or intercalary   **40. Lophocolea (p. 204)**
  Mature leaves rounded to emarginate, dorsal margin decurrent, perianth short with calyptra protruding, male inflorescence intercalary   **41. Chiloscyphus (p. 211)**
5 Underleaves simple, female inflorescence developing a perianth, marsupium lacking
                                                                         42. Harpanthus (p. 213)
  Underleaves bilobed almost to base, inflorescence developing a marsupium but no perianth   **43. Geocalyx (p. 215)**

SUBFAMILY LOPHOCOLEOIDEAE

Vegetative branches often lateral. Rhizoids in tufts at base of underleaves. Male inflorescence lateral and spicate or intercalary, if arising ventrally not concealed by lateral leaves. Antheridial wall cells ± isodiametric. Female inflorescence terminal on main axis or branch, bracts as large as or larger than leaves. Perianth present, wide-mouthed, perigynium or marsupium lacking. Capsule wall 4–5-stratose.

39. LEPTOSCYPHUS MITT., *J. Bot. Kew Gard. Misc.* 3: 358. 1851

Dioecious. Plants minute to robust, often brownish. Shoots prostrate, stems branching irregularly, innovating from below female inflorescence, flagelliform shoots lacking. Rhizoids in tufts. Leaves succubous, alternate, suborbicular to ovate or rarely obovate or obdeltoid, entire or bluntly 2–3-lobed or obscurely toothed, insertion very oblique, dorsal margin often decurrent, ventral margin sometimes connate with underleaves. Underleaves usually patent, 1–2-lobed with 1–2 ciliate teeth, rarely simple. Vegetative propagation, where present, by caducous leaves. Male bracts similar to leaves or smaller. Female bracts of similar size to or larger than leaves, bracteole present. Perianth inflated below, laterally compressed above, mouth truncate, entire to toothed, bilabiate with rudimentary third lip. Capsule ellipsoid, wall usually 2- or 4–5-stratose. 18–20 species.

1. L. cuneifolius (Hook.) Mitt., *J. Bot. Kew Gard. Misc.* 3: 358. 1851   (Fig. 87)
Anomylia cuneifolia (Hook.) Schust., *Mylia cuneifolia* (Hook.) S.F. Gray

Plants minute, in reddish-brown patches on or scattered and creeping over other bryophytes. Shoots prostrate to ascending, 2–7 mm, sparingly branched. Rhizoids in bunches at base of underleaves. Leaves caducous, often leaving lengths of stem bare, distant to approximate, patent to spreading, obovate to obdeltoid, slightly concave, apex rounded-truncate, truncate or retuse, insertion oblique, not decurrent, 120–220 μm long, 130–220 μm wide, ± as long as wide, mid-leaf cells 16–26 μm wide, walls brownish, thickened, trigones distinct. Underleaves patent, usually lanceolate or subulate, 80–100 μm long on sterile stems, 100–200 μm long on fertile stems. Vegetative propagation by caducous leaves. Male bracts larger than leaves, *ca* 260 μm long, 300

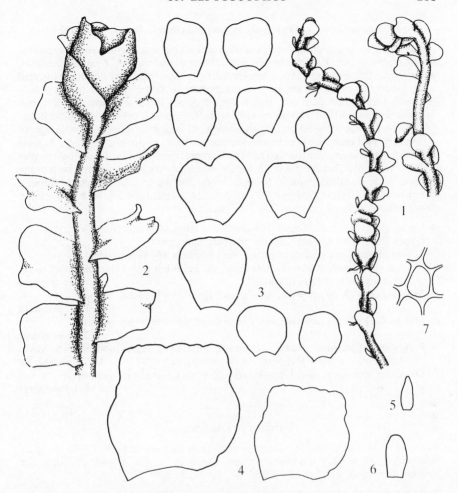

Fig. **87**. *Leptoscyphus cuneifolius*: 1, sterile shoots (× 45); 2, female shoot (× 45); 3, leaves from sterile stems (× 84); 4, leaves from female stems (× 84); 5 and 6, underleaves from sterile and female stems (× 84); 7, mid-leaf cell (× 450).

μm wide. Leaves below female inflorescence large, 360–400 μm long, 360–440 μm wide. Female bracts similar to upper leaves. Perianth protruding beyond bracts at maturity, laterally compressed, wide-mouthed, mouth bilabiate, toothed. Capsules unknown in Britain or Ireland. Tree trunks, especially *Betula*, often growing with *Frullania* or on boulders in sheltered, western sites, usually at low altitudes but ascending to 300(–760) m, rare, Merioneth and W. Scotland from Dumbarton and Perth north to W. Sutherland, Outer Hebrides, W. Ireland. 14, H8. S.W. Norway, Spain, Azores, Madeira, Virginia, N. Carolina, Tennessee and C. America south to Straits of Magellan, W. Indies, Juan Fernandez, Tristan da Cunha.

For a detailed description of *L. cuneifolius* see Paton, *Trans. Br. bryol. Soc.* **5**, 232–6. 1967.

40. LOPHOCOLEA (DUM.) DUM., *Recueil Observ. Jungerm.*: 17. 1835

Plants small to large, whitish- to yellowish-green or green. Stems usually creeping, irregularly branched. Rhizoids in bunches at bases of underleaves. Leaves succubous, alternate (in European species), often spreading horizontally, very obliquely inserted with dorsal margin decurrent or not, emarginate to deeply bilobed or irregularly spinosely toothed at apex; cells thin-walled, trigones small or absent. Underleaves bilobed, with or without a tooth on either side. Gemmae, if present, 1–multicellular. Male inflorescence terminal on branch, sometimes becoming intercalary with age, or intercalary or below female, bracts saccate at base, antheridia solitary. Female inflorescence terminal on main axis or branch, bracts usually bilobed, bracteole similar to or smaller than bracts. Perianth longly emergent, trigonous above, keels often winged, wide-mouthed, mouth 3-lobed, lobes dentate or ciliate-dentate. Capsule ovoid, wall 4–6-stratose. Spores 8–22 μm. A mainly tropical and southern hemisphere, taxonomically difficult genus of 200–300 species.

1 Upper leaves bilobed, or apex 2–3-dentate or irregularly and coarsely dentate   2
  Upper leaves with rounded, retuse or emarginate apex   4
2 Leaves bilobed, margins not toothed, mid-leaf cells 16–50 μm wide   3
  At least some leaves with 3-dentate apex, margin often with 1 or more small teeth,
    mid-leaf cells 12–20 μm wide   **5. L. fragrans**
3 Mid-leaf cells 25–50 μm wide, deciduous flagelliform branches lacking
                                                                     1. L. bidentata
  Mid-leaf cells 16–24 μm wide, deciduous flagelliform branches present
                                                                     2. L. bispinosa
4 Paroecious, perianths usually abundant, lowermost leaves and branch leaves
    bilobed to ⅓, upper leaves retuse or emarginate   **4. L. heterophylla**
  Dioecious, perianths rare, lower leaves and branch leaves similar to upper leaves
                                                                     3. L. semiteres

Subgenus *Lophocolea*

Plants of variable size. Leaves bilobed or not, rarely toothed; leaf cells 12–50 μm wide, thin-walled, all lacking mamillose spines. Gemmae present or not. Perianth not inflated, trigonous, keels often winged.

1. L. bidentata (L.) Dum., *Recueil Observ. Jungerm.*: 17. 1835

Whitish, pale green or yellowish-green to dark green patches or growing through vegetation. Shoots 1–6 cm, prostrate or procumbent, simple or sparsely to much branched. Leaves approximate to subimbricate, diverging to ± spreading horizontally, symmetrical to asymmetrical, insertion very oblique, dorsal margin decurrent, mature leaves 0.7–2.0 mm long, (0.4–)0.6–1.8 mm wide, 1.1–1.5(–1.7) times as long as wide, bilobed (⅛–)¼–⅖, very rarely 3-lobed, lobes often divergent, acuminate, sinus angular to lunate; mid-leaf cells (20–)24–44(–48) μm wide, thin-walled, trigones very small or absent. Underleaves 0.28–0.64 mm long, bilobed to *ca* ¾, lobes subulate, 0–1 acute to subulate teeth on each side. Gemmae lacking. Male inflorescence spicate, terminal on main axis or branch, becoming intercalary with age, bracts 2–5 pairs, smaller than leaves, closely imbricate, 2–3-lobed, saccate at base with inflexed lobe on dorsal basal margin. Female inflorescence terminal on main axis or branch, bracts much larger than leaves, bilobed ¼–⅓, lobes longly acuminate, margins with 0–1 teeth, bracteole ± similar but smaller. Perianth trigonous above, keels winged or not, wings entire to ciliate-dentate, mouth deeply 3-lobed, lobes acuminate, margins dentate to ciliate-dentate.

Key to varieties of *L. bidentata*

Stems much branched, leaf lobes ending in (2–)4–8 uniseriate cells, mid-leaf cells
24–40(–48) μm wide, perianths very common   var. **bidentata**
Stems hardly branched, leaf lobes ending in 1–5(–6) uniseriate cells, mid-leaf cells
20–36(–40) μm wide, perianths rare   var. **rivularis**

Var. **bidentata**   **(Fig. 88)**
Chiloscyphus cuspidatus (Nees) Engel & Schust., *L. alata* (Nees) Schiffn., *L.
cuspidata* (Nees) Limpr.

Autoecious. Shoots 1–3 cm, stems much branched. Mature leaves bilobed $\frac{1}{5}$–$\frac{2}{5}$, lobes
ending in (2–)4–8 uniseriate cells, mid-leaf cells 24–40(–48) μm wide. Perianths very
common. Capsules very common, winter to summer. $n = 18*$. On tree boles, exposed
roots, logs, stumps and rotting wood in woodland, on shaded rocks and on exposed soil
in turf, in lowland habitats, common or very common. 112, H38, C. Europe north to S.
Fennoscandia, Faroes, China, Japan, Macaronesia, montane parts of Africa from S.
Africa to Tunisia and Morocco, Fernando Po, Madagascar, Réunion, N. America,
Mexico, Cuba, New Zealand.

Var. **rivularis** (Raddi) Warnst., *Kryptogamenfl. Mark Brandenburg* 1: 243. 1902
                                                                          (Fig. 88)
Chiloscyphus latifolius (Nees) Engel & Schust., *L. bidentata* auct. non (L.) Dum., *L.
latifolia* Nees

Apparently dioecious. Shoots 2.5–6.0 cm, sparsely branched. Mature leaves $\frac{1}{5}$–$\frac{1}{4}$
bilobed, lobes ending in 1–5(–6) uniseriate cells, mid-leaf cells (20–)24–36(–40) μm
wide. Perianths rare. Capsules very rare, summer. $n = 9*, 18*$. On humus and soil in
woodland, in damp turf and in marshes, common in lowland habitats but ascending to
850 m. 112, H39, C. Europe north to S. Fennoscandia and W. Russia, Faroes, Iceland,
Asia Minor, Himalayas, Macaronesia, Tunisia, N. America.

    L. bidentata is extremely variable morphologically and it has been shown experimentally (Steel,
J. Bryol. **10**, 49–59, 1978) that much of this variation is environmentally induced. Steel also points
out that the characters used to separate *L. alata* are too plastic to be of use; the strongly winged
perianth of *L. alata* can occur at times in *L. bidentata*.

    Sex has been used in the past to separate var. *bidentata* (as *L. cuspidata*) and var. *rivularis* (as *L.
bidentata*). According to Steel this is an unreliable character as antheridia and archegonia are not
necessarily produced at the same time so that material of var. *bidentata* may appear to be
dioecious. Further, D. Vogelpoel (pers. comm.) says that the production of antheridia or
archegonia or both is affected by environmental factors and that in some apparently dioecious
specimens abortive gametangia of the other sex occur. Thus it is not certain that var. *rivularis* is
really dioecious. Steel treats the two taxa as species, and Vogelpoel recognises *L. alata* as a variety
of *L. bidentata* (var. *alata* Nees). Because of the uncertainty of the sex of plants and because of the
degree of environmentally induced variation I have treated *L. bidentata* as consisting of only two
varieties in Britain and Ireland but whether the two are taxonomically distinct is debatable.

    Small forms of *L. bidentata* tend to have small, more deeply bilobed leaves with smaller cells and
determinations should be based only on mature shoots. Because of the overlap in distinguishing
characters specimens can often only be named on the basis of a combination of characters from
several shoots.

2. L. bispinosa (Hook. f. & Tayl.) Gott. et al., *Syn. Hep.*: 162. 1845   (Fig. 89)
Chiloscyphus bispinosus (Hook. f. & Tayl.) Engel & Schust.

Dioecious. Yellowish-green to pale brownish-green, dense turfs, thin patches or
scattered shoots. Shoots procumbent to erect, to 1.5 cm, usually irregularly branched,
caducous flagelliform microphyllous branches sometimes present. Leaves variable,

Fig. **88**, 1–9, *Lophocolea bidentata* var. *bidentata*: 1, shoot with male and female branches (× 8); 2, perianth from vigorous plant (× 8); 3, male branch (× 8); 4, leaves (× 15); 5, underleaves (× 21); 6, female bract (× 8); 7, male bract (× 23); 8, leaf lobe tips (× 84); 9, mid-leaf cell (450); 10–12, *L. bidentata* var. *rivularis*: 10, shoot (× 8); 11, leaves (× 15); 12, leaf lobe tips (× 84).

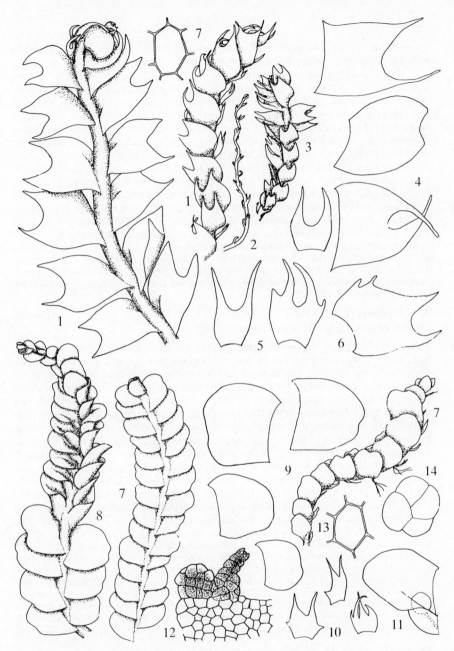

Fig. **89**. 1–7, *Lophocolea bispinosa*: 1, shoots (× 21); 2, flagelliform shoot (× 21); 3, male shoot (× 21); 4, leaves (× 63); 5, underleaves (× 63); 6, male bract (× 63); 7–14, *L. semiteres*: 7, shoot (× 13); 8, male shoot (× 13); 9, leaves (× 21); 10, underleaves (× 33); 11, male bract (× 21); 12, leaf margin with developing gemmae (× 130); 13, mid-leaf cell (× 450); 14, gemma (× 450).

erect and imbricate to approximate and spreading horizontally, concave, insertion oblique, dorsal margin decurrent or not, 0.35–0.9(–1.0) mm long, 0.3–0.7 mm wide, (1.1–)1.3–1.6(–2.0) times as long as wide, bilobed $\frac{1}{5}$–$\frac{1}{2}$, occasionally only emarginate, lobes divergent, parallel or connivent, acute to longly acuminate or occasionally obtuse; mid-leaf cells 16–24 μm wide, thin-walled, trigones small or lacking, marginal row of cells sometimes elongated and with fewer chloroplasts and forming a translucent, poorly defined border. Underleaves 0.24–0.65 mm long, bilobed $\frac{2}{5}$–$\frac{4}{5}$, lobes acuminate, outer margins with 0–1(–2) teeth on one or both sides. Gemmae lacking but vegetative propagation by deciduous flagelliform branches. Male inflorescence spicate, becoming intercalary with age, bracts of similar size to leaves, saccate, dorsal margin toothed towards base. Female inflorescence terminal, bracts and bracteole similar, bilobed to *ca* $\frac{1}{2}$, margins entire or with tooth on one or both sides. Perianth sharply trigonous, wide-mouthed, mouth irregularly ciliate. Capsules occasional, spring. $n = 8 + m^*$. Peat, earth and sand on banks and tracks, on soil on rocks and walls, in moist or dry, open or shaded situations. Scilly Is., Colonsay. 2. Australasia.

Almost certainly an introduction with horticultural plants from the southern hemisphere. The plant is exceedingly variable and may be mistaken for *L. fragrans* or slender forms of *L. bidentata*. The latter has larger cells and larger shoots of normal size are also usually present. In *L. fragrans* at least some of the leaves are 3-lobed or spinosely toothed. For an account of *L. bispinosa* in Britain see Paton, *J. Bryol.* **8**, 191–6, 1974.

3. L. semiteres (Lehm.) Mitt., *J. Linn. Soc. Bot.* 16: 188. 1877   (Fig. 89)
Chiloscyphus semiteres (Lehm.) Engel & Schust.

Dioecious. Pale green or yellowish-green, sometimes extensive mats. Shoots procumbent, to 1(–3) cm, occasionally branched. Leaves imbricate, dorsally secund, spreading horizontally or reflexed, apical erect and appressed, ovate-quadrate to rounded-quadrate, apex emarginate or occasionally rounded, margin sometimes irregular from gemma production, insertion very oblique, dorsal margin not or only slightly decurrent, 0.4–1.1 mm long, 0.4–1.0 mm wide, 1.0–1.1(–1.2) times as long as wide; mid-leaf cells 20–32 μm wide, thin-walled, trigones small. Underleaves 0.30–0.55 mm long, bilobed $\frac{1}{2}$–$\frac{3}{4}$, lobes acuminate, with large tooth on each outer margin. Gemmae sometimes present on leaf margins, irregular, multicellular, sometimes germinating *in situ*. Male inflorescence terminal, becoming intercalary with age, bracts numerous, imbricate, smaller and narrower than leaves, reflexed from sac-like base. Perianths rare, trigonous with dentate wings. Capsules rare, spring. $n = 8 + m^*$. Tree bases and rotting logs, especially under conifers, in turf and on peaty soil and banks, very rare but locally common, Scilly Is., Argyll. 2. Cape Province, S. Africa, S. Chile, Australia, Tasmania, New Zealand, New Hebrides.

Most likely to be mistaken for a form of *Chiloscyphus polyanthos* but may be distinguished by the non- or scarcely decurrent leaves and multicellular gemmae. It also differs from *Chiloscyphus* in being dioecious, in the inflated sac-like basal part of the male bracts and in the large, trigonous, keeled perianths. *L. semiteres* may also be mistaken for *L. heterophylla* but the latter is paroecious and the lowermost leaves are bilobed and the male bracts differ in position and shape. *L. semiteres* is almost certainly an introduction from the southern hemisphere and it is very likely that the Scilly Isles and Argyll populations were introduced separately. For an account of the occurrence of *L. semiteres* in Britain see Paton, *Trans. Br. bryol. Soc.* **4**, 775–9, 1965 and Long, *J. Bryol.* **12**, 113–5, 1982.

4. L. heterophylla (Schrad.) Dum., *Recueil Observ. Jungerm.*: 18. 1835   (Fig. 90)
Chiloscyphus profundus (Nees) Engel & Schust.

Usually paroecious, rarely autoecious. Pale green to yellowish-green patches. Shoots procumbent, adhering to substrate, to 2(–3) cm, irregularly branched. Leaves very variable even on a single shoot, lower leaves and sometimes branch leaves bilobed to up to $\frac{1}{3}$ with acute to acuminate lobes, mature leaves subimbricate or imbricate, rarely

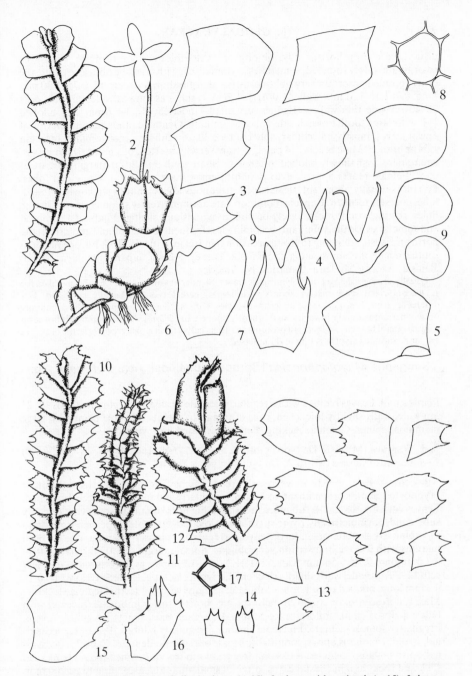

Fig. **90**. 1–9, *Lophocolea heterophylla*; 1, shoot (× 15); 2, shoot with perianth (× 15); 3, leaves (× 23); 4, underleaves (× 23); 5, male bract (× 23); 6, female bract (× 23); 7, bracteole (× 23); 8, mid-leaf cell (× 450); 9, gemmae (× 450). 10–17, *L. fragrans*: 10–12, sterile, male and female shoots (× 15); 13, leaves (× 24); 14, underleaves (× 33); 15, bract (× 24); 16, bracteole (× 24); 17, mid-leaf cell (× 450).

distant, spreading horizontally, oblong or ovate-oblong, apex truncate, obtuse or emarginate, rarely rounded, sometimes asymmetrical with one lobe obtuse to acute, the other rounded, insertion very oblique, dorsal margin shortly decurrent, 0.75–1.30 mm long, 0.7–1.1(–1.3) mm wide, (1.0–)1.1–1.2(–1.3) times as long as wide; mid-leaf cells 28–40 μm wide, thin-walled, trigones very small. Underleaves 0.3–0.6 mm long, bilobed ½–⅔, with large tooth on each side, lobes acuminate. Gemmae rarely present on tips of apical leaves, mostly unicellular, spherical, ca 30 μm. Inflorescence terminal on main axis or branch. Male bracts 2–4 pairs, ± transversely inserted, of similar size to leaves, emarginate or shallowly bilobed, often with ciliate teeth on ventral margin, saccate at base. Female bracts appressed to perianth, emarginate or with 3–4 rounded lobes. Perianths usually abundant, urn-shaped, trigonous above, mouth shallowly 3-lobed, lobes coarsely toothed. Capsules common, late autumn to early summer. $n = 9$. On tree boles, rotting wood, rocks, rarely on soil, in woods and sheltered habitats, common and sometimes locally abundant, becoming less frequent in C. Scotland and rare in the north and west, extending to Ross, occasional in Ireland. 104, H27, C. Europe north to southern half of Fennoscandia and Russia, Turkey, Siberia, Japan, Himalayas, India, Tunisia, Azores, Madeira, Canaries, N. America.

Usually distinguished by the polymorphic leaves. Slender or young shoots may be mistaken for *L. bidentata* but more mature shoots with ± emarginate leaves can usually be found. For differences from *L. semiteres* see under that species. Differs from *Chiloscyphus* in the commonly present perianths and the lower leaves frequently bilobed. In *Chiloscyphus* the perianth is borne on a dwarf lateral branch and there is a protuberant fleshy calyptra; in *L. heterophylla* fertile branches are long, and the calyptra is thin and immersed.

Subgenus *Microlophocolea* (Spruce) Vogelpoel, *Acta Bot. Neerl.* 26: 494. 1977

Plants small. Leaves bilobed with irregularly dentate or ciliate margins; leaf cells 12–25 μm wide, walls slightly thickened, cells of perianths, bracts or whole plant with large mamillose spines. Gemmae lacking. Perianth inflated, not winged.

5. L. fragrans (Moris & De Not.) Gott. et al., *Syn. Hep.*: 166. 1845   (Fig. 90)
Chiloscyphus fragrans (Moris & De Not.) Engel & Schust.

Autoecious. Pale green or yellowish-green patches or growing through other bryophytes. Shoots procumbent, to 1.5 cm, irregularly and sometimes freely branched. Leaves variable, distant to subimbricate, spreading horizontally to somewhat dorsally subsecund, asymmetrically ovate-quadrate to ovate-rectangular, apex emarginate to 2–3-spinosely dentate or, in large leaves, irregularly spinosely dentate, one or both margins with 0–2 teeth, insertion very oblique, dorsal margin slightly decurrent, 0.3–0.8 mm long, 0.20–0.56 mm wide, (1.0–)1.2–1.5(–1.6) times as long as wide; mid-leaf cells 12–20 μm wide, walls slightly thickened, trigones lacking. Underleaves mostly 0.1–0.3 mm long, bilobed to ca ⅔, lobes subulate, margins with 0–1 teeth. Gemmae lacking. Male inflorescence on lateral branch, spicate, bracts closely imbricate, saccate, with reflexed tips. Female inflorescence terminal on main axis, bracts irregularly ovate, irregularly spinose-dentate, bracteole smaller, irregularly bilobed. Perianths occasional, ovoid, trigonous above, mouth 3-lobed, lobes spinose-dentate. Capsules occasional, spring. Damp sheltered rocks and tree trunks in woods, ravines and on banks, on sheltered rocks and in coastal caves, often in small quantity, ascending to ca 200 m in mainland habitats but to 300 m on St. Kilda, very rare in Dorset, I. of Wight, E. Sussex and Derby and rare to occasional in W. Britain from Cornwall north to Sutherland and Outer Hebrides, occasional in W. Ireland, rare elsewhere. 40, H23, C. Corsica, Spain, Portugal, France, Italy, S.W. Norway, Azores, Madeira, La Palma, Tenerife, S. Africa, Tanzania.

41. CHILOSCYPHUS CORDA IN OPIZ CORR. DUM., *Sylloge. Jung. Europ. Indig.*:
67. 1831

Autoecious or dioecious. Plants medium-sized to large, pale to dark green. Shoots procumbent to erect, branching laterally. Rhizoids in tufts arising from underleaf bases. Leaves alternate, succubous, very obliquely inserted, entire, rounded-quadrate to oblong, dorsal margin decurrent, cells thin-walled, trigones very small or absent; oil-bodies 2–8 per cell. Underleaves usually bilobed with a tooth on one or both sides. Gemmae lacking. Male bracts intercalary, resembling leaves but with 1(–2) antheridia in small inflexed lobe at base on dorsal side. Female inflorescence on dwarf lateral branch, bracts several, smaller than leaves, 2–3-lobed. Perianth short, campanulate to cylindrical, wide-mouthed, mouth 3-lobed. Calyptra extending beyond perianth. Seta stout. Capsule ovoid, wall 4–6-stratose. Spores 12–18 μm. A mainly southern hemisphere tropical, subtropical and temperate genus which because of taxonomic difficulties is of uncertain size (possibly of about 100 spp.).

1. C. polyanthos (L.) Corda in Opiz corr. Dum., *Sylloge. Jung. Europ. Indig.*: 67. 1831

Autoecious. Pale green, yellowish-green, dark green, brownish- or blackish-green tufts or thin patches or mixed with other bryophytes. Shoots procumbent to ascending, to 5.5 cm, stems occasionally to frequently branched. Leaves distant to subimbricate, very obliquely inserted and spreading horizontally, oblong to rectangular, apex rounded to emarginate, dorsal margin decurrent, 0.9–2.2 mm long, 0.8–2.7 mm wide, (0.7–)1.0–1.5 times as long as wide, mid-leaf cells 20–60 μm wide, thin-walled, trigones very small or absent. Underleaves 0.13–0.90 mm long, bilobed ½–⅔, lobes subulate, subulate tooth often present on one or both sides. Male bracts intercalary, numerous, similar to leaves but with small inflexed lobe at base. Female inflorescence on dwarf ventral branch, bracts much smaller than leaves, bilobed. Perianth short, campanulate, wide-mouthed, mouth deeply 3-lobed, calyptra protruding at maturity. Capsules rare to occasional, late winter, spring.

Key to varieties of *C. polyanthos*

Lobes of perianth mouth entire, marginal cells at middle of leaf 18–30 μm long, 25–39 μm wide, branch leaves rarely bilobed   var. **polyanthos**
Lobes of perianth mouth dentate to ciliate, marginal cells at middle of leaf 24–36 μm long, 30–50 μm wide, branch leaves often bilobed   var. **pallescens**

Var. **polyanthos** (Fig. 91)
C. polyanthos var. *rivularis* (Schrad.) Nees

Plants dull green to blackish-green. Leaves 0.9–2.0 mm long, apex rounded to emarginate, branch leaves rarely bilobed, apical marginal cells 18–20 μm long, 22–34 μm wide, marginal cells at middle of leaf 18–30 μm long, 25–39 μm wide, mid-leaf cells 22–30 μm wide. Lobes of perianth mouth entire or crenulate. *n* = 8 + m*, 9, 18. On stones, tree roots and soil in and by streams and pools, wet soil in woods, in flushes, fens, bogs, wet turf and rotting logs, mainly at low altitudes but ascending to 800 m in Scotland, frequent or common. 108, H36, C. Europe north to Spitzbergen, Faroes, Iceland, Turkey, Himalayas, Siberia, Japan, Azores, Madeira, Morocco, Tunisia, N. America.

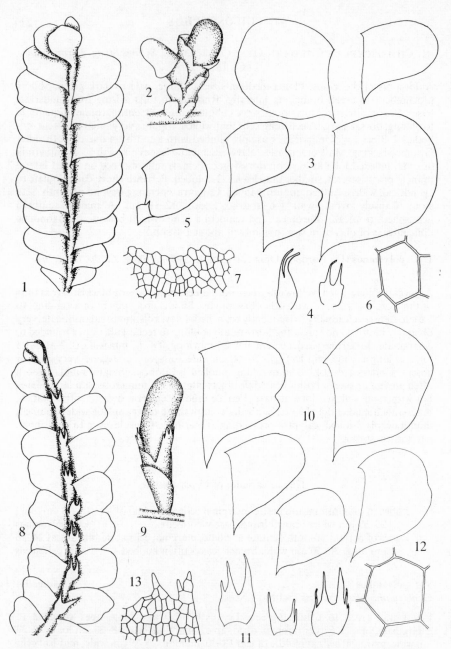

Fig. **91**. 1–7, *Chiloscyphus polyanthos* var. *polyanthos*: 1, shoot (× 8); 2, female shoot with mature sporophyte (× 8); 3, young female shoot (× 8); 4, leaves (× 15); 5, underleaves (× 21); 6, mid-leaf cell (× 330); 7, part of perianth mouth (× 84). 8–12, *C. polyanthos* var. *pallescens*: 8, ventral side of shoot (× 8); 9, female shoot with mature sporophyte (× 8); 10, leaves (× 15); 11, underleaves (× 21); 12, mid-leaf cell (× 330); 13, margin of lobe of perianth (× 84).

Var. **pallescens** (Ehrh. ex Hoffm.) Hartm., *Skand. Fl.* ed. 10, 2: 145. 1871 (Fig. 91)
C. pallescens (Ehrh. ex Hoffm.) Dum., *C. polyanthos* var. *fragilis* (Roth) K. Müll.

Pale yellowish-green to pale to dark green, often translucent, thin patches or tufts or mixed with other bryophytes. Leaves 1.00–2.25 mm long, rounded to emarginate, branch leaves sometimes bilobed; apical marginal cells 24–34 μm long, 27–45 μm wide, marginal cells at middle of leaf, 24–36 μm long, 30–50 μm wide, mid-leaf cells 27–42 μm wide. Lobes of perianth mouth dentate to spinose-dentate or ciliate. $n = 18^*$. In similar but sometimes more basic habitats than var. *polyanthos*, occasional to frequent. 100, H28. Europe north to Spitzbergen, Faroes, Iceland, Caucasus, Novaya Zemlya to Sakhalin, Japan, Azores, Madeira, N. America.

Although there are reported biochemical and chromosomal differences between *C. polyanthos* var. *polyanthos* and var. *pallescens* there is an overlap in leaf characters measured on specimens identified on the basis of the perianth. Further, perianths of an intermediate character do occur, and the oil-body characters given by Schuster (1980) also intergrade. As the two taxa tend to occur in similar habitats and have similar geographical ranges they cannot be regarded as more than varieties of a single species. Not all specimens can be named when sterile. Cell measurements given are based on observations by M.J.M. Yeo (pers. comm.). He points out that the characters given by some authors (e.g. Macvicar, 1926) to distinguish var. *rivularis* from *polyanthos* intergrade to such an extent that the variety cannot be maintained. However, it is possible that var. *fragilis*, on the basis of its larger size, may be a distinct variety but this requires experimental investigation.

SUBFAMILY GEOCALYCOIDEAE

Vegetative branches arising ventrally. Rhizoids usually scattered along ventral side of stem. Fertile branches very small, arising ventrally and concealed under lateral leaves. Male inflorescence spicate, of determinate length, antheridia with elongated, partially tiered jacket cells. Female bracts smaller than leaves or vestigial, apex of branch developing into perigynium bearing reduced perianth narrowed to mouth or marsupium with a few scales at mouth and no perianth. Capsule wall bistratose.

42. HARPANTHUS NEES, *Naturg. Europ. Leberm.* 2: 351. 1836

Dioecious. Small to medium-sized plants. Shoots prostrate to erect, stems irregularly and sparingly branched, branches arising in axils of underleaves. Rhizoids arising in branches from underleaf bases and also scattered along stem. Leaves succubous, alternate, obliquely or very obliquely inserted, $\pm$ ovate, bilobed to $\frac{1}{3}$ or apex retuse or emarginate; cells with trigones. Underleaves ovate-lanceolate or lanceolate, apex entire or notched, margins entire or toothed, joined at base to adjacent lateral leaf. Two-celled gemmae present or not. Fertile branches very short, arising on ventral side of stem. Male inflorescence spicate, concealed by lateral leaves, bracts 2–5 pairs, imbricate, very concave, bilobed, antheridia 1–2. Female bracts smaller than leaves. Sporophyte protected by tubular upgrowth of stem apex (perigynium) and short perianth, the whole forming a fleshy ovoid structure, perianth narrowed at mouth, mouth lobed or not, crenulate. Capsule ovoid, wall 2-stratose. Spores 8–12 μm. Three species, the third, *H. drummondii* (Tayl.) Grolle, occurring in N. America.

Shoots mostly more than 2 cm long, leaves 0.6–1.5 mm long, retuse to emarginate
1. H. flotovianus
Shoots not more than 1.5 cm long, leaves 0.28–0.56 mm long, bilobed
2. H. scutatus

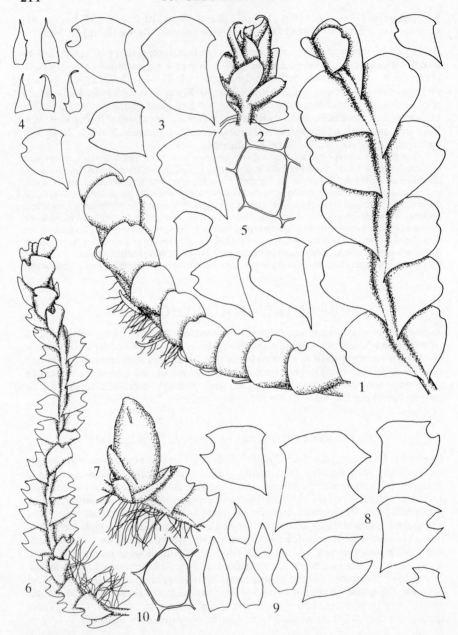

Fig. **92**. 1–5, *Harpanthus flotovianus*: 1, shoots (× 15); 2, male shoot (× 21); 3, leaves (× 15); 4, underleaves (× 38); 5, mid-leaf cell (× 450). 6–10, *H. scutatus*: 6, shoot (× 15); 7, perianth (× 15); 8, leaves (× 38); 9, underleaves (× 38); 10, mid-leaf cell (× 450).

1. H. flotovianus (Nees) Nees, *Naturg. Europ. Leberm. 2*: 353. 1836   (Fig. 92)

Dioecious. Yellowish-green, green or yellowish-brown, soft tufts or creeping through other bryophytes. Shoots procumbent to ascending, to 5 cm, stems sparsely branched, brittle. Leaves soft, distant to imbricate, spreading horizontally to dorsally subsecund, obliquely ovate-quadrate, apex retuse to emarginate, lobes rounded or obtuse, very rarely acute or acuminate, very obliquely inserted, dorsal margin longly decurrent, 0.6–1.5 mm long, 0.6–1.4 mm wide, about as long as wide; mid-leaf cells (24–)28–40 µm wide, walls yellowish-brown, thin to somewhat thickened, trigones small. Underleaves lanceolate-triangular, acute to acuminate, margins with 0–2 teeth on either side, 0.4–0.8(–1.0) mm long. Gemmae lacking. Inflorescences bud-like on ventral side of stem, to 1.5 mm long. Male bracts imbricate, bilobed. Female bracts concave, 2–3-lobed, lobes obtuse to acuminate, inflexed at inflorescence apex. Perianths and capsules unknown in Britain. Basic flushes and edges of streams in montane habitats, up to *ca* 1000 m but descending to sea-level in the Shetlands, very rare in Westmorland, formerly in Kirkcudbright, rare or occasional from C. Scotland northwards. 17. France, N. Italy and Yugoslavia north to Fennoscandia, Russia and Spitzbergen, Iceland, Siberia, N. America, S. Greenland.

Only likely to be mistaken for a *Lophocolea, Chiloscyphus polyanthos* or a large *Leiocolea* from all of which it differs in underleaf shape. According to Macvicar (1926) *H. flotivianus* is commonly fertile but this does not appear to be so; when present, however, the bud-like, ventrally borne inflorescences are distinctive.

2. H. scutatus (Web. & Mohr) Spruce, *Ann. Mag. Nat. Hist.* ser. 2, 4: 114. 1849
    (Fig. 92)

Dioecious. Pale green patches or growing through other bryophytes. Shoots ascending to prostrate with ascending tips, to 1.0(–1.5) cm, stems sparsely branched. Leaves approximate to imbricate, more rarely distant, spreading, erecto-patent to dorsally secund, obliquely ovate, obliquely inserted, dorsal margin decurrent, (0.20–)0.28–0.56 mm long, (0.20–)0.24–0.56 mm wide, (0.8–)1.0–1.2 times as long as wide, bilobed $\frac{1}{6}$–$\frac{1}{4}$(–$\frac{3}{8}$), lobes acute, sometimes somewhat connivent, sinus usually rounded; mid-leaf cells 20–32 µm wide, walls thin or thickened, yellowish-brown, trigones medium-sized to large and bulging. Underleaves lanceolate or ovate-lanceolate, 0.25–0.55 mm long, acute, entire or with an obscure tooth on one or both sides. Gemmae very rare, 2-celled, 20–28 µm long (Müller, 1957). Inflorescences bud-like on ventral side of stem. Male bracts imbricate, bilobed, very concave. Female bracts smaller than leaves, 2–3-lobed. Perianths very rare, ovoid, bluntly trigonous above, narrowed to shallowly lobed crenulate mouth. Capsules very rare, late spring, early summer. Sandstone or siliceous rocks, damp peat or soil and rotting logs in sheltered humid situations at low altitudes but ascending to 300 m in the Wicklow Mts, often in small quantities, occasional in N. Wales and the extreme west of Scotland, north to Shetland, very rare elsewhere, E. Cornwall, S. Devon, E. Sussex, W. Kent, Hereford, W. Lancashire and Yorkshire rare in Ireland. 32, H14. Europe, Iceland, Caucasus, Japan, eastern N. America.

May be mistaken, at least in W. Scotland and W. Ireland, for *Acrobolbus wilsonii* or *Geocalyx graveolens* from which it differs in underleaf shape and size and the usual presence of large trigones. Small Lophoziaceae may also be distinguished by shape of or absence of underleaves.

43. GEOCALYX NEES, *Naturg. Europ. Leberm. 1*: 97. 1833

Autoecious. Shoots prostrate, sparingly branched. Rhizoids arising in bunches from underleaf bases and also scattered along stem. Leaves succubous, alternate, insertion very oblique, bilobed; cell walls thin, trigones small. Underleaves bilobed almost to base, lobes entire. Fertile branches very short, arising ventrally. Male inflorescence

16. GEOCALYCACEAE

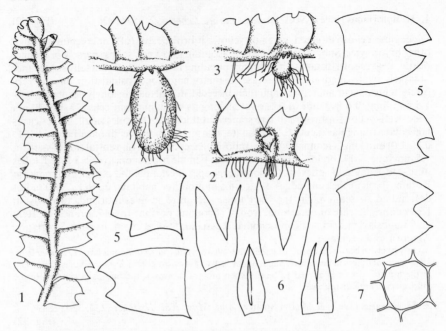

Fig. **93**. *Geocalyx graveolens*: 1, shoot (× 11); 2, side-view of shoot with male branch (× 15); 3, shoot with young marsupium (× 15); 4, shoot with mature marsupium (× 15); 5, leaves (× 23); 6, underleaves (× 23); 7, mid-leaf cell (× 450).

spicate, hidden by lateral leaves, bracts 4–8 pairs, bilobed. Female inflorescence with small bracts at apex, initially spherical, developing into fleshy, rhizoid-covered marsupium, perianth absent. Capsule cylindrical, wall bistratose. Spores 10–15 μm. A small genus with, in addition to *G. graveolens*, 2 southern hemisphere and 2 Japanese species and one species from Réunion.

1. G. graveolens (Schrad.) Nees, *Naturg. Europ. Leberm.* 2: 337. 1836   (Fig. 93)

Thin, pale green or yellowish-green patches or growing through other bryophytes. Shoots prostrate, to 1.5(–2.0) cm, very sparsely branched. Leaves approximate to subimbricate, ±spreading horizontally, ovate-quadrate, very obliquely inserted, dorsal margin slightly decurrent, 0.5–1.2 mm long, 0.4–1.2 mm wide, 1.0–1.4 times as long as wide, bilobed ⅙–⅓, lobes acute, sinus acute to rounded; mid-leaf cells 20–32 μm wide, thin-walled, trigones small. Underleaves 0.20–0.36 mm long, bilobed almost to base, lobes lanceolate, acute, entire. Male inflorescence ventral, *ca* 1.5 mm long. Perianths lacking, marsupia initially ± spherical, becoming cylindrical, rhizoidal, *ca* 2.5 mm long at maturity of capsule. Capsules occasional. On damp, shaded peat and humus at low altitudes near the sea, often in small quantity, very rare, W. Inverness, Skye, W. Ross, S. Kerry, W. Mayo. 3, H2. Europe from Spain, Italy and Romania north to Fennoscandia and Russia, Faroes (?), Caucasus, Siberia, Azores, Madeira, N. America.

Curiously rare and restricted in distribution in Britain and Ireland for a species so widespread, albeit rare, in Europe. May be mistaken in the field for *Acrobolbus wilsonii* or a small *Lophozia* or *Lophocolea* but differing in underleaf shape and the very small fertile branches.

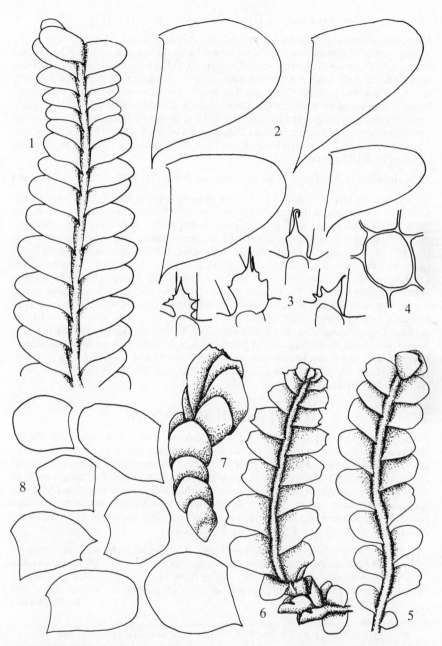

Fig. **94**. 1–4, *Saccogyna viticulosa*: 1, shoot (× 8); 2, leaves (× 15); 3, underleaves (× 21); 4, mid-leaf cells (× 450). 5–8, *Pedinophyllum interruptum*: 5, sterile shoot (× 8); 6, male shoot (× 8); 7, female shoot (× 8); 8, leaves (× 15).

44. SACCOGYNA DUM., *Commentat. Bot.*: 113. 1822

Dioecious. Plants medium-sized to robust. Shoots procumbent, branches occasional, arising ventrally. Rhizoids in tufts at underleaf bases. Leaves succubous, subopposite, entire, dorsal margin decurrent, ventral margin joined to underleaves or not, insertion very oblique; cells with a few large chloroplasts. Underleaves simple or bilobed, toothed. Gemmae lacking. Male branches dwarf, arising ventrally, bracts very small, bilobed, saccate at base; antheridia large, shortly stalked, stalks uniseriate. Female branches dwarf, arising ventrally. Perianth lacking, fleshy rhizoid-covered marsupium developing after fertilisation, mouth of marsupium with a few small bracts. Seta stout. Capsule cylindrical, wall bistratose. Except for *S. viticulosa*, a southern hemisphere genus of about 12 species.

1. S. viticulosa (L.) Dum., *Sylloge Jung. Europ. Indig.*: 74. 1831   (Fig. 94)

Plants medium-sized, in pale to olive- or brownish-green flat patches, sometimes extensive, or creeping through other bryophytes. Shoots procumbent, to 8 cm, branching occasionally. Leaves succubous, subopposite, imbricate, very obliquely inserted and spreading horizontally, plane or convex, entire, oblong, ovate-oblong or ovate-triangular, apex rounded, 0.8–1.7 mm long, 0.7–1.6 mm wide, (0.8–)1.0–1.7 times as long as wide; mid-leaf cells 20–32 μm wide, thin-walled, trigones small. Underleaves about at wide as stem, 0.25–0.70 mm long, triangular with acuminate apex, margins sharply toothed, often joined at base to one or both adjoining lateral leaves. Gemmae lacking. Male inflorescence very small, concealed under lateral leaves. Perianth lacking, marsupium cylindrical, horizontal, *ca* 2 mm long at maturity of capsule. Capsules very rare, throughout year. $n = 8 + m^*$. Rocks, tree roots and soil in humid woodland, in turf and on cliffs by the sea, in scree and on montane slopes, common and sometimes locally abundant in W. Britain and W. Ireland, rare or very rare elsewhere. 66, H33, C. Corsica, Sardinia, Portugal, W. Spain, W. France, S.W. Norway, Macaronesia.

17. PLAGIOCHILACEAE

Usually dioecious. Plants small to robust, greenish to brownish but not reddish. Shoots procumbent to erect, markedly dorsiventral, branches sparse to frequent, arising laterally or ventrally, ± leafless rhizomatous system sometimes present. Rhizoids continuous or in tufts. Leaves succubous, alternate or rarely opposite, margin entire to ciliate-dentate, decurrent or not; cells usually ± thin-walled, with or without trigones. Underleaves usually present, small, restricted to new growth. Male bracts smaller than leaves. Perianth laterally compressed, mouth truncate. Seta many cells of ± similar size in diameter. Capsule wall 4–10-stratose. Seven genera.

> Shoots usually with rhizoids ± to apex, not arising from rhizomes, leaves ± plane, margin entire, retuse or with 2–4 apical teeth   **45. Pedinophyllum**
> Shoots with few or no rhizoids, arising from horizontal rhizomatous system, leaves usually convex, usually dentate or bilobed or shoots apparently laterally compressed   **46. Plagiochila**

45. PEDINOPHYLLUM (LINDB.) LINDB., *Acta Soc. Sci. Fenn.* 10: 508. 1875

A small genus, with 4 poorly defined species, characterised by the characters given in the key to the Plagiochilaceae and for *P. interruptum*.

1. P. interruptum (Nees) Kaal., *Nyt. Mag. Naturv.* 33: 190. 1893   (Fig. 94)
P. interruptum var. *pyrenaicum* (Spruce) Kaal.

Autoecious. In flattish, yellowish-green or greenish-brown patches or growing through other bryophytes. Shoots prostrate, 1–4 cm, branching occasionally, not producing flagelliform branches or stolons. Rhizoids often presesnt ± to apex. Leaves succubous, distant to subimbricate, suberect to spreading horizontally, apical sometimes erect-appressed, ± plane, ovate-oblong to rounded-quadrate, very obliquely inserted, not or only slightly decurrent, margin entire or apex retuse or occasionally with 2–3(–4) blunt to spinose-ciliate teeth, 0.8–2.0(–3.2) mm long, 0.6–1.6(–1.9) mm wide, (0.8–)1.0–1.3(–1.7) times as long as wide; mid-leaf cells 24–32 μm wide, walls not or scarcely thickened, trigones small or absent. Underleaves small, filiform, limited to new growth. Male inflorescence terminal, becoming intercalary, on main or lateral branches, bracts smaller than leaves, imbricate, with saccate base with single tooth. Female bracts of similar size to leaves, emarginate or toothed. Perianths rare, terminal on main or lateral branches, laterally compressed, mouth truncate, ± toothed. Capsules very rare. *n* = 8 + m*, 9. Sheltered limestone rocks and ledges, locally frequent in M.W. and N.W. Yorkshire, rare or very rare elsewhere, Derby, Westmorland, Argyll, S. Kerry, Clare, Sligo, Leitrim, Fermanagh, W. Donegal. 5, H6. Spain, Corsica, Italy and Greece north to C. Europe, Belgium, France, Caucasus, Algeria, eastern N. America.

Much confused with entire-leaved or sparsely dentate forms of *Plagiochila porelloides* but differs in the absence of rhizomatous stems and stolons and in the presence of rhizoids ± to stem apex. Also, leaf bases do not reach the mid-line on the dorsal side of the stem, leaving a bare strip of stem, and the oil-bodies are smokey. In *P. porreloides* the leaf bases reach the stem mid-line and the oil-bodies are translucent.

46. PLAGIOCHILA (DUM.) DUM., *Recueil Observ. Jungerm.*: 14. 1835

Dioecious. Plants small to robust. Procumbent to erect, leafy shoots arising from rhizoid-covered rhizomatous stems, shoots sparsely to freely branched, often producing flagelliform branches and occasionally stolons, innovating from below perianth, rhizoids sparse or lacking. Leaves often varying in shape and size along stem, obliquely inserted, succubous, dorsal margin frequently reflexed, with or without teeth, decurrent or not, ventral margin usually toothed, apex rounded or truncate, toothed, rarely deeply bilobed; cell walls not or only slightly thickened, trigones small to large and bulging, oil-bodies 4–15 per cell. Underleaves where present small and usually subulate. Gemmae lacking but vegetative propagation by caducous leaves and leaf and shoot fragments. Male bracts smaller than leaves, closely imbricate, saccate. Female bracts usually 2, similar to leaves but with more numerous and coarser teeth; bracteole usually lacking. Perianth strongly laterally compressed, often with a keel on ventral side, mouth truncate, dentate or ciliate. Capsules usually rare, sometimes unknown, wall 4–10-stratose. Spores 12–20(–40) μm. A ± cosmopolitan, taxonomically exceedingly difficult genus with 1200–1300 described species.

1 Leaves concave, erect-appressed, shoots appearing laterally compressed
                                                                    1. P. carringtonii
  Leaves usually convex, variously spreading   2
2 Dorsal margin of leaf markedly decurrent   3
  Dorsal margin of leaf not or scarcely decurrent   8
3 Leaf margin with numerous teeth, those at apex of ± similar size to others, margin
    rarely obscurely dentate or entire, mid-leaf cells 24–52 μm wide, leaves to 1.4
    times as long as wide*   4

*Leaf dimensions include teeth (see Fig. 1).

Margin with 2-several teeth, those at apex usually larger than others, rarely apex retuse and otherwise entire, mid-half cells 20–32 μm wide, if teeth numerous then leaves 1.4–2.0 times as long as wide   6

4 Mid-leaf cells 36–52 μm wide, epidermal cells on dorsal side of stem 25–40 μm wide
                                                                                       5. P. britannica
   Mid-leaf cells 24–48 μm wide, epidermal cells on dorsal side of stem 16–24 μm wide
                                                                                       5

5 Shoots 1.6–5.6 mm wide, leaves mostly 1.3–2.5 mm long, mid-leaf cells 24–44 μm wide   **3. P. porelloides**
   Shoots 5–9 mm wide, leaves 2.5–4.2 mm long, mid-leaf cells 32–48 μm wide
                                                                                       4. P. asplenioides

6 Leaves 1.4–2.0 times as long as wide, shoots 3–7 mm wide   **8. P. atlantica**
   Leaves 1.0–1.5 times as long as wide, mature shoots 1.2–3.5 mm wide   7

7 Median basal cells elongated, ± sharply defined from isodiametric cells on either side and forming distinct vitta, leaves not caducous   **2. P. killarniensis**
   Median basal cells, if elongated, not sharply demarcated and not forming distinct vitta, leaves caducous   **6. P. spinulosa**

8 Leaves, at least on mature stems, not bilobed to ⅓, ventral margin usually with several teeth   **7. P. punctata**
   Leaves bilobed to *ca* ⅓, ventral margin with 0–1 teeth   **9. P. exigua**

Section *Carringtoniae* Inoue, *Bull. Natl. Sci. Mus.* 8: 377. 1965

Leaves erect, appressed so that stems appear laterally compressed, concave, widest at about middle, usually entire, dorsal margin decurrent, vitta lacking; trigones very large. Caducous propaguliferous leaves lacking.

1. P. carringtonii (Balfour) Grolle, *Trans. Br. bryol. Soc.* 4: 656. 1964   (Fig. 95)
Jamesoniella carringtonii (Balfour) Schiffn.

Robust, yellowish-green tufts. Shoots erect, 3–10(–13) cm, flagelliform shoots sometimes present. Leaves somewhat translucent, lower approximate, upper subimbricate to imbricate, erect-appressed so that shoots appear laterally compressed, overlapping stem by *ca* ⅓ on dorsal side, by *ca* ⅔ on ventral side, orbicular-reniform, somewhat concave, margin entire, somewhat sinuose, or with an occasional ciliate tooth on ventral side, rarely leaves near apex of stem bilobed to ⅙ with margin of sinus dentate, insertion oblique, dorsal margin decurrent, 0.9–1.6(–2.1) mm long, 1.2–2.0(–2.9) mm wide, 0.60–0.85 times as long as wide; mid-leaf cells 20–28 μm wide, walls thin or slightly thickened, trigones large, sometimes strongly bulging. Underleaves occasionally present on new growth, filiform, simple or bilobed almost to base, to 0.7 mm long. Only male plants known. In turf on montane heaths, rocks, rock ledges and cliffs and in scree, 150–1000 m, occasional to frequent in W. Scotland, rare in the eastern Scottish highlands, Arran, Dumbarton, Mid Perth and Angus north to Shetland, Outer Hebrides, W. Galway, W. Mayo. 19, H2. Faroes, Japan.

  The British plant belongs to ssp. *carringtonii*; ssp. *lobuchensis* Grolle, which occurs in Nepal, has been found with perianths. *P. carringtonii* bears a superficial resemblance to *Nardia compressa* but the two species differ in colour, habitat and leaf cells.

Section *Zonatae* Carl, *Ann. Bryol. Suppl.* 2: 97. 1931

Leaves ventrally secund, not concave, widest near base, margin spinose-dentate, dorsal margin decurrent, elongated cells forming vitta near base; oil-bodies numerous, homogeneous. Caducous propaguliferous leaves lacking.

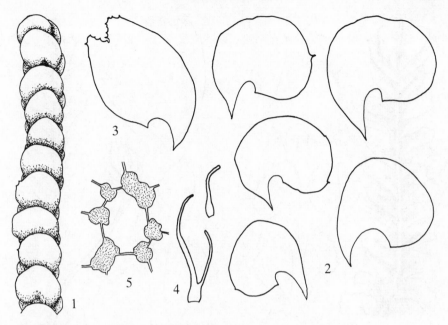

Fig. **95.** *Plagiochila carringtonii*: 1, shoot (× 8); 2, leaves (× 15); 3, leaf from shoot apex (× 15); 4, underleaf (× 38); 5, mid-leaf cell (× 450).

2. P. killarniensis Pears., *J. Bot. Br. Foreign* 43: 281. 1905   (Fig. 96)
P. spinulosa var. *inermis* Carring., *P. spinulosa* var. *killarniensis* (Pears.) Macv.

Similar to small plants of *P. spinulosa*, loose to dense, pale yellowish-green to brownish-green patches or mixed with other bryophytes. Shoots procumbent to ascending. 1–5 cm long, 1.2–2.6 mm wide, stems 120–240(–320) μm wide, occasionally branching, producing flagelliform branches. Leaves not caducous, approximate to imbricate, patent to spreading, variable in shape and size along stem, convex, markedly decurrent on dorsal side, mature leaves 0.8–1.5 mm long. 0.7–1.2 mm wide, 1.0–1.5 times as long as wide, apex often truncate with two teeth, thus appearing bilobed, dorsal margin entire, ventral margin and apex with 2-several spinose teeth, teeth on ventral margin 50–120(–160) μm long, apical cell of teeth 20–30(–40) μm long; mid-leaf cells 20–28 μm wide, cells in median basal part of leaf elongated, bounded on either side by isodiametric cells and forming a distinct vitta at least in larger leaves. Fertile plants occasional. Male bracts imbricate, appressed, saccate. Female bracts longer than wide, with more numerous and coarser teeth than leaves, dorsal as well as ventral margin usually spinose-dentate. Perianth usually without dorsal wing. Capsules not known in Britain or Ireland. On acidic to basic banks, boulders and fallen trees in woods, ravines and also coastal and montane habitats in sheltered or exposed sites, ascending to 430 m, occasional but overlooked, W. Britain and Ireland, from Cornwall north to W. Ross, Jersey, Guernsey, formerly also S. Hants, E. Sussex. 27, H13, C. Atlantic coastal Europe from S. Portugal north to Normandy, Belgium, Luxemburg, Italy, Macaronesia.

The three species, *P. spinulosa*, *P. punctata* and *P. killarniensis* are likely to be confused. The latter is distinctive in the presence of a distinct vitta at least in larger leaves. Also the leaf apices are

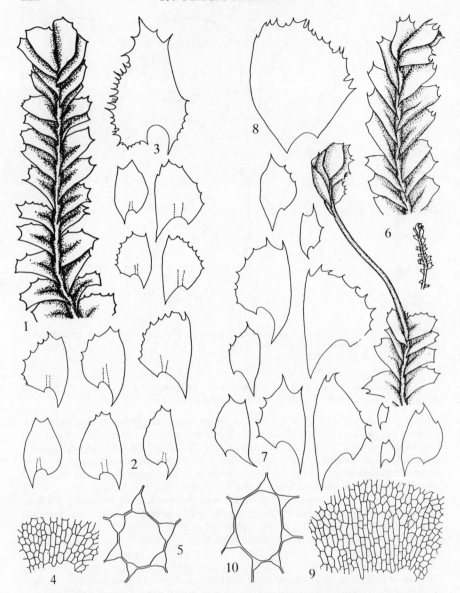

Fig. **96**. 1–5, *Plagiochila killarniensis*: 1, shoot (× 8); 2, leaves (× 11); 3, female bract (× 11); 4, median basal cells (× 60); 5, mid-leaf cell (× 450). 6–10, *P. spinulosa*: 6, shoots (× 6); 7, leaves (× 11); 8, female bract (× 11); 9, median basal cells (× 60); 10, mid-leaf cell (× 450).

often truncate, a feature not found in the other two species except sometimes in leaves of attenuated shoots. Also the leaves are not caducous. In *P. punctata* the dorsal leaf bases are not decurrent. For an account of the differences between the three species see Paton, *J. Bryol.* **9**, 451–9, 1977; for a further account of the differences between *P. killarniensis* and *P. spinulosa* see also Grolle, *J. Bryol.* **12**, 215–25, 1982.

Section *Plagiochila*

Leaves spreading horizontally to ventrally secund, not concave, widest ¼(–½) from base, margin entire to dentate with numerous teeth, dorsal margin decurrent, vitta lacking; oil-bodies few, segmented. Caducous propaguliferous leaves lacking, trigones small.

3. P. porelloides (Torrey ex Nees) Lindenb., *Spec. Hep. Fasc.* 1: 61. 1840   (Fig. 97)
P. asplenioides auct., *P. asplenioides* (L.) Dum., ssp. *porelloides* (Torrey ex Nees) Schust.

Green to dark green tufts or patches, sometimes extensive, dull green when dry. Shoots hardly pellucid when moist, ascending or spreading, 1.5–7.0 cm long, (1.6–)2.4–5.6 mm wide, simple or occasionally branched, horizontal to downward-growing leafless branches usually present, epidermal cells on dorsal side of stem 16–24 μm wide. Leaves not caducous, distant to subimbricate, patent to spreading horizontally, sometimes secund, somewhat shrunken but not longitudinally inrolled when dry, variable in shape, convex with dorsal margin usually reflexed, markedly decurrent, apex usually rounded, mature leaves 1.3–2.4(–3.0) mm long, 0.9–2.4(–3.2) mm wide, 0.9–1.2(–1.4) times as long as wide, margin dentate to spinose-dentate with numerous teeth, ventral margin more strongly toothed than dorsal which may be without teeth, teeth to 130 μm long, becoming eroded with age, sometimes rudimentary, rarely totally lacking; mid-leaf cells 24–44 μm wide, walls slightly thickened, trigones small to medium-sized and slightly bulging. oil-bodies translucent, longest median basal cells to 90 μm long. Underleaves very small, usually restricted to new growth, or lacking. Male bracts closely imbricate, appressed, saccate. Female bracts similar to leaves but larger and more coarsely toothed. Perianths rare. Capsules rare, spring. *n* = 8 + m*, 9. On soil, rocks and tree boles in sheltered habitats in woods, turf, on banks, cliffs and by streams in both acidic and basic habitats, ascending to 1230 m, occasional in lowland areas, common and sometimes locally abundant elsewhere. 105, H40, C. Europe north to Fennoscandia, Russia, Caucasus, Turkey, Madeira, N. America, Greenland.

   P. porelloides and *P. asplenioides* are ± identical in appearance but differ in overall dimensions. However, intermediate plants may be encountered which are difficult to name and for this reason Schuster (1980) treats them as subspecies of a single species. Cultivation experiments and biochemical studies have, however, shown that they are specifically distinct. *P. asplenioides*, which is the most luxuriant British species of the genus, tends to be the more frequent species in southern and eastern Britain, whilst *P. porelloides* is the commoner in western and northern Britain and Ireland.

   P. britannica differs from *P. porelloides* and *P. asplenioides* in larger stem epidermal and leaf cells. Although there is an overlap in cell size between the two species, the smallest cells of *P. porelloides* are smaller than those of *P. britannica* and similarly the largest cells of *P. britannica* exceed the width of the largest of *P. porelloides*. The width of the epidermal cells is best determined by averaging the width of 3 or 4 cells measured as a block.

4. P. asplenioides (L. emend. Tayl.) Dum., *Recueil Observ. Junger.*: 14. 1835.
(Fig. 98)

P. asplenioides var. *major* Nees, *P. major* (Nees) S. Arn.

Similar to *P. porelloides* but larger in all its parts. Shoots 3–12 cm long, 5–9 mm wide, not producing leafless flagelliform shoots, dorsal epidermal cells of stem 3–12 μm long.

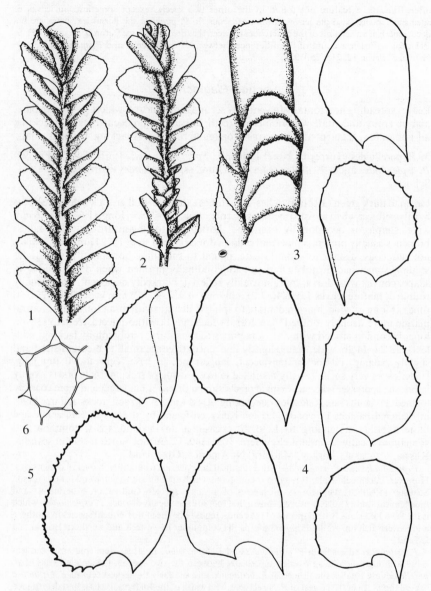

Fig. **97**. *Plagiochila porelloides*: 1 and 2, sterile and male shoots (× 8); 3, shoot with perianth in side view (× 8); 4, leaves (× 15); 5, female bract (× 15); 6, mid-leaf cell (× 450).

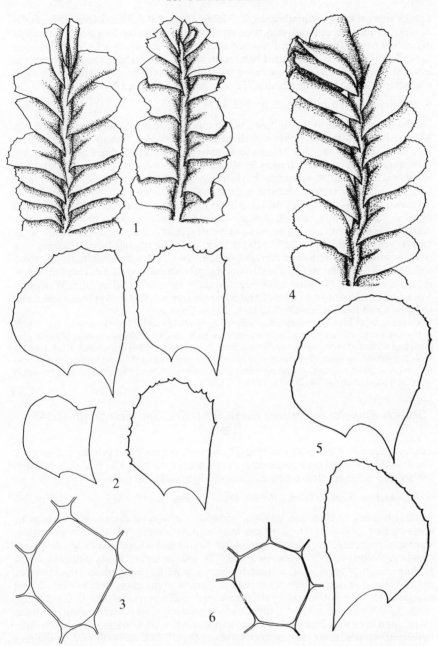

Fig. **98**. 1–3, *Plagiochila britannica*: 1, shoots (× 6); 2, leaves (× 11); 3, mid-leaf cell (× 450); 4–6, *P. asplenioides*: 4, shoot (× 6); 5, leaves (× 11); 6, mid-leaf cell (× 450).

Leaves approximate to subimbricate, 2.5–4.2 mm long, 2.0–3.7 mm wide; mid-leaf cells 32–48(–52) μm wide, cell walls thin, trigones very small or lacking, longest cells at middle of leaf base to 90(–105) μm long. Perianths and capsules very rare. $n = 8 + m^*$, 9. On basic to acidic soil and rocks in sheltered habitats in woods, turf, on banks and by streams at low altitudes, occasional to frequent, extending north to Orkney. 109, H32, C. Europe, extending north to S. Fennoscandia, Turkey, Madeira, western N. America (?).

5. P. britannica Paton, *J. Bryol.* 10: 245. 1979   (Fig. 98)

Similar in appearance to *P. porelloides* but shoots pellucid when moist, in tufts, patches or mixed with other bryophytes including *P. porelloides*. Shoots procumbent to ascending, 2–7 cm long, 2.5–7.0 mm wide, simple or branched, flagelliform branches rare, dorsal epidermal cells mostly 25–40 μm wide. Leaves not caducous, distant to subimbricate, patent to spreading horizontally or frequently ventrally secund, very variable in shape, convex, decurrent on dorsal side, apex rounded or truncate or occasionally retuse, mature leaves 1.2–3.0 mm long, 0.9–2.9 mm wide, 0.85–1.25(–1.40) times as long as wide, margin dentate to spinose-dentate, teeth to 170 μm long, becoming eroded with age, sometimes rudimentary; mid-leaf cells (32–)36–52 μm wide, longest cells at middle of base to 110 μm long, walls slightly thickened, trigones very small or absent. Perianths and capsules rare. $n = 16 + 2m^*$, $17 + m^*$. On usually basic soil and rocks in sheltered situations in woods and on banks, 30–305 m, rare but under-recorded in W. and N. Britain, scattered localities from W. Gloucester, S. Wales and Yorkshire north to Moray and N. Ebudes, rare in Ireland. 32, H8. Apparently endemic to Britain and Ireland but likely to be found in Europe.

   This plant is probably a polyploid derivative of *P. porelloides* and is likely to be found elsewhere in Europe. Forms of this plant with retuse leaves, which, because of the positioning of teeth, give the leaves a bilobed appearance, are distinct from *P. porelloides* but in other forms the leaves are indistinguishable in shape. For differences from *P. porelloides* and *P. asplenioides* see under the former. A detailed description and account of the differences from the previous two species is given by Paton, *J. Bryol.* **10**, 245–56, 1979.

Section *Spinulosae* Spruce, *Trans. Proc. Bot. Soc. Edinburgh* 15: 454. 1885

Leaves spreading horizontally to ventrally secund, not concave, usually widest at or above middle, margin spinose-dentate, dorsal margin decurrent or not, vitta lacking; oil-bodies few, segmented. Caducous propaguliferous leaves present.

6. P. spinulosa (Dicks.) Dum., *Recueil Observ. Jungerm.*: 15. 1835   (Fig. 96)

Yellowish-green to brownish patches, sometimes extensive. Shoots procumbent to spreading or ascending, (1–)2–10 cm long, mature shoots (1.0–)2.0–3.5 mm wide, sparingly branched, producing flagelliform branches and sometimes whole plants consisting largely of such branches, stems 120–360 μm wide. Leaves caducous, often leaving length of stem bare, distant to approximate or occasionally subimbricate, erecto-patent to patent, often very variable in shape and size along stem, convex, dorsal margin reflexed or not, markedly decurrent on dorsal side, mature leaves 0.8–2.0 mm long, 0.7–1.8 mm wide, 1.0–1.5 times as long as wide, dorsal margin entire, ventral margin and apex with 2–several spinose teeth, teeth on ventral margin 70–230 μm long, apical cell of teeth 24–60(–80) μm long; leaves on flagelliform shoots distant, sometimes with only 2 apical teeth and hence appearing bilobed; mid-leaf cells 24–32 μm wide, thin-walled, trigones large, bulging, rarely small, cells ± isodiametric throughout or median basal cells elongated, intergrading with isodiametric cells on either side and

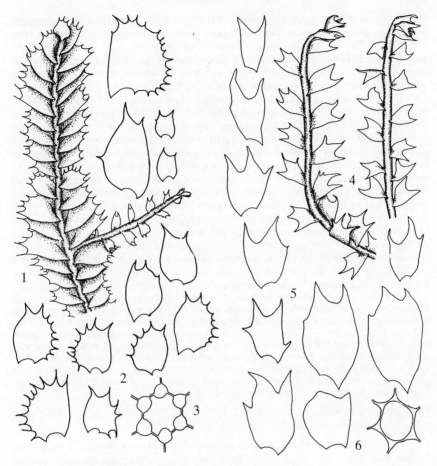

Fig. **99**. 1–3, *Plagiochila punctata*: 1, shoot (× 11); 2, leaves (× 15); 3, mid-leaf cell (× 450). 4–6, *P. exigua*: 4, shoots (× 15); 5, leaves (× 27); 6, mid-leaf cell (× 450).

forming an ill-defined vitta. Underleaves if presesnt small and subulate. Vegetative propagation by caducous leaves. Male plants not known in Britain. Female bracts as long as or slightly longer than wide, with more numerous and coarser teeth than leaves, dorsal margin without teeth. Perianths rare, with dorsal wing, mouth spinose-dentate. On and amongst acidic to basic rocks and on bark in sheltered, humid habitats in woods, on cliffs and in scree, ascending to 1250 m, widespread and sometimes locally abundant in W. and N. Britain and Ireland, from Cornwall to Somerset, Wales and the Welsh Borders and Yorkshire north to Shetland, Jersey. 57, H23, C. A few localities in France, Belgium, Luxemburg and S.W. Norway, Faroes, Macaronesia.

7. P. punctata Tayl., *Trans. Bot. Soc. Edinburgh* 1: 179. 1844   (Fig. 99)

Yellowish-green to yellowish-brown tufts or patches or mixed with other bryophytes, becoming pale brown when dry. Shoots procumbent to spreading. 1.0–2.5 cm long, mature shoots 0.7–2.0 mm wide, branching freely and producing frequently abundant flagelliform shoots and sometimes whole plant consisting entirely of such shoots, stem 120–200 (–280) mm wide. Leaves caducous, often leaving lengths of stem bare, approximate to subimbricate, erecto-patent to patent, often varying in shape and size along stem, convex, insertion arched so that base is concave, dorsal margin not or scarcely decurrent, leaves on mature stems 0.5–1.5 mm long, 0.4–1.1 mm wide, 1.0– 1.4(–1.6) times as long as wide, dorsal margin entire or very rarely with one tooth, ventral margin with 2–several spinose-ciliate teeth, teeth 50–185 μm long, apical cell 30–45 μm long, 2 apical teeth often larger than others; leaves on flagelliform shoots distant, sometimes only with two apical teeth or even emarginate, very rarely leaves on mature shoots similar; mid-leaf cells 20–28 μm wide, walls slightly thickened, trigones large, bulging, those of leaves of flagelliform shoots sometimes very small, median basal cells elongated or not, intergrading with isodiametric cells on either side and not forming a vitta. Underleaves if present small and subulate. Vegetative propagation by caducous leaves. Fertile plants rare. Male bracts imbricate, appressed, saccate. Female bracts with spinose-dentate ventral and dorsal margins. Capsules not known in Britain or Ireland. On acidic rocks and occasionally on bark in woods, ravines, scree and on cliffs, ascending to 600 m, occasional to frequent and sometimes abundant in W. Britain from Cornwall, Devon, S. Somerset, Wales, W. Yorkshire, Lake District and W. Scotland north to Shetland, Ireland. 38, H22. Portugal, Pyrenees, Belgium, Luxemburg, Rhineland, Norway, Sweden, Azores, Madeira, Gomera, Tenerife.

Likely to be confused with small forms of *P. spinulosa*. Typically it differs in the more freely branched shoots, the finer teeth, the dorsal margins of the leaf not or hardly decurrent and the dorsal margin of the female bracts spinose-dentate. Both species are, however, extremely variable and some forms may be very difficult to separate. *P. punctata* is a more markedly atlantic species than *P. spinulosa* and less frequent. Flagelliform shoots may be confused with *P. exigua* with which it sometimes grows. The leaves of *P. exigua* are more deeply bilobed (to about ⅓) and have a much narrower insertion giving the leaves a V-shaped appearance not found in *P. punctata* and the cell walls are often brownish; in flagelliform plants of *P. punctata* careful search will usually reveal larger leaves with several teeth on the ventral margin.

8. P. atlantica F. Rose, *J. Bryol.* 8: 417. 1975   (Fig. 100)
P. ambagiosa auct.

Dark green to pale brownish-green tufts or patches. Shoots procumbent, 3–9 cm long, 3.7 mm wide, stems 200–360 μm wide, branching and producing occasional horizontal or downward-growing flagelliform shoots, innovating successively from beneath female inflorescences. Leaves not caducous, distant to approximate or rarely subimbricate, when dry longitudinally inrolled, convex, markedly decurrent on dorsal side, mature leaves 1.5–3.1 mm long, 1.6–2.2 mm wide, 1.4–2.0 times as long as wide, becoming relatively wider below perianth, dorsal margin entire, ventral margin and apex usually with several, often spinose teeth, teeth on ventral margin 40–140(–210) μm long, apical cell of teeth 16–30(–40) μm long; mid-leaf cells (20–)24–28(–32) μm wide, thin-walled, trigones large, bulging, median basal cells elongated, ± grading on either side into isodiametric cells, sometimes forming an ill-defined vitta. Male plants not known. Female bracts with more numerous teeth than leaves, dorsal margin entire. Perianths frequent. On acidic, vertical or steeply sloping rock or more rarely tree boles in sheltered, humid, habitats, ascending to 450 m, very rare but sometimes locally abundant in W. Britain and S.W. Ireland, Merioneth, W. Inverness, Argyll, Kintyre, Mull, Skye, W. Ross, Kerry, W. Cork. 7, H3. Finisterre, France.

When dry the longitudinally rolled leaves give this plant a characteristic appearance. The leaves are relatively narrower than in other British species and the teeth have shorter apical cells. *P. atlantica* is more frequently fertile although male plants are unknown. For an account of the species see Jones & Rose, *J. Bryol.* **8**, 417–22, 1975.

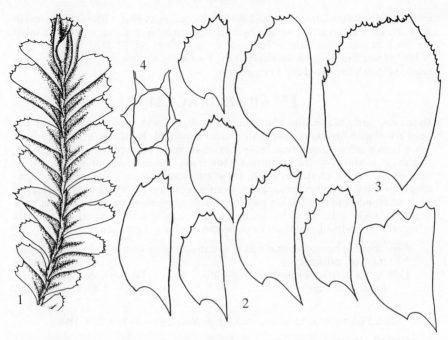

Fig. **100**. *Plagiochila atlantica*: 1, shoot (× 6); 2, leaves (× 11); 3, female bract (× 11); 4, mid-leaf cell (× 450).

Section *Caducilobae* Inoue, *J. Hattori bot. Lab.* 39: 339. 1975

Leaves spreading, not concave, widest at about middle, bilobed, with or without a few marginal teeth, dorsal margin not decurrent, vitta lacking; oil-bodies numerous, homogeneous. Vegetative reproduction by caducous leaves.

9. P. exigua (Tayl.) Tayl., *London J. Bot.* 5: 265. 1846   (Fig. 99)
P. corniculata auct., *P. stableri* Pears., *P. tridenticulata* auct.

Dull green patches or growing through other bryophytes, becoming brownish to blackish when dry. Shoots procumbent to ascending, 0.5–2.0 cm long, 0.8–1.8 mm wide when mature, stem 110–160 μm wide, branching sparingly and producing frequently abundant flagelliform shoots. Leaves caducous, leaving lengths of stem bare, distant, erecto-patent to patent, rarely spreading horizontally, varying in shape and size along stem, convex, markedly narrowed to insertion, margin not or slightly decurrent on dorsal side, leaves on mature stems 0.5–1.2 mm long, 0.3–0.6 mm wide, (1.0–)1.3–2.1 times as long as wide, bilobed to *ca* ⅓, lobes acuminate, spreading so that leaves sometimes appear V-shaped, rarely more shallowly lobed or emarginate, or 3-lobed, dorsal margin entire or very rarely with a tooth near base, ventral margin with 0–1 or very rarely 2 teeth; mid-leaf cells 20–28 μm wide, walls slightly thickened, often brownish, trigones small to medium-sized. Underleaves small and subulate or absent. Vegetative propagation by caducous leaves. Male plants rare, female unknown in Britain or Ireland. On acidic to mildly basic rocks and ledges and more rarely on bark, in sheltered woodland, in ravines and in scree, ascending to 300(–610) m, occasional in N.W. Wales, Lake District, western Scottish highlands, W. Ireland, very rare in other western localities, from Carmarthen north to W. Sutherland. 22, H10. S.W. Norway, N.W. France, Pyrenees, Switzerland, Italy, Faroes, Macaronesia, southern Appalachians (U.S.A), Mexico, Juan Fernandez Is.

18. ARNELLIACEAE

Dioecious, rarely monecious. Plants small to robust, green, sometimes brownish but never red-tinged. Shoots dorsiventral, simple or sparingly branched, innovating from below female inflorescence, stem fleshy. Leaves opposite, often connate on dorsal side, succubous, ovate to orbicular, entire, ± transversely inserted; cells thin-walled, with or without trigones. Underleaves small, sometimes restricted to below female inflorescence or absent. Vegetative propagules sometimes present on dorsal surface of leaves. Male bracts intercalary. Perianth short or absent; perigynium present, usually at right angles to main axis, after fertilisation marsupium usually developing. Capsule spherical to cylindrical, wall bistratose. Spores 8–12 μm. Three genera.

Cells towards base of ventral part of leaf to 4 times as long as wide, female bracts toothed, calcicole   **47. Southbya**
Cells towards base of ventral part of leaf 4–7 times as long as wide, female bracts entire, calcifuge   **48. Gongylanthus**

47. SOUTHBYA SPRUCE, *Ann. Mag. Nat. Hist.* ser. 2, 3: 501. 1849

Plants small. Underleaves present below female inflorescence. Gemmae present on leaf surface or not. Female bracts toothed. Archegonia terminal. Perianth immersed to barely emergent, mouth wide, toothed; perigynium present, rudimentary marsupium developing after fertilisation or not. Capsule globose to ellipsoid. Four species.

Dioecious, pale brown to brown when dry, stem T.S. less than twice as wide as thick, male bracts entire   **1. S. tophacea**
Paroecious, dark brown to blackish when dry, stem T.S. at least twice as wide as thick, male bracts bluntly toothed   **2. S. nigrella**

1. S. tophacea (Spruce) Spruce, *Ann. Mag. Nat. Hist.* ser. 2, 3 501. 1849   (Fig. 101)
S. stillicidiorum (Raddi) Lindb.

Dioecious. Bright green to yellowish-green patches or scattered plants, pale brown to brown when dry. Shoots prostrate to ascending, 2–7 mm, occasionally branching, innovating from below female inflorescence, stem in section circular to oval, less than twice as wide as thick. Rhizoids abundant, colourless to very pale brown. Leaves opposite, both surfaces of similar colour, often connate on dorsal side, distant to imbricate, erecto-patent, not incurved when dry, ovate to ± orbicular, concave, entire, lacking appendage on outer face, ± transversely inserted, not decurrent, 0.4–1.0 mm long, 0.35–0.80 mm wide, 0.9–1.4 times as long as wide; mid-leaf cells (24–)28–40 μm wide, thin-walled, trigones very small or absent, cuticle slightly papillose, cells becoming longer towards base on ventral side, longest 60–90 μm long, 3–4 times as long as wide. Underleaves only present below female inflorescence, subulate, to *ca* 180 μm long. Male bracts to 12 pairs, entire, very saccate. Female bracts and immediately preceding leaves somewhat undulate, with irregularly toothed margins and apex. Perianths frequent, immersed to barely emergent; perigynium present, in same plane as main axis, no marsupium developing after fertilisation. Capsules frequent, September

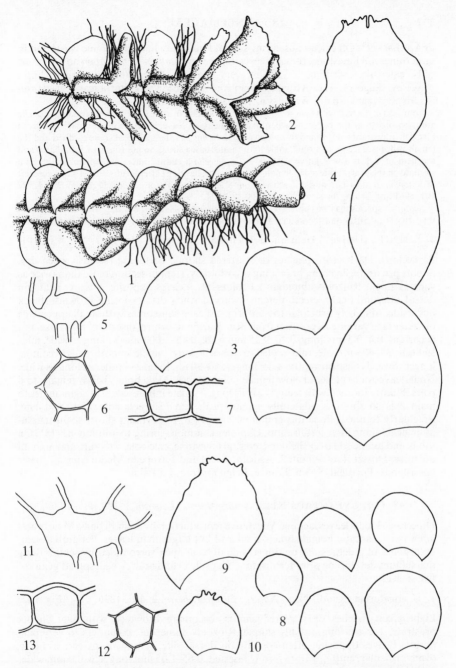

Fig. **101**. 1–7, *Southbya tophacea*: 1, sterile shoot (× 21); 2, fertile shoot (× 21); 3, leaves (× 38); 4, female bract (× 38); 5, т.s. stem (× 63); 6, mid-leaf cell (× 450); 7, т.s. leaf cells (× 450). 8–13, *S. nigrella*: 8, leaves (× 38); 9, female bract (× 38); 10, male bract (× 38); 11, т.s. stem (× 63); 12, mid-leaf cell (× 450); 13, т.s. leaf cell (× 450).

to April. $n = 9*$. On calcareous clayey soil and on sandy humus on mine waste and in sand-dunes, on limestone, tufa and growing over *Eucladium verticillatum* on limestone cliffs, calcicole, ascending to 120 m, very rare, W. Cornwall, Glamorgan, Flint, Anglesey, Sligo. 4, H1. African, Asian and European Mediterranean coasts and islands, Portugal, Spain, W. Russia, Caucasus, Macaronesia.

Confused in the past with *S. nigrella* which is very similar in appearance. The leaves of *S. nigrella* are dark brown at the back and when dry the incurved leaves give the plant a dark brown to blackish appearance not found in *S. tophacea*; in *S. nigrella* the male bracts are toothed and the female inflorescence set at a right angle to the main axis, a situation not found in *S. tophacea* and the stem in T.S. is also relatively wider. In *S. nigrella* a rudimentary marsupium, visible as a swelling on the ventral side of the female inflorescence, develops after fertilisation. For an account of *S. tophacea* in Britain and differences from *S. nigrella* see Paton, *Trans. Br. bryol. Soc.* **4**, 98–101, 1961 and *Trans. Br. bryol. Soc.* **6**, 328–30, 1971.

Southbya species may be confused with *Gongylanthus ericetorum* but they are calcicole whilst the latter is calcifuge and the longest basal cells of the leaves are shorter than in *Gongylanthus*.

2. S. nigrella (De Not.) Henriques, *Boletim Soc. Broter.* 4: 244. 1887   (Fig. 101)

Paroecious. Dark green patches or scattered shoots, becoming blackish when dry. Shoots prostrate, 2–5 mm, branching occasionally, stem in section more than twice as wide as thick. Rhizoids abundant, ±colourless. Leaves opposite, often connate on dorsal side, imbricate, erecto-patent, incurved when dry, exposing brownish-black underside, obliquely orbicular to reniform, concave sometimes with small appendage on outer face varying in position from near margin to centre, insertion transverse, not decurrent, 0.4–0.5 mm long, 0.40–0.55 mm wide, 0.85–1.00 times as long as wide, mid-leaf cells 32–40 μm wide, thin-walled, trigones lacking, cuticle smooth, cells becoming longer towards base on ventral side, longest 60–80 μm long, 2–4 times as long as wide. Underleaves only present below female inflorescence. Male bracts below female, to 6 pairs, bluntly toothed with teeth 1–2 cells long. Female inflorescence at right angles to main axis so shoot is bilaterally symmetrical, female bracts somewhat undulate, irregularly toothed. Perianths common, immersed; perigynium short, slight marsupium developing after fertilisation. Capsules common, spring to autumn. $n = 18*$. On sandy and humic soil over sheltered, coastal limestone, calcicole, very rare and in small quantity, Dorset, I. of Wight. 2. African, Asian and European Mediterranean coasts and islands, Portugal, Spain, France, Belgium, Iraq, La Palma.

48. GONGYLANTHUS NEES, *Naturg. Europ. Leberm.* 2: 388, 405. 1836

Dioecious. Underleaves lacking. Vegetative reproduction by axillary buds. Male bracts a few pairs, saccate. Female bracts similar to but larger than leaves. Perianth absent, archegonia in a depression on dorsal side of stem apex surrounded by scales. Long marsupium developing after fertilisation. Capsule cylindrical. A widespread genus of five species.

1. G. ericetorum (Raddi) Nees, *Naturg. Europ. Leberm.* 2: 407. 1836   (Fig. 102)

Light green patches or scattered shoots, becoming brownish with age. Shoots prostrate, 1–2 cm long, usually simple. Rhizoids numerous, colourless to very pale purple. Leaves opposite, often connate on dorsal side, imbricate, patent, ovate to ovate-orbicular, entire, ± transversely inserted, 0.65–1.05 mm long, 0.6–0.9 mm wide, 0.95–1.25 times as long as wide; mid-leaf cells 28–36 μm wide, thin-walled, trigones distinct, cells becoming longer ± from middle of leaf to ventral margin, longest cells 80–120 μm long, 4–7(–11) times as long as wide. Underleaves lacking. Vegetative propagation and possibly perennation by axillary buds. Only female plants known in

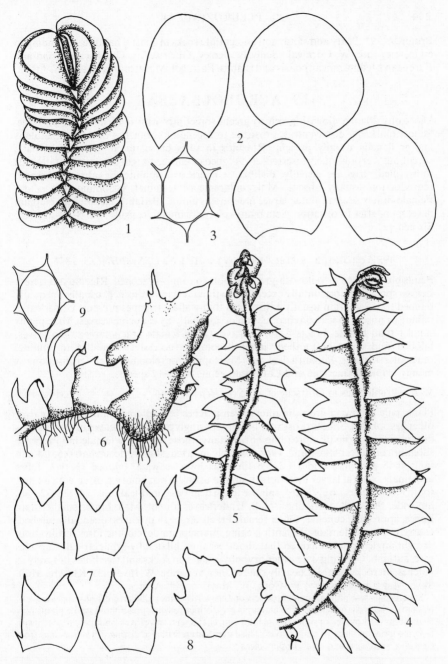

Fig. **102**. 1–3, *Gongylanthus ericetorum*: 1, shoot (× 21); 2, leaves (× 38); 3, mid-leaf cell (× 450). 4–9, *Acrobolbus wilsonii*: 4, sterile shoot (× 15); 5, male shoot (× 15); 6, female shoot with marsupium containing mature capsule, (× 15); 7, leaves (× 15); 8, female bract (× 15); 9, mid-leaf cell (× 450).

Britain. $n = 9^*$. Peaty and sandy soil and granite rocks in coastal habitats, ascending to 60 m, very rare, W. Cornwall, Scilly Is., Jersey, Guernsey. 1, C. African, Asian and European Mediterranean coasts and islands, Portugal, Macaronesia, W. and S. Africa.

19. ACROBOLBACEAE

Dioecious. Plant yellowish-green to green, sometimes with reddish or brown tinge. Shoots markedly dorsiventral, prostrate to ascending, branches occasional, stolons and/or flagella usually present. Rhizoids in tufts or scattered. Leaves alternate, succubous, very obliquely inserted, 2(–5)-lobed or entire, margin entire to dentate; cells thin-walled, trigones usually distinct. Underleaves minute or absent. Asexual reproduction usually absent. Male inflorescence terminal, becoming intercalary. Female bracts several, outer large, innermost minute. Perianth lacking, marsupium developing after fertilisation. Seta many cells in diameter; capsule wall 5–10-stratose. Six genera.

49. ACROBOLBUS NEES IN GOTTSCHE ET AL., *Syn. Hep.*: 5. 1844

Plants bluish-green or yellowish-green. Shoots sparingly branched. Rhizoids scattered. Leaves spreading horizontally except beneath female inflorescence, usually unequally bilobed to $\frac{1}{3}-\frac{2}{3}$, margin usually entire; cells thin-walled, trigones present. Underleaves minute or absent. Asexual reproduction occasionally by caducous leaves. Male plants smaller than female, bracts 4–10, concave. Outer female bracts larger than leaves, bilobed and coarsely toothed, inner becoming successively smaller to minute; bracteoles lacking. Perianth lacking. Marsupium developing after fertilisation. A mainly tropical American and Caribbean genus of 10–12 species.

1. A. wilsonii Nees in Gottsche et al., *Syn. Hep.*: 5. 1844   (Fig. 102)

Plants pale yellowish-green, in interwoven patches or scattered shoots growing over other bryophytes. Shoots prostrate, 1–2 cm, sparingly branched. Leaves of sterile stems distant to approximate, spreading horizontally, those below the female inflorescence subimbricate, erect-spreading, very obliquely inserted, slightly decurrent, 0.6–1.4 mm long, 0.45–1.20 mm wide, 1.2–1.6 times as long as wide, bilobed $(\frac{1}{6}-)\frac{1}{3}-\frac{1}{2}$, lobes acuminate, ventral larger than dorsal, sinus ± acute, ventral margin often with a blunt tooth near middle, rarely with a spinose tooth near base; mid-leaf cells (24–)28–32(–36) μm wide, thin-walled, trigones small. Underleaves absent. Male bracts smaller than leaves, appressed, concave. Outer female bracts larger than leaves, unequally bilobed, coarsely spinose-dentate. Perianth lacking; marsupium developing after fertilisation, at right angles to main axis so that shoot becomes bilaterally symmetrical. Capsules rare, autumn. Growing over other bryophyte mats on rocks and tree boles or rarely as patches, on rocks in moist, heavily shaded sites, very rare, W. Inverness, Kintyre, Mull, Skye, Mid Ebudes, Kerry, W. Cork, W. Mayo. 5, H4. Azores.

Superficially resembling a small *Lophozia* or *Leiocolea* but the pallid green, fleshy appearance of fresh material is characteristic. The marsupia are distinctive when present and sterile plants may be distinguished by one or more of the following characters: leaves very obliquely inserted with tapering lobes and lack of underleaves. This latter feature will also distinguish *Acrobolbus* from *Geocalyx graveolens* and small Lophocoleas.

20. PLEUROZIACEAE

A small family of two genera, the second, *Eupleurozia* Schust., differing from *Pleurozia* in leaf morphology and branching pattern.

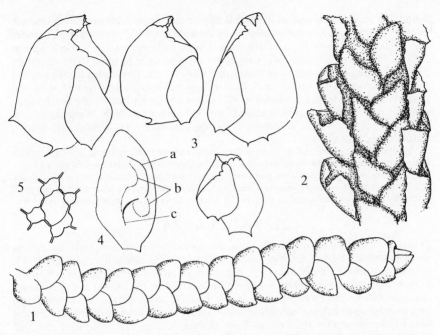

Fig. **103**. *Pleurozia purpurea*: 1, shoot, dorsal side (× 8); 2, shoot, ventral side (× 13); 3, leaves, ventral side (× 13); 4, inner face of ventral lobe, *a* = fold, *b* = internal flaps, *c* = slit (× 15); 5, mid-dorsal lobe cell (× 450).

50. PLEUROZIA DUM., *Recueil. Observ. Jungerm.*: 15. 1835

Plants robust, often purplish. Stems ascending, arising from rhizomatous base. Leaves incubous, bilobed, dorsal lobe large, entire or dentate, apex emarginate or not, ventral lobe smaller, saccate, attached at base to stem, opening by a slit on inner face, with 2 flaps on interior wall. Underleaves absent. Inflorescences on short lateral branches. Male inflorescence spicate with 6–12 pairs of imbricate bracts. Perianth elongate, plicate above, narrowed to mouth. Tubular cylindrical structures, 1–3 mm long, wide-mouthed and smooth, referred to as 'sterile perianths' sometimes borne in position of true perianths. About 15 mainly tropical species.

   The structure of the ventral lobe of the leaf and the presence of 'sterile perianths' are unique to the Pleuroziaceae. The family is also unusual in the Jungermanniales in having an apical cell with only 2 cutting faces.

1. P. purpurea Lindb., *Acta Soc. Fauna Fl. Fenn.* 1(2): 27. 1877   (Fig. 103)

Autoecious. Plants robust, reddish-yellow to reddish-purple or deep purple, in tufts or growing with other bryophytes. Shoots ascending with ventrally secund tips, stems simple or with 2–3 branches. Leaves closely imbricate, incubous, ventrally secund, conduplicate bilobed almost to base, dorsal lobe broadly ovate, widely crossing stem, convex, apex emarginate, dentate, dorsal margin ± semicircular with 2–3 teeth near base, (1.1–)1.6–2.4 mm long, (0.9–)1.2–2.1 mm wide; ventral lobe about ¼–⅓ size of dorsal, saccate, pyriform to ovoid or broadly ovoid, (0.9–)1.2–1.8 mm long, (0.5–)0.7–1.0 mm wide, inner face with a curved elongated ridge ending in a curved slit with a round flap on one side and an elongated flap on the other, both projecting into the interior; cells in middle of dorsal lobe 20–24 μm wide, walls slightly thickened, trigones very large. Underleaves lacking. Gemmae lacking. Fertile plants unknown in Britain or Ireland. Female inflorescence on a short branch near apex of main stem; immature perianth oblong-ovoid with three folds from below middle, contracted to ciliate-lobed mouth (Macvicar, 1926). $n = 8 + m^*$. Blanket bog, wet heath, montane turf, scree and peat-covered rock ledges, common in W. Scotland from Kirkcudbright northwards, common in W. Ireland, curiously absent from N. England and N.W. Wales, ascending to *ca* 1000 m. 15, H24. S.W. Norway, Faroes, Jan Meyen Is., Himalayas, Alaska, Guadeloupe, Hawaii.

The ventral lobe of the leaf of *Pleurozia* has excited considerable interest. The sac-like lobe opens to the exterior by a slit on the inner face, and on either side of the interior of the slit is a flap of thin-walled cells. It is said that when moist the flaps close the opening in such a way as to allow the entry of water but preventing its egress. These flaps collapse when dry. It is also reported that small invertebrates become entrapped in the sacs but whether this is of any significance is unknown.

21. RADULACEAE

Dioecious or rarely monoecious. Plants mostly medium-sized to large and forming thin patches, yellowish-green to olive-green or brown. Stems prostrate, sparingly 1(–2)-pinnate, branching of the *Radula* type (a form of terminal branching in which the branch appears to arise slightly in front of the leaf base on the ventral half of the stem, the branches usually being less vigorous than the main axis). Leaves incubous, approximate to imbricate, spreading horizontally, conduplicate-bilobed, the ventral lobe markedly smaller than dorsal lobe, margins entire, keel region inflated and rhizoids arising only from inflated part of ventral lobe near keel, cells with or without trigones, oil-bodies 1(–3), very large and rendering cells opaque. Underleaves absent. Discoid gemmae or caducous leaves sometimes present. Male inflorescence in dioecious species spike-like with few to numerous bracts, bracts imbricate, smaller than leaves, erect, saccate; in monoecious species bracts few, ± similar to vegetative leaves but saccate. Female inflorescence usually with 1–2-subinvolucral innovations, female bracts usually 2, smaller than vegetative leaves with relatively larger ventral lobe. Perianth ± terete at base, dorsiventrally compressed above, smooth, mouth wide, entire to dentate. Capsule wall bistratose. Spores 15–55 μm. One genus.

51. RADULA DUM., *Commentat. Bot.*: 12. 1822

With the characters of the family. A mainly tropical and subtropical genus, probably with about 150–210 species.

1 Ventral lobe of leaf widely crossing stem   **3. R. voluta**
  Ventral lobe not crossing stem   2
2 Plants reddish-brown, dorsal lobe of leaf convex, keel region strongly inflated
                                                                                      5. R. aquilegia
  Plants greenish, rarely brownish, dorsal lobe plane or slightly convex, keel region not
  or only slightly inflated   3
3 Dorsal lobe of leaf ovate-orbicular, crossing stem, gemmae often present   4
  Dorsal lobe of leaf ovate, not or scarcely crossing stem, gemmae lacking   5
4 Paroecious, perianths common   **1. R. complanata**
  Dioecious, perianths very rare   **2. R. lindenbergiana**
5 Paroecious, perianths common   **4. R. holtii**
  Dioecious, perianths very rare   **6. R. carringtonii**

Subgenus *Radula*

Female inflorescence terminal on short or long leafy branch, often apparently lateral because of development of subinvolucral innovations, with 1–2 innovations from below female inflorescence.

1. R. complanata (L.) Dum., *Sylloge Jung. Europ. Indig.*: 38. 1831   (Fig. 104)

Paroecious. Thin, yellowish-green patches or growing over other bryophytes. Shoots prostrate, to *ca* 2 cm, stems irregularly pinnately branched. Leaves on main stems approximate to imbricate below, imbricate to closely imbricate above, spreading, margins entire or irregular if gemmiferous, dorsal lobe ovate-orbicular, crossing stem, ± flat to slightly convex with inflexed margin towards rounded apex, 0.7–2.0 mm long, 0.54–1.70 mm wide; ventral lobe appressed to dorsal, obliquely rectangular, not crossing stem, 0.4–1.2 mm long, 0.2–0.9 mm wide, 0.5–0.6 length of dorsal lobe, apex pointed to rounded; keel region slightly inflated; leaves on branches smaller but otherwise similar; cells in middle of dorsal lobe 16–24(–30) μm wide, walls thin or slightly thickened, trigones small or absent. Underleaves absent. Gemmae frequent on leaf margins, ± discoid, 20–70 μm or more in diameter. Plants usually fertile, inflorescences terminal on stems and branches; female bracts erect, somewhat saccate at base and sometimes with antheridia; male bracts below female, (1–)2–4 pairs, ± erect, saccate at base. Perianths common, compressed, ± rectangular, contracted at base, mouth truncate, entire. Capsules common, most of the year, spores 24–40 μm. $n = 6, 8, 12$. Epiphytic on deciduous trees, less commonly on rocks, mainly at low altitudes but ascending to *ca* 850 m, occasional to common throughout Britain and Ireland. 111, H40, C. Europe north to Fennoscandia and N. Russia, Iceland, extending east to China, Sakhalin and Japan, Algeria, Azores, Madeira, Tenerife, N. America, Greenland.

R. complanata and R. lindenbergiana are very closely related and can only be named when fertile. In R. complanata there are always 1–4 pairs of saccate male bracts below the female inflorescence which is subtended by two saccate bracts. In R. lindenbergiana there are only vegetative leaves immediately below the female inflorescence and the female bracts are not saccate; female plants without perianths may appear sterile but archegonia, if present, may be detected by clearing or dissecting a stem or branch apex. In R. lindenbergiana lowland plants are mainly female and montane ones male. In the past there has been a tendency to assume that montane or gemmiferous plants are R. lindenbergiana but R. complanata is often abundantly gemmiferous and may ascend to at least 850 m. Some authorities regard R. lindenbergiana as a subspecies or variety or even a synonym of R. complanata but the male inflorescence and female bracts are very different and there are some differences in their geographical distribution so such treatment seems unjustified.

Lowland forms of R. lindenbergiana are almost invariably gemmiferous but some montane plants lack gemmae and may belong to a distinct taxon. This, together with R. complanata which is very variable in cell and spore size and chromosome number, requires further study.

2. R. lindenbergiana Gott ex Hartm. f., *Skand. Fl.* 9: 98. 1964   (Fig. 104)
R. complanata ssp. *lindbergiana* (Gott.) Schust., *R. lindbergiana* Gott.

Dioecious. Vegetatively similar to *R. complanata*. Male plants rare, inflorescence terminal, becoming intercalary, spike-like, bracts numerous, closely imbricate, erect, ± equally bilobed, strongly saccate. Female plants frequent, bracts erect, ± similar to vegetative leaves, not saccate at base. Perianths rare, similar as those of *R. complanata*. Capsules very rare. $n = 8$. On damp sheltered rocks, especially where basic, from sea-level to *ca* 1200 m, occasional in western and montane areas, Cornwall, S. Devon, and N.W. Wales, Shropshire and N.W. Yorkshire north to Orkney, very rare in Ireland. 41,

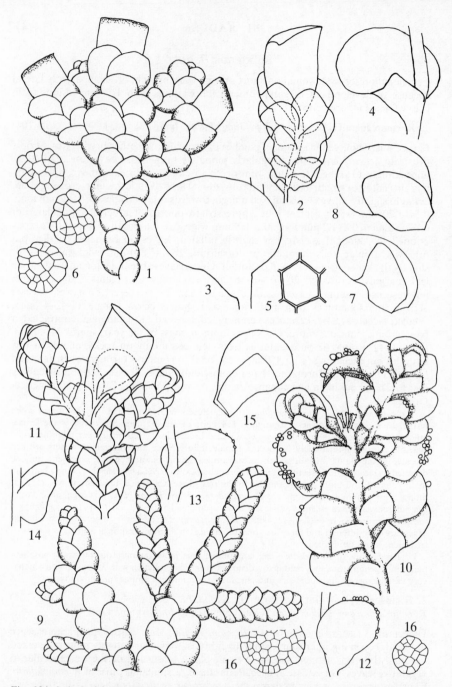

Fig. **104**. 1–6, *Radula complanata*: 1, shoot with perianths, dorsal side (× 16); 2, outline of paroecious inflorescence in ventral view (× 16); 3 and 4, leaves in dorsal and ventral view (× 23); 5, cell from middle of dorsal lobe of leaf (× 450); 6, gemmae (× 150); 7, male bract (× 23); 8, female bract (× 23). 9–16, *R. lindenbergiana*: 9, male shoot with inflorescences (× 15); 10, female shoot with archegonia, ventral view (× 15); 11, outline of female inflorescence with perianth (× 16); 12 and 13, dorsal and ventral views of leaves (× 23); 14, male bract (× 38); 15, female bract, ventral view (× 23); 16, gemmae (× 150).

H8, C. Europe north to Fennoscandia and N. Russia, Faroes, Iceland, Caucasus, Turkey, Iran, Shensi, Korea, Japan, Taiwan, Himalayas, Tunisia, Morocco, Algeria, S. Africa, Macaronesia, Tennessee, Greenland.

3. R. voluta Tayl. ex Gott et al., *Syn. Hep.*: 255. 1845   (Fig. 105)

Dioecious. Thin, pale green or yellowish-green patches or growing over other bryophytes. Shoots prostrate, to 3 cm, stems irregularly 1(–2)-pinnately branched. Leaves on main stems flaccid, approximate to loosely subimbricate, spreading, margins entire, dorsal lobe ovate-rectangular, overlapping stem, ± flat or somewhat undulate, 0.7–1.4 mm long, 0.6–1.1 mm wide, apex obtuse to rounded; ventral lobe broadly ovate-cordate, widely crossing stem, 0.40–0.64 mm long, 0.3–0.5 mm wide, 0.4–0.6 length of dorsal lobe, apex acute to rounded; keel region not inflated; leaves on branches smaller but otherwise similar; cells in middle of dorsal lobe (16–)20–24 μm wide, walls very thin, trigones very small. Underleaves absent. Gemmae rare, disciform. Male plants common, inflorescence spike-like, terminal on or constituting whole of branch, bracts 2–4(–6) pairs, smaller than leaves, closely imbricate, erect, saccate. Female plants unknown in Britain, inflorescence terminal, becoming pseudolateral by growth of an innovation (Schuster, 1980). Perianths and capsules unknown. Usually on basic rocks and on other bryophytes in deeply shaded, humid habitats, especially in the spray zone of waterfalls, ascending to 200 m in Britain and to 750 m in S.W. Ireland, rare in N.W. Wales, Cumberland, Kirkcudbright, Ayr, N.W. Scotland north to Skye and W. Ross, occasional in S.W. Ireland, very rare elsewhere. 11, H10. N. Carolina, Tennessee.

Distinct from other native *Radula* species in the ventral lobe of the leaf widely crossing the stem.

4. R. holtii Spruce, *J. Bot. Br. Foreign* 25: 209. 1887   (Fig. 106)

Paroecious. Thin dark green to olive-green patches or growing over other bryophytes. Shoots prostrate, to *ca* 1.5 cm, stems sparsely irregularly pinnately branched and innovating from below female inflorescence, medullary cells of stem in section thin-walled with small trigones. Leaves on main stem distant below, approximate to subimbricate above, spreading, margins entire, ventral edge of leaf often angled at junction or margin and keel, dorsal lobe ovate, not crossing stem, slightly convex, 0.6–1.1 mm long, 0.4–0.9 mm wide, apex rounded; ventral lobe obliquely ovate to ovate-triangular, not crossing stem, 0.24–0.48 mm long, 0.16–0.36 mm wide, 0.35–0.50 length of dorsal lobe, ± rounded; keel region slightly inflated; leaves on branches smaller but otherwise similar; cells in middle of dorsal lobe 16–24 μm wide, walls thin or very thin, trigones absent. Underleaves absent. Gemmae absent. Inflorescences terminal on stems, sometimes becoming pseudolateral by growth of one innovation, often produced successively, female bracts erect, smaller than leaves, not saccate at base; male bracts 2–4 below female, spreading, similar to leaves but saccate at base, male bracts sometimes also produced at base of subfloral innovations. Perianths usually present, trumpet-shaped, from very narrow base gradually widening to shallowly lobed mouth. Capsules rare, early summer. Shaded wet rocks in ravines and by waterfalls, ascending to *ca* 300 m, rare in S. and W. Ireland, Kerry, W. Cork, W. Mayo. H4. Portugal, N.W. Spain, Azores, Madeira, Tenerife.

Readily recognised by the usually present paroecious inflorescence and trumpet-shaped perianths and distinguished from other superficially similar *Radula* species except *R. carringtonii* by the dorsal lobe not crossing the stem. Sterile material of *R. carringtonii* may be separated from *R. holtii* by leaf morphology and the stem cells in T.S. with large trigones. Although *R. carringtonii* is more widespread, in the extreme S.W. of Ireland *R. holtii* is considerably commoner. The two species may grow together and the two have been confused as Macvicar (1926) says, incorrectly,

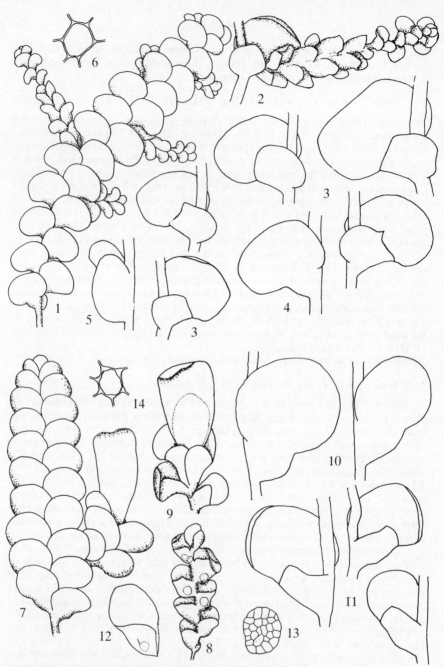

Fig. **105**. 1–6, *Radula voluta*: 1, male shoot, dorsal view (× 13); 2, male branch, ventral view (× 23); 3 and 4, leaves, ventral and dorsal views (× 23); 5, male bract (× 38); 6, cell from middle of dorsal lobe (× 450). 7–14, *R. aquilegia*: 7, shoot with perianth, dorsal side (× 15); 8, male shoot, ventral side (× 33); 9, shoot with perianth, ventral side (× 15); 10 and 11, leaves, dorsal and ventral sides (× 38); 12, male bract, ventral side (× 63); 13, gemma (× 150); 14, cell from middle of dorsal lobe (× 450).

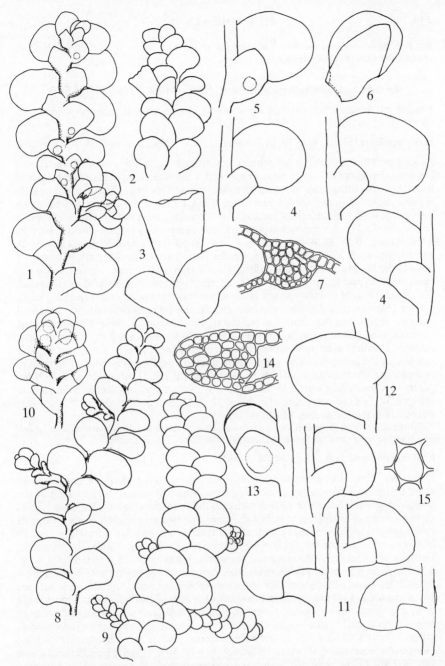

Fig. **106**. 1–8, *Radula holtii*: 1, paroecious shoot, ventral side (× 15); 2, shoot, dorsal side, with perianth (× 15); 3, perianth, dorsal side (× 38); 4, leaves, ventral side (× 63); 5, male bract, ventral side (× 63); 6, female bract, ventral side (× 63); 7, т.s. stem (× 150). 8–15, *R. carringtonii*: 8 and 9, male shoots, dorsal side (× 15); 10, male shoot, ventral side (× 21); 11 and 12, leaves, ventral and dorsal side (× 38); 13, male bract, ventral side (× 63); 14 т.s. stem (× 150); 15, cell from middle of dorsal lobe (× 450).

that *R. holtii* is dioecious, and the key character he gives for separating them, the dorsal leaf lobe crossing the stem in *R. carringtonii*, is unreliable.

Subgenus *Cladoradula* Spruce, *Hep. Amaz. And.*: 315. 1885

Female inflorescence terminal on a very short lateral branch, subfloral innovations absent.

5. R. aquilegia (Hook. f. & Tayl.) Gott. *et al., Syn. Hep.*: 260. 1845   (Fig. 105)

Dioecious. Thin reddish-brown patches or scattered shoots. Shoots prostrate, to *ca* 5 cm, stems irregularly 1(–2)-pinnately branched. Leaves on main stems subimbricate to imbricate, spreading, convex, margins entire, dorsal lobe broadly ovate, not or only slightly crossing stem, 0.6–1.5 mm long, 0.4–1.1 mm wide, apex rounded, often inflexed; ventral lobe obliquely rectangular, ±crossing stem or not, concave towards keel so that keel region appears strongly inflated, margin appressed to dorsal lobe, 0.2–0.8 mm long, (0.16–)0.24–0.40 mm long, (0.45–)0.50–0.60 length of dorsal lobe, apex acute to rounded; leaves on branches smaller but otherwise similar; cells in middle of dorsal lobe 16–24 μm wide, walls reddish-brown, thin to thickened, trigones lacking to medium-sized and bulging. Underleaves absent. Gemmae rare, discoid or elongated, 80–120 μm long. Male inflorescence terminal on branch, ±spike-like, bracts 2–5 pairs, smaller than leaves, imbricate, erect, saccate. Female inflorescence terminal on lateral branch, bracts smaller than leaves, erect, not saccate at base. Perianths rare, ±rectangular above, tapering to very narrow base, mouth entire. Capsules rare, summer. Damp shaded rocks and rock faces, often where basic, in woods and wooded ravines, on sheltered rocks by the sea and epiphytic in the spray zone of waterfalls, ascending to 920 m, occasional in N.W. Wales and the Lake District, frequent in N.W. Scotland from Stirling and Clyde Is. north to W. Sutherland and Shetland, frequent in extreme W. Ireland, very rare elsewhere. 22, H12. Spain, Portugal, S.W. Norway, Faroes, Himalayas, Azores, Madeira, Tenerife.

Distinguished from other British species of the genus by its reddish-brown colour, inflated keel region of the leaves and the frequently somewhat thickened leaf cell walls.

6. R. carringtonii Jack, *Flora* 64: 385. 1881   (Fig. 106)

Dioecious. Thin olive-green to brownish-patches, scattered plants or growing over other bryophytes. Shoots prostrate to *ca* 4 cm, stems irregularly and sparsely 1(–2)-pinnately branched, stems in section with cells thick-walled with large trigones. Leaves on main stems subimbricate to imbricate, spreading, plane or slightly convex, margins entire, ventral edge ± straight or somewhat curved but not angular, dorsal lobe ovate to broadly ovate, crossing stem or not, 0.5–1.5 mm long, 0.4–1.1 mm wide, apex rounded; ventral lobe obliquely ovate-quadrate, not overlapping stem, 0.25–0.80 mm long, 0.24–0.52 mm wide, 0.5–0.6 length of dorsal lobe, apex acute; keel region not or only slightly inflated; leaves on branches smaller but otherwise similar; cells in middle of dorsal lobe 16–28 μm wide, walls colourless to brownish, thin or slightly thickened, trigones small or absent. Underleaves absent. Gemmae absent. Male plants frequent, inflorescence terminal on lateral branch, spike-like, becoming intercalary, sometimes successional, bracts 2–7 pairs, smaller than leaves, subimbricate to imbricate, erect, saccate at bases. Female inflorescence on short lateral branch, bracts smaller than leaves. Perianths very rare, narrow below, widened above and wide-mouthed (Macvicar, 1926). Capsules unknown. On moist shaded vertical rocks in ravines at low altitudes, rare, W. Inverness, Kintyre, Mid and N. Ebudes, Kerry, W. Cork, W. Mayo, Fermanagh. 4, H5. N.W. Spain, Azores, Madeira.

22. PTILIDIACEAE

A monogeneric family with the characters of the genus.

52. PTILIDIUM NEES, *Naturg. Europ. Leberm.* 1: 95. 1836

Dioecious. Plants moderately robust, brown or reddish-brown. Stems irregularly 1–2-pinnately branched, branching of the *Frullania*-type, stems in section with cortical and central cells differentiated. Leaves incubous, insertion ± transverse, bilobed, dorsal lobe larger than ventral, lobes 1–3 subdivided giving leaves appearance of being 2–6-lobed, margins with uniseriate cilia. Cells usually with large trigones, oil-bodies small, numerous. Underleaves smaller than leaves, bilobed, margins ciliate. Gemmae lacking. Male plants smaller than female. Male inflorescence terminal, becoming intercalary, on lateral branch. Female inflorescence terminal on main axis, bracts and bracteoles larger and more densely ciliate than leaves and underleaves. Perianth cylindrical to obovate, plicate above, narrowed to ciliate mouth. Capsule wall 4–5-stratose. Spores 25–33 μm. Three species, the third occurring in the North Pacific.

Leaves lobed to 0.4–0.5, largest lobe 11–33 cells wide at base, this base wider than length of marginal cilia **1. P. ciliare**
Leaves lobed to 0.6–0.8, largest lobe 5–10(–12) cells wide at base, this base narrower than length of many marginal cilia **2. P. pulcherrimum**

1. P. ciliare (L.) Hampe, *Prodrom. Fl. Hercyn.*: 76. 1836 (Fig. 107)

Dioecious. Plants reddish-brown, rarely olive-green or brown, scattered amongst other bryophytes, gregarious, in lax patches or erect tufts. Shoots procumbent, ascending or erect, to 0.75 cm, not adhering to substrate, stems laxly irregularly 1(–2)-pinnate. Leaves approximate to loosely imbricate, concave, ± ovate-quadrate, 1.5–2.5 cm long including cilia, leaves unequally bilobed but apparently 3–5-lobed with large dorsal lobe 2(–3)-lobed to (0.3–)0.4–0.5(–0.6), lobes lanceolate to triangular and 11–23 cells wide at base, ventral lobe small, simple or bilobed, margins ciliate, or one or more margins of dorsal lobe entire, cilia shorter than width of base of divisions of dorsal lobe except sometimes at base; cells at base of divisions of dorsal lobe (24–)28–40 μm wide, walls somewhat thickened, trigones medium-sized to large, bulging. Underleaves to 1.2 mm long including cilia, shallowly bilobed, margins ciliate. Gemmae lacking. Perianths rare, obpyriform, shallowly plicate above, narrowed to lobed, ciliate mouth (Schuster, 1966). Capsules very rare. $n = 8, 8 + m^*, 9$. Bogs, heaths and moorland, damp rocks and dunes, ascending to 1310 m, widely distributed but sometimes only occasional and in small quantity, rare in Ireland. 97, H10. Europe north to Spitzbergen, Faroes, Iceland, Caucasus, Novaya Zemlya, Siberia, N. America, Greenland, Tierra del Fuego, New Zealand.

Although there is great variation in leaf morphology in *P. ciliare*, the genus *Ptilidium* is sufficiently distinct from other liverworts that the only confusion likely to arise is between *P. ciliare* and *P. pulcherrimum*. The former is a usually larger, laxer plant than the latter. Rarely the marginal cilia of the leaves may be longer than the width of the lobe bases and occasional leaves have lobes no wider at the base than in *P. pulcherrimum* but leaves from mature stems of *P. ciliare* are distinctive.

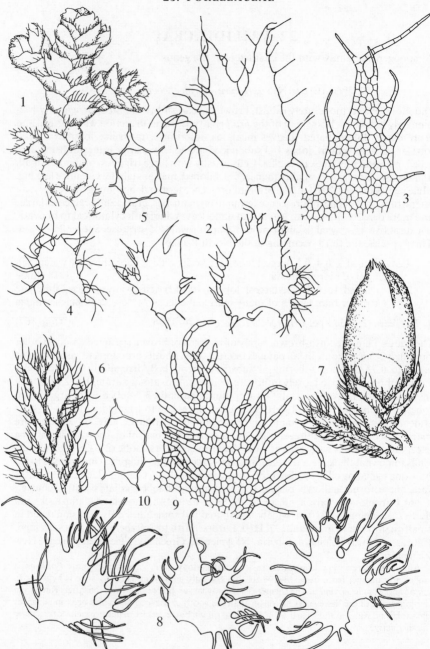

Fig. **107**. 1–5, *Ptilidium ciliare*: 1, shoot (× 15); 2, leaves (× 23); 3, leaf lobe (× 84); 4, underleaf (× 23); 5, cell from middle of leaf (× 450). 6–10, *P. pulcherrimum*: 6, shoot (× 21); 7, shoot with perianth (× 21); 8, leaves (× 33); 9, leaf lobe (× 84); 10, cell from middle of leaf (× 450).

2. P. pulcherrimum (G. Webb). Vainio, *Meddeland. Soc. Faun. Flor. Fenn.* 3: 88. 1878

(Fig. 107)

Dioecious. Yellowish-green to light brown or reddish-brown, dense patches. Shoots procumbent, to *ca* 2 cm, adhering to substrate, stems irregularly 1–2-pinnate, leaves imbricate, often closely so, concave, ±ovate-quadrate, 0.7–1.0(–1.7) mm long including cilia, unequally bilobed but apparently 3–4-lobed, dorsal lobe 2(–3)-lobed to 0.6–0.8 with narrowly lanceolate or lanceolate divisions, 5–10(–12) cells wide at base, ventral lobe very small, 1(–2)-lobed, margins with long curved cilia longer than width of bases of divisions of dorsal lobe; cells at base of divisions of dorsal lobe (24–)28–40 μm wide, walls somewhat thickened, trigones large, bulging, more rarely walls thin and trigones small. Underleaves to *ca* 0.4 mm long, shallowly bilobed, margins with long cilia. Gemmae lacking. Female inflorescence terminal on main axis, becoming pseudolateral with age, bracts larger and less deeply lobed than leaves, margins densely ciliate. Perianths occasional, narrowly obovoid, shallowly plicate above, narrowed to ciliate mouth. Capsules occasional, summer. $n = 8 + m^*$, 9. Trunks, branches and exposed roots of usually deciduous trees or shrubs, rotting logs, rarely on rocks, in sheltered situations, occasional and apparently increasing in England and C. and E. Scotland, north to E. Ross and W. Sutherland, rare in Wales, Louth. 71, H1. Europe north to Fennoscandia and N. Russia, Iceland, Caucasus, Siberia, Shensi, N. America.

23. PORELLACEAE

Dioecious. Plants robust, yellowish-green, green or brown. Stems with rhizomatous basal part, 1–3-pinnate, branching of *Frullania*-type. Leaves conduplicate-bilobed with very short or vestigial keel, dorsal lobe incubous, orbicular to lanceolate, entire, dentate or ciliate, ventral lobe to *ca* ½ size or dorsal, parallel to stem, both lobes with or without water-sacs; cells with or without hyaline papillae. Underleaves large, entire, dentate or ciliate. Gemmae and other vegetative propagules unknown. Inflorescences on dwarf lateral branches. Male inflorescence spicate, bracts imbricate with solitary axillary antheridia. Perianth often dorsiventrally compressed, at least above, narrowed or not to entire to ciliate mouth. Wall of capsule 5(–6)-stratose, dehiscing somewhat irregularly. Spores 24–80 μm. Three genera.

53. PORELLA L., *Sp. Plant.*: 1106. 1753

Leaf lobes without water-sacs, margins entire or dentate but not ciliate; cells without hyaline papillae. An actively evolving genus, probably with more than 100 species, centred mainly in eastern Asia.

1 Plants acrid, dorsal lobes of leaves acute, margins of ventral lobes and underleaves dentate at least on branches **2. P. arboris-vitae**
  Plants not acrid, dorsal lobes rounded, margins of ventral lobes and underleaves (except sometimes on decurrent part) entire throughout 2
2 Ventral lobes of leaves mostly 0.3–0.6 width of dorsal lobe, to as wide as or wider than stem 3
  Ventral lobes of leaves to *ca* 0.2 width of dorsal lobe, narrower than stem 4
3 Ventral lobes much wider than stem, of ± similar size to underleaves **3. P. obtusata**
  Ventral lobes not wider than stem, to *ca* ½ width of underleaves **5. P. platyphylla**
4 Underleaves scarcely decurrent, entire at base, cells in mid-dorsal lobe 16–18 μm **1. P. pinnata**
  Underleaves longly decurrent, dentate at base, cells 28–40 μm **4. P. cordaeana**

Section *Porella*

Leaves distant to approximate, lobes mostly ± plane, ventral lobe not wider than stem, plane, not decurrent, insertion not J-shaped. Underleaves not or scarcely wider than stem, not decurrent, margins not reflexed.

1. P. pinnata L., *Sp. Plant.*: 1106. 1753   (Fig. 108)
Madotheca porella (Dicks.) Nees

Plants robust, in lax, dull, olive-green patches or tufts. Shoots procumbent, often with leaves silt-covered and eroded, to 7 cm, stems irregularly 2(–3)-pinnately branched. Leaves on main stem approximate to slightly subimbricate, margins entire, dorsal lobe ovate-rectangular to rectangular not or scarcely crossing stem, convex, rounded apex often deflexed and ventral margin sometimes decurved, 1.2–2.6 mm long, 0.8–2.0 mm wide, (1.1–)1.3–1.6 times as long as wide; ventral lobe relatively very small, appressed to stem and dorsal lobe, lanceolate to narrowly lanceolate or lingulate, frequently slightly falcate, apex obtuse, margins entire, not decurrent, 0.4–1.0 mm long, 0.16–0.40 mm wide, 0.13–0.20(–0.22) width of dorsal lobe, narrower than stem, half or less than width of underleaves; cells in middle of dorsal lobe of leaf 16–28 μm wide, walls slightly thickened, trigones small to large and bulging. Underleaves distant, ± orbicular to ovate-oblong or lingulate, slightly decurrent, entire, apex rounded to truncate, 0.4–0.9 mm long, 0.4–0.9 mm wide. Male inflorescence spicate on short lateral branch, bracts 3–6 pairs, closely imbricate (Schuster, 1980). Female inflorescence on short lateral branch, bracts smaller than leaves, entire; perianths unknown in Britain or Ireland, obovate, narrowed to ciliate-dentate mouth (Schuster, 1980). On rocks, tree roots and tree boles subject to flooding, by streams and rivers and at water level by lakes, at low altitudes, very rare in S.W. England, S. and Mid Wales, Westmorland, occasional in N. Wales, rare but widespread in Ireland. 14, H14. Extreme oceanic Europe from Portugal north to Germany, Madeira, eastern N. America, Cuba.

Section *Platyphyllae* Schust., *Hep. Anthoc. N. Amer.* 4: 690. 1980

Leaves usually imbricate on main stems, lobes often convex with decurved apices and margins, ventral lobes as wide as or wider than stem, longly decurrent, insertion J-shaped. Underleaves wider than stem, longly decurrent, margins reflexed.

2. P. arboris-vitae (With.) Grolle, *Trans. Br. bryol. Soc.* 5: 770. 1969   (Fig. 108)
Madotheca laevigata (Schrad.) Dum; *P. laevigata* (Schrad.) Pfeiff.

Plants robust in glossy or occasionally dull, yellowish-green to dark green or brownish-green, dense to lax patches, acrid to the taste. Shoots procumbent, to 7(–13) cm, irregularly 1–2-pinnately branched, older branches becoming attenuated. Leaves on main stems imbricate to closely imbricate, dorsal lobe broadly ovate with obliquely cordate base to obliquely rounded-cordate, crossing stem or not, convex with decurved tip and ventral margin, apex acute, less commonly obtuse or rounded and with or without an apiculus, dorsal margin undulate or not, sinuous or not, entire or with 1–3 sparse teeth towards base or with obscure to spinose teeth near apex, 1.3–2.6 mm long, 1.0–2.0 mm wide, 1.0–1.3(–1.6) times as long as wide; ventral lobe lanceolate to ovate or oblong, apex obtuse to rounded or truncate, decurved or not, not decurrent, spinose-dentate, to dentate or entire, 0.8–1.6 mm long, 0.36–0.80 mm wide, 0.30–0.45 width of dorsal lobe, to as wide as or wider than stem, 0.5–1.0 width of underleaves; cells in middle of dorsal lobe 20–32 μm wide, walls slightly thickened, trigones small to medium-sized. Leaves on branches smaller, always acute, ventral lobes and under-

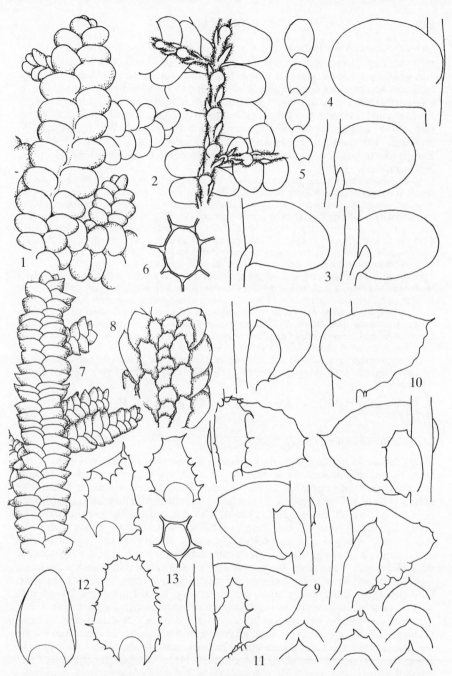

Fig. **108**. 1–6, *Porella pinnata*: 1, shoot, dorsal side (× 5); 2, shoot, ventral side (× 8); 3, leaves, ventral side (× 15); 4, leaf, dorsal side (× 15); 5, underleaves (× 15); 6, mid-dorsal lobe cell (× 450). 7–13, *P. arboris-vitae*: 7, shoot, dorsal side (× 5); 8, shoot, ventral side (× 8); 9, leaves, ventral side (× 13); 10, leaf, dorsal side (× 13); 11, dorsal lobe apices (× 13); 12, underleaves (× 21); 13, mid-dorsal lobe cell (× 450).

leaves usually with toothed margins. Underleaves orbicular to oblong or lingulate, not recurved, entire to spinose-dentate, longly decurrent, 0.44–1.00 mm long, 0.44–0.80 mm wide. Female bracts similar to but larger than leaves, ciliate (Macvicar, 1926). Perianths and capsules unknown in Britain or Ireland. $n = 8 + m*$. On usually dry, at least partly sheltered, basic rock, in stony turf and on nearby tree boles, rarely in deeply shaded humid ravines or woodlands ascending to *ca* 700 m, very rare in S.E. England, rare in S.W. England and S. Wales, occasional in N. Wales, N.W. England and W. Scotland north to Shetland, rare in W. Ireland and very rare elsewhere. 68, H20, C. Europe north to Fennoscandia, Faroes, Iceland, Caucasus, Turkey, Morocco, Macaronesia.

An extremely variable species particularly in the shape of the dorsal leaf lobe apex and the degree and nature of the toothing of the leaf lobe and underleaf margins. It is evident from the examination of specimens from a variety of habitats that these characters are very susceptible to environmental modification. The characters defining var. *obscura* (Nees) Corley occur either singly or in various combinations within the same and in different plants. Var. *killarniensis* (Pears.) Corley is at first sight very distinctive but all the characters except shoot length exhibit the same phenomenon as those of var. *obscura*; the considerable shoot length (to 13 cm) can be attributed to the very sheltered and humid habitats in which the plant grows.

Forms with the dorsal leaf lobe rounded and the margins of the ventral lobes and underleaves entire may be confused with *P. platyphylla*. *P. arboris-vitae* is usually glossy and the acrid taste when fresh is characteristic. Plants with leaves and underleaves of the main stem resembling those of *P. platyphylla* do, however, have branches with acute leaves and ventral lobes and underleaves with at least some sparse teeth. The underleaves of *P. arboris-vitae* are about as wide as long (length measured from middle of insertion to apex), whereas those of *P. platyphylla*, which are more longly decurrent, are mostly 0.5–0.7 times as long as wide.

I have found that the acrid taste, characteristic of *P. arboris-vitae*, persists strongly in herbarium specimens up to 40 years old and still persists in 80 year old gatherings: I have not sampled older specimens.

3. P. obtusata (Tayl.) Trev., *Mem. Real. Istit. Lombardo Sci. Mat. Nat.* ser. 3, 4: 497. 1877   (Fig. 109)

Madotheca thuja auct., *P. thuja* auct.

Plants robust, in dense, glossy, dark green to olive-brown patches, not acrid to the taste. Shoots procumbent, to 4(–15) cm, stems irregularly 1–2-pinnately branched, older branches not attenuated. Leaves on main stems closely imbricate, dorsal lobe obliquely ovate or ovate-orbicular, crossing stem, convex with deflexed margins, apex rounded or very obtuse, margins entire or with 1–3 small teeth near undulate and auriculate ventral base, 1–2 mm long, 0.9–1.8 mm wide, slightly longer than wide; ventral lobe oblong or oblong-orbicular, margins entire, recurved, auriculate and undulate near base, apex rounded, 0.6–1.4 mm long, 0.4–1.0 mm wide, 0.4–0.6 width of dorsal lobe, wider than stem, of ± similar size to underleaves, cells in middle of dorsal lobe 24–36 μm wide, walls slightly thickened, trigones large, sometimes bulging. Underleaves orbicular with entire deflexed margins, 0.6–0.9 mm long, 0.6–1.0 mm wide. Female bracts similar to leaves but with ciliate-dentate margins (Macvicar, 1926). Perianths and capsules unknown in Britain or Ireland. $n = 9*$. On sheltered or exposed, acidic or basic rocks, rarely on trees, ascending to 400 m, rare but sometimes locally frequent on the coast from Cornwall and Berwick north to Shetland, I. of Wight, E. Kent, very rare inland, rare along Irish coast. 39, H12, C. Mediterranean islands and coast of Europe, Portugal, S.W. Norway, Tunisia, Algeria, Macaronesia.

10, leaves, ventral side (× 13); 11, leaf, dorsal side (× 13); 12, underleaves (× 23); 13, mid-dorsal cell (× 450). 14–20, *P. platyphylla*: 14, shoot with perianth (arrowed), dorsal side (× 5); 15, shoot with male branch, ventral side (× 8); 16, perianth, ventral side (× 15); 17, leaves, ventral side (× 15), 18, leaf, dorsal side (× 15); 19, underleaf (× 38); 20, mid-dorsal lobe cell (× 450).

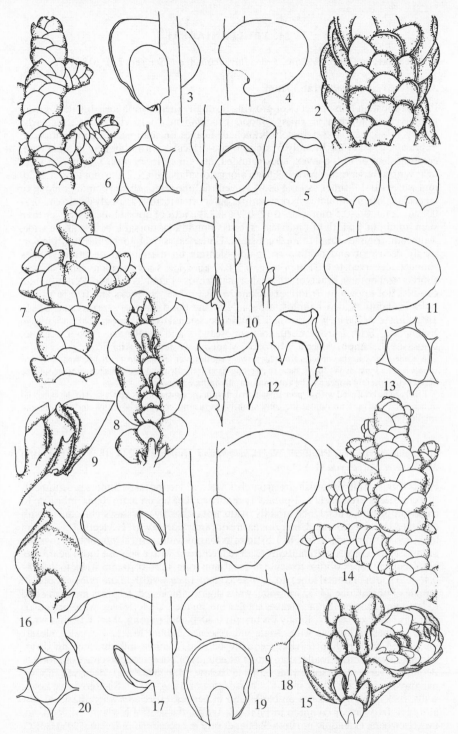

Fig. **109**. 1–6, *Porella obtusata*: 1, shoot, dorsal side (× 3.5); 2, shoot, ventral side (× 8); 3, leaves, ventral side (× 13); 4, leaf, dorsal side (× 13); 5, underleaves (× 13); 6, mid-dorsal lobe cell (× 450). 7–13, *P. cordaeana*: 7, shoot, dorsal side (× 5); 8, shoot, ventral side (× 8); 9, female branch (× 21);

4. P. cordaeana (Hüb.) Moore, *Proc. Roy. Irish Acad. Sci.* ser. 2, 2: 618. 1976

(Fig. 109)

Madotheca cordaeana (Hüb.) Dum.

Plants robust, in loose patches or growing through other bryophytes, dull dark green, rarely yellowish-green or glossy. Shoots procumbent, to 10 cm, stems irregularly, closely to sparsely 1–2-pinnately branched. Leaves on main stems approximate to imbricate, dorsal lobe broadly ovate to ovate-orbicular with obliquely cordate base, crossing stem or not, convex, apex rounded, margin entire or with 1–3 spinose teeth near ventral base, rarely with 1–2 teeth along dorsal margin, 1.2–2.6 mm long, 1.2–2.0 mm wide, 1.0–1.4 times as long as wide; ventral lobe relatively very small, oblong to lanceolate, often undulate, margins entire, outer often recurved, longly decurrent, 0.5–0.9 mm long, 0.2–0.5 mm wide, 0.12–0.20(–0.23) width of dorsal lobe, narrower than stem, to about ½ width of underleaves; cells in middle of dorsal lobe 28–40 μm wide, walls thin, trigones small to medium-sized. Underleaves ovate-orbicular to orbicular, longly decurrent and dentate or spinose-dentate on margins of decurrency, apex rounded, decurved, 0.4–0.8 mm long, 0.5–0.8 mm wide. Female inflorescence on very short lateral branch, bracts much smaller than leaves, ± equally bilobed. Perianths and capsules not known in Britain or Ireland. $n = 8 + m^*$, 8. On rocks and tree roots and boles in damp shaded habitats, especially in the flood zone of streams and rivers, ascending to *ca* 700 m, very rare in S. Britain, occasional to frequent in Scotland, rare in Ireland. 62, H17. Europe north to Fennoscandia and N. Russia, Faroes, Iceland, Caucasus, Lebanon, Algeria, Morocco, Madeira, western N. America.

A somewhat variable species for which two varieties, var. *simplicior* (Zett.) S. Arn. and var. *simplicior* (C. Jens.) E.W. Jones, have been recognised in the British Isles. These are not recognised here as they merely appear to be environmentally induced forms

Likely to be confused with *P. pinnata* which differs in the ventral lobes appressed to the stem and dorsal lobe and not undulate, the only slightly decurrent, completely entire underleaves and smaller cells.

5. P. platyphylla (L.) Pfeiff. *Fl. Neiderhessen und Munchen* 2: 234. 1835   (Fig. 109)

Madotheca platyphylla (L.) Dum.

Plants robust, dull, yellowish-green patches, not acrid to the taste. Shoots procumbent, to 8 cm, stems irregularly 2–3-pinnately branched. Leaves on main stems imbricate to closely imbricate, dorsal lobe broadly ovate, not or hardly crossing stem, convex with decurved tip, apex rounded, margins entire or rarely dorsal with 1–3 teeth, 1.3–2.7 mm long, 0.9–2.0 mm wide. 1.2–1.4(–1.5) times as long as wide; ventral lobe oblong to ovate or lanceolate, convex with margins recurved, entire or with a spinose tooth near base, not decurrent, apex obtuse to acute, 0.5–1.1 mm long, 0.24–0.70 mm wide, (0.2–)0.3–0.4(–0.5) width of dorsal lobes, to as wide as stem, to *ca* ½ width of underleaves; cells in middle of dorsal lobe 24–32 μm wide, walls slightly thickened, trigones medium-sized to large and bulging. Branch leaves similar but smaller. Underleaves ovate-orbicular, margins recurved, entire, longly decurrent, 0.40–0.64 mm long, 0.64–1.12 mm wide, 0.5–0.7 times as long as wide. Male inflorescence obloid, bracts 4–5 pairs, closely imbricate, smaller than leaves, ± equally bilobed. Female inflorescence on dwarf lateral branch, bracts smaller than leaves, ± equally bilobed, margins entire or toothed. Perianths very rare, somewhat inflated below, flattened, and decurved above, narrowed to dentate-ciliate mouth. Capsules very rare. $n = 8$, 9^*. Sheltered tree boles, walls, rocks, soil and stones on banks and in hedges, especially where basic, at low altitudes, frequent or common in England, Wales, the central lowlands of Scotland, rare elsewhere, extending north to Shetland, rare to occasional in Ireland. 99, H33, C. Europe north to Spitzbergen, Turkey, N. and C. Asia, Madeira, N. America, Greenland.

24. FRULLANIACEAE

Dark green to reddish-brown, frequently glossy plants. Stems in section with or without distinct cortex, irregularly 1–3-pinnately branched, branches of the *Frullania*-type (i.e. arising terminally, each replacing ventral lobe of associated leaf), never intercalary. Rhizoids arising from underleaf bases. Leaves sometimes caducous, bilobed, dorsal lobes large, incubous, imbricate, overlapping stem on dorsal side, margin usually entire, ventral lobe lanceolate or helmet-shaped, initial branch leaves bilobed, insertion narrow, ± transverse, insertion of leaves subtending branches not extending on to branch. Underleaves usually with 0–1 teeth on either side. Male branches of *Frullania*-type, usually without bracteoles, antheridia 1–2 per bract. Female inflorescence usually without innovations, bracts and bracteoles in (1–)2–3 whorls, larger than the leaves, bracts asymmetrically bilobed, entire to toothed, bracteoles bilobed, entire to toothed, archegonia 1–4. Perianth ovoid to obovoid, 3–5-gonous with 1–3 keels on ventral side, abruptly narrowed to beak-like mouth. Seta in section of many irregularly arranged cell rows. Capsule wall bistratose, inner with irregular, hyaline, thin walls. Spores irregular in shape because of precocious division, with irregular rosette-like ornamentations. *Frullania* and two small genera, *Schusterella* and *Steeria*.

54. FRULLANIA RADDI, *Jungerm. Etrusca*: 9. 1818

Dioecious, rarely autoecious or paroecious. Plants dull green, reddish-brown or brownish-black. Stems irregularly 1–3-pinnately branched, female plants usually more branched than male. Leaves spreading ± horizontally, bilobed almost to base, dorsal lobes incubous, imbricate, orbicular to ovate or obliquely so, crossing stem on dorsal side, sometimes with discoloured cells (ocelli) scattered or in a median row, margin entire, ventral lobe much smaller, helmet-shaped or sometimes evolute, often with linear or circular stylus between lobe and insertion; cell walls thickened, trigones often present. Underleaves usually bilobed, usually with 0–1 teeth on one or both margins. Vegetative propagation by gemmae from leaf margins or surface of perianth sometimes present. Male inflorescence on short lateral branch, bracts closely imbricate, ± symmetrical. Female inflorescence terminal on lateral branch, rarely on main axis, bracts 2–5 pairs, asymmetrically bilobed, margins entire to laciniate, innermost joined at base and to smaller, bilobed bracteole. Perianth extending beyond bracts, ovoid or obovoid, 3–4-gonous, abruptly narrowed to beak-like mouth. Calyptra fleshy. Spores 40–50 μm. Female plants mostly with $n = 9$, male with $n = 8$. A large, mainly tropical and subtropical genus for which *ca* 800 species have been described but many of these may be synonymous.

1 Dorsal lobes of leaves without ocelli, plants usually dull   2
  Dorsal lobes of leaves, at least of ultimate branches, with ocelli, plants usually glossy
                                                                                                 3
2 Ventral lobes of leaves ± twice as long as wide, perianth smooth   **1. F. teneriffae**
  Ventral lobes of leaves ± as long as wide, perianth tuberculate   **5. F. dilatata**
3 Underleaves with recurved margins and auricles   **2. F. tamarisci**
  Underleaves with plane margins, auricles lacking   4
4 Leaves not caducous, ocelli in a median row, underleaf lobes acute **3. F. microphylla**
  Leaves sometimes caducous, ocelli scattered, underleaf lobes obtuse **4. F. fragilifolia**

Subgenus *Frullania*

Leaves often with ocelli, ventral lobe narrowly helmet-shaped, narrowed at base, sometimes evolute. Perianth trigonous, wall smooth.

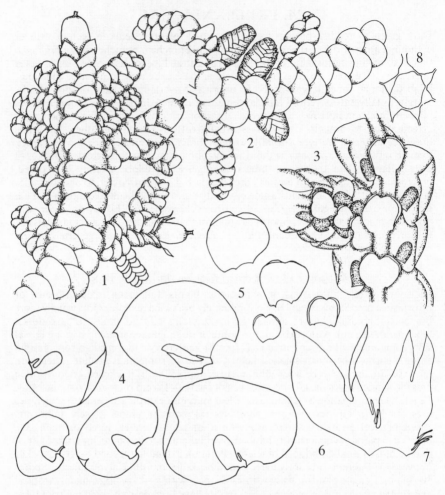

Fig. **110**. *Frullania teneriffae*: 1 and 2, female and male shoots (× 8); 3, shoot, ventral side (× 21); 4, leaves, ventral side (× 21); 5, underleaves (× 21); 6 and 7, bract and bracteole (× 21); 8, mid-leaf cell (× 450).

1. F. teneriffae (Web.) Nees, *Naturg. Europ. Leberm.* 3: 239. 1838   (Fig. 110)
F. germana (Tayl.) Gottsche et al.

Dioecious. Dull, yellowish-brown to reddish-brown patches. Shoots procumbent, to 6(–8) cm, stems irregularly 1–2-pinnately branched. Leaves imbricate, dorsal lobe obliquely broadly ovate, lacking ocelli, apex obtuse to acute, deflexed, on main axis 0.5–1.4 mm long; ventral lobe narrowly helmet-shaped, narrowed at base, *ca* twice as long as wide, narrower than underleaves, on main axis frequently evolute, linear or rounded stylus often present, branch leaves smaller with ventral lobe always narrowly helmet-shaped; mid-leaf cells 16–20 μm wide, walls thickened, trigones large. Underleaves 0.30–0.65 mm long, *ca* 3 times as wide as stem, bilobed to *ca* ⅓, lobes rounded, sinus obtuse, margin recurved above and sometimes ± to base, entire, with or without small auricles. Male inflorescence globular to spicate, bracts 6–numerous,

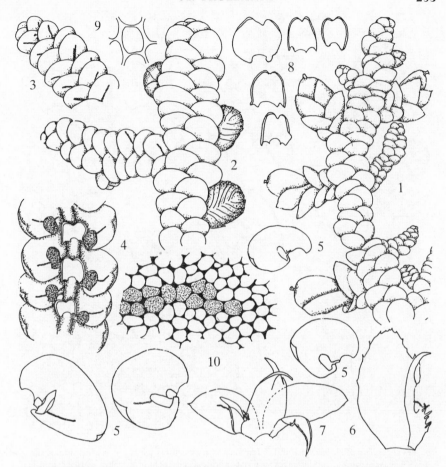

Fig. **111**. *Frullania tamarisci*: 1 and 2, female and male shoots (× 6); 3, ultimate branch (× 21); 4, ventral side of shoot (× 27); 5, leaves, ventral side, (× 21); 6, female bract (× 21); 7, bracts and bracteole (× 21); 8, underleaves (× 21); 9, mid-leaf cell (× 450; 10, mid-leaf cells including ocelli (× 210).

closely imbricate, bilobed, saccate. Female bracts bilobed to *ca* ½, dorsal lobe broadly ovate, acute, entire or with 1–3 cilia near base, ventral lobe smaller, lanceolate, acuminate, with 0–1 coarse teeth at about middle and 0–3 cilia at base, bracteole bilobed to *ca* ½, lobes lanceolate, acuminate, margin with 1–2 cilia at base. Perianths frequent, obovoid, trigonous, beaked, wall smooth, lacking ocelli. Capsules occasional, summer, autumn. On acidic to slightly basic rocks and on bark, especially near the sea, ascending to *ca* 600 m, frequent or common in W. Scotland and W. Ireland, rare in S.W. and N.W. England, N. Wales and S. Scotland. 40, H25, C. Faroes, France, Portugal, Sardinia, Morocco, Macaronesia.

2. F. tamarisci (L.) Dum., *Recueil Observ. Jungerm.*: 13. 1835   (Fig. 111)

Dioecious. Glossy, reddish-brown, rarely dark green patches or loose wefts. Shoots procumbent to ascending, to 10 cm, stems irregularly pinnately branched. Leaves imbricate, dorsal lobe obliquely broadly ovate, at least branch leaves with 1(–2) rows of

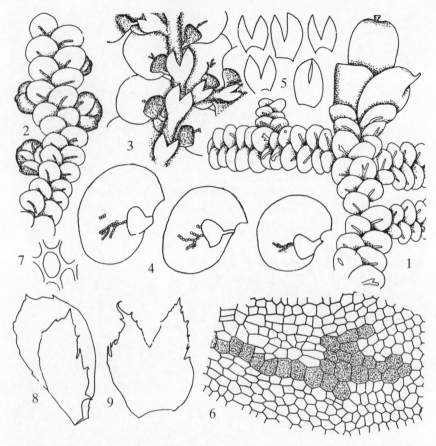

Fig. **112**. *Frullania microphylla*: 1 and 2, female and male shoots (× 21); 3, ventral side of shoot (× 38); 4, leaves (× 63); 5, underleaves (× 63); 6, mid-leaf cells including ocelli (× 210); 7, mid-leaf cells (× 450); 8, bract (× 63); 9, bracteole (× 63).

median ocelli and sometimes also with a few scattered ocelli, very rarely ocelli completely lacking, apex rounded to obtuse or acute or obtuse and apiculate, deflexed, on main axis 0.5–0.9(–1.2) mm long; ventral lobe narrowly helmet-shaped, narrowed at base, usually about twice as long as wide, narrower than underleaves, sometimes evolute on upper part of main axis, linear or rounded stylus present or not, branch leaves smaller with ventral lobes always narrowly helmet-shaped; mid-leaf cells 16–24 μm wide, walls usually thickened and trigones usually distinct. Underleaves 0.25–0.40 mm long, 2.0–(2.5) times as wide as stem, bilobed $\frac{1}{8}$–$\frac{1}{4}$, lobes obtuse to acute, sinus obtuse, margin entire, usually recurved from apex ± to base, rarely only at apex or very rarely plane, entire, base undulate with prominent auricles. Male inflorescence ± glob-ular, bracts 4–6, closely imbricate, bilobed, saccate. Female bracts bilobed, dorsal lobe broadly ovate, acute, margin entire or toothed, ventral lobe smaller, lanceolate, acuminate, with cilia at base, bracteole bilobed to *ca* $\frac{1}{2}$, lobes lanceolate, acuminate, with large tooth on either side, otherwise entire to ciliate-dentate. Perianths occasional to frequent, trigonous, beaked, wall smooth, without ocelli. Capsules occasional, spring to autumn. $n = 8, 9$. On rocks and bark in sheltered or open habitats, in scree, on cliffs and in usually basic grassland, ascending to 1000 m, rare to occasional in S.E. England, E. Anglia and the Midlands, frequent or common and sometimes locally abundant elsewhere. 102, H40, C. Europe north to Fennoscandia and Russia, Faroes, Iceland, Siberia, China, Macaronesia, Cape Verde Is.

A variable species for which three varieties have been described from Britain but these intergrade to such an extent that they are not worthy of taxonomic recognition. Distinguished in the field from *F. germana* and *F. dilatata* by its glossy appearance and usually reddish-brown coloration. Small forms may be confused with *F. fragilifolia* which, however, has readily caducous leaves. In some forms the ocellia may be lacking from most leaves but the glossy colour and underleaf morphology will distinguish *F. tamarisci* from other British species of the genus.

It has recently been shown experimentally that *F. tamarisci* consists of four genetically and geographically distinct species (Crandall-Stotler *et al.*, *Bryologist* **90**, 287–308, 1987). No information is given about distribution but European material belongs to *F. tamarisci* s.s. so the world distribution given above is likely to be innaccurate.

3. F. microphylla (Gottsche) Pears., *J. Bot. Br. Foreign* 32: 328. 1894 (Fig. 112)

Dioecious. Thin, glossy, reddish-brown patches. Shoots procumbent, to 5 cm, irregularly 1–3-pinnately branched. Leaves approximate to closely imbricate, dorsal lobe obliquely orbicular to obliquely broadly ovate, row of median ocelli, sometimes branching at distal end, and sometimes a few scattered ocelli present on most leaves, apex rounded, not or hardly deflexed, on main axis 0.25–0.55 mm long; ventral lobe helmet-shaped, narrowed at base, 1.3–1.6 times as long as wide, narrower than underleaves, occasionally evolute, linear stylus often present, branch leaves smaller; mid-leaf cells (10–)12–16 μm wide, walls strongly thickened, trigones not developed. Underleaves 0.1–0.2(–0.3) mm long, 1.5–2.0 times as wide as stem, bilobed to *ca* ½, lobes and sinus acute, margin plane, entire or very rarely with a single tooth, auricles lacking. Male inflorescence ± globular, bracts 2–5 pairs, closely imbricate, bilobed, saccate. Female bracts bilobed, dorsal lobe ovate, ventral lobe smaller, lanceolate, margin usually irregularly dentate or ciliate-dentate, rarely ± entire, bracteole bilobed to *ca* ½, lobes lanceolate, dentate or ciliate-dentate. Perianths occasional, trigonous, beaked, wall smooth with scattered ocelli. Capsules occasional, winter, spring. On sheltered, ± vertical rocks in coastal areas, rarely inland, at low altitudes, rare to frequent in W. Britain from Scilly Is. and Cornwall north to Orkney and St. Kilda, occasional in W. Ireland, S. Tipperary. 28, H11. Portugal, Spain, France, Germany, Faroes, Morocco, Azores.

A smaller plant than *F. tamarisci* and *F. teneriffae* with which it sometimes grows. It differs from both in the tips of the leaves not being deflexed and the underleaf margins not reflexed. It may also grow with *F. fragilifolia* (q.v.).

4. F. fragilifolia (Tayl.) Gottsche et al., *Syn. Hep.*: 437. 1843 (Fig. 113)
F. maritima Steph.

Dioecious. Thin, glossy, deep red to reddish-brown patches. Shoots procumbent, to 2.5 cm, stems irregularly 1–2(–3)-pinnately branched. Leaves approximate to imbricate, sometimes caducous and leaving bare lengths of stem, dorsal lobe orbicular to obliquely ovate, ocelli scattered, rarely in a median row, present in most leaves, apex rounded, plane to deflexed, on main axis 0.28–0.52 mm long; ventral lobe helmet-shaped, narrowed at base, 1.2–1.6 times as long as wide, narrower than underleaves, rarely evolute, linear stylus often present, branch leaves smaller; mid-leaf cells 16–20 μm wide, walls heavily thickened, trigones poorly developed. Underleaves 0.17–0.24 (–0.31) mm long, 1.4–2.2(–3.0) times as wide as stem, bilobed, lobes obtuse, sometimes appearing obliquely truncate, margin plane, sometimes with blunt tooth on one or both sides, auricles lacking. Male inflorescence ± globular, bracts 3–4 pairs, closely

Fig. **113**. *Frullania fragilifolia*: 1 and 2, female and male shoots (× 15); 3, ventral side of shoot (× 38); 4 leaves, ventral side (× 63); 5, underleaves (× 63); 6, bract (× 27); 7, bracteole (× 27); 8, mid-leaf cells including ocelli (× 236); 9, mid-leaf cell (× 450).

imbricate, bilobed, saccate. Female bracts bilobed, dorsal lobe broadly ovate, ventral lobe smaller, lanceolate, margins very coarsely and irregularly toothed, bracteole bilobed $ca \frac{1}{3}$, lobes lanceolate, coarsely toothed. Perianths rare, trigonous, beaked, wall smooth, without ocelli. Capsules rare, spring, summer. $n = 9^*$. Sheltered, ± vertical rocks and tree trunks, ascending to *ca* 1200 m, occasional to common in W. Britain from Scilly Is. to Shetland, very rare elsewhere in England, S. Hants, E. Sussex, W. Gloucester, rare elsewhere in Scotland, rare in Ireland. 49, H16, C. Europe north to Fennoscandia and N. Russia, Faroes, Azores, Canaries.

   Forms with the ocelli in a median row may be mistaken for *F. microphylla* but differ in the larger leaf cells and shape of the underleaves. May be recognised in the field by the frequently caducous leaves. Multicellular gemmae have been reported occurring on leaf margins (Bisang, *Bot. Helvetica* **97**, 111–14, 1987).

Subgenus *Trachycolea* Spruce, *Hep. Amaz. And.*: 7 and 31. 1884

Leaves without ocelli, ventral lobe of leaf about as long as wide, wide-mouthed. Perianth 4–5-gonous, tuberculate, papillose or scaly.

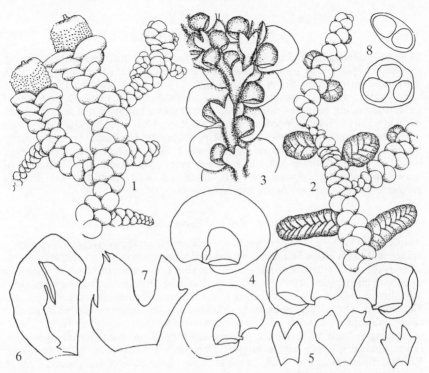

Fig. **114**. *Frullania dilatata*: 1 and 2, female and male shoots (× 9); 3, ventral view of shoot (× 21); 4, leaves, ventral side (× 38); 5, underleaves (× 38); 6, bract (× 38); 7, bracteole (× 38); 8, gemmae (× 450).

5. F. dilatata (L.) Dum., *Recueil Observ. Jungerm.*: 13. 1835   (Fig. 114)

Dioecious. Thin, dull, reddish-brown to brownish-black or occasionally greenish patches. Shoots procumbent, to *ca* 4 cm, stems irregularly 1–3-pinnately branched. Leaves imbricate, dorsal lobe obliquely orbicular, ocelli lacking, apex rounded, deflexed, on main axis 0.36–0.72 mm long; ventral lobe helmet-shaped, wide-mouthed, 1.0–1.2 times as long as wide, ± as wide as or wider than underleaves, rarely evolute, linear stylus often present, branch leaves smaller; mid-leaf cells 16–20 μm wide, walls ± thickened, trigones large. Underleaves 0.20–0.36 mm long, 1.5–2.0(–2.5) times as wide as stem, bilobed, lobes acute, margin plane, with 0–1 blunt teeth on one or both sides, auricles lacking. Gemmae 3–6-celled, sometimes present on leaf margins and perianths. Male inflorescence spicate, bracts 3-numerous pairs, closely imbricate, bilobed, saccate. Female bracts bilobed, dorsal lobe broadly ovate with rounded apex, margin entire, ventral lobe smaller, lanceolate, with large acute tooth at middle and sometimes small teeth near base, bracteole somewhat asymmetrically bilobed, with large acute tooth on either side. Perianths common, trigonous, beaked, wall tuberculate, without ocelli. Capsules frequent, autumn to spring. $n = 8$ (♂), 9(♀). On trees and rocks, common. 112, H40, C. Europe north to C. Sweden, S. Finland, N. Russia, Cyprus, Turkey, Iran, Siberia, Shensi, Macaronesia.

258   25. JUBULACEAE

25. JUBULACEAE

Close to Frullaniaceae and possibly only a subfamily of it, differing in the following characters. Plants dull green, lacking accessory pigments. Stem in section with unistratose cortex, branching of *Frullania*-type but intercalary branches also frequent. Leaves never caducous, toothed, insertion ± longitudinal and J-shaped, insertion of leaves subtending branches extending to branch base, initial branch leaves simple; trigones small or absent. Underleaves entire to ciliate. Male branches intercalary, with bracteoles. Female inflorescence with (1–)2 innovations of the *Radula*-type. Bracts and bracteoles in 1, rarely 2, whorls. Archegonia 2. Seta in section with 16 rows of irregularly arranged outer cells and 4 rows of inner cells. Inner cells of capsule wall brown and reticulate. Spores spherical, 22–28 μm, finely papillose, not dividing precociously. Two genera, *Jubula* and *Neohattoria*.

55. JUBULA Dum., *Recueil Observ. Jungerm.*: 12. 1835

Autoecious. Plants light to dark green or blackish-green. Stems irregularly pinnately branched. Leaves bilobed almost to base, dorsal lobes incubous, imbricate, asymmetrically ovate, not or hardly crossing stem, lacking ocelli, margin entire to spinosely toothed or ciliate, ventral lobe helmet-shaped, with or without a stylus, or triangular to subulate and reflexed; cells thin-walled, trigones inconspicuous, oil-bodies 4–7 per cell. Underleaves bilobed ⅓–½, lobes acute to ciliate, margins entire to spinose-dentate or ciliate. Male inflorescence spicate, bracts much smaller than leaves, imbricate, saccate, subtending 1–2 antheridia. Female inflorescence terminal on main axis or long branch, bracts bilobed, margins entire to ciliate, bracteole bilobed to ½, entire to ciliate. Perianth trigonous with one keel ventral, abruptly narrowed to beaked mouth. Capsule wall bistratose. Spores 22–28 μm. Four species.

1. J. hutchinsiae (Hook.) Dum., *Sylloge Jung. Europ. Indig.*: 38. 1831   (Fig. 115)

Autoecious. Soft, dark green patches. Shoots procumbent, to 4 cm, irregularly 1(–2)-pinnately branched, two innovations usually arising immediately below perianth. Leaves approximate to imbricate, spreading horizontally, dorsal lobe obliquely ovate, dorsal margin curved but not crossing stem, ventral margin ± straight, apex rounded, apiculate, marginal toothing variable, spinose-dentate or teeth restricted to dorsal margin and apex or apex only, or teeth poorly developed or wanting; ventral lobe variable, helmet-shaped, 1.4–2.0 times as long as wide, sometimes with a cilium projecting from base, or small, lanceolate, acuminate, with 0–2 spinose teeth or very small and appearing as a reflexed tooth at base of dorsal lobe; cells 20–32 μm wide in mid-leaf, thin-walled, trigones lacking. Underleaves ovate to ± orbicular, bilobed ½–⅔, lobes emarginate, margins with 0–2 spinose teeth. Male branches lateral, shorter than to hardly longer than leaves, bracts 3–4 pairs, bilobed, saccate. Female inflorescence terminal on main axis or long branch, often successional, bracts conduplicate, asymmetrically bilobed, lobes acuminate, margins spinose-dentate, bracteole smaller, symmetrically bilobed, spinose-dentate. Perianths occasional, obovate, strongly trigonous above, abruptly narrowed to shortly beaked mouth. Capsules occasional, spring. *n* = 9. Deeply shaded wet or dripping rocks in streams, on stream banks, in rock crevices and caves, rarely on wet tree roots, rare to occasional in W. Britain from Cornwall north to W. Ross and Outer Hebrides, very rare elsewhere, Yorkshire, rare to occasional in Ireland. 35, H16. W. France and Spain, Faroes, Azores, Madeira, La Palma, Tenerife, Cape Verde Is.

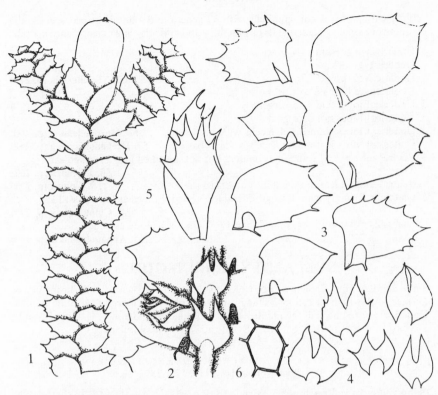

Fig. **115**. *Jubula hutchinsiae*: 1, shoot with perianth (× 15); 2, portion of shoot with male branch, ventral side (× 21); 3, leaves, ventral side (× 21); 4, underleaves (× 21); 5, female bract (× 21); 6, mid-leaf cell (× 450).

European and Macaronesian material belongs to ssp. *hutchinsiae*, while material from elsewhere belongs to ssp. *javanica* (Steph.) Verd. Bearing a superficial resemblance to other British spinose-toothed hepatics but readily distinguished by the leaves being bilobed with large, incubous dorsal lobes.

26. LEJEUNEACEAE

Plants minute to medium-sized, yellowish to green, occasionally brown-tinged. Branching usually of *Lejeunea*-type. Rhizoids restricted to underleaf bases. Leaves unequally bilobed, margins entire, ventral lobe folded under dorsal and much smaller, joined to both stem and dorsal lobe with the latter forming a conspicuous keel region, ventral lobe inflated, terminating in a single apical cell with a hyaline papilla at base. Underleaves usually present, much smaller than lateral leaves. Vegetative propagation frequently by caducous leaves, microphyllous small branches or discoid gemmae. Female bracts 2, archegonia one per inflorescence, long-necked. Perianth with 3–5 keels, abruptly narrowed at apex to conspicuous beak. Seta cells in tiers so that seta appears articulated. Capsules spherical, wall 2-stratose. Spores 15–65 μm long, germinating before release. A rapidly evolving family of predominantly tropical and subtropical epiphytes and epiphylls with 80 genera and 1600 described species, the taxonomy of which is such that its terminology and difficulty makes it a science in itself.

1 Underleaves present   2
   Underleaves lacking   6
2 Underleaves simple   **62. Marchesinia (p. 279)**
   Underleaves bilobed to ⅓ or more   3
3 Underleaf lobes widely divergent   4
   Underleaf lobes not divergent   5
4 Underleaf lobes rounded, 4–6 cells wide at base   **56. Harpalejeunea (p. 260)**
   Underleaf lobes acuminate, 2 cells wide at base   **57. Drepanolejeunea (p. 262)**
5 Ventral lobe of leaf sometimes inflated but not modified into a beaked sac
                                                                           58. Lejeunea (p. 262)
   Ventral lobe of leaf modified into a beaked sac   **59. Colura (p. 272)**
6 Leaf cells strongly mamillose or dorsal lobe 1.1–1.4 times as long as wide
                                                                           60. Cololejeunea (p. 273)
   Leaf cells not mamillose, dorsal lobe 1.6–2.4 times as long as wide
                                                                           61. Aphanolejeunea (p. 278)

SUBFAMILY LEJEUNEOIDEAE

Plants small to minute. Stem in T.S. with 7 cortical and 5–25 central cells, branching of *Lejeunea*-type. Leaf insertion J-shaped; ventral lobe with marginal hyaline papilla. Underleaves present, half the number of the lateral leaves. Gemmae lacking. Perianth 5-keeled, not compressed. Seta in T.S. with 12 outer and 4 inner cells.

56. HARPALEJEUNEA (SPRUCE) SCHIFFN. IN ENGLER & PRANTL., *Natürl. Pflanzenfam.* 1(3): 119, 126. 1893

Plants yellow or yellowish-green. Stem in section with 7 cortical and 3–10 central cells, cell walls somewhat thickened, branching of *Lejeunea*-type. Leaves distant to subimbricate, spreading, obliquely ovate, falcate and deflexed, apex of dorsal lobe acute; ventral lobe strongly inflated, about ½ length of dorsal lobe, ± forming a right angle with free margin of dorsal lobe, apex a single-celled tooth with hyaline papilla at base of inner face, enlarged cells (ocelli) at base each containing a single large oil-body; stylus lacking. Underleaves distant, obdeltoid or obcordate, lobes divergent with rounded apices. Male inflorescence with bracteoles only at base. Female inflorescence with 2 subinvolucral innovations, bracteole ± free from bracts. Perianth 5-keeled, not compressed. An almost entirely tropical genus for which about 60 species have been described but of which probably only half are genuine.

1. H. ovata (Hook.) Schiffn. in Engler & Prantl., *Näturl. Pflanzenfam.* 1(3): 127.
1893   (Fig. 116)

Dioecious. Plants minute, yellowish-green, solitary, gregarious or in thin patches. Shoots procumbent, to *ca* 1 cm long, sparsely irregularly branched. Leaves on mature stems distant to subimbricate, spreading, margins entire, dorsal lobe obliquely ovate, subfalcate, convex with deflexed acute or subacute apex, 280–420 μm long, junction of ventral margin and keel forming an angle of 90–120°; ventral lobe inflated, 160–250 μm long, 0.4–0.6 length of dorsal lobe, apex a unicellular tooth often obscured because of strongly inrolled free margin; keel rounded, not or only slightly crenulate with bulging cells; cells in mid-dorsal lobe 16–24 μm wide, smooth, walls slightly thickened, trigones small to medium-sized, oil-bodies 4–6 per cell, compound. Underleaves distant, obdeltoid or obcordate, 70–190 μm long, 100–200 μm wide, base very narrow, apex

Fig. **116**. 1–7, *Harpalejeunea ovata*: 1 and 2, shoots dorsal and ventral sides (× 33); 3, leaves, ventral side (× 63); 4, leaf, dorsal side (× 63); 5, dorsal lobe apex (× 275); 6, underleaves (× 150); 7, mid-dorsal lobe cell (× 450). 8–13, *Drepanolejeunea hamatifolia*: 8 and 9, shoots, dorsal and ventral sides (× 63); 10, leaves, ventral side (× 84); 11, leaf, dorsal side (× 84); 12, dorsal lobe apex (× 275); 13, underleaf (× 150).

emarginate, 2–3 times width of stem, lobes 4–6 cells wide at base, apices rounded. Gemmae lacking. Perianths not known in Britain or Ireland, obovate or obovate-clavate, 5-keeled, keels entire (Schuster, 1980). On vertical damp rock, bark and epiphytic on other bryophytes in shaded humid ravines and cliffs, on boulders in scree, ascending to 460 m, very rare in S.W. England, S. Wales and S.W. Scotland, occasional in N.W. Wales and the Lake District, occasional to common from Kintyre north to W. Sutherland and Outer Hebrides, frequent in extreme west of Ireland, very rare elsewhere. 27, H16. Corsica, Spain, Portugal, France, Italy, S.W. Norway, Faroes, Macaronesia.

The plant that occurs in Europe and Macaronesia is ssp. *ovata*.

57. DREPANOLEJEUNEA (SPRUCE) SCHIFFN. IN ENGLER & PRANTL., *Natürl. Pflanzenfam.* 1(3): 119, 126. 1893

Plants usually yellowish-green. Stem in section with 7 cortical and 3–4 central cells, cell walls thickened, branching of *Lejeunea*-type. Leaves distant to imbricate, dorsal lobe ovate to lanceolate, falcate, apex acute to acuminate, margin crenulate to toothed; ventral lobe strongly inflated, apex with 1-celled tooth with hyaline papilla on inner face at base, line of ocelli present, keel apex often with a 1-celled tooth; stylus lacking. Underleaves obdeltoid, bilobed, lobes widely divergent, acute, 2–3(–4) cells wide at base. Vegetative propagation by deciduous dwarf branches. Male inflorescence only with bracteoles at base. Female inflorescence with one subinvolucral innovation, bracts sharply toothed and at least one connate with bracteole. Perianth 5-keeled above, keels often winged and variously dentate. About 150 almost exclusively tropical and subtropical species.

1. D. hamatifolia (Hook.) Schiffn. in Engler & Prantl. *Natürl. Pflanzenfam.* 3(1): 126. 1893
(Fig. 116)

Dioecious. Plants minute, yellowish-green, solitary, gregarious or in thin patches. Shoots procumbent, to *ca* 1 cm long, sparsely irregularly branched. Leaves on mature stems distant to approximate, erecto-patent, dorsal lobe convex, from broadly ovate base abruptly narrowed to deflexed acuminate apex ending in 1–2 uniseriate cells, dorsal margin with (0–)1–3 blunt to sharp unicellular teeth, 200–340 µm long; ventral lobe inflated, 140–190 µm long, 0.5–0.7 length of dorsal lobe, apex a unicellular straight or curved tooth, free margin strongly incurved; keel rounded, crenulate with bulging cells; stylus lacking; cells in mid-dorsal lobe 16–25 µm wide, smooth, thin-walled, trigones small or lacking, oil-bodies compound, 4–9 per cell. Underleaves distant, obdeltoid, 60–120 µm long, 100–170 µm wide, (2.0–)2.5–3.5 times width of stem, lobes widely diverging, acuminate, 2 cells wide at base, terminating in 2 uniseriate cells. Female bracts ± conduplicate, bracts and bracteole spinosely toothed. Perianths very rare, 5-winged above, wings irregularly spinose (Macvicar, 1926). Capsules very rare, spring. On vertical damp rock, bark and creeping over other bryophytes in shaded, humid ravines and cliffs, on boulders in scree, ascending to 430 m, rare to frequent from the Lake District north to Sutherland and Outer Hebrides, very rare elsewhere, N. Cornwall, Merioneth, Caernarfon, W. Yorkshire, occasional to frequent in the extreme west and in N.W. Ireland, Wicklow, Dublin. 21, H24. Atlantic coast of France, W. Pyrenees, Portugal, Azores, Madeira, Tenerife, Natal.

58. LEJEUNEA LIBERT CORR. HAMPE, *Linnaea* 11: 92. 1837

Plants small to minute, often pale or yellowish. Stem in T.S. with 7 cortical cells and 3–14 much smaller central cells; branches arising immediately behind ventral base of leaf

(*Lejeunea*-type branching). Leaves distant to imbricate, ± spreading laterally, apex of dorsal lobe rounded or obtuse; ventral lobe small, sometimes vestigial, usually inflated, apex a blunt 1-celled tooth with hyaline papilla at base; leaf cells not mamillose or papillose, trigones small or lacking. Underleaves ± orbicular, bilobed, 1–4 times width of stem. Gemmae and vegetative propagules lacking. Male inflorescence on short branch, bracts imbricate, ± symmetrically bilobed, saccate, bracteoles present at base of inflorescence only. Female inflorescence terminal on main axis or long or short branch, rarely on a dwarf branch, often becoming pseudolateral by overgrowth of a subinvolucral innovation, bracts unequally bilobed, bracteole bilobed, joined at base to one or both bracts. Perianth usually 5-keeled at least above. A mainly tropical and subtropical genus with innumerable species (Schuster, 1980).

1 Ventral lobe more than half the length of dorsal lobe at least in well developed leaves, perianths very rare   2
   Ventral lobe less than half the length of dorsal lobe, perianths frequent   3
2 Leaves not varying much in size along stem, ventral lobe ± constantly 0.7–0.9 length of dorsal lobe, oil-bodies compound   **8. L. ulicina**
   Leaves variable in size along stem, ventral lobe variable, 0.0–0.8 length of dorsal lobe, oil-bodies simple   **5. L. hibernica**
3 Dorsal lobe broadly ovate   4
   Dorsal lobe elliptical to lingulate or ovate   6
4 Underleaves on mature stems contiguous to subimbricate, 1.2–2.0 length of ventral lobe, oil-bodies numerous, simple, glistening, perianth winged in upper ⅓
                                                                                1. L. cavifolia
   Underleaves on mature stems distant, 0.9–1.2 length of ventral lobe, oil-bodies 4–6 per cell, compound, opaque, perianth winged to below middle   5
5 Angle between dorsal lobe and keel 120–150°, margin of ventral lobe at an angle with margin of dorsal lobe (Fig. **117**, 14), apical tooth of ventral lobe straight, directed ± to stem apex   **2. L. lamacerina**
   Angle between dorsal lobe and keel 70–90°, margins of ventral lobe and dorsal lobes forming a ± straight line, (Fig. **118**, 7), apical tooth of ventral lobe curved ± towards leaf apex   **3. L. patens**
6 Underleaves contiguous to subimbricate, 1.1–1.3 times as long as wide, oil-bodies compound   **4. L. flava**
   Underleaves distant, 0.8–1.1 times as long as wide, oil-bodies simple   7
7 Dorsal lobes mostly more than 500 μm long, cells 24–32 μm wide, underleaf sinus wide, perianth 5-winged   **6. L. holtii**
   Dorsal lobes less than 500 μm long, cells 20–24 μm wide, underleaf sinus narrow, perianth smooth   **7. L. mandonii**

Subgenus *Lejeunea*

Autoecious. Plants small, often forming tight thin patches. Stems in section with 8–15 central cells. Leaves patent to spreading, contiguous to imbricate; ventral lobe not usually more than 0.25 area of dorsal lobe; ocelli lacking. Male inflorescence with bracteoles only at base, bracts with 1–2 antheridia. Female bracts with keel region unwinged. Perianths frequent, 5-angled or winged above.

1. L. cavifolia (Ehrh.) Lindb., *Acta Soc. Sci. Fenn.* 10: 43. 1871   (Fig. 117)

Autoecious. Plants very small, yellowish-green to green, in small to extensive thin patches. Shoots to *ca* 2 cm, stems irregularly branched. Leaves on mature stems contiguous to imbricate, patent, dorsal lobe obliquely broadly ovate, slightly convex to convex, apex rounded, margin entire, (560–)600–900(–1100) μm long, 220–480(–600) μm wide, 1.2–1.6(–1.8) times as long as wide; ventral lobe 180–240(–280) μm long, 0.2–0.4 length of dorsal lobe but on some stems very small or ± vestigial, inflated in keel region, margin plane to inrolled, apex a single, forwardly directed cell; keel convex, not crenulate, forming an angle of 120–160° with margin of dorsal lobe; cells in mid-dorsal lobe 24–36(–40) μm wide, walls thin but sometimes with intermediate thickenings, trigones small, oil-bodies simple, glistening, 25–50 per cell, 1–2 μm long, often persisting in dried specimens. Underleaves on mature stems broadly ovate to ± orbicular, 176–400(–480) μm long, 160–400(–480) μm wide, 0.8–1.0 times as long as wide, 1.2–2.0 times length of adjacent ventral lobes, bilobed to ca 0.5, sinus narrow, acute to obtuse at base. Gemmae lacking. Male inflorescence terminal on short branch, imbricate, 2–4 pairs, smaller than leaves, saccate. Female inflorescence terminal on main axis or long or short branch, becoming pseudolateral with age, bracts smaller than leaves, ± conduplicate, unequally bilobed. Perianths frequent, narrowly obpyriform, 5-winged in upper ⅓, wings not or hardly crenulate, beaked. Capsules occasional, summer to winter. Damp shaded rocks and boulders in ravines, on cliffs, in scree, often creeping over other bryophytes, also on tree trunks and shrub stems, usually in basic habitats, ascending to 580 m, frequent or common in W. and N. Britain, rare to occasional elsewhere, widespread in Ireland. 96, H30. Europe north to Fennoscandia and N. Russia, Faroes, Iceland, Caucasus, Turkey, Siberia, Tunisia, Algeria, Azores, Madeira, Gran Canaria, Tenerife.

2. L. lamacerina (Steph.) Schiffn. *Hedwigia* 41: 278. 1902   (Fig. 117)
L. azorica Steph., *L. lamacerina* var. *azorica* (Steph.) Greig-Smith; *L. planiuscula* (Lindb.) Buch

Autoecious. Plants very small, pale to yellowish-green, in thin patches. Shoots to *ca* 15 mm, irregularly branched. Leaves on mature stems contiguous to imbricate, patent, dorsal lobe obliquely broadly ovate, slightly convex, apex rounded, margin entire, 400–600(–760) μm long, 280–480(–600) μm wide, 1.3–1.5 times as long as wide; ventral lobe 150–210 μm long, 100–160 μm wide, 0.28–0.37(–0.40) length of dorsal lobe, but in some leaves very small or vestigial, apex a single cell ± pointing to stem apex but often obscured by inrolled margin, margin between apex and keel at an angle to margin of dorsal lobe; keel strongly convex, not crenulate, forming angle of 120–150° with margin of dorsal lobe; cells in mid-dorsal lobe 20–28(–32) μm wide, walls thin to thickened, trigones absent or minute, oil-bodies compound, opaque (but see footnote), 4–6 per cell, 4–8 μm long. Underleaves distant, broadly ovate to suborbicular, (120–)150–220 μm long, 140–230 μm wide, 0.9–1.0(–1.1) times as long as wide, 0.9–1.2 length of adjacent ventral lobes, bilobed to 0.4–0.5, sinus narrow, acute to rounded at base, margin sometimes with a tooth on one or both sides. Male inflorescence on short branch, bracts imbricate, 2–4 pairs, smaller than leaves, saccate at base. Female inflorescence terminal on long or short branch, becoming pseudolateral with age, bracts ± conduplicate, unequally bilobed. Perianths frequent, obpyriform or clavate, 5-winged in upper ⅔ or more, wings not crenulate, beaked. Capsules occasional, ± throughout year. Usually growing over other bryophytes on damp shaded rocks and boulders in ravines, in woodland and by waterfalls, especially where base-rich, occasionally on sheltered stream banks and on tree trunks, ascending to *ca* 500 m, frequent or common in W. Britain from Cornwall to Shetland, rare elsewhere but extending east to Surrey, scattered in Ireland. 61, H29, C. Portugal, N.W. Spain, S.W. France, Elba, Faroes, Azores, Madeira, Canaries, Cape Verde Is.

When fresh *L. lamacerina* and *L. patens* have a small number of relatively large, compound, opaque oil-bodies in the leaf cells, providing a ready distinction from *L. cavifolia* which has numerous, small, glistening oil-bodies. The oil-bodies in the first two species, when the plants are drying, either due to drought or after collection, tend to round off, lose their opacity and glisten but differ from those of *L. cavifolia* in their small number and relatively large size (about 5 μm diameter). The oil-bodies of *L. cavifolia* often persist for many years and become visible on rehydration (I have seen them in an 81 year old herbarium specimen). In *L. lamacerina* and *L. patens* the oil-bodies often vanish on drying. However, in some specimens the compound oil-bodies may be seen briefly immediately after rehydration even after several years.

L. cavifolia is a generally larger plant with larger cells than the other two species and the relatively larger contiguous to subimbricate underleaves are diagnostic; on juvenile or depauperate stems and branches the underleaves are smaller and distant. The perianth is also only winged in the upper one-third.

L. patens differs from *L. lamacerina* in the acute angle between the keel and the margin of the dorsal lobe, the edge of the dorsal and ventral lobes forming a ± straight line, and the nature of the apical tooth of the ventral lobe. For a statistical analysis of the differences between these two species see Greig-Smith, *Trans. Br. bryol. Soc.* **2**, 458–69, 1954.

L. mandonii is a smaller plant and like *L. holtii* and *L. flava* has relatively narrower leaves.

3. L. patens Lindb., *Acta Soc. Sci. Fenn.* 10: 482. 1875   (Fig. 118)

Autoecious. Plants very small, pale to whitish-green, shiny when dry, in thin patches. Shoots to *ca* 12 mm, irregularly branched. Leaves on mature stems contiguous to imbricate, patent, dorsal lobe very obliquely broadly ovate to almost broadly reniform, convex, or very convex, apex rounded, margin entire, 360–600 μm long, 280–400 μm wide, 1.3–1.6 times as long as wide; ventral lobe 180–240 μm long, 120–170 μm wide, 0.35–0.40(–50) length of dorsal lobe, inflated, apex a single cell curved towards leaf apex but often obscured by inrolled margin, margin between apex and keel forming a ± straight line with margin of dorsal lobe; keel strongly convex, not crenulate, forming angle of (60–)70–90° with margin of dorsal lobe; cells in mid-dorsal lobe 20–28(–32) μm wide, walls thin, trigones small, oil-bodies compound, opaque (but see footnote to *L. lamacerina*), 4–6 per cell, 4–8 μm long. Underleaves distant, broadly ovate to suborbicular, 140–240 μm long, 150–260 μm wide, 0.9–1.0 times as long as wide, 0.7–1.1 length of adjacent ventral lobes, bilobed to 0.4–0.5, sinus narrow with obtuse or rounded base. Gemmae lacking. Male inflorescence on short branch, bracts imbricate, 2–4 pairs, smaller than leaves, saccate at base. Female inflorescence terminal on long or short branch, becoming pseudolateral with age, bracts ± conduplicate, unequally bilobed. Perianths occasional, narrowly obpyriform, 5-winged in upper ⅔ or more, wings not crenulate, beaked. Capsules rare, winter, spring, late summer. Damp shaded rocks and boulders in ravines, on cliffs, in scree, often creeping over other bryophytes, especially where base-rich, also on tree trunks and shrub stems, ascending to 650 m, rare in S.W. Britain and S. Wales, frequent or common in the west, from Mid Wales and Yorkshire north to Shetland, very rare elsewhere. 56, H33, C. W. and Mediterranean Europe from Spain and Portugal north to S.W. Norway and east to the Crimea and Caucasus, Turkey, Macaronesia.

4. L. flava (Sw.) Nees, *Naturg. Europ. Leberm.* 3: 277. 1838   (Fig. 118)

Autoecious. Plants very small, yellowish-green to green, in thin patches. Shoots to 2.0 (–2.5) cm, stems irregularly branched. Leaves on mature shoots contiguous to imbricate, patent, dorsal lobe elliptical to ovate-lingulate, ± convex, apex rounded, margin entire, 480–800 μm long, 265–480 μm wide, 1.6–2.0 times as long as wide; ventral lobe 120–200 μm long, (0.2–)0.3(–0.4) length of dorsal lobe but in some plants very small or vestigial, slightly inflated in keel region, margin inrolled, apex a single cell directed ± to leaf apex; keel convex, not crenulate, forming angle of 120–160° with margin of dorsal lobe; cells in mid-dorsal lobe (16–)20–24 μm wide, walls ± thickened, trigones medium-sized, oil-bodies compound, opaque, mostly 5–12 per cell. Under-leaves contiguous to subimbricate, broadly ovate, 240–480 μm long, 240–440 μm wide, (1.0–)1.1–1.3(–1.4) times as long as wide, 1.6–2.0(–3.0) times length of adjacent ventral

Fig. **117**. 1–7, *Lejeunea cavifolia*: shoot, dorsal side (×24); 2, shoot with perianth and male branches, dorsal side (×24); 3, shoot with perianth, ventral side (×24); 4, leaves, ventral side (×38); 5, leaf, dorsal side (×38); 6, underleaves with adjoining ventral lobes (×63); 7, mid-dorsal lobe cell with oil-bodies (×450). 8–15, *L. lamacerina*: 8, shoot with perianths, dorsal side (×38); 9, shoot with perianth, ventral side (×38); 10, male branch, ventral side (×38); 11, leaves, ventral side (×45); 12, leaf, dorsal side (×45); 13, underleaves with adjoining ventral lobes (×63); 14, apex of ventral lobe (×275); 15, mid-dorsal lobe cell (×450).

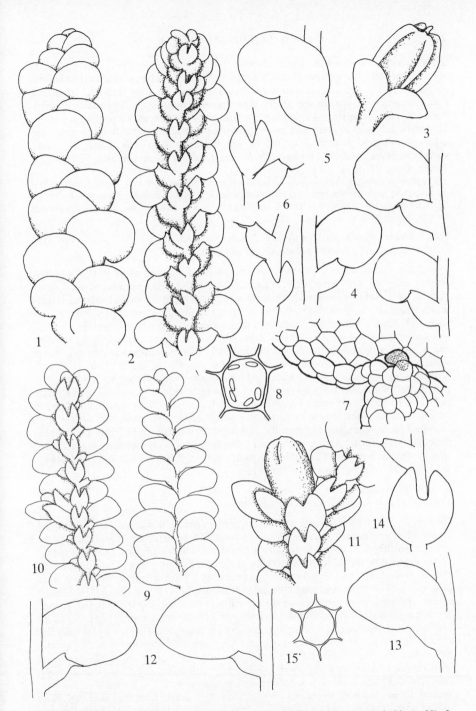

Fig. **118**. 1–8, *Lejeunea patens*: 1, shoot, dorsal side (× 38); 2, shoot, ventral side (× 38); 3, perianth, dorsal side (× 63); 4, leaves, ventral side (× 63); 5, leaf, dorsal side (× 63); 6, underleaves with adjacent ventral lobes (× 63); 7, apex of ventral lobe, apical tooth stippled (× 275); 8, mid-dorsal lobe cell with oil-bodies (× 450). 9–15: *L. flava*: 9, shoot ventral side (× 27); 10, shoot with female inflorescence, ventral side (× 27); 11, perianth, ventral side (× 27); 12, leaves, ventral side (× 63); 13, leaf, dorsal side (× 63); 14, underleaf with adjacent ventral lobe (× 63); 15, mid-dorsal lobe cell (× 450).

268 26. LEJEUNEACEAE

lobes, bilobed to 0.3–0.5, sinus narrow, base obtuse. Gemmae lacking. Male inflorescence terminal on short branch, bracts imbricate, 2–4 pairs, smaller than leaves, saccate. Female inflorescence terminal on main axis or long branch, becoming pseudolateral with age, bracts larger than leaves, conduplicate, unequally bilobed. Perianths rare, obovate, 5-angled but not winged towards apex. Capsules rare. $n = 9$. Dry sheltered rocks in glens and ravines, by streams and in scree, often creeping over other bryophytes, ascending to 300 m, occasional in Kerry, W. Cork, very rare elsewhere, Waterford, W. Galway. H5. Azores, Madeira, Gran Canaria, Tenerife, La Palma, Cape Verde Is.

The plant described here belongs to ssp. *moorei* (Lindb.) Schust. (Schuster, 1980). It differs from the preceding species in the relatively narrower dorsal lobe and underleaves longer than wide. In the latter character it also differs from *L. holtii* and the much smaller *L. mandonii*, both of which have simple oil-bodies.

5. L. hibernica Bischl. et al. ex Grolle, *Lindbergia* 3: 48. 1975   (Fig. 119)
L. diversiloba auct., *Microlejeunea diversiloba* auct.

Dioecious. Plants minute, yellowish-green, scattered, gregarious or in thin patches. Shoots to *ca* 15 mm, irregularly branched, branches sometimes ascending. Leaves distant, very variable in size along same stem, erecto-patent, dorsal lobe of largest leaves ovate or elliptical, slightly convex, apex rounded, margin entire, 260–380 µm long, 150–220 µm wide, 1.5–1.9 times as long as wide; ventral lobe in largest leaves 160–240 µm long, 0.6–0.8 length of dorsal lobe, but often smaller or even lacking in many leaves, apex a unicellular tooth, margin inrolled; cells in mid-dorsal lobe 16–20 µm wide, walls thickened, trigones lacking, oil-bodies simple, 1–2 µm (2–)3–4(–8) per cell. Underleaves very small, 60–110 µm long, 40–80 µm wide, bilobed to 0.6–0.8, sinus rounded at base, lobes broadly lanceolate, (3–)4 cells wide at base, ending in a single cell. Gemmae lacking. Female bracts larger than leaves, conduplicate, dorsal lobe with sinuose margin, apex obtuse or rounded, ventral lobe much narrower, with a large marginal tooth, apex acute. Perianths and capsules not known in Ireland. On damp rock and on leaf litter and creeping over other bryophytes on rocks in deeply shaded sites by streams and waterfalls, at low altitudes, very rare, Kerry, W. Cork, W. Galway, W. Mayo. H5. Azores, Madeira, La Palma.

6. L. holtii Spruce, *J. Bot. Br. Foreign* 25: 33. 1887   (Fig. 119)
Potamolejeunea holtii (Spruce) Greig-Smith

Autoecious. Plants very small, yellowish-green to green, sometimes reddish-tinged, glossy when dry, in thin patches. Shoots to about 2.5 cm, irregularly branched below, unbranched above. Leaves on mature stems distant to contiguous or subimbricate, spreading widely, dorsal lobe lingulate or more usually elliptical and narrowing from middle to rounded apex, ± plane, margin entire, 480–960 µm long, 280–560 µm wide, (1.4–)1.5–1.8(–1.9) times as long as wide; ventral lobe 120–180 µm long, 0.2–0.3 length of dorsal lobe but on some stems very small or ± vestigial, slightly inflated, margin inrolled, apex a single cell directed ± to leaf apex; keel convex, forming angle of *ca* 130–160° with margin of dorsal lobe; cells in mid-dorsal lobe (20–)24–32 µm wide, walls thin or with intermediate thickenings, trigones very small or absent, oil-bodies simple, glistening, 30–50 per cell, 2.0–3.5 µm long, often persisting in dried specimens. Underleaves distant, standing away from stem, ± orbicular, 160–280(–340) µm long, 200–300(–320) µm wide, 0.8–1.0 times as long as wide, 1.2–2.0 length of adjacent ventral lobes, bilobed to *ca* ⅓, sinus widely V-shaped. Gemmae lacking. Male inflorescence terminal on short branch, bracts 3–4 pairs, imbricate, much smaller than leaves, saccate. Female inflorescence on very short lateral branch, bracts smaller than leaves, ± conduplicate, unequally bilobed. Perianths rare, obovate-pyriform, 5-winged

Fig. **119**. 1–7, *Lejeunea hibernica*: 1, shoot, dorsal side (× 63); 2, shoot, ventral side (× 63); 3, portion of shoot with female inflorescence, ventral side (× 63); 4, leaves, ventral side (× 84); 5, leaves, dorsal side (× 84); 6, underleaves (× 230); 7, mid-dorsal lobe cell (× 450). 8–13, *L. holtii*: 8, shoot, dorsal side (× 18); 9, shoot with perianth and female inflorescence, ventral side (× 18); 10, leaves, ventral side (× 38); 11, leaf, dorsal side (× 38); 12, underleaves with adjoining ventral lobes (× 63); 13, mid-dorsal lobe cell (× 450).

to below middle, wings crenulate, beaked. Capsules unknown. On wet shaded rocks in ravines and by waterfalls, ascending to 330 m, rare in S.W. Ireland, very rare elsewhere, Islay, Kerry, W. and Mid Cork, Waterford, N. Tipperary, W. Galway. 1, H7. N.W. Spain, Azores, Madeira, La Palma, Tenerife.

Differs from other *Lejeunea* species in the dorsal leaf lobe narrowing towards the apex, the underleaves standing away from the stem instead of appressed and the female inflorescence borne on a dwarf branch with 1–2 vestigial leaves. This latter feature should not be confused with inflorescences of other species which appear sessile and lateral because of the overgrowth of a single subinvolucral innovation.

7. L. mandonii (Steph.) K. Müll., *Rabenh. Krypt.-Fl. Deutschl.* ed. 3: 1281. 1958

(Fig. 120)

L. macvicari Pears.

Autoecious. Plants minute, yellowish-green, sometimes reddish-tinged, scattered stems or small thin patches. Shoots to *ca* 1 cm, stems irregularly branched. Leaves on mature stems sometimes variable, distant to approximate, largest leaves with dorsal lobe obliquely broadly ovate to lingulate, plane or slightly convex, apex rounded to rounded-obtuse or obtuse, margin entire, 240–480 μm long, 160–280 μm wide, 1.6–1.9 times as long as wide; ventral lobe 110–200 μm long, 0.4(–0.5) length of dorsal lobe but in some leaves small or vestigial, inflated, margin plane to inrolled, apex a single cell directed ± to stem apex; keel convex, not crenulate, forming an angle of 130–160° with margin of dorsal lobe; cells in mid-dorsal lobe 20–24 μm wide, walls thin, trigones small or absent, oil-bodies simple, *ca* 15 per cell, 2–3 μm, often persisting in dried specimens. Underleaves distant, ± orbicular, 120–200(–260) μm long, 110–220 μm wide, 0.9–1.1 times as long as wide, (0.7–)0.9–1.1 length of adjacent ventral lobes, bilobed to *ca* 0.5, sinus narrow to wide, rounded at base. Gemmae lacking. Male inflorescence terminal on short branch, bracts imbricate, 2–4 pairs, smaller than leaves, saccate. Female inflorescence terminal on main axis or branch, bracts larger than leaves, conduplicate, unequally bilobed. Perianths occasional, obovate-clavate, completely smooth, beaked. Capsules rare, spring. Among other bryophyte on sheltered, humid rocks especially where basic, occasionally on tree trunks in humid woods and sheltered ravines, at low altitudes in Britain but ascending to 330 m in W. Ireland, very rare, W. Cornwall, W. Inverness, Skye, Kerry, Leitrim. 3, H3. Portugal, Madeira, La Palma, Tenerife.

When present the completely smooth perianth will distinguish *L. mandonii* from all other British and Irish Lejeuneaceae. The simple oil-bodies may persist for at least 88 years and other species with simple oil-bodies have larger leaves except for *L. diversiloba* which differs in leaf morphology.

Subgenus *Microlejeunea* Spruce, *Trans. Proc. Bot. Soc. Edinburgh* 15: 286. 1884

Dioecious. Plants minute, often scattered. Stems in section with 3 central cells. Leaves erect, distant to contiguous; dorsal lobe with 1–3 ocelli at base, cuticle smooth. Male inflorescence with bracteoles only at base, bracts with solitary antheridia. Female bracts with keel region winged. Perianths very rare, 5-winged above.

8. L. ulicina (Tayl.) Gott. et al., *Syn. Hep.*: 387. 1845   (Fig. 120)
Microlejeunea ulicina (Tayl.) Evans

Dioecious. Plants minute, green, solitary, gregarious or in thin patches. Shoots to *ca* 8 mm, sparsely branched. Leaves on mature stems distant to contiguous, rarely subimbricate, erecto-patent, dorsal lobe ovate to rounded-cordate, slightly convex, apex obtuse to rounded, margin entire, (160–)180–250(–280) μm long, 100–180(–210)

Fig. **120**. 1–6, *Lejeunea mandonii* 1, shoot, dorsal side (× 45); 2, shoot, ventral side (× 45); 3, shoot with perianth (× 45); 4, leaves, ventral side (63); 5, leaf, dorsal side (× 63); 6, underleaves and adjacent ventral lobes (× 63); 7, mid-dorsal lobe cell (× 450). 8–14, *L. ulicina*: 8, shoot, dorsal side (× 63); 9, shoot, ventral side (× 63); 10, shoot with pseudolateral female inflorescence (× 63); 11, leaves, ventral side (× 84); 12, leaf, dorsal side (× 84); 13, underleaves (× 84); 14, mid-dorsal lobe cell (× 450).

μm wide, (1.1–)1.2–1.5(–1.8) times as long as wide; ventral lobe inflated, ovate-quadrate, 130–200 μm long, (0.6–)0.7–0.9 length of dorsal lobe, apex a unicellular tooth, free margin inrolled; cells in mid-dorsal lobe 16–24 μm wide, walls slightly thickened, trigones very small, oil-bodies 3–8, compound. Underleaves 60–100 μm long, 50–80 μm wide, bilobed to 0.5–1.7, sinus acute, lobes sometimes ± connivent, lanceolate, 2 cells wide at base, ending in 1–2 uniseriate cells. Gemmae lacking. Male inflorescence on short branch, bracts 2–3 pairs, ± equally bilobed, saccate (Macvicar, 1926). Female inflorescence terminal on main axis, bracts unequally bilobed, conduplicate, margins bluntly crenulate-dentate, dorsal with apex obtuse to subacute, ventral acute, keel winged. Perianth obpyriform, 5-keeled above (Macvicar, 1926). Capsules unknown (?) in Britain and Ireland. Epiphytic on usually deciduous trees and shrubs, rarely on sheltered rock or compressed peat, at low altitudes, occasional to common in S. and W. Britain from Cornwall east to Kent and north to W. Ross and Outer Hebrides, scattered localities in Ireland. 53. H36, C. S. and W. Europe north to Norway, Japan, Turkey, Macaronesia, eastern N. America.

The plant described above belongs to ssp. *ulicina* (Schuster, 1980). Differs from other British and Irish Lejeuneaceae except *L. hibernica* in the very large ventral lobe and/or the non-crenulate keel and leaf margins. *Harpalejeunea* and *Drepanolejeunea* also differ in underleaf shape and *Cololejeunea* in lack of underleaves. *L. hibernica* differs in the variable leaf and ventral leaf lobe size and simple oil-bodies. As pointed out by Schuster (1980) this plant is remarkably common and widespread for a species with no known method of propagation.

The subgenus *Microlejeunea* is sufficiently different from the other subgenera of *Lejeunea* to merit generic status (E.W. Jones, pers. comm.)

SUBFAMILY COLOLEJEUNEOIDEAE

Plants small to minute. Stem in T.S. with 5–7 cortical and (0–)1–3 central cells, branching of *Lejeunea* or *Radula*-type. Leaf insertion narrow, ± transverse. Underleaves present, or where absent, rhizoid bunches same number as lateral leaves. Vegetative reproduction by discoid gemmae. Perianths usually 5-keeled, rarely flattened. Seta in T.S. with 12 outer and 4 inner cells.

59. COLURA (DUM.) DUM., *Recueil Observ. Jungerm.*: 12. 1835

Plants minute. Stem in section with 7 cortical and 3 inner cells, branching of *Lejeunea*-type. Leaves erect or ascending, insertion very narrow, dorsal lobe reduced and shorter than ventral; ventral lobe modified in upper part into ovoid to tubular sac, margin near base inrolled, forming channel leading to sac, the opening of which is guarded by a moveable flap; stylus lacking. Underleaves present, bilobed almost to base. Male inflorescence terminal on short branch, bracts unlobed, bracteoles present throughout. Female bracts shallowly bilobed, bracteoles reduced or absent. Perianth from very narrow base clavate, not beaked, 5-keeled above, each keel ending in a horn-like process. A genus of about 50 species almost entirely restricted to tropical and subtropical Asia, the Americas and Pacific islands.

Unique (except for *Diplasiolejeunea*) in the underleaves and lateral leaves equal in number and (except for *Pleurozia*) in the water-sac modification of the leaf with a moveable flap guarding the orifice.

1. C. calyptrifolia (Hook.) Dum., *Recueil Observ. Jungerm.*: 12. 1835   (Fig. 121)

Autoecious. Plants minute, pale green, scattered shoots, or very small patches. Shoots procumbent, to *ca* 4 mm, sparsely branched, branches sometimes ascending and arcuate. Leaves ascending to erect, crowded, from a very narrow base flask-shaped, 720–1400 μm long, 240–480 μm wide, dorsal lobe smaller than ventral lobe, ovate to

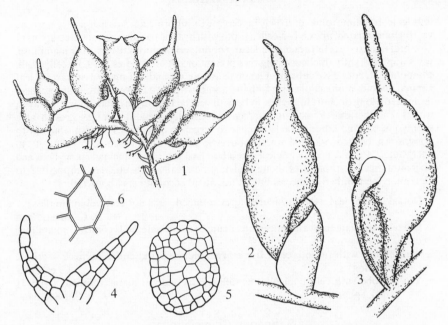

Fig. **121**. *Colura calyptrifolia*: 1, shoot with perianth (× 21); 2, leaf, dorsal side (× 55); 3, leaf, ventral side (× 55); 4, underleaf (× 150); 5, gemma (× 150); 6, cell from middle of sac (× 450).

orbicular, overlapping stem and decurved; ventral lobe ovoid and sac-like above, abruptly narrowed into beak constituting ¼–⅓ total leaf length, below free margin strongly incurved and forming channel to sac, the entrance to which is guarded by a moveable flap; cells very thin-walled, in middle of sac 16–24(–32) μm wide, trigones lacking. Underleaves as many as lateral leaves, bilobed almost to bases, lobes subulate, to 240 μm long. Gemmae rare, multicellular, discoid. Male inflorescence on short branch, bract 3–5 pairs, emarginate. Female inflorescence terminal on main axis, often produced successively from single innovations from below female inflorescences, bracts very small, to 500 μm long, channelled, emarginate. Perianths frequent, from very narrow base obovoid, crenulate with bulging cells, apex truncate, 5-keeled, each keel terminating in a horn-like process. Capsules frequent, spring. On twigs and branches of *Calluna*, *Ulex*, etc., damp to dry shaded rock in humid ravines and on cliffs, mostly where basic, and epiphytic on other bryophytes, especially *Frullania tamarisci*, ascending to 430 m, usually in small quantity but sometimes locally frequent near the coast, W. Cornwall, N. Somerset, E. Sussex, Glamorgan, N.W. Wales, N.W. Yorkshire, Lake District, W. Scotland from Dumfries north to W. Sutherland and Outer Hebrides, frequent in extreme W. Ireland, rare elsewhere. 22, H22. N.W. Spain, N.W. France, Azores, Madeira, Kilimanjaro, Brazil, Peru, Chile, Patagonia, Juan Fernandez.

60. Cololejeunea (Spruce) Schiffn. in Engler & Prantl., *Natürl. Pflanzenfam.* 1(3): 117, 121. 1893

Plants minute. Stem in t.s. with 5–6 cortical and 1 central cell, branching of the *Lejeunea*-type, branches arising internally, with collar round base. Leaves distant to imbricate, not dimorphic, narrowly lanceolate to suborbicular, insertion very narrow; ventral lobe inflated, apex a 1–2-celled tooth with hyaline papilla on inner face; stylus of 1-several uniseriate cells present, persisting or not; cells smooth to conically mamillose, walls thin or slightly thickened, trigones present or not, oil-bodies 2–10 per cell, small, compound. Underleaves absent. Gemmae arising on and flat against surface of leaf lamina, discoid, multicellular, smooth in outline, without adhesive cells. Inflorescences terminal on long or short branches, male with ventral lobe more inflated than in leaves, with 1–2 antheridia. Female inflorescence becoming pseudolateral by growth of a subinvolucral innovation, bracts smaller to larger than leaves. Perianth obovoid to obpyriform, beaked, 5-keeled at least above. Spores 12–20 μm (Schuster, 1980), 30–40 μm (Müller, 1957). A mainly tropical and subtropical genus of epiphytic, epiphyllous and saxicolous plants; rapidly evolving and taxonomically difficult, it is impossible to determine the number of species but an uncritical account lists about 150.

1 Dorsal lobe of leaf ovate-orbicular, apex rounded, cells not conically mamillose
                                                                    3. C. minutissima
   Dorsal lobe ovate to ovate-lanceolate, acute to acuminate, cells conically mamillose
                                                                                      2
2 Ventral lobe with smooth cells, free margin incurved and entire, stylus 2–4-celled, persisting   **1. C. calcarea**
   Ventral lobe with cells conically mamillose, free margin plane, toothed, stylus 1-celled, ephemeral   **2. C. rossettiana**

Subgenus *Cololejeunea*

Plants not strongly compressed, closely adhering to substrate. Leaves never with hyaline margin or lobe apex; cells in part or all of plant except stems conically mamillose. Perianth inflated, 5-keeled above.

1. C. calcarea (Libert) Schiffn. in Engler & Prantl., *Natürl. Pflanzenfam.* 3(1): 122. 1893   (Fig. 122)

Autoecious. Plants very minute, yellowish-green, scattered, gregarious or in small thin patches. Shoots procumbent, to *ca* 8 mm, irregularly branched. Leaves on mature stems distant, approximate or imbricate, erect or erecto-patent, dorsal lobe convex, ovate or ovate-lanceolate, apex decurved, acute to acuminate, margin entire to sharply serrulate-crenulate, sometimes obscurely to sharply toothed, (200–)250–400(–460) μm long, (110–)140–200(–220) μm wide, (1.6–)1.8–2.3 times as long as wide, cells conically mamillose on both sides; ventral lobe 170–210(–280) μm long, 0.6–0.7 length of dorsal lobe, free margin inrolled, terminating in a 1–2-celled tooth, cells not conically mamillose; stylus persisting, uniseriate, 2–4 cells long; cells in mid-dorsal lobe 12–16 μm wide, walls thin or thickened, trigones small or medium-sized. Underleaves absent. Gemmae occasional, discoid, multicellular, 40–64 μm. Male inflorescence terminal on long branch, bracts imbricate, 3–4 pairs, similar to leaves (Macvicar, 1926). Female inflorescence terminal on long branch, becoming pseudolateral with age, bracts of ± similar size to leaves, unequally bilobed, ± conduplicate, mamillose. Perianths occasional, obpyriform, 5-keeled above, mamillose in upper half. Capsules rare, summer. $n = 9$. On other bryophytes or on steeply sloping to vertical rocks, especially limestone, in humid ravines on cliffs and by waterfalls, calcicole, ascending to 800 m, rare to occasional in Gloucester, Hereford, Derby and Wales, occasional to frequent from the Pennines north to Sutherland, widely distributed but rare in Ireland. 48, H17. Europe north to Norway and Sweden, Faroes, Caucasus, Crimea, Turkey.

The commonest of the small Lejeuneaceae (*Colura, Drepanolejeunea, Harpalejeunea, Cololejeunea* and *Aphanolejeunea*) in the British Isles. All of these, except *Cololejeunea minutissima*, the

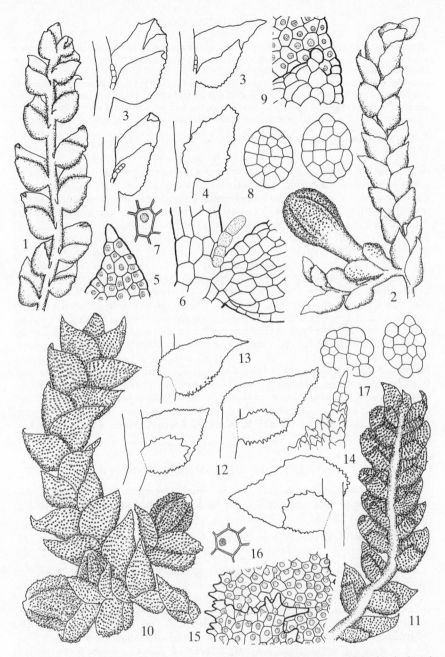

Fig. **122**. 1–8, *Cololejeunea calcarea*: 1, shoot, ventral side (× 63); 2, shoot with perianth, dorsal side (× 63); 3, leaves, ventral side (× 84); 4, leaf, dorsal side (× 84); 5, dorsal lobe apex (× 380); 6, base of ventral lobe with stylus stippled (× 380); 7, mid-dorsal lobe cell, mamilla stippled (× 450); 8, gemmae (× 380); 9, apex of ventral lobe (× 380). 10–17, *C. rossettiana*: 10, shoot with perianths, dorsal side (× 63); 11, shoot, ventral side (× 63); 12, leaves, ventral side (× 84); 13, leaf, dorsal side (× 84); 14, dorsal lobe apex (× 380); 15, apex of ventral lobe (× 380); 16, mid-dorsal lobe cell (× 450); 17, gemmae (× 380).

second most frequent, are each recorded from about 45 vice-counties although this is not a reflection of their frequency in those vice-counties. *C. calcarea* is superficially similar to *Drepanolejeunea hamatifolia* in leaf shape but differs in lack of underleaves and the mamillose leaf cells. *Cololejeunea rossettiana* differs in the ventral lobe of the leaf being plane, coarsely mamillose and having a toothed margin, and the lack of a persistent stylus. This latter character will also separate forms of *Aphanolejeunea microscopica* with a smaller than usual ventral lobe and this species, and also *Cololejeunea minutissima*, differ in the leaf cells lacking conical mamillae.

2. C. rossettiana (Mass.) Schiffn. in Engler & Prantl., *Natürl. Pflanzenfam.* 3(1): 122. 1893   (Fig. 122)

Autoecious. Plants very minute, yellowish-green, scattered, gregarious or in small thin patches. Shoots procumbent, to *ca* 6 mm, stems sparsely and irregularly branched. Leaves on mature stems distant to imbricate, patent, ventrally subsecund, coarsely mamillose throughout, margins crenulate to serrulate, dorsal lobe convex, inflated towards keel, ovate to ovate-lanceolate, acute to acuminate, 240–290 μm long, 150–290 μm wide, (1.5–) 1.7–2.1 times as long as wide; ventral lobe not inflated, 120–230 μm long, (0.4–)0.5–0.6 length of dorsal lobe, cells coarsely mamillose, margin plane, toothed; keel rounded, coarsely crenulate with projecting mamillae; stylus unicellular, ephemeral; cells in mid-dorsal lobe 12–20 μm wide, walls thin, trigones very small or absent, each cell with a tall conical mamilla on either face. Underleaves absent. Gemmae frequent, multicellular, 40–80 μm. Male inflorescence terminal on short branch, spicate, bracts imbricate, 2–5 pairs, of similar size to leaves but saccate. Female inflorescences 1(–2) terminal on long branch, becoming pseudolateral with age, bracts larger than leaves, unequally bilobed, ± conduplicate, coarsely mamillose throughout. Perianths common, obovate to obpyriform, 5-keeled almost to bases, coarsely mamillose throughout except at base. Capsules common, spring, summer. On rock faces, boulders and bark and epiphytic on other bryophytes in humid, shaded habitats, at low altitudes, calcicole, rare or very rare in W. Britain from S. Devon north to Argyll and Dumbarton, very rare elsewhere, I. of Wight, Stafford, Roxburgh, Berwick, Kincardine, rare in Ireland. 28, H12. S., W. and C. Europe, Crimea, Caucasus, Turkey, Iran, Szechuan, Morocco.

Subgenus *Protocolea* Schust., *Hep. Anthoc. N. Amer.* 4: 1239. 1980

Plants not strongly compressed nor closely adhering to substrate. Leaves never with hyaline margin or lobe apex; cells never conically mamillose. Perianth inflated or compressed, not keeled.

3. C. minutissima (Sm.) Schiffn. in Engler & Prantl., *Natürl. Pflanzenfam.* 3(1): 122. 1893   (Fig. 123)

Autoecious. Plants very minute, pale green, scattered, gregarious or in small thin patches. Shoots procumbent but sometimes with sharply ascending tips, to *ca* 8 mm, irregularly branched. Leaves on mature stems distant to approximate, rarely subimbricate, patent, margins and keel crenulate with bulging cells, dorsal lobe convex, suborbicular to ovate-orbicular, apex rounded, 200–260(–300) μm long, 150–190 (–210) μm wide, 1.1–1.4 times as long as wide; ventral lobe inflated, 160–250 μm long, 0.8–0.95 length of dorsal lobe, free margin incurved with an obscure tooth towards apex, apex a uniseriate 2-celled tooth; keel rounded, crenulate with bulging cells; stylus unicellular, ephemeral; cells in mid-dorsal lobe (16–)20–24 μm wide, not conically mamillose, thin-walled, trigones lacking. Underleaves absent but small structures composed of 2 elongated collateral cells, present towards stem tips, one to each lateral leaf. Gemmae frequent, discoid, multicellular, 45–100 μm. Male inflorescence terminal on short lateral branch, spicate, bracts imbricate, 2–4 pairs, similar to leaves but

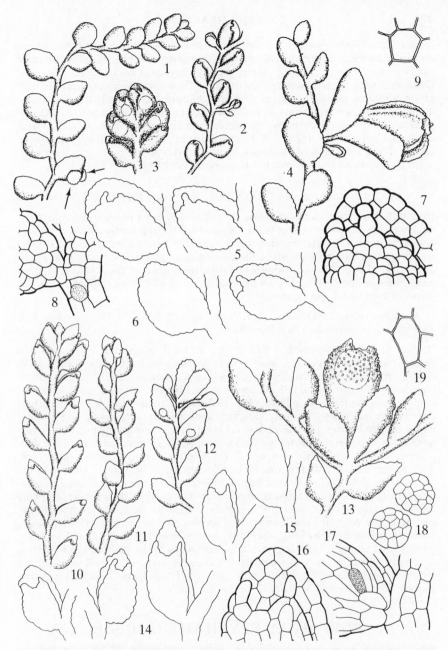

Fig. **123**. 1–9, *Cololejeunea minutissima*: 1, shoot, dorsal side, gemmae on leaf arrowed (× 45); 2, shoot, ventral side (× 45), 3, male branch, ventral side (× 63); 4, shoot with perianth, dorsal side (× 63); 5, leaves, ventral side (× 93); 6, leaf, dorsal side (× 93); 7, leaf apex, ventral side (× 208); 8, base of young leaf, ventral side, with stylus stippled (× 208); 9, mid-dorsal lobe cell (× 450). 10–19, *Aphanolejeunea microscopica*: 10, shoot, ventral side (× 45); 11, shoot, dorsal side (× 45); 12, fertile shoot with paroecious inflorescence and developing subinvolucral innovations (× 45); 13, perianth, dorsal side (× 63); 14, leaves ventral side (× 63); 15, leaf, dorsal side (× 130); 16, leaf apex, ventral side (× 270); 17, base of young leaf, ventral side, with stylus stippled (× 270); 18, gemmae (× 270); 19, mid-dorsal lobe cell (× 450).

saccate. Female inflorescences 1–2, terminal on long branch, bracts larger than leaves, unequally bilobed, ±conduplicate, margins crenulate with bulging cells. Perianths common, obpyriform, 5-keeled above, slightly roughened with bulging cell walls. Capsules common, spring, summer. On trunks and branches, mainly of deciduous trees at low altitudes, mostly in coastal situations but also occasionally inland, rare or occasional in the south and west, from Cornwall east to E. Kent and north to Anglesey, (Westmorland), very rare in W. Scotland, extending north to Skye and Harris, occasional in Ireland. 27, H25, C. W. France, Portugal, N.W. Spain, Italy, Yugoslavia, Greece, Tunisia, Algeria, Macaronesia, south-east U.S.A.

The plant described above belongs to ssp. *minutissima* (Schuster, 1980).

61. APHANOLEJEUNEA EVANS, *Bull. Torrey Bot. Club* 88: 272. 1911

Doubtfully distinct from *Cololejeunea*, the differences being of uncertain constancy. All branching somewhat similar to the *Radula*–type but apparently differing in a way unique to *Aphanolejeunea*, branches without collar at base. Gemmae usually angular with projecting marginal cells, with adhesive cells, arising on leaf lobe margins at a right angle to the margin. A mainly tropical and subtropical genus of about 40 species, these being amongst the smallest known liverworts and mostly being epiphyllous on the leaves of ferns and flowering plants.

1. A. microscopica (Tayl.) Evans, *Bull. Torrey Bot. Club* 38: 273. 1911   (Fig. 123)
Cololejeunea microscopica (Tayl.) Schiffn.

Paroecious. Plants very minute, pale green, scattered, gregarious or in small thin patches. Shoots procumbent, to *ca* 5 mm, sparsely irregularly branched. Leaves distant, erecto-patent, dorsal lobe very convex, apex obtuse to acute, margin smooth to crenulate with bulging cells, 190–280 μm long, 90–130(–140) μm wide, 1.6–2.4 times as long as wide; ventral lobe inflated, (140–)160–200(–250) μm long, 0.7–0.9 length of dorsal lobe, free margin strongly incurved, apex a 1(–2)-celled tooth with another barely discernible tooth beside it; keel rounded, crenulate with bulging cells; stylus unicellular, ephemeral; cells in mid-dorsal lobe 12–20 μm wide, smooth, walls thin, trigones lacking; oil-bodies compound. Underleaves absent. Gemmae occasional, discoid, multicellular, 60–110 μm. Male inflorescence of two bracts, similar to the leaves, immediately below female. Female inflorescence terminal, becoming pseudo-lateral by growth of a subinvolucral innovation, bracts larger than leaves, unequally bilobed, ±conduplicate, margins crenulate. Perianths rare, obpyriform, terete, smooth below, coarsely mamillose above. Capsules rare, late summer. On rock faces, boulders, bark and epiphytic on other bryophytes in humid shaded habitats in ravines, on cliffs and in scree, ascending to 520 m, rare in N. Wales and the Lake District, occasional to frequent in W. Scotland from Kirkcudbright north to W. Ross and Outer Hebrides, occasional to frequent in W. Ireland. 27, H17. Belgium, Luxemburg, Spain, Faroes, Borneo, Azores, Madeira, Tenerife, Tanzania.

SUBFAMILY PTYCHANTHOIDEAE

Plants medium-sized or large. Stem in section with cortex 12 or more cells in circumference, central cells numerous, branching of *Frullania* and/or *Lejeunea*-type. Leaf insertion J-shaped, ventral lobe with 1–several marginal teeth towards base and hyaline papilla of the lobe apex often displaced from the margin to the inner surface; cell walls often pigmented. Underleaves about half the number of lateral leaves, entire or emarginate. Vegetative propagules lacking. Male inflorescence with bracteoles

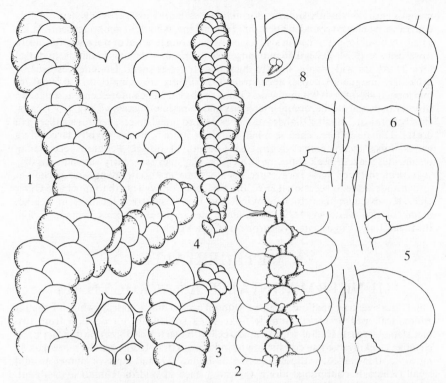

Fig. **124**. *Marchesinia mackaii*: 1, shoot, dorsal side (× 8); 2, shoot, ventral side (× 8); 3, portion of female shoot with perianth (× 8); 4, male shoot (× 8); 5, leaves, ventral side (× 23); 6, leaf, dorsal side (× 23); 7, underleaves (× 21); 8, male bract with antheridia, ventral side (× 23); 9, mid-dorsal lobe cell (× 450).

throughout. Perianth with 3 keels or plicate. Seta in T.S. with 16 outer and 4 inner cells. Wall thickenings of inner layer of cells of capsule wall with 16 large pits (fenestrated).

62. MARCHESINIA S.F. GRAY, *Nat. Arr. Br. Pl.* 1: 679, 689, 817. 1821

Plants medium-sized to large, often pigmented brownish or blackish. Stems irregularly pinnately branched. Leaves imbricate, dorsal lobe ovate to broadly elliptical, crossing stem, margin entire, ventral lobe oval with 0–3 teeth, hyaline papilla at about middle of free margin. Underleaves orbicular to reniform, entire or emarginate. Male bracts bilobed, lobes of similar or slightly differing size, with 1–2 axillary antheridia. Female bracts with ventral lobe about half size of dorsal. Perianth obcordate, truncate, beaked, compressed, plane on both surfaces or with obscure keel on ventral surface. Spores 35–40 μm. A mainly tropical genus with a single species in Europe.

1. M. mackaii (Hook.) S.F. Gray, *Nat. Arr. Br. Pl.* 1: 689. 1821 (Fig. 124)

Autoecious. Plants medium-sized, in olive-green to blackish-green or more rarely brownish flat patches. Shoots procumbent, to *ca* 3 cm, stems irregularly pinnately branched. Leaves imbricate or closely imbricate, incubous, spreading horizontally, margins entire or slightly sinuose, dorsal lobe ovate-orbicular, crossing stem, slightly convex with deflexed rounded apex, 0.8–1.8 mm long, 0.56–1.30 mm wide; ventral lobe small, ovate-oblong, inflated in keel region with margin inflexed or not and with 2 teeth towards base, (0.24–)0.28–0.56 mm long, 0.24–0.48 mm wide; cells in middle of dorsal lobe 24–32 μm wide; walls slightly thickened, trigones small. Underleaves broadly orbicular, margins ± sinuose, apex straight or slightly emarginate, 0.32–0.60(–0.76) mm long, 0.48–0.80(–0.96) mm wide. Gemmae absent. Male inflorescence terminal on short branch, spicate, becoming intercalary and sometimes successive, bracts 3–6 pairs, closely imbricate, smaller than leaves, lobes with rounded apices, ventral smaller than dorsal. Female inflorescence terminal on short branch, bracts slightly larger than leaves, lobes ovate, entire, ventral *ca* ⅔ length of dorsal. Perianths occasional, obcuneate, beaked, beak short, mouth entire. Capsules rare, early summer. On dry, ± vertical, sheltered, basic rock, especially limestone, in woods, ravines and on cliffs, at low altitudes, rare to occasional in W. Britain from Cornwall north to Inner and Outer Hebrides, extending east to I. of Wight and Derby, scattered localities in Ireland, mainly in the west. 44, H29. An atlantic-mediterranean species, France, Spain, Portugal, Italy, Yugoslavia, Macaronesia, Cape Verde Is.

3. METZGERIALES

(JUNGERMANNIALES ANACROGYNAE)

Plants dorsiventral, thalloid, with or without well defined midrib, wings of thallus entire, scalloped or incised into leaf-like lobes (which do not develop from two meristematic cells). Thallus with an apical cell with two (rarely three) cutting faces; air-chambers lacking; median conducting strand present or not, cells usually more than 35 μm wide, thin-walled, without trigones, oil-bodies few and large or numerous and small. Rhizoids continuous along ventral surface of midrib. Underleaves absent. Archegonia produced behind growing point, growth of which is continuous (i.e. monopodial or anacrogynous), on dorsal surface or on short lateral or ventral branches. Developing sporophyte protected by flap-like or cylindrical pseudoperianth or by scales (such structures never being leafy in origin) or by a fleshy calyptra. Capsule spherical, wall 2–6-stratose, usually dehiscing by 4 valves; elaterophore present at base or apex of capsule. Spores (except in Aneuraceae) usually more than 30 μm diameter. About 32 genera.

ARTIFICIAL KEY TO FAMILIES OF METZGERIALES

1 Plants with very obliquely inserted crisped or finely divided leaves, or dorsal surface of thallus with lamellae radiating from midrib to margin, rhizoids frequently deep purple or violet   **27. Codoniaceae (p. 281)**
  Plants thalloid, lacking radiating dorsal lamellae, rhizoids colourless to brown   2
2 Thallus margin scalloped, flask-shaped gemma-receptacles towards thallus tips, *Nostoc*-cavities visible as dark spots at bases of margin lobes
                                                        30. Blasiaceae (p. 296)
  Thallus margin plain or crisped but not scalloped, gemma-receptacles and *Nostoc*-cavities lacking   3
3 Thallus 0.3–1.8 mm wide, often with marginal hairs, regularly dichotomously branched, gametangia and sporophytes in dwarf globular ventral branches
                                                        32. Metzgeriaceae (p. 302)

Thallus without marginal hairs, if dichotomously branched then more than 2 mm
    wide, sporophytes on dorsal side of thallus or on lateral branches   4
4 Thallus simple, pinnate or palmate but not dichotomous, margin not crisped,
    gametangia and sporophytes on short lateral branches **31. Aneuraceae (p. 297)**
Thallus simple with crisped margins or dichotomously branched, gametangia and
    sporophytes on dorsal side of thallus   5
5 Thallus dichotomous, without central strand, antheridia immersed in dorsal surface
    of thallus, archegonia and sporophyte protected by flap-like or tubular
    pseudoperianth, dorsal scales lacking   **28. Pelliaceae (p. 289)**
Thallus dichotomous with longitudinal conducting strand or simple with strongly
    undulate or crisped margins, antheridia and archegonia subtended by scales
    29. Pallaviciniaceae (p. 292)

27. CODONIACEAE

Plants thalloid with lamellae on dorsal surface radiating from midrib to margin or with
main axis with obliquely inserted succubous leaves. Oil-bodies small, numerous.
Gametangia on upper surface of midrib towards apex. Pseudoperianth campanulate or
cup-shaped. Calyptra shorter than pseudoperianth. Seta cells numerous, uniform in
section. Capsule wall 2–4-stratose. Capsule dehiscing irregularly by fragmentation of
wall of upper side of capsule. Spores areolate or with irregular lamellae or spinose on
convex surface. Elaterophore lacking, elaters 2–3-spiral. A small family of three
genera, the third *Sewardiella* Kashyap occurring in India.

Plants with obliquely inserted succubous leaves, rhizoids usually purple to violet
    63. Fossombronia
Plants thalloid with erect lamellae on dorsal surface radiating from midrib to
margin, rhizoids colourless or brown   **64. Petalophyllum**

63. FOSSOMBRONIA RADDI, *Jungerm. Etrusca*: 29. 1818

Monoecious or dioecious. Solitary or gregarious plants, usually of open, temporary or
disturbed habitats. Shoots prostrate or ascending. Stem in section usually semicircular.
Rhizoids usually violet to purple. Leaves in two ranks, obliquely inserted, succubous,
wider than long, margin usually sinuose or lobed, two cells thick at base, elsewhere
unistratose. Archegonia naked near apex on dorsal side of stem. Pseudoperianth
developing after fertilisation, campanulate, wide-mouthed, often plicate, mouth
usually sinuose or lobed. Capsule globose, wall two cells thick, inner layer with
variously developed semi-annular thickenings. Spores (20–)40–70 μm, ornamented
with spines or lamellae on convex face. Antheridia orange-yellow, naked or protected
by bracts on dorsal side of stem.

1 Leaves dissected into irregular fimbriate lobes giving plant a fluffy appearance
    9. F. fimbriata
Leaves ± entire to shallowly lobed, never fimbriate, plants not fluffy in appearance 2
2 Rhizoids colourless or pale brown   **4. F. husnotii**
Rhizoids violet to purple   3
3 Spores 20–27 μm   **8. F. incurva**
Spores (34–)40–64(–76) μm   4
4 Plants 10–17 × 2–5 mm, forming dense patches, dioecious, spores in profile with
    translucent wings joining spines, convex face of spore with several areolae
    2. F. angulosa

Plants 2–10 mm long, solitary or gregarious, paroecious, spores in profile without
    translucent wings or if wings present then only 1–3 areolae on convex face 5
5 Convex face of spores with several ± hexagonal areolae 6
   Convex face of spores with spines or lamellae, areolae 0–3, irregular in shape 7
6 Areolae 11–16 μm across, spores in profile with 18–22 papillae **1. F. foveolata**
   Areolae 3–4 μm across, spores in profile with *ca* 40 papillae **7. F. crozalsii**
7 Spores with flattened truncate papillae **3. F. caespitiformis**
   Spores with lamellae 8
8 Spores in side view with 5–7 lamellae 6–10 μm apart **5. F. pusilla**
   Spores in side view with 10–13 lamellae 3–5 μm apart **6. F. wondraczekii**

Subgenus *Fossombronia*

Paroecious, rarely dioecious. Spores 35–76 μm, not adhering in tetrads.

1. F. foveolata Lindb., *Helsingfors Dagblad* 1873: 2. 1873   (Fig. 125)
F. dumortieri Hüb. & Genth. ex Lindb.

Autoecious. Plant solitary or gregarious, green, 3–7 mm long, prostrate or ascending.
Rhizoids violet-purple. Leaves entire or shallowly lobed. Pseudoperianth campanulate
with irregularly lobed mouth. Semi-annular thickenings of capsule wall mostly
incomplete. Spores brown, (39–)44–48 μm, lamellae anastomosing on convex face to
form ± hexagonal areolae 11–16 μm across, in profile with 18–22 rounded papillae 2–4
μm high, not winged. Elaters (100–)120–180 × 8–10 μm, 2(–3)-spiral. Capsules spring,
late summer, autumn. $n = 9$. On moist soil or peat on heaths, stream banks and by
pools, occasional throughout Britain and Ireland. 39, H11. W. and C. Europe, Finland,
Italy, Turkey, N. Africa, Macaronesia, N. America.

2. F. angulosa (Dicks.) Raddi, *Jungerm. Etrusca*: 29. 1818   (Fig. 125, 127)

Dioecious. Plants perennial, forming green to bright green dense patches, older parts
becoming yellowish-brown; shoots 10–17 mm long, 2–5 mm wide. Rhizoids deep
purple. Leaves crowded towards stem apex, crisped, irregularly and shallowly lobed.
Pseudoperianth campanulate. Inner wall of capsule with complete semi-annular
thickenings. Spores reddish-brown, 33–49 μm, lamellae anastomosing on convex face
to form ± hexagonal areolae 8–15 μm across, in profile with 13–20(–24) spines 3–6 μm
high, joined by translucent wings. Elaters 2(–3)-spiral, 140–310 × 7–12 μm. Capsules
frequent, winter, spring. $n = 9*$. A salt-tolerant plant of peaty soil in sheltered places on
banks, stream-sides, rock crevices, cliffs and heathy ground near the sea, locally
abundant in Cornwall, rare elsewhere, Pembroke, Merioneth, Caernarfon, St Kilda,
scattered localities along the west coast of Ireland. 6, H6, C. W. France, Iberian
Peninsula, Mediterranean region, Macaronesia.
    Readily recognised by the large size of the plants which sometimes form extensive patches.

3. F. caespitiformis De Not. ex Rabenh., *Hep. Eur. Exs.* no. 123, 1860   (Fig. 125)

Paroecious. Plants solitary or gregarious, pale green, shoots 5–10 mm, prostrate.
Rhizoids violet-purple. Leaves entire or shallowly lobed, strongly crisped near stem
apex. Pseudoperianth campanulate with bluntly lobed mouth. Inner layer of capsule
wall with poorly developed semi-annular thickenings. Spores brown, very variable in
size within and between capsules, typically 48–56(–64) μm with flattened truncate
papillae sometimes joining on convex face to form short lamellae, in profile with 19–29

spines 4–8 μm high; occasional plants with smaller spores, 32–42 μm, with shorter spines or spines sometimes reduced to bumps. Capsules autumn to early spring. $n = 9$. On fine, compressed soil in fields, on disturbed ground and by footpaths, rare, scattered localities in S. and E. England, extending north to W. Lancashire and Westmorland. 19. Belgium, France, Spain, Portugal, Switzerland, Italy, Sardinia, Sicily, Yugoslavia, Greece, N. Africa, Ethiopia, Macaronesia.

Some spores of *F. husnotii* very closely resemble those of *F. caespitiformis* but the colourless to pale brown rhizoids of *F. husnotii* are distinctive. Also, spores of *F. husnotii* can usually be found which have well developed lamellae.

4. F. husnotii Corb., *Mem. Soc. Natl. Sci. Nat. Math. Cherbourg* 26: 353. 1889
(Fig. 125)
F. husnotii var. *anglica* Nich.

Paroecious or dioecious. Plants solitary or gregarious, light green, shoots 2–5 mm long, arising from bulbous perennating stem embedded in soil. Rhizoids colourless or pale brown. Leaves shallowly and irregularly lobed, crisped. Pseudoperianth campanulate. Inner layer of capsule wall with well developed semi-annular thickenings. Spores yellowish-brown, 40–53(–58) μm, with papillae or short wavy lamellae, sometimes anastomosing but not forming areolae on convex face, in side view usually with parallel lamellae, in profile with 24–36 flattened truncate papillae 3–6 μm long. Elaters 2–3-spiral, 140–192 × 6–10 μm. Capsules frequent, late summer to spring. $n = 8 + m^*$. Wood rides and soil on paths, roadside verges and rocky slopes in open maritime situations, rare, Cornwall, S. Devon, N. Wilts, W. Sussex, S. Wales, W. Lancs, S. Kerry, W. Cork, Clare, W. Galway, W. Mayo, Channel Is. 9, H5, C. N.W. France, Mediterranean region of Europe, Macaronesia, N. Africa.

5. F. pusilla (L.) Nees, *Naturg. Europ. Leberm.* 3: 319. 1838.

Paroecious. Plants solitary or gregarious, stems fleshy or not. Rhizoids violet-purple. Leaves ± entire to lobed, sometimes acutely so, crisped towards stem apex. Pseudo-perianth campanulate, mouth ± entire to lobed. Inner wall of capsule with incomplete semi-annular thickenings. Spores brown to dark brown, (40–)44–64(–76) μm, lamellae straight to sinuose, sometimes anastomosing on convex face, ± parallel, mostly 5–7 in number and 6–10 μm apart in side view of spores, in profile with 18–28 spines often joined by translucent wings. Elaters 100–250 × 6–13(–16) μm, mostly bispiral.

Key to varieties of *F. pusilla*

Stem not thick and fleshy, spores (40–)44–58(–64) μm with 18–22(–26) spines often joined by translucent wings   var. **pusilla**
Stems often thick and fleshy, spores (52–)56–64(–68) μm with 21–28 spines, translucent wings absent   var. **maritima**

Var. **pusilla**   (Fig. 126)
F. pusilla var. *decipiens* Corb.

Plants prostrate, shoots 3–10 mm, stem not becoming thick and fleshy towards apex. Spores (40–)44–58(–64) μm, lamellae sometimes anastomosing on convex face but not forming areolae, spines 18–22(–26), 4–8 μm high, often joined by translucent wings. Elaters 120–200(–212) × 6–8(–12) μm, bispiral. Capsules produced all year round but especially autumn to spring. $n = 8 + m^*$. On fine-textured or loamy soil in temporary habitats, in arable fields, by ditches, streams, pools and paths, frequent to common in

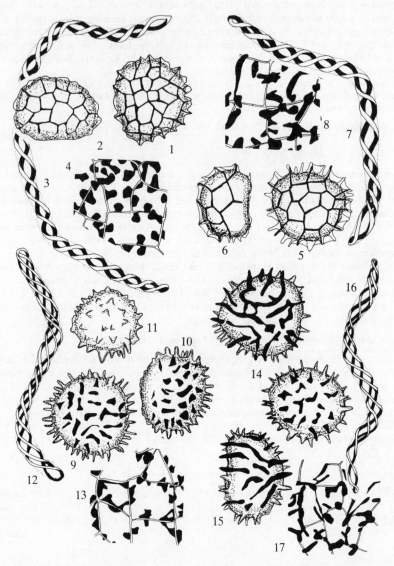

Fig. **125**. 1–4, *Fossombronia foveolata*: 1 and 2, convex face and side of spores; 3, elater; 4, inner wall cells of capsule. 5–8, *F. angulosa*: 5 and 6, convex face and side of spores; 7, elater; 8, inner wall cells of capsule. 9–12, *F. caespitiformis*: 9 and 10, convex face and side of spores; 11, spore from small-spored form; 12, elater; 13, inner wall cells of capsule. 14–17, *F. husnotii*: 14 and 15, convex faces and side of spores; 16, elater; 17, inner wall cells of capsule. Spores and elaters × 520, cells × 280.

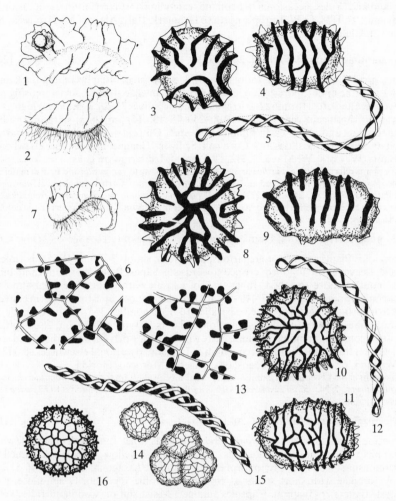

Fig. **126**. 1–6, *Fossombronia pusilla* var. *pusilla*: 1 and 2, plants in dorsal and side view; 3, 4, convex face and side of spores; 5, elater; 6, inner wall cells of capsule. 7–9, *F. pusilla* var. *maritima*: 7, plant in side view; 8 and 9, convex face and side of spores. 10–13, *F. wondraczekii*: 10 and 11, convex face and side of spores; 12, elater; 13, inner wall cells of capsule. 14–15, *F. incurva*: 14, single spore and tetrad 15, elater. 16, *F. crozalsii*: convex face of spore, redrawn from Müller, 1954(× 600). Plants × 6; spores and elaters × 520, cells × 280.

England and Wales, occasional in Scotland, extending north to Caithness, occasional in Ireland. 97, H28, C. W. Europe, north to Denmark, Italy, Algeria, Macaronesia, N. America, Chile.

Var. **maritima** Paton, *J. Bryol.* 7: 244. 1973   (Fig. 126)

Plants prostrate to ascending, shoots mostly 3–5 mm, stem often becoming thick and fleshy near the apex. Spores (52–)56–64(–68) μm, lamellae sinuose, anastomosing on convex face and often forming 1–3 irregular areolae, spines 21–28, 3–5(–6) μm high, not joined by translucent wings. Elaters 100–250 × 8–13(–16) μm, 2(–3)-spiral. Capsules common, late autumn to early summer. *n*-8 + m*. On peaty soil on heaths in coastal habitats, in several localities in S. Cornwall, Scilly Is., Channel Is., very rare elsewhere, S. Hants, W. Cork, W. Mayo. 2, H2, C. French Mediterranean coast.

The plants referred to as *F. pusilla* var. *decipiens* Corb. belong to var. *pusilla* and British material named *F. loitlesbergeri* Schiffn. and most of that named *F. crozalsii* Corb. belongs to var. *maritima* (see Paton, *J. Bryol.*, 7, 243–52, 1973). Like *F. husnotii, F. pusilla* var. *maritima* perennates by means of its tuberous stems; the two may grow together but *F. husnotii* is distinct in rhizoid colour and smaller spores.

6. F. wondraczekii (Corda) Dum., *Recueil Observ. Jungerm.*: 11. 1835   (Fig. 126)

Paroecious. Plants solitary or gregarious, pale green, shoots 3–7 mm. Rhizoids violet-purple. Leaves entire to lobed, crisped towards stem apex. Pseudoperianth campanulate, mouth ± entire to lobed. Inner wall of capsule with incomplete semi-annular thickenings. Spores brown, (37–)40–54 μm, lamellae sinuose, anastomosing on convex face to form 0–1 irregular areolae, ± parallel, 10–13 in number, 3–5 μm apart on side of spores, spines 35–40 in profile, 2–4 μm long. Elaters 100–160 × 8–10 μm, 2(–3)-spiral. Capsules common, summer, autumn, early winter. *n* = 8 + m*. On damp disturbed soil in fields, on tracks, ditch-sides, frequent in S. England, occasional elsewhere. 88, H23, C. W. and C. Europe, Italy, Yugoslavia, Algeria, Morocco, Florida.

The plant named *F. mittenii* Tindall is in all probability an aberrant form of *F. wondraczekii* (see Paton, *J. Bryol.*, 7, 243–52). *F. wondraczekii* has a similar distribution and habitat to *F. pusilla* but is less frequent.

7. F. crozalsii Corb., *Rev. Bryol.* 30: 13. 1903   (Fig. 126)

Paroecious. Plants similar to *F. wondraczekii*, rhizoids violet. Inner wall of capsule with incomplete semi-annular thickenings. Spores reddish-yellow, 34–41 μm, lamellae anastomosing on convex face to form mostly complete hexagonal areolae 3–4(–5) μm across, areolae with short spines at corners, in profile very shortly spinulose, not winged. Elaters 2(–3)-spiral. Capsules summer. Moist soil on woodland rides, very rare, N. Wilts, Berks. 2. Germany, Algeria, Canaries.

I have seen no authentic material of *F. crozalsii* and the above description is based upon those of Müller (1954) and Paton, *J. Bryol.* 7, 243–52, 1973. The illustration (Fig. **126**) is based upon that of Müller, 1954, p. 537.

Subgenus *Simodon Lindb., Rev. Bryol.* 12: 37. 1885

Dioecious. Spores 20–38 μm, adhering in tetrads.

8. F. incurva Lindb., *Bot. Notis.* 1873: 152. 1873   (Fig. 126)

Dioecious. Plants pale green, solitary or in small patches, shoots 2–6 mm, prostrate or more usually ascending. Rhizoids purplish. Leaves 2–4-lobed. Vegetative propagation by buds produced on leaves. Pseudoperianth pyriform, ± plicate, mouth with small uncurved teeth. Inner layer of capsule wall with well developed semi-annular thickenings. Spores yellowish-brown, 20–27 μm, often adhering in tetrads, bluntly papillose in outline, lamellae anastomosing to form irregular to hexagonal areolae. Elaters 2–3-spiral, 90–240 × 8–10 μm. Capsules common, summer, early autumn. $n = 8 + m^*$. On damp soil, often where sandy, by water, on waste ground, paths and in dune-slacks, usually at low altitudes but ascending to 500 m. Rare in S. Britain and Ireland, occasional in Scotland, extending from Cornwall and Sussex to Shetland. 28, H7, C. N. Germany and Poland, S. Sweden and Finland.

Distinct from other British species in the small size of the spores and the perianth narrowed at the mouth before elongation of the seta. For the discovery of this plant in Britain see Crundwell, *Trans. Br. bryol. Soc.*, **4**, 767–74, 1965.

9. F. fimbriata Paton, *J. Bryol.* 8:1. 1974   (Fig. 127)

Dioecious. Plants yellowish-green, solitary or in small patches. Shoots to 3–4 mm. Rhizoids purplish. Leaves very fragile, divided almost to base into 2 or more lobes, each branching dichotomously two or more times and often ending in uniseriate branches of ± quadrate cells, giving a fimbriate appearance to the plants. Vegetative propagation by buds produced on leaves. Pseudoperianth campanulate, fimbriate. Spores dark brown, in tetrads, 36–42 μm diameter. Moist, ± neutral, sandy or gravelly soil by lakes, roads and paths, sea-level to 400 m, rare. Caernarfon, N. Northumberland, Peebles, Selkirk, Mid Perth, Inverness, Argyll, Kintyre, E. Ross, Sligo, W. Mayo. 9, H2. Endemic to Britain and Ireland.

The above description is based in part upon the account of the species by Paton, *J. Bryol.* **8**,, 1–4, 1974.

64. PETALOPHYLLUM NEES & GOTT. EX LEHM., *Novat. Min. Cogn. Stirp. Pugillus* 8: 29. 1844

Dioecious. Thallus simple or branched, sometimes producing swollen perennating proliferations from the underside, midrib semicircular in section. Rhizoids hyaline. Dorsal surface of thallus with 1-cell thick erect lamellae radiating from midrib to margin. Antheridia ± globose, protected by bracts. Archegonia in groups, protected by laciniate bracts. Pseudoperianth campanulate, mouth lobed. Seta 11–15 mm. Capsule globose, wall 3–4 cells thick, inner layer of wall with semi-annular thickenings. Spores large, areolate. Elaters 2–3-spiral. A small genus of three species, the two non-British species coming from Australia and Bolivia, respectively.

1 P. ralfsii (Wils.) Nees & Gott. ex Lehm., *Novat. Min. Cogn. Stirp. Pugillus* 8: 20. 1844   (Fig. 127)

Plants solitary or gregarious, pale green. Rhizoids colourless or brown. Tuberous portion of thallus circular in section, embedded obliquely in substrate, prostrate portion semicircular in section and winged, 5–10(–15) mm long, simple or once or twice dichotomously branched; upper surface of wings with erect unistratose lamellae radiating from midrib to margin. Pseudoperianth widely campanulate, mouth lobed. Spores 48–64 μm with 4–6 hexagonal areolae on convex face. Capsules occasional, spring. $n = 9$. A maritime plant of moist shallow calcareous dune-slacks or damp turf, rare, W. Cornwall north to S. Lancashire, W. Ross, W. Norfolk, S.E. Yorkshire, N. Northumberland, S. Kerry north to Londonderry, Dublin. 16, H8. European Mediterranean coast and islands, Algeria, Texas, Arkansas.

Superficially resembling a *Fossombronia* but differing in having a winged thallus with lamellae. The oblique, cylindrical part of the stem embedded in the substrate serves as a perennating structure. It may be produced from the undersurface of the prostrate thallus midrib or it may arise from an old thallus which has been buried by several millimetres of sand.

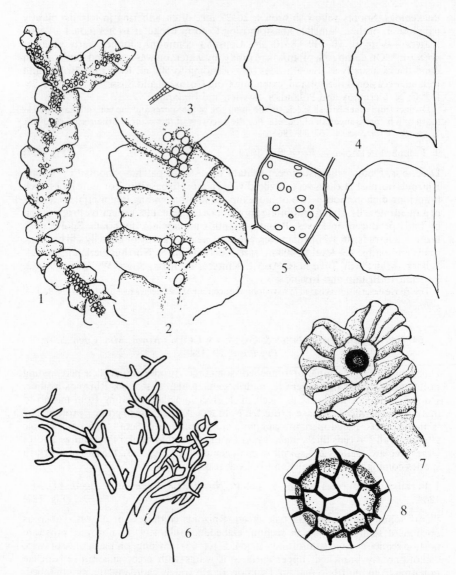

Fig. **127**. 1–5, *Fossombronia angulosa*: 1, male shoot with antheridia (× 2.5); 2, portion of same (× 7.5); 3, antheridium (× 63); 4, leaves (× 8); 5, mid-leaf cell with oil-bodies (× 385). 6, *F. fimbriata*: leaf (× 50). 7–8, *Petalophyllum ralfsii*: 7, thallus with pseudoperianth and capsule (× 5); 8, spore(× 630).

28. PELLIACEAE

Plants thalloid with ovoid spores (*Pellia*) or leafy with spherical spores (*Noteroclada*). European species with the characters of *Pellia*.

65. PELLIA RADDI, *Jungerm. Etrusca:* 38. 1818

Plants thalloid. Thallus to 1 cm wide, dichotomously branched, midrib poorly defined, 8–15 cells thick, without central strand but sometimes with transverse and vertical thickening bands, grading into wings 1–3 cells thick; thallus apices emarginate with small ventral swelling. Rhizoids numerous on ventral surface of midrib. Inner thallus cells with numerous persistent oil-bodies. Ventral scales lacking. Gemmae absent but propaguliferous branches produced by *P. endiviifolia*. Antheridia solitary in antheridial cavities on dorsal side of midrib. Archegonia several in cavity on dorsal side of midrib, protected by flap-like or tubular pseudoperianth (involucre). Calyptra thick-walled, fleshy, emergent or not from pseudoperianth. Capsule spherical, wall 2–3-layered, dehiscing by 4 valves. Elaterophore basal, elaters 2–4-spiral. Spores ovoid, large, germinating before and hence multicellular at time of release. A northern hemisphere genus of 5 species, thought to be of very ancient origin but still showing considerable genetic variation.

1 Monoecious, antheridial cavities situated behind flap-like pseudoperianth (invo-
    lucre)   (*P. epiphylla* agg.) 3
  Dioecious, pseudoperianth shortly cylindrical or tubular   2
2 Mouth of pseudoperianth toothed, thallus in section with thickening bands,
    marginal cells mostly 60–140 μm long   **3. P. neesiana**
  Mouth of pseudoperianth ± fimbriate, thallus in section without thickening bands,
    marginal cells mostly 40–80 μm long   **4. P. endiviifolia**
3 Interphase nuclei with one large heterochromatin body*, cells of outer layer of
    capsule wall 34–63 μm long, spores 60–117 μm long   **1. P. epiphylla**
  Interphase nuclei with two large heterochromatin bodies, cells of outer layer of
    capsule wall 43–81 μm long, spores 104–144 μm long   **2. P. borealis**
    *For detection of heterochromatin bodies see note following the description of P. epiphylla
    agg.

P. epiphylla agg.

Monoecious. Plants green to dark green with pale green young parts, often reddish-tinged, forming patches, sometimes extensive, or as scattered thalli. Thallus dichotomously branched, to 10 mm wide, apices expanded and lobed, margins ± undulate. Midrib 10–15 cells thick in T.S., cells with transverse and vertical colourless to brown thickenings, these sometimes sparse or rarely absent. Marginal row of cells of thallus wing very variable in size, mostly (40–)60–180(–200) μm long, 20–50 μm wide, adjacent epidermal cells 30–90 μm long, 30–80 μm wide. Antheridial cavities behind pseudoperianth, their position marked by small reddish protuberances. Pesudoperianth flap-like, entire. Calyptra protruding beyond pseudoperianth at maturity. Capsule blackish-green, wall 2–3-layered, cells of outer layer hexagonal with nodular thickenings, cells of inner walls elongated with semi-annular thickenings.

    The *P. epiphylla* agg. contains two taxa, one haploid, the other diploid. There has been much dispute as to whether or not these are distinct taxa. Messe (*Bull. Soc. Roy.*

Bot. Belgique **114**, 3–14, 1981) has shown that there are differences, albeit overlapping, in sporophyte dimensions between the two taxa, and Newton (*J. Bryol.* **14**, 215–30, 1986) has shown that there are cytological differences and that gametophytes of the two taxa may be distinguished by the number of heterochromatic bodies in the interphase nuclei. The technique for detecting the heterochromatin bodies given by Newton is as follows:

(1) Immerse in a stoppered glass tube of freshly mixed 3:1 absolute alcohol:glacial acetic acid 5mm lengths of growing thallus apices for at least 3 h but preferably overnight at 15–20°C.
(2) Dealing with one apex at a time, dissect out a 1 mm³ meristematic segment in 45% acetic acid. Cut it into three or four pieces. The material must not be allowed to dry out at any stage.
(3) Transfer the pieces to a small drop (*ca* 3 mm diameter) of acetic orcein or lacto-propionic orcein stain on a clean slide and apply a clean cover-slip. All particles of dust must be scrupulously excluded.
(4) Hold the cover-slip firmly in place with the thumb and forefinger of one hand while tapping briskly over the area of tissue with a blunt mounted needle. Lateral movement of the cover-slip, however slight, must be prevented at all times.
(5) Invert the slide on a thick pad of blotting paper on a smooth bench. Hold the slide steady with one hand and apply firm and heavy pressure to the slide with the other. The slide is ready for immediate inspection but will remain satisfactory for several days if sealed with rubber solution and stored in the freezing compartment of a refrigerator.

1. P. epiphylla (L.) Corda in Opiz. (edn), *Beiträge zur Naturgesch.* 12: 1654. 1829
(Fig. 128)

Interphase nuclei of thallus cells with one large heterochromatin body. Cells of outer layer of capsule wall 34–63 μm long. Spores 80–117 μm long. Capsules very common, late winter, early spring. *n* = 9*. Patches, sometimes extensive, on damp soil in woods and sheltered habitats, by ditch and stream banks and in block scree, usually in non-basic localities, common in suitable habitats except in calcareous areas. 111, H40, C. Europe, Faroes, Himalayas, Algeria, Tunisia, Azores, Madeira, N. America, Greenland.

2. P. borealis Lorbeer, *Jahrb. Wiss. Bot.* 80: 698. 1934   (Fig. 128)
P. epiphylla ssp. *borealis* (Lorbeer) Messe

Interphase nuclei of thallus cells with two large heterochromatin bodies. Walls of outer layer of capsule wall 43–81 μm long. Spores 100–144 μm long. Capsules not known in Britain. *n* = 18*. Margins of Scottish Lochs, 60–390 m, very rare. Mid Perth, W. Sutherland. 2. Germany, Belgium, Poland, Russia.

According to Müller (1954) *P. borealis* and *P. epiphylla* are separable by the length of the marginal cells of the thallus and on this basis *P. borealis* was reported from Scotland and Ireland by Müller (*Beitr. Kryptogamen-Flora Schweiz* **10**, pt 2, 8–55, 1947). It has since been shown conclusively by several authors that gametophytes of the two taxa cannot be separated without recourse to cytology. According to De Sloover & Messe (*J. Hattori bot. Lab.* **53**, 153–5, 1982) Belgian plants which have spores more than 115 μm long, and 75 μm wide and capsule wall cells more than 60 μm long and 45 μm wide can be referred to *P. borealis* with a determination error of 8%. They do not, however, give dimensions below which plants can be referred to *P. epiphylla*. It would appear that in Britain *P. borealis* is a plant of high altitudes but its distribution is imperfectly known. In Belgium, however, *P. borealis* descends to 300m and *P. epiphylla* does not ascend above 590m. For an account of *P. borealis* in Britain see Newton, *J. Bryol.* **14**, 215–30, 1986.

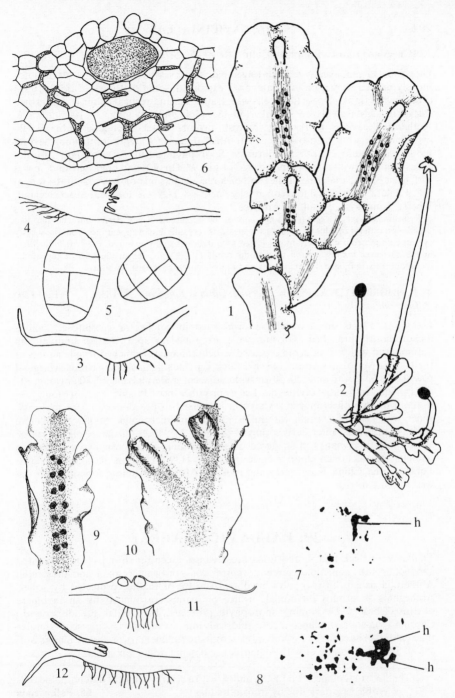

Fig. **128**. 1–6, *Pellia epiphylla* agg.: 1, thallus with antheridia and pseudoperianths (× 3.5); 2, thallus with mature sporophytes (× 1); 3, T.S. thallus (× 15); 4, L.S. through pseudoperianth (× 11); 5, spores (× 230); 6, T.S. through thallus with antheridium, antheridium and thickening bands stippled (× 210). 7 and 8, nuclei of *P. epiphylla* s.s. and *P. borealis* with heterochromatin bodies (*h*) (× 2500). 9–12, *P. neesiana*: 9, male thallus with antheridia (× 3.5); 10, female thallus with pseudoperianths (× 3.5); 11, T.S. thallus through antheridial cavities (× 15); 12, L.S. pseudoperianth (× 7).

3. P. neesiana (Gott.) Limpr. in Cohn., *Krypt.-Fl. Schles.* 1: 329. 1876   (Fig. 128)

Dioecious. Green, usually reddish-tinged patches or scattered thalli. Thallus dichotomously branched, margin undulate, apex expanded, somewhat lobed, Midrib 10–14 cells thick in transverse section, cells with transverse and vertical colourless to brown thickenings. Marginal row of cells (40–)60–140(–200) μm long, 30–100 μm wide. Antheridial cavities marked by reddish protuberances. Pseudoperianth shortly cylindrical, very low on proximal side, toothed at mouth. Calyptra protruding beyond mouth of pseudoperianth at maturity. Sporophyte as in *P. epiphylla*. Capsules occasional, late winter, early spring. $n = 8 + m^*$, 9^*. On peat and boggy ground, in wet turf, dune-slacks, marshes, flushes and on stream banks, in non-basic habitats; rare in S.E. England, occasional to frequent elsewhere. 58, H27, C. Europe, Faroes, Iceland, Siberia, Himalayas, Japan, N. America.

   P. neesiana and *P. epiphylla* agg. cannot be separated when sterile but the flap-like pseudoperianth (or involucre) of the latter species is very different in appearance from the shortly cylindrical toothed pseudoperianth of *P. neesiana* and can usually be found. *P. endiviifolia* differs in the narrower thallus lacking thickening bands in the cells, the production of repeatedly dichotomously branched, fragile shoots and in its occurrence in basic habitats.

4. P. endiviifolia (Dicks.) Dum., *Recueil. Observ. Jungerm.*: 27. 1835   (Fig. 129)
P. fabbroniana Raddi

Dioecious. Plants green, rarely reddish-tinted in patches or caespitose. Thallus dichotomously branched, 3–8 mm wide, expanded at apex, margin undulate; in autumn and winter producing repeatedly dichotomously branched fragile shoots at thallus tips. Midrib in section 8–13 cells thick, cells lacking thickening bands. Marginal row of cells 50–80 μm long, 20–50 μm wide, adjacent epidermal cells 40–80 μm long, 20–50 μm wide. Antheridial cavities marked by protuberances. Pseudoperianth tubular, 3–4 mm long, mouth ± fimbriate. Calyptra immersed in pseudoperianth at maturity. Inner wall cells of capsule without semi-annular thickenings. Spores 56–80 μm. Capsules rare, late winter, early spring. $n = 8 + m^*(\male)$, $9^*(\female)$. On damp soil, wet rocks, stream and ditch banks in sheltered places and dune-slacks, usually in calcareous habitats, frequent or common in basic areas. 111, H39, C. Europe, Iceland, Caucasus, Japan, Korea, China, Kamchatka and India (?), Morocco, Tunisia, Algeria, Madeira, western N. America.

29. PALLAVICINIACEAE

Thallus with thick midrib and unistratose wings; a central strand of differentiated conducting cells sometimes present; ventral scales absent or small and uniseriate. Antheridia usually in two rows on dorsal surface of thallus, protected by scales. Archegonia in groups on dorsal surface of thallus surrounded by an involucre of ± fused bracts. Developing sporophyte protected by tubular pseudoperianth. Capsule long-stalked, ellipsoid to cylindrical, inner wall cells lacking semi-annular thickenings, dehiscence by 2–4 valves; elaterophore rudimentary or absent, elaters 2–3-spiral. Spores reticulate or with sometimes coalescing ridges. Six genera.

Thallus wings plane or undulate, midrib with a central strand of differentiated cells
visible as a dark line by transmitted light   **66. Pallavicinia**
Thallus wings usually crisped, midrib lacking a central strand   **67. Moerckia**

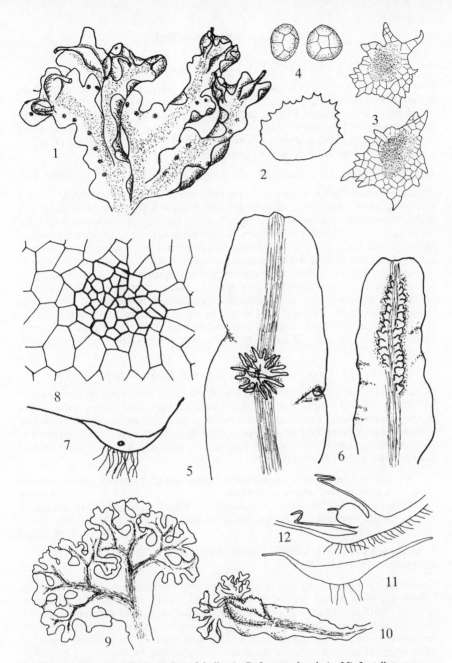

Fig. **129**. 1–4, *Blasia pusilla*: 1, portion of thallus (× 7); 2, ventral scale (× 35); 3, stellate gemmae (× 84); 4, gemmae from gemma-receptacle (× 84). 5–8, *Pallavicinia lyellii*: 5, female thallus (× 8); 6, male thallus (× 8); 7, т.s. thallus (× 10); 8, т.s. conducting strand of thallus (× 210). 9–12, *Pellia endiviifolia*: 9, thallus in spring (× 2.5); 10, female thallus with pseudoperianth (× 2.5); 11, т.s. thallus (× 15); 12, l.s. pseudoperianth with young sporophyte (× 7).

66. PALLAVICINIA S.F. GRAY, *Nat. Arr. Br. Pl.* 1: 175. 1821

Dioecious. Thallus prostrate or with a prostrate portion and an erect portion, ± simple, unistratose wings entire or lobed. Midrib with a central strand of small elongated conducting cells. Ventral scales lacking. Antheridia biseriate, subtended by forwardly directed laciniate scales. Archegonia on dorsal surface of thallus or on ventral branches. Capsule shortly cylindrical, wall of two cell layers only. About 30 species, mostly tropical or southern hemisphere.

1. P. lyellii (Hook.) Carruth., *J. Bot. Br. Foreign* 3: 203. 1865   (Fig. 129)

Plant pale green. Thallus simple or occasionally branched, to 4(–5) cm long, 2–4 mm wide, wings plane or undulate. Midrib *ca* 12 cells thick, well defined, with central strand of small elongated thickened conducting cells visible as a dark line by transmitted light; wings unistratose. Rhizoids pale brown. Antheridia in two rows, one on either side of midrib, protected by lobed scales. Archegonia in groups surrounded by involucre of basally fused laciniate scales. Pseudoperianth cylindrical, 5–7 mm long. Calyptra ± protruding beyond mouth of pseudoperianth. Capsules rare. $n = 8$. Patches or scattered thalli amongst tussocks or on detritus on marshy ground or in bogs, or rarely on damp shaded sandstone, at low altitudes, rare, from N. Somerset, I. of Wight and Sussex north to N.W. Yorkshire and Westmorland, scattered localities in Ireland. 16, H9. W. Europe from Spain and Portugal to Denmark, Hungary, China and Japan, Madeira, Azores, N. and S. America, Bermuda, Polynesia, Philippines, New Zealand.

67. MOERCKIA GOTT. IN RABENH., *Hepat. Europ. Exs.* no. 121. 1860

Dioecious. Thallus midrib very convex below, gradually narrowing into unistratose, undulate or crisped wings. Ventral scales uniseriate, concealed amongst rhizoids at margin of midrib, soon vanishing. Forwardly directed scales present on dorsal surface of thallus of male plants and, in some species, of female plants also. Capsule ellipsoid to shortly cylindrical, wall 3–6 cells thick, outer layer in section as wide as or wider than all inner layers together. Four species.

Rhizoids colourless to straw-coloured, dorsal surface of female thallus without
    scales, bracts irregularly laciniate   **1. M. hibernica**
Rhizoids pale reddish-brown, dorsal surface of both male and female thalli with
    scales, female bracts irregularly lobed, scarcely differing from thallus scales
                                                    2. M. blyttii

1. M. hibernica (Hook.) Gott. in Rabenh., *Hep. Eur. Exs.* no. 121. 1960   (Fig. 130)
M. flotoviana (Nees) Schiffn.

Plants solitary or gregarious, pale green. Thallus simple or furcate, to 15(–25) mm long, 2–4(–6) mm wide, wings usually strongly crisped, rarely only undulate. Midrib in section (10–)15–21 cells thick, usually with 2 lateral strands of differentiated cells, gradually narrowing into unistratose wings. Rhizoids colourless or straw-coloured. Antheridia numerous on dorsal surface, each subtended by an irregularly toothed scale, scales coalescing to form chambers. Female thalli lacking scales on dorsal surface, archegonia surrounded by irregularly laciniate fused bracts on dorsal side of thallus. Pseudoperianth narrowly ellipsoid, *ca* 5 mm long. Spores 35–55 μm, with anastomosing ridges giving an irregular outline in profile. Capsules rare, early spring. $n = 9$. Occasional in dune-slacks, base-rich flushes, rarely on ditch banks, silt or marshy ground, usually at low altitudes, rarely ascending to 500(–750) m, from Cornwall and

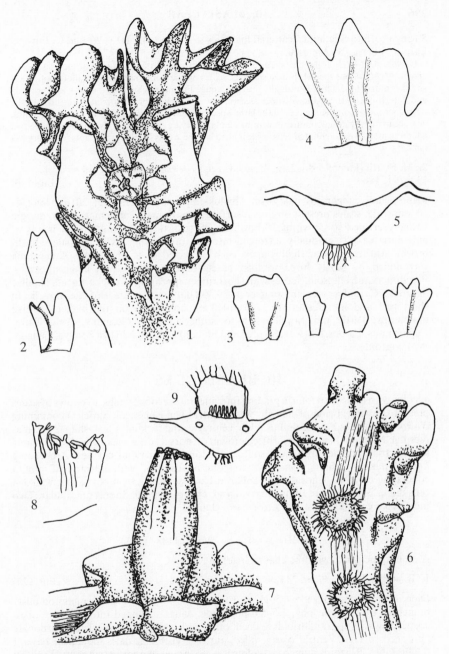

Fig. **130**. 1–5, *Moerckia hibernica*: 1, thallus with male bracts (× 7.5); 2, male bracts (× 17); 3, female bracts (× 17); 4, involucral bract (× 17); 5, т.s. thallus (× 12). 6–9, *M. blyttii*: 6, thallus with young involucres (× 7.5); 7, portion of thallus with mature pseudoperianth (× 7.5); 8, portion of involucre (× 17); 9, т.s. thallus through young involucre (× 12).

Surrey north to Shetland, scattered localities in Ireland. 42, H15. W. and C. Europe, Faroes, Siberia, N. America.

Usually only likely to be mistaken for *M. blyttii* but the altitudinal range of the two species differs, *M. blyttii* has reddish-brown rhizoids, scales on the dorsal surface of thalli of both sexes and the midrib tapers more gradually into the wings. Forms in which the thallus wings are only undulate, formerly treated as a distinct species, may be mistaken for *Pallavicinia lyellii* but lack the central strand of conducting cells of the latter species. Sand-dune plants appear drought tolerant, new shoots innovating from the midrib of plants surviving the summer. Most plants have two lateral strands of differentiated cells in the midrib but these may be absent in plants from very wet habitats.

2. M. blyttii (Moerch) Brockm., *Archiv. Vereins Freunde Naturg. Mecklenburg* 17: 190, 191. 1863   (Fig. 130)

Plants usually gregarious, pale green. Thallus simple or furcate, 7–15(–20) mm long, 4–10 mm wide, scales present on dorsal surface of both male and female thalli, margin strongly crisped to convolute. Midrib 20–25 cells thick, lacking lateral bands of differentiated cells, gradually narrowing into unistratose wings. Rhizoids pale reddish-brown. Antheridia near thallus apex, each protected by a lobed scale. Archegonia surrounded by partly fused toothed bracts differing little from adjacent scales. Pseudoperianth ellipsoid, spores with short ridges appearing as thick truncate papillae in profile. Capsules rare, late summer. $n = 9$. On damp soil in areas of late snow-lie, in short turf and in scree above (750–)1000 m; occasional on the higher Scottish mountains; Perth and Argyll north to Sutherland. 13. Germany, Alps, Tatra, Caparthians, Faroes, Norway, Sweden and Finland north to Novaya Zemlya, Canada, Washington.

30. BLASIACEAE

Thallus with ill-defined midrib grading into unistratose lobed wings; two rows of scales present on ventral side of midrib; small ventral hemispherical 'auricles' containing *Nostoc* colonies at base of thallus lobes; central strand lacking. Flask-shaped gemma-receptacles and stellate gemmae often present on dorsal surface near apex. Male plants smaller than female; antheridia solitary in dorsal antheridial cavities. Archegonia dorsal, naked, behind apex. Pseudoperianth fusiform with constricted mouth concealing calyptra. Capsule with collar at base, dehiscing by 4(–6) valves; wall 3–4 cells thick. Elaterophore rudimentary, basal; elaters bispiral. Spores unicellular. Two monotypic genera, *Blasia* and *Cavicularia* (Japan).

68. BLASIA L., *Sp. Plant.*: 1138. 1753

A monotypic genus with the characters of the species.

1. B. pusilla L., *Sp. Plant.*:, 1139. 1753   (Fig. 129)

Dioecious. Green to yellowish-green rosettes or patches. Thallus dichotomously branched, to *ca* 2 cm long, 1.5–5.0 mm wide, wings scalloped into leaf-like lobes, crisped towards apex; midrib ill-defined, often with a white line of calcium carbonate on dorsal surface. Ventral *Nostoc*-filled 'auricles' visible as dark spots near base of thallus lobes. Rhizoids numerous, colourless, restricted to midrib; two rows of ventral scales on underside of midrib. Stellate, ± unistratose gemmae, to 500(–700) μm long, often present on dorsal surface near apex; flask-shaped gemma-receptacles containing stalked, slightly flattened, rounded to elliptical gemmae, 80–112 μm, usually present on dorsal surface near apex. Capsule narrowly ellipsoid, enclosed by decumbent pseudoperianth. Spores ellipsoid, 35–48 μm. Capsules occasional, late summer to spring. $n = 9*$, 18. On moist, frequently clayey soil on disturbed ground, tracks, ditch and stream banks, old quarries; rare to occasional in lowland areas, occasional to frequent elsewhere. 104, H28. Europe, Faroes, Iceland, Caucasus, Siberia, Kamchatka, Japan, Java, Himalayas, Madeira, Canaries, N. America, Greenland.

31. ANEURACEAE

Thallus several to many cells thick, midrib poorly defined, without conducting strand or thickening bands, branching irregularly to regularly 1–3-pinnate. Oil-bodies 1–30 per cell, at least in internal cells, rarely absent. Gemmae frequent, 1–many-celled. Antheridia in loculi in short lateral branches. Archegonia 2–8, on short laciniate lateral branches. Pseudoperianth lacking. Calyptra fleshy, cylindrical to clavate, papillose above. Capsule oblong-cylindrical, dehiscing by four valves, 2-layered wall cells variously thickened, elaterophore apical. Elaters unispiral. Spores unicellular. Three genera.

1 Subterranean, plants whitish, cells lacking chloroplasts **70. Cryptothallus (p. 299)**
  Plants not subterranean, green, cells with chloroplasts 2
2 Thalli (2–)3–7 mm wide, greasy in appearance, 7–12 cells thick, 5–40 oil-bodies* per
  cell **69. Aneura (p. 297)**
  Thalli 0.3–1.2(–2.6) mm wide, not greasy in appearance, 5–8(–9) cells thick, 0–3 oil-
    bodies per cell **71. Riccardia (p. 299)**
*N.B. Oil-bodies are lost on drying.

69. ANEURA DUM., *Comment. Bot.*: 115. 1822

Dioecious. Plants yellowish-green to dark green, greasy in appearance. Thallus ± simple to bipinnate, mostly 2–6 mm wide; plano-convex to deeply concave in section, mostly 7–12 cells thick, apical slime papillae non-persistent. Oil-bodies usually 5–40 per cell. Gemmae absent in most species, where present exogenous, multicellular. Calyptra 5–12 cells thick, smooth, tuberculate or ciliate. Seta 8–16 cells diameter. Capsule wall cells with thickenings on inner and outer walls and with thickenings on inner tangential walls of inner layer.

1. A. pinguis (L.) Dum., *Sylloge Jung. Europ. Indig.*: 86. 1831 (Fig. 131)
Riccardia pinguis (L.) S.F. Gray

Plants yellowish-green to dark green, greasy in appearance. Thallus brittle, midrib very poorly defined, with abundant rhizoids, simple or sparsely and irregularly pinnately branched, (2–)3–6(–7) cm long, (2–)3–7 mm wide, apices rounded, margins often ± crisped. Midrib 10–12(–13) cells thick, epidermal cells smaller than inner cells. Oil-bodies 4–6(–9) μm, 6–30 per cell, present in all cells. Capsules frequent, winter. $n = 10*$, 20. Patches or scattered thalli amongst other bryophytes in ± permanently wet habitats, on wet rocks, wet heath, marshy ground, ditch and stream banks, wet ground in woods and in dune-slacks, ascending to 1000 m or more, frequent or common. 111, H40, C. Cosmopolitan.

    Small forms may be mistaken for *Riccardia* species but may be readily distinguished when fresh by the greasy appearance and numerous oil-bodies. *Pellia* species differ in the emarginate apices.

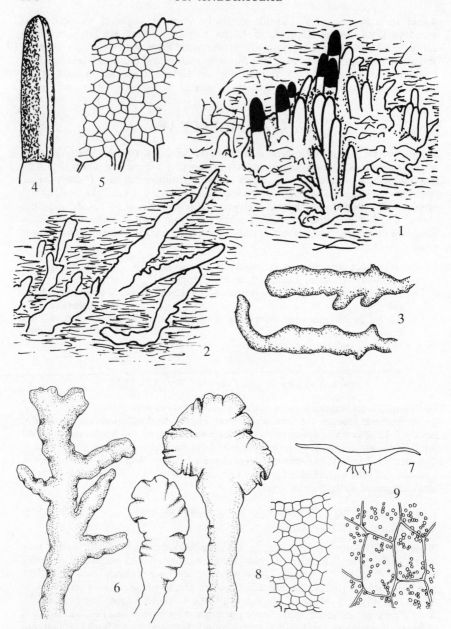

Fig. 131. 1–5, *Cryptothallus mirabilis*; 1 and 2, thalli with and without capsules after removal of covering mosses (× 1.5); 3, thalli (× 3.5); 4, capsule (× 14); 5, т.s. thallus (× 84). 1, 2 and 4 drawn from photographs by Dr S.R. Edwards. 6–9, *Aneura pinguis*: 6, thalli (× 2); 7, т.s. thallus (× 8); 8, т.s. midrib region of thallus (× 60); 9, dorsal epidermal cells (× 300).

70. CRYPTOTHALLUS MALMB., *Ann. Bryol.* 6: 122. 1933

A monotypic genus with the character of the species.

1. C. mirabilis Malmb., *Ann. Bryol.* 6: 122. 1933 (Fig. 131)

Dioecious. Plants creamy white, lacking chlorophyll. Thallus prostrate or occasionally ascending, simple or irregularly pinnate or sometimes coralloid in appearance, female thalli to 4 cm long, to 5 mm wide, male thalli smaller. Thallus in section circular to semicircular or concavo-convex, to *ca* 20 cells thick, upper cells with starch grains, lower with mycorrhizal fungal hyphae. Antheridia and archegonia on short lateral branches. Calyptra fleshy, cylindrical. Capsule cylindrical with apical elaterophore, dehiscing below or above ground level. Spores *ca* 30 μm, in tetrads, probably without chlorophyll. Capsules common, throughout the year. $n = 10^*$. Under humus, leaf litter or bryophytes, in dry to damp sites in *Betula* or *Pinus* woodland and *Molinia* heath; rare but probably much under-recorded. Scattered localities from E. Cornwall and Surrey north to Ross, N. Kerry, W. Galway, Wicklow. 35, H3. Norway, Sweden, Finland, western U.S.S.R., W. Greenland (possibly also France and Austria).

This remarkable liverwort was first recorded in Britain in 1948 (see Williams, *Trans. Br. bryol. Soc.* 1, 357–66). Its ecology is varied (see Williams, op. cit.; Crundwell, *Trans. Br. bryol. Soc.* 1, 485–6: Scott, *Trans. Br. bryol. Soc.* 4, 336–7). *Cryptothallus* tends to grow at the soil/litter or covering bryophyte interface. In wet habitats, especially under *Sphagnum*, the thalli tend to be vigorous and horizontal, in drier habitats such as under leaf litter the thalli are thinner and ascending. *Cryptothallus* has been found under several *Sphagnum* species, *Pellia epiphylla*, *Mnium hornum*, *Pleurozium schreberi* and *Hylocomium splendens* and under *Pinus*, *Betula*, *Vaccinium* and *Molinia* litter. Usually said to be saprophytic, *Cryptothallus* derives its nutrition from the mycorrhizal fungi with which it is associated and is therefore parasitic upon these fungi (see Pocock & Duckett, *J. Bryol.* 13, 227–33, 1984).

71. RICCARDIA S.F. GRAY, *Nat. Arr. Br. Pl.* 1: 679. 1821

Monoecious or dioecious. Plants pale to dark green, not greasy in appearance. Thallus 1–4-pinnate or palmately or irregularly branched, mostly 0.5–2.0 mm wide, circular to concavo-convex in section, 5–15(–40) cells thick, apical slime papillae persisting. Oil-bodies 0–15 per cell. Gemmae often present at thallus apices, 1–3-celled, endogenous. Calyptra 3–12 cells thick, smooth, papillose or ciliate above. Seta 4 cells in diameter. Capsule wall cells with semi-annular thickenings.

The species of *Riccardia* are phenotypically variable and at times difficult to separate. The oil-bodies have proved of use in naming British material (see Little, *Trans. Br. bryol. Soc.* 5, 536–40, 1968). The thickenings of the outer and inner cell layers of the capsule wall are characteristic of particular species but are not of practical use (see Macvicar, 1926).

1 Thalli 1–3 pinnate, in section 5–8 cells thick with dorsal side flat or convex   2
  Thalli not pinnately branched or if so then in T.S. 5–6 cells thick and dorsal side concave   3
2 Branches with ± parallel sides, oil-bodies mostly absent from epidermal and marginal cells   **1. R. multifida**
  Branches narrowed towards base, oil-bodies present in epidermal and marginal cells   **2. R. chamedryfolia**
3 Thalli irregularly pinnately branched, concave, semi-lunar in section   **3. R. incurvata**
  Thalli simple to palmate or irregularly branched, biconvex in section   4
4 Branches ± parallel-sided, epidermal cells 24–56 × 16–48 μm, oil-bodies present in inner cells   **4. R. palmata**

Branches narrowed at base, epidermal cells 60–120 × 30–50 μm, oil-bodies lacking
5. R. latifrons

1. R. multifida (L.) S.F. Gray, *Nat. Arr. Br. Pl.* 1: 684. 1821   (Fig. 132)
Aneura multifida (L.) Dum.

Autoecious. Plants dark green to greenish-brown. Thallus prostrate, ± regularly 2–3-
pinnate, to 30 mm long, 0.3–1.0(–1.2) mm wide. Branches ± linear, at least ultimate
with translucent margin of 2–3 rows unistratose cells. Midrib in section ± biconvex, 5–7
cells thick, epidermal cells much smaller than inner cells. Oil-bodies rare or absent in
epidermal and marginal cells, 1–2 per inner cell of thallus, 10–25 × 8–12 μm. Gemmae
2-celled, rare. Female branches laciniate with short uniseriate hairs. Capsules frequent,
winter, spring. *n* = 10, *ca* 19*, 20. Patches on wet rocks, in flushes, bogs and fens, on
damp soil, peat and in dune-slacks, ascending to 1000 m or more, occasional to
frequent. 105, H37, C. Europe north to Fennoscandia and N. Russia, Faroes, Iceland,
Himalayas, Japan, Tunisia, Azores, Madeira, Tenerife, S. Africa, N. America,
Greenland, Hawaii, Tierra del Fuego, Falkland Is.

In similar habitats to but usually less common than *R. chamedryfolia* with which it may be
confused, the two species being variable morphologically and sometimes resembling one another.
Plants of *R. chamedryfolia* lacking a unistratose margin to the thallus are distinctive but some
forms with 1–4 rows of unistratose cells may be confused with *R. multifida* in dry ground forms of
which the unistratose margin may be only one cell wide. In *R. multifida* the branches
have ± parallel sides and often tend to overlap whereas those of *R. chamedryfolia* widen upwards
and do not overlap. The distribution of oil-bodies is very distinctive in fresh material.

2. R. chamedryfolia (With.) Grolle, *Trans. Br. bryol. Soc.* 5: 772. 1960   (Fig. 132)
Aneura sinuata Dum., *R. sinuata* (Hook.) Trev.

Autoecious. Plants pale green. Thallus prostrate, irregularly 1–3-pinnate, to 30(–40)
mm long, 0.3–1.2 mm wide. Branches narrowed towards base, ultimate branches
lacking translucent margin but sometimes with 1–4 rows unistratose cells. Thallus in
section ± plane on dorsal side, ± convex on ventral side, 5–8 cells thick, epidermal cells
much smaller than inner cells. Oil-bodies in most cells including marginal and
epidermal, 1–2(–3) per cell, 9–20(–25) × 8–15 μm. Gemmae 2-celled, rare. Female
branches shortly laciniate. Capsules occasional, winter, early spring. *n* = 20*, 30.
Patches on wet rocks and tree roots, sometimes submerged, in flushes, bogs, fens,
± permanently damp soil, rotting logs and dune-slacks, ascending to *ca* 1000 m,
frequent. 109, H36, C. Europe north to Fennoscandia and N. Russia, Faroes, Iceland,
Shensi, Japan, Tunisia, Macaronesia, N. America.

May be confused with *R. multifida* (q.v.) and *R. incurvata*. Rarely the thalli may be semi-lunar in
section as in *R. incurvata*, but the epidermal cells in the latter are similar in size to the inner cells
and the plants are dioecious. The plant referred to as *Aneura sinuata* var. *major* (Lindb.) by
Macvicar (1926) is a habitat form with simple or simply pinnate thalli from moist ground.

3. R. incurvata Lindb., *Bot. Notis.* 1878: 187, 1878   (Fig. 132)
Aneura incurvata Steph.

Dioecious. Plants pale green. Thalli prostrate to ascending, irregularly pinnately
branched, to 20 mm long, 0.5–1.0 mm wide, concave, especially towards apices.
Branches ± lingulate with translucent margin of a single row of unistratose cells.
Thallus semi-lunar in section, 5–6 cells thick, epidermal cells hardly differing in size
from inner cells. Oil-bodies present in most cells, 1–2 per cell, 9–12 × 9 μm. Gemmae 2–
3-celled, often abundant. Female branches shortly laciniate. Capsules rare. *n* = 10*.
Patches or mixed with other bryophytes on damp sandy soil by ponds, in dune-slacks
and damp hollows in open habitats, ascending to *ca* 350 m, very rare in S. England, rare

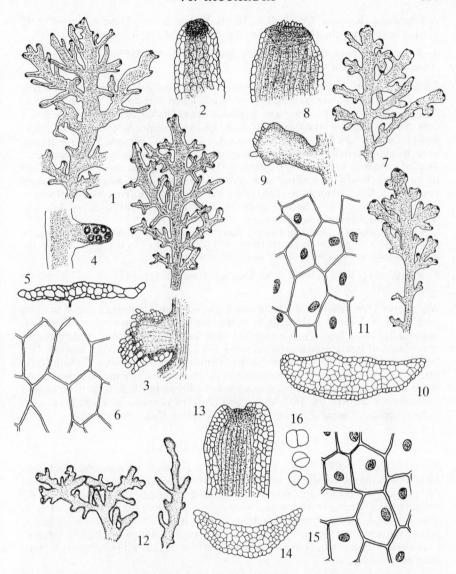

Fig. **132**. 1–6, *Riccardia multifida*: 1, thalli (× 4); 2, apex of secondary branch (× 38); 3 and 4, female and male branches (× 21); 5, T.S. main axis of thallus (× 38); 6, dorsal epidermal cells of thallus (× 210). 7–11, *R. chamedryfolia*: 7, thalli (× 4); 8, apex of secondary branch (× 38); 9, female branch (× 21); 10, T.S. main axis of thallus (× 38); 11, dorsal epidermal cells of thallus (× 210). 12–16, *R. incurvata*: 12, thalli (× 4); 13, apex of secondary branch (× 38); 14, T.S. main axis of thallus (× 38); 15, dorsal epidermal cells of thallus (× 210); 16, gemmae (× 84).

to occasional elsewhere, Cornwall, S. Hants, Sussex, Surrey, Carmarthen and S.W. Yorkshire northwards. 43, H5. W. Europe, Switzerland, N. America.

4. R. palmata (Hedw.) Carruth., *J. Bot. Br. Foreign* 3: 302. 1865   (Fig. 133)
Aneura palmata Dum.

Dioecious. Plants dark green, reddish-brown below. Thallus to 10 mm long, adhering to substrate, with numerous simple or palmately lobed branches to 3 mm high, 0.15–0.30(–0.70) mm wide, branches with ± parallel sides. Thallus biconvex in section, 6–9 cells thick, in ultimate branches 4 cells thick, epidermal cells smaller than inner, thick-walled, $24–56 \times 16–48$ μm in surface view. Oil-bodies 1–2 per cell, $9–15 \times 7–10$ μm, present in inner cells but mostly absent from epidermal and marginal cells. Gemmae 1–2-celled, common. Capsules rare, spring. $n = 10$. On rotting logs, rarely on damp rocks or humus, in humid woodland and ravines, very rare in S. England, occasional elsewhere. S. Devon, Dorset, S. Hants, Bucks, Brecon and N.E. Yorkshire northwards. 45, H23. Europe, N. Asia, Shensi, Japan, Madeira, Azores, Madagascar, N. America, Mexico, Cuba, Bermuda, S. America.

  R. latifrons sometimes occurs on decaying wood but may be distinguished from *R. palmata* by its paler colour, larger size, larger epidermal cells and lack of oil-bodies. In *R. latifrons* the epidermal cells in section are of similar size to the inner cells.

5. R. latifrons (Lindb.) Lindb., *Acta Soc. Sci. Fenn.* 10: 513. 1875   (Fig. 133)
Aneura latifrons Lindb.

Autoecious. Plants pale green. Thallus prostrate, adhering to substrate or not, to 8 mm long, 0.5–1.0(–2.0) mm wide, irregularly branched, branches prostrate to ascending, narrowed towards base. Thallus biconvex in section, 5–6 cells thick, epidermal cells thin-walled, of similar size to inner cells, $(40–)60–120 \times 30–50$ μm in surface view. Oil-bodies lacking in all cells. Gemmae 2-celled. Female branches laciniate. Capsules rare, spring. $n = 20$. Growing over *Sphagnum* or in patches on peat or more rarely on rotting wood, mainly at low altitudes but ascending to *ca* 400 m, rare in S. Britain, occasional to frequent elsewhere. 58, H33, C. Europe north to Fennoscandia and N. Russia, Faroes, Iceland, Siberia, Tunisia, Azores, N. America, Cuba, Bermuda.

32. METZGERIACEAE

Plants thalloid, thallus linear with well defined midrib several cells thick and unistratose wings; wings with or without marginal hairs and hairs on ventral and dorsal surfaces; oil-bodies absent or very small. Fertile branches dwarf, involute, borne on ventral side of midrib. Male branches with midrib, not hairy (ventral surface hairy in *Apometzgeria*). Female branches lacking midrib, hairy. Calyptra becoming fleshy, hairy, pyriform. Pseudoperianth lacking. Seta short, in section of 28 similar-sized cells. Capsule with apical elaterophore, wall two cells thick. Elaters with single broad spiral. Four genera.

  Thallus without hairs on dorsal surface, hairs sparse or absent on ventral surface
                                                                   72. Metzgeria
  Thallus densely hairy on both surfaces   **73. Apometzgeria**

72. METZGERIA RADDI, *Jungerm. Etrusca*: 34, 1818

Thallus linear, dichotomously branched and with branches arising from ventral surface of midrib. Midrib sharply defined from unistratose thallus wings, with small elongate inner cells and large short cells on dorsal and ventral surface; rhizoids present at least

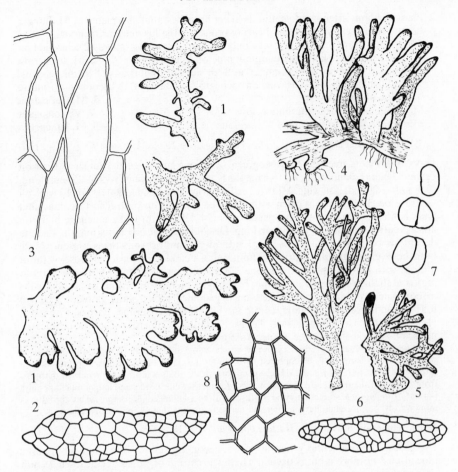

Fig. **133**. 1–3, *Riccardia latifrons*: 1, thalli (× 7); 2, т.s. main axis of thallus (× 84); 3, dorsal epidermal cells of thallus (× 230). 4–8, *R. palmata*: 4, thalli (× 7); 5, thallus with sporophyte (× 7); 6, т.s. thallus (× 84); 7, gemmae (× 84); 8, epidermal cells of thallus (× 230).

on older parts. Thallus wings usually with marginal hairs, with or without scattered rhizoids on ventral surface, dorsal surface usually without hairs; cells ± hexagonal. Gemmae sometimes borne on wing margins or on modified branches. Male branches dwarf, involute, hairless, with a midrib. Female branches dwarf, involute, subglobose, with numerous hairs, midrib lacking, sometimes later continuing growth as vegetative branches. Calyptra pyriform, fleshy, hairy above. Seta 4–6 cells in diameter. Capsule dehiscing by 4 valves. A world-wide genus of about 120 species.

1 Marginal hairs of thallus where present usually single, plants with or without
      gemmae   2
   Marginal hairs usually in pairs, gemmae lacking   4

2 Plants without attenuated gemma-bearing branches, thallus ± plane **1. M. furcata**
   Plants with attenuated gemma-bearing branches, thallus margins recurved   3
3 Attenuated branches with gemmae only on margins, midrib mostly 2 cells wide on
      each surface, plant not becoming blue on long drying   **2. M. temperata**
   Attenuated branches with gemmae on both sides of thallus as well as on margins,
      midrib mostly 4–5 cells wide on each surface, plants often becoming turquoise
      blue on long drying   **3. M. fruticulosa**
4 Marginal hairs ± straight, monoecious   **4. M. conjugata**
   Marginal hairs strongly curved, dioecious   **5. M. leptoneura**

1. M. furcata (L.) Dum., *Recueil Observ. Jungerm.*: 26. 1835   (Fig. 134)

Dioecious. Plants green to yellowish-green, becoming grey-green to whitish with age on drying. Thallus dichotomously to irregularly branched, ± plane, to 20(–25) mm long, (0.4–)0.6–1.0(–1.2) mm wide. Midrib 2 cells wide on dorsal side, (3–)4(–5) cells wide on ventral side. Thallus wings often with scattered rhizoids on ventral surface, marginal hairs solitary, straight, usually sparse; cells (24–)28–48 μm wide; mucilage hairs on apical ventral surface (28–)40–64 μm long. Discoid gemmae on thallus margin, ligulate deciduous branches from the ventral side of the midrib sometimes present. Male branches ± globose, with midrib, without hairs. Female branches ± globose, without midrib, with hairs. Capsules occasional, at most times of year. *n* = 8. Dense mats or scattered thalli growing through other bryophytes on bark of various trees and shrubs in both sheltered and exposed sites, on dry shaded rocks, walls, boulder scree, in lowland and montane habitats, common. 112, H40, C. Europe, Faroes, Caucasus, Himalayas, Shensi, Japan, Korea, Tunisia, Ethiopia, Cameroun, S. Africa, Macaronesia, N. America, W. Indies, Chile, Australasia.

   M. furcata sometimes produces gemmae on the wings of the thallus, especially in exposed situations. Such forms may be confused with immature plants of the next two species lacking attenuated branches. *M. furcata* differs from both in the thallus not having recurved margins, from *M. temperata* in the midrib mostly 4 cells wide on the ventral side; and longer mucilage hairs than *M. fruticulosa*. *M. furcata* may be confused with small plants of *M. conjugata* but the latter is monoecious, has marginal hairs in pairs, larger cells and the thallus convex in section.

2. M. temperata Kuwah., *J. Hattori bot. Lab.* 40: 219. 1976   (Fig. 134)

Dioecious. Plants green to yellowish-green, becoming yellowish to whitish but never blue after 2 or more months drying. Thallus prostrate or procumbent, to *ca* 25 mm long, 0.6–1.5 mm wide, margins recurved, with ascending or erect attenuated gemmiferous branches to 5(–7) mm long. Midrib mostly 2 cells wide on each side of thallus. Thallus wings with usually single straight marginal hairs; rhizoids very rare on ventral side except submarginally; cells 30–44 × 32–56 μm; mucilage hairs on apical ventral surface 32–72 μm long, visible also at apices of attenuated branches. Attenuated branches usually with at least 1–2 rows of wing cells on either side of midrib right to apex; gemmae produced only from wing margins, unistratose, discoid to *ca* twice as long as wide, rarely with a few short marginal hairs. Capsules very rare, spring. Dense mats or growing through other bryophytes, usually on the more acid types of bark, in shaded or damp situations, rarely on shaded rock, usually at low altitudes but ascending to 360 m, frequent in S. and W. Britain, extending north to Midlothian, Mid Perth and Kintyre, S. and W. Ireland. 57, H17, C. France, Germany, Japan, Madeira, N. America.

   Until recently not distinguished from *M. fruticulosa* (see Paton, *J. Bryol.* 9, 441–9, 1977). *M. temperata* differs from that species in its usually larger size, the gemmae arising only on the wings of the thallus of attenuated branches, the apical portions of which in *M. fruticulosa* often consists only of midrib and are narrower, the longer mucilage cells which are visible at the apices of

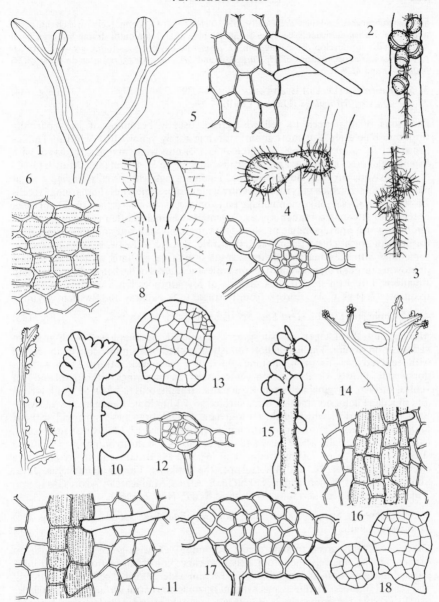

Fig. **134**. 1–8, *Metzgeria furcata*: 1, thallus, dorsal side, 2–4, ventral side of thalli with male branches, female branches and female branch containing mature sporophyte (× 17); 5, thallus margin; 6, ventral side of thallus midrib; 7, T.S. midrib; 8, mucilage hairs on ventral surface of thallus apex (× 310). 9–13, *M. temperata*: 9, thallus; 10, apex of attenuated branch (× 42); 11, ventral side of thallus midrib; 12, T.S. midrib; 13, gemma. 14–17, *M. fruticulosa*: 14, thallus; 15, apex of attenuated branch (× 42); 16, ventral side of thallus midrib; 17, T.S. midrib; 18, gemmae. Thalli × 8, cells and gemmae × 170.

attenuated branches, and there are usually only 2 instead of 3–4 epidermal cells on each side of the midrib. *M. fruticulosa* usually becomes turquoise-blue after 2–6 months drying but occasional specimens do not develop this coloration. *M. temperata* has not been found growing with *M. furcata* but mixed populations of *M. fruticulosa* and *M. temperata* do occur as do those of *M. fruticulosa* and *M. furcata*.

3. M. fruticulosa (Dicks.) Evans, *Ann. Bot.* 24: 293, 296. 1910.			(Fig. 134)
M. furcata var. *fruticulosa* (Dicks.) Lindb.

Dioecious. Plants green to yellowish-green, usually becoming at least partially turquoise blue after 2 or more months drying, rarely yellowish to whitish. Thallus prostrate, to *ca* 10 mm long, (0.3–)0.6–0.8(–1.0) mm wide, margins recurved, with ascending or erect attenuated gemmiferous branches to 3(–4) mm long. Midrib (3–)4–5(–6) cells wide on each side of thallus. Thallus wings with usually single, straight marginal hairs, rhizoids very rare on ventral side except sometimes submarginally; cells 24–40(–44) × 32–48(–56) μm; mucilage hairs on apical ventral surface 16–24 μm long, obscure on attenuated branch apices. Attenuated branches often consisting only of midrib towards apex, gemmae produced on wings and both sides of midrib and often forming an apical cluster; gemmae unistratose, discoid to *ca* twice as long as wide, sometimes with a few marginal hairs. Capsules very rare, autumn, winter. Dense mats or growing through other bryophytes on the less acid types of bark in sheltered or damp situations, rarely on shaded rock, usually at low altitudes but ascending to 430 m, frequent. 77, H33, C. W. Europe from Portugal to S. Norway and Sweden, Azores.

4. M. conjugata Lindb., *Acta Soc. Sci. Fenn.* 10: 495. 1875			(Fig. 135)

Monoecious. Plant green or pale green, ±pellucid, becoming yellowish-green to yellowish on drying. Thallus dichotomously branched or ± simple, convex in section with recurved margins, to 30 mm long, (0.6–)1.0–1.6 mm wide. Midrib 2 cells wide on dorsal side, mostly 4 cells wide on ventral side. Thallus wings without rhizoids on ventral surface, marginal hairs usually in pairs, straight; cells (32–)40–56 × (40–)48–64 μm, mucilage hairs on apical ventral surface 56–72 μm long. Gemmae absent. Male branches ± spherical, with midrib, without hairs. Female branches ± spherical, without midrib, with hairs. Capsules rare, autumn to spring. *n* = 9, 17. Mats on shaded rocks, ledges, cliffs, especially where basic and on tree trunks and shrub branches in sheltered sites, ascending to *ca* 750 m, frequent in W. and N. Britain and in Ireland, rare elsewhere. 73, H29. N. and W. Europe, the Balkans, Faroes, Himalayas, Java, Sumatra, Celebes, Korea, Japan, Sakhalin, S. Africa, Madagascar, Azores, Madeira, Tenerife, N. America, Bermuda, W. Indies, Brazil, New Zealand.

5. M. leptoneura Spruce, *Trans. Proc. Bot. Soc. Edinburgh* 15: 555. 1885	(Fig. 135)
M. hamata Lindb.

Dioecious. Plants green to pale green, becoming yellowish-green or yellowish on drying. Thallus sparsely dichotomously branched, convex in section with strongly reflexed margins, to 6 cm long, (0.8–)1.2–1.8 mm wide. Midrib 2 cells wide on dorsal and ventral surfaces. Thallus wings without rhizoids on ventral surface, marginal hairs usually in pairs, sometimes single or in threes, strongly curved; cells 44–56 × 32–56 μm; mucilage hairs on apical ventral surface 60–100 μm long. Gemmae lacking. Frequently fertile but capsules not known in Britain or Ireland. *n* = 9. Patches, sometimes extensive, on steep moist rocks, or creeping over other bryophytes in moist, shaded ravines and on cliff ledges, especially where calcareous, rarely on the ground or on trees, from sea-level to *ca* 1000 m, occasional in N.W. Wales, the Lake District and W. Scotland, from Cardigan north to Outer Hebrides and Orkney, rare in W. Ireland. 26,

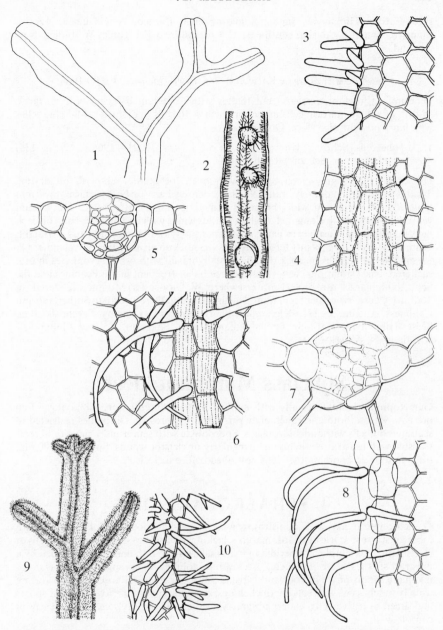

Fig. **135**. 1–5, *Metzgeria conjugata*: 1, thallus, dorsal side; 2, ventral side of thallus with fertile branches; 3, thallus margin; 4, ventral side of thallus midrib; 5, т.s. midrib. 6–8, *M. leptoneura*: 6, ventral side of thallus midrib; 7, т.s. midrib; 8, thallus margin. 9–10, *Apometzgeria pubescens*: 9, thallus; 10, thallus margin. Thalli × 17, cells × 170

H10. Faroes, Himalayas, Japan, Sumatra, Java, Borneo, New Guinea, Azores, Madeira, Tenerife, Alaska, south-east U.S.A. south to Patagonia, W. Indies, New Zealand.

73. APOMETZGERIA KUWAH., *Rev. Bryol. Lichén*. 34: 212. 1966

Dioecious. Close to *Metzgeria* but thallus with abundant hairs on both surfaces. Midrib of nearly uniform cells in section. Seta 8–10 cells in diameter. Male branches with hairs on ventral surface. One species.

1. A. pubescens (Schrank) Kuwah., *Rev. Bryol. Lichén*. 31: 214. 1966   (Fig. 135)
Metzgeria pubescens (Schrank) Raddi

Dioecious. Plants grey-green to yellowish-green, becoming yellowish on drying. Thallus irregularly pinnately branched, plane, densely hairy on both surfaces, to 30 mm long, (0.6–)0.8–1.6 mm wide, branches sometimes longly attenuated. Dorsal and ventral cells of midrib elongated. Cells of thallus wings with one or more hairs on each face, marginal hairs usually in pairs, sometimes singly or in threes, all hairs ± straight. Gemmae lacking. Frequently fertile but capsules not known in Britain or Ireland. $n = 8$, $8 + m^*$. Dense patches on rock or more usually mixed with other bryophytes in dry, sheltered, basic situations, very rarely on tree boles, frequent in the Pennines and the Scottish highlands, rare or very rare elsewhere, W. Gloucester, Monmouth, Hereford, Stafford, Caernarfon, Denbigh, Flint, Nottingham and Derby north to Sutherland and Caithness, Antrim. 38, H1. Belgium, Denmark, Norway, Germany, Czechoslovakia, Sicily, Himalayas, Kashmir, China, Japan, Korea, Sakhalin, Borneo, Philippines, Aleutians, N. America.

Subclass Marchantiideae

Gametophyte thalloid, usually with marked tissue differentiation, air-cavities often present; both smooth and tuberculate rhizoids often present. Oil-bodies restricted to scattered oil-cells without chloroplasts. Sporophyte with seta usually short or absent. Capsule wall unistratose, dehiscing irregularly or cleistocarpous. Spores 40–200 μm, often in tetrads; spore mother cells not lobed before meiosis.

4. SPHAEROCARPALES

Gametophyte an axis with a unistratose wing or leaf-like lobes or a lobed thallus with lobes unistratose at least towards margins, lacking internal air-spaces, mucilage cells or pores. Rhizoids simple. Antheridia in marginal cavities in thallus or surrounded by a unistratose sheath or involucre. Archegonia solitary in a unistratose sheath or involucre. Sporophyte of foot and capsule only. Capsule ± globose with unistratose usually indehiscent wall lacking thickenings, containing at maturity tetrads of spores and green nutritive cells, elaters absent. Spore tetrads usually released by decay of capsule wall.

33. SPHAEROCARPACEAE

Dioecious. Gametophyte a lobed thallus, multistratose in the middle, becoming unistratose in the lobes, perennating tuber present (*Geothallus*) or not. Gametangia in

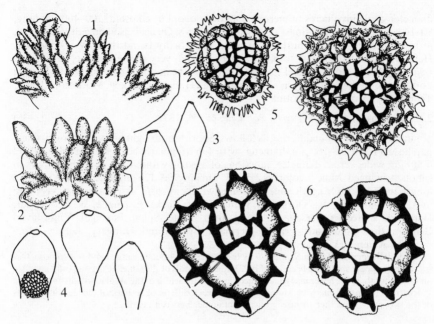

Fig. **136**. 1–5, *Sphaerocarpos michelii*: 1, male thallus (× 15); 2, portion of female thallus (× 4); 3, male involucres (× 63); 4, female involucres (× 38); 5, spore tetrads (× 275); 6, *S. texanus*: spore tetrads (× 275).

involucres on dorsal surface of thallus. Capsule indehiscent or dehiscing irregularly (*Geothallus*). Two extant genera, the second, *Geothallus*, occurring in southern California.

74. Sᴘʜᴀᴇʀᴏᴄᴀʀᴘᴏs Bᴏᴇʜᴍᴇʀ in G. Ludwig, *Definit. Gen. Pl.* ed. 3: 501. 1760

Gametophyte annual or ephemeral, lacking perennating tuber. Capsule indehiscent. A ± world-wide genus of about 12 species.

Spore tetrads spinulose in silhouette, with 8–12 areolae, 10–14 μm wide, across diameter **1. S. michelii**
Spore tetrads winged in silhouette, with 4–6 areolae, 20–36 μm wide, across diameter **2. S. texanus**

1. S. michelii Bellardi, *Appendix ad Fl. Piedmont.*: 52. 1792 (Fig. 136)

Dioecious. Female plants in pale or glaucous green, rarely reddish-purple rosettes, mostly 4–12 mm diameter. Thalli irregularly lobed, several cells thick at centre of rosette, unistratose towards edge. Female involucres numerous, at maturity obovoid or obpyriform with uniseriate wall, narrowed to small mouth, 1.2–1.6 mm long, 0.8–1.1 mm diameter, 1.4–2.3 times as long as wide. Spore tetrads dark brown, 70–150 μm diameter, with 8–12 irregular to ± hexagonal areolae, 10–14 μm wide, across diameter, with tall papillae at corners of areolae rendering tetrads spinulose in silhouette. Capsules very common, autumn to spring. Male rosettes often reddish, to *ca* 5 mm diameter, male involucres numerous, narrowly ovoid or ellipsoid, 200–400 μm long, 80–160 μm diameter, 2–3 times as long as wide. $n = 8$. On acidic soil on cultivated land, at low altitudes, rare in southern England from Scilly Is. north to W. Gloucester, Hereford and Oxford, east to Kent and E. Anglia, very rare elsewhere, Worcester, Leicester, Yorkshire, Channel Is. 22, C. Mediterranean region and islands of Europe north to the Netherlands, Germany, Austria and Hungary, Turkey, Morocco, Algeria, Tenerife, N. America, Argentina.

2. S. texanus Aust., *Bull. Torrey Bot. Club* 6: 158. 1877   (Fig. 136)

Similar to *S. michelii* but differing as follows. Spore tetrads light brown to brown, finely papillose, (120–)135–160 μm diameter, with 4–6 irregularly hexagonal areolae, 20–32 (–36) μm wide, across diameter, junctions of areolae without tall papillae, tetrads in silhouette winged. Male involucres 215–275 μm long, 90–120 μm wide, 2.0–2.6 times as long as wide. $n = 8$. On acidic soil, sometimes with *S. michelii*, at low altitudes, very rare in S. England, Scilly Is., Dorset, N. Hants, Kent, Bucks, E. Anglia, W. Gloucester, Hereford, Worcester, Jersey. 13, C. Sardinia, Crete, Spain, Portugal, Yugoslavia, Belgium, the Netherlands, France, Germany, Switzerland, Hungary, Turkey, Morocco, N. America, Uruguay.

Readily distinguished from *S. michelii* by the spores in silhouette winged and not spinose. Old spores may appear spinulose because of breakdown of walls of the lamellae leaving only lamellae junctions but there are fewer marginal projections and the width of the areolae will distinguish the spores from those of *S. michelii*. Macvicar (1926) gives a difference in shape of the male involucres of the two species but there is no significant difference between them.

5. MARCHANTIALES

Gametophyte a dichotomously branching thallus. Thallus with a dorsal epidermis usually lacking chloroplasts but usually with pores opening into air-chambers in a subepidermal layer of chlorenchyma analogous to spongy mesophyll of leaves of higher plants, cells beneath chlorenchyma parenchymatous and containing starch grains with occasional cells with single large compound oil-bodies or sometimes mucilage. Ventral surface with smooth and tuberculate rhizoids in midrib region and two or more rows of ventral scales. Antheridia and archegonia borne on modified branches with essentially similar internal structure to thallus or immersed in dorsal surface of thallus. Capsules with unistratose walls, with or without elaters.

The Marchantiales have a more elaborate gametophyte structure than other liverworts, this being related to the frequent occurrence of species in xeric habitats. Compared with the rest of Europe the Marchantiales are poorly represented in the British Isles (21 of a total of 64 European species). They represent about 7.5% of the British hepatic flora compared with about 85% in Mediterranean Europe, this being a reflection of climate and the xeromorphic nature of many Marchantiales.

ARTIFICIAL KEY TO THE FAMILIES OF MARCHANTIALES

1 Plants medium-sized to large, thalli (2–)5–12 mm wide, terrestrial or saxicolous, with pores on dorsal surface surrounded by 2 or more rows of distinct concentric or superposed cells, gametangia borne on sessile or stalked receptacles   2
  Plants small, thalli 0.3–3.0 mm wide (but sometimes larger if aquatic), pores not

surrounded by differentiated cells or pores lacking, gametangia and capsules immersed in thallus **40. Ricciaceae (p. 321)**
2 Gemma-cups present on dorsal surface of thallus 3
  Gemma-cups absent 4
3 Gemma-cups semilunar with entire margins, male receptacles sessile, female receptacles cruciately 4-rayed **35. Lunulariaceae (p. 312)**
  Gemma-cups cup-shaped, with frilled margins, female receptacles stellately rayed with about 9–11 rays, male receptacles stalked **39. Marchantiaceae (p. 317)**
4 Thallus translucent with obvious midrib, without pores or air-chambers
  36. Weisnerellaceae (p. 314)
  Thallus opaque, midrib not obvious, with pores and air-chambers 5
5 Pores not conspicuous, thallus parchment-like when dry, in section air-chambers in several layers **38. Aytoniaceae (p. 316)**
  Pores conspicuous, thallus not parchment-like when dry, air-chambers in a single layer 6
6 Ventral surface of thallus purplish-black, female receptacle sessile at ventral apex of thallus with conspicuous purplish-black involucre **34. Targioniaceae (p. 311)**
  Ventral surface of thallus green or reddish-brown, female receptacle stalked, involucres membranous and not conspicuous 7
7 Dorsal surface of thallus conspicuously reticulate, pores simple
  37. Conocephalaceae (p. 316)
  Dorsal surface of thallus faintly reticulate, pores barrel-shaped
  39. Marchantiaceae (p. 317)

SUBORDER MARCHANTIINEAE

Thallus with pores surrounded by 2 or more rings of differentiated concentric or superposed rings of cells; air-chambers usually well defined with columns of chlorophyllose cells. Male branches arising at thallus apex, often becoming lateral with age, consisting of a sessile or stalked discoid receptacle with antheridia, naked at first but embedded in receptacle tissue at maturity, on dorsal side, sperm often dispersed by a splash-cup mechanism. Female branches arising similarly, consisting of a stalked, (sessile in *Corsinia* and *Targionia*) entire or variously lobed receptacle, mature receptacle entire, lobed or rayed; archegonia in radiating rows, initially dorsal, becoming ventral by centrifugal growth of cells at centre of receptacle, protected by fragile hyaline to tough coloured involucres, membranous pseudoperianths sometimes present. Sporophyte, consisting of capsule, seta and foot, on ventral surface of receptacle, capsule wall unistratose, dehiscing regularly or irregularly. Elaters present. Twenty-five genera and about 230 species.

34. TARGIONIACEAE

Thallus pores simple, air-chambers in one layer. Male receptacles on short lateral branches. Female receptacles sessile on underside of thallus apex, archegonia and later single sporophyte surrounded by a 2-lobed involucre, pseudoperianth lacking. Capsule dehiscing irregularly, wall cells with annular thickenings. Elaters sometimes branched. Spores large, unicellular. Two genera, the second, *Cyathodium* Kunze, occurring in tropical Asia and differing from *Targionia* in mode of dehiscence of the capsule.

75. TARGIONIA L., *Sp. Plant*.: 1136. 1753

Three species, the two non-British being *T. lorbeeriana* K. Müll. from Europe and *T. elongata* Bisch. from Ethiopia.

1. T. hypophylla L. *Sp. Plant*.: 1136. 1753   (Fig. 137)

Autoecious. Plants dull green in small to extensive patches. Thalli 2–3 mm wide, to 1.5 cm long, branching from ventral side, faintly reticulate, pores ± conspicuous, midrib not obvious, margins becoming strongly incurved when dry revealing glossy blackish ventral scales, ventral surface purplish-black. Dorsal epidermal cells with distinct trigones, pores hardly elevated, simple, surrounded by 3–4 concentric rings of thin-walled cells, these often lost with age. Thallus in T.S. with midrib quickly merging into wings, air-chambers with filaments 3–4 green cells long, chlorenchyma layer narrow. Ventral scales large, imbricate, in one row on either side of midrib, purplish-black, becoming black and glossy when dry, semilunar-triangular with a toothed appendage. Male receptacles at the ends of ventral lateral branches. Female receptacles sessile on ventral side of thallus apex, archegonia and, later, solitary capsule surrounded by a conspicuous laterally compressed involucre of 2 convex purplish-black scales. Capsules common, late winter, early spring, dehiscing within involucres. Spores brown, coarsely crenulate, with coarse areolae and fine reticulations, 55–75 μm. *n* = 9. On shallow soil overlying usually basic rock at low altitudes, occasional in S. and W. England and Wales, from E. Cornwall east to N. Hants and Surrey, north to Derby and Cumberland, rare in C. Scotland, very rare in Ireland, Kerry, Mid Cork, Antrim. 32, H4, C. Mediterranean and Atlantic Europe north to the Netherlands, S.W. Russia, Crimea, Turkey, Cyprus, Korea, Szechuan, India, Macaronesia, N. and S. Africa, Madagascar, N., C. and S. America, Australia (Victoria).

Müller (1954) gives details of a triploid species, *T. lorbeeriana* K. Müll. which he distinguishes from *T. hypophylla* on the basis of various cell dimensions and characteristic smell but there seems some doubt about the reliability of these characters (Düll, 1983; Grolle, 1983a). The triploid is said to occur in S. Europe and Macaronesia.

35. LUNULARIACEAE

Thallus pores simple, air-chambers in one layer. Male receptacle sessile, discoid. Stalk of female receptacle without rhizoid furrows, pseudoperianths lacking. Capsule dehiscing by 4 valves, wall cells without annular thickenings. Elaters simple. Spores small, unicellular. One genus.

76. LUNULARIA ADANS., *Fam. Pl.* 2: 15. 1763

Two species, the secund, *L. tarteri* Ev. & Herz., occurring in Chile.

1. L. cruciata (L) Dum. ex Lindb., *Not. Sällsk. Fauna Fl. Fenn.* 9: 298. 1868
                                                                              (Fig. 137)

Dioecious. Plants glossy, light green to green, in patches. Thalli 5–10 mm wide, to 2.5 cm long, irregularly dichotomously branched, faintly reticulate, pores obvious, midrib not obvious, margins somewhat undulate, semilunar gemma-cups with lenticular gemmae always present; ventral surface pale green. Dorsal epidermal cells thin-walled with small trigones, pores elevated, simple, surrounded by 4–5 concentric rings of thin-walled cells. Thallus in T.S. with midrib very gradually merging into wings, air chambers with simple or branched filaments 2–3 green cells long, chlorenchyma layer narrow.

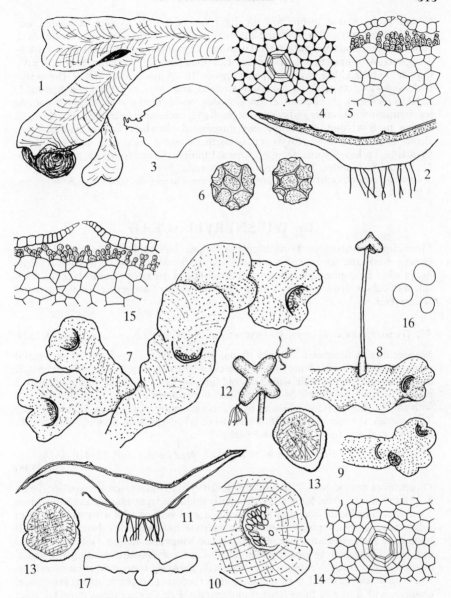

Fig. **137.** 1–6, *Targionia hypophylla*: 1, thallus with mature involucre (× 5); 2, T.S. thallus, chlorenchyma stippled (× 23); 3, ventral scale (× 21); 4, pore in surface view (× 150); 5, T.S. pore, chlorenchyma stippled (× 84); 6, spores (× 230). 7–17. *Lunularia cruciata*: 7, sterile thallus (× 1.5): 8, thallus with female receptacle (× 1.5); 9, thallus with male receptacle (× 1.5); 10, gemma cup (× 4); 11, T.S. thallus, chlorenchyma stippled (× 33); 12, female receptacle with two dehisced capsules (× 4); 13, gemmae (× 38); 14, pore in surface view (× 50); 15, T.S. pore, chlorenchyma stippled (× 84); 16, spores (× 230); 17, ventral scale (× 21).

Ventral scales sparse, in one row on either side of midrib, hyaline, elongated, one margin with teeth and a ± circular appendage. Male receptacles rare, sessile, oval, concave, becoming apparently lateral with age. Female receptacles occasional, stalk 2–3 cm long, with uniseriate hairs, receptacles green, cruciate at maturity with 4 rays. Capsules very rare, late summer. Spores smooth, 18–24 µm. $n = 8 + m^*$, 9. Damp soil, rocks and walls, on paths, roadsides, by streams, and rivers, in natural and man-made habitats including glasshouses, at low altitudes, very common in S. Britain, becoming less frequent northwards and rare in N. Scotland, occasional in Ireland. 112, H38, C. Native in S.W. and S. Europe but introduced elsewhere and extending to S. Fennoscandia and S.W. Russia (at its northern extent often only in glasshouses), Caucasus, Turkey, Cyprus, Iran, N. Africa, Ethiopia, C. and S. Africa, Macaronesia, N., C. and S. America, Juan Fernandez, Australia.

For an account of sexual reproduction in British *Lunularia* see Goodman, *Trans. Br. bryol. Soc.* 3, 98–102, 1956.

36. WIESNERELLACEAE

Monoecious or dioecious. Pores simple or lacking, air-chambers in a single layer, or lacking. Male and female receptacles stalked, stalks with 2 rhizoid furrows. Female receptacle ± hemispherical, lobed, involucre 2-lipped, pseudoperianths lacking. Capsule wall cells with annular thickenings. Elaters simple. Two genera, *Dumortiera* and *Wiesnerella*.

77. DUMORTIERA NEES IN REINW. ET AL., *Nova Acta Leop. Carol.* 12: 410. 1824

Monoecious or dioecious. Mature thallus without pores, air-chambers or ventral scales, smooth rhizoids abundant but tuberculate rhizoids very sparse. Male receptacle discoid with very short stalk with 2 rhizoid furrows. Stalk of female receptacle with 2 rhizoid furrows, receptacle 6–10-lobed, pseudoperianths lacking. Capsule wall cells with annular thickenings, dehiscing by 4 or occasionally more valves. Elaters simple. Two species, the second being *D. nepalensis* (Tayl.) Nees. with a similar range to *D. hirsuta* but absent from Europe and Africa.

1. **D. hirsuta** (Sw.) Nees in Reinw. et al., *Nova Acta Leop. Carol.* 12: 410. 1824

(Fig. 138)

Dioecious or monoecious. Plants in dark green in sometimes extensive patches. Thalli 6–12 mm wide, to 10 cm long, margins slightly undulate, apex emarginate, wings often somewhat translucent and midrib obvious, dorsal surface with a network of faint ridges, pores absent, ventral surface with numerous smooth but no tuberculate rhizoids on midrib, scattered short spine-like rhizoids on wings and margin. Dorsal epidermal cells very thin-walled, lacking trigones. Thallus in T.S. with midrib grading into wings, air-chambers lacking. Ventral scales very small, narrowly triangular, soon vanishing. Male receptacles terminal, with very short stalk rendering receptacle apparently sessile, concave, with a ring of short hairs round margin. Female receptacles terminal, stalk 1.5–2.0 cm long, receptacle ca 6-lobed. Capsules very rare, late summer (?). Spores brown, coarsely papillose, 24–35 µm. Damp soil banks and dripping rocks by streams and waterfalls in deeply shaded humid sites, at low altitudes but ascending to 320 m in S.W. Ireland, very rare in Britain, Cornwall, Devon, E. Sussex, I. of Man, Jura, occasional in S.W. Ireland, very rare elsewhere, Wicklow, Leitrim, Antrim. 8, H13. Portugal, N.W. Spain to Pyrenees, Apennines (Italy), Greece (?), Nepal, Sri Lanka,

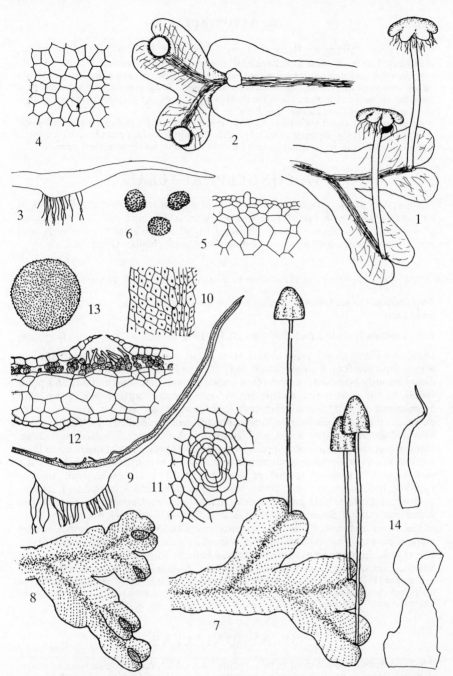

Fig. **138**. 1–6, *Dumortiera hirsuta*: 1, thallus with female receptacles (× 1.5); 2, thallus with male receptacles (× 1.5); 3, т.s. thallus (× 21); 4, dorsal epidermis in surface view (× 150); 5, т.s. dorsal part of thallus (× 84); 6, spores (× 230). 7–14, *Conocephalum conicum*: 7, thallus with female receptacles (× 1.5); 8, thallus with male receptacles (× 1.5); 9, т.s. thallus, chlorenchyma stippled (× 13); 10, part of thallus, dorsal side (× 3); 11, pore in surface view (× 150); 12, т.s. pore, chlorenchyma stippled (× 84); 13, spore (× 230); 14, ventral scales (× 21).

China, Japan, Philippines, Réunion, Kenya, Cameroon, Fernando Po, Macaronesia, south-east U.S.A., C. and S. America, W. Indies, Bermuda, Hawaii, Tahiti.

Sterile dried specimens may be confused with *Pellia* as, after resoaking, the network of fine ridges on younger parts of the thallus are not always visible. It differs in the presence of the short spine-like rhizoids on the underside of the thallus wings and the dorsal epidermal cells ± hexagonal near the margin (very elongated in *Pellia*).

According to Macvicar (1926) air-chambers are laid down at the thallus apex but disappear later and Müller (1954) illustrates such an air-chamber together with a pore. In serial transverse and longitudinal sections of a thallus apex I could find neither air-chambers nor pores.

37. CONOCEPHALACEAE

Thallus pores simple, air-cavities in one layer. Male receptacle sessile, discoid. Stalk of female receptacle with one rhizoid furrow, receptacle ± entire with 5–8 involucres, pseudoperianths lacking. Capsule dehiscing by lid and 4–8 valves, wall cells with annular thickenings. Elaters simple. Spores large, multicellular. One genus.

78. CONOCEPHALUM HILL CORR. WIGGERS, *Primit. Fl. Holsat.* 82: 1780

Two species, the second, *C. supradecompositum* (Lindb.) Steph., occurring in China and Japan.

1. C. conicum (L.) Underw., *Bot. Gaz.* 20: 67. 1895   (Fig. 138)

Dioecious. Plants glossy, green to dark green in flat, sometimes very extensive patches with a strongly fungal smell. Thalli 6–12 mm wide, to 10–15 cm long, regularly dichotomously branched, margin often somewhat undulate, reticulations and pores visible to the naked eye, midrib not obvious, ventral surface pale green with conspicuous midrib. Dorsal epidermal cells thin-walled, lacking trigones, pores elevated, simple, surrounded by 4–5 concentric rings of thin-walled cells. Thallus in T.S. with midrib grading into wings, air-chambers with filaments 2–3 green cells long, filaments on floor beneath pores ending in a colourless projection, chlorenchyma layer narrow. Ventral scales sparse, in one row on either side of midrib, narrowly triangular with a violet, ± rounded appendage. Male receptacle violet, sessile, oval, concave, papillose, becoming apparently lateral with age. Female receptacles frequent, stalk 2–5(–6) cm, receptacle pale green, conical, margin ± entire. Capsules occasional, late winter, early spring. Spores brown, papillose, (65–)80–110 μm. $n = 8 + m^*$, 9. Sheltered banks, rocks and walls by ditches, streams, rivers and waterfalls, in flushes and on marshy ground, common and sometimes locally abundant, ascending to *ca* 1000 m. 112, H40, C. Europe north to Swedish Lapland and N. Russia, Faroes, Iceland, east to Siberia, China, Japan, Himalayas, Macaronesia, Algeria, Tunisia, N. America.

Macvicar (1926) and Müller (1954) report tubers being produced on the ventral surface of the midrib but I have not seen these and it is possible that they were referring to *C. supradecompositum* which has such tubers (see Wanstall, *Trans. Br. bryol. Soc.* **4**, 464–5, 1963).

38. AYTONIACEAE

Thallus pores simple, air-cavities in 2 or more layers. Male receptacles arising on dorsal surface of midrib. Stalk of female receptacle with one rhizoid furrow, receptacle entire or lobed, pseudoperianths lacking. Capsule dehiscing by a lid (except in *Reboulia*), wall cells lacking annular thickenings. Elaters simple. Spores large. Five genera.

79. REBOULIA RADDI, *Opusc. Sci. (Bologna)* 2: 357. 1818

Thallus pores surrounded by 3–5 concentric rings of cells, dorsal epidermal cells thin-walled with bulging trigones. Two or possibly four species.

1. R. hemisphaerica (L.) Raddi, *Opusc. Sci. (Bologna)* 2: 357. 1818   (Fig. 139)

Monoecious or dioecious. Plants dull, light green, often with purplish margins, becoming smooth and parchment-like when dry, in rosettes or patches. Thalli 3–6(–8) mm wide, to 2(–3) cm long, dichotomously branched, margins crenulate, dorsal surface not reticulate, pores scarcely visible, midrib not obvious; ventral surface deep purple. Dorsal epidermal cells with large, often bulging trigones, pores hardly elevated, simple, surrounded by 3–5 concentric rings of cells which may have thickened radial walls. Thallus in T.S. with midrib very gradually merging with wings, chlorenchyma composed of air-cavities in several layers, occupying about ⅔ of midrib region and almost all of wings, cavities irregularly rounded in midrib region, elongated in wings. Ventral scales in one row on either side of midrib, purple, obliquely semilunar, with two elongated appendages at least when young. Male receptacles sessile on dorsal surface of midrib, concave. Female receptacles common, arising from thallus apex, stalk 1–2 cm with hair-like scales at base and apex, receptacle with 4–7 rounded lobes. Capsules frequent, late winter to early summer. Spores with crenulate margin, areolate and papillose, 70–100 μm. $n = 8 + m^*$, 9. Rocks and soil on banks, amongst rocks, in ravines, on cliffs, in sheltered situations especially where basic, rarely in turf, mainly at low altitudes, rare to frequent from Cornwall and Kent north to Sutherland and Caithness, widely distributed but rare in Ireland. 85, H21, C. More or less cosmopolitan except for arctic and subarctic climatic regions.

39. MARCHANTIACEAE

Thallus pores barrel-shaped; air-chambers in one layer. Receptacles arising at thallus apices, stalked in both sexes, stalks with two rhizoid furrows. Female receptacles lobed or palmately or stellately rayed. Pseudoperianths present. Capsules dehiscing irregularly, wall cells with annular thickenings. Elaters simple. Four genera.

Thalli without gemma-receptacles, lacking a dark line in midrib region, ventral scales in 2 rows   **80. Preissia**

Thalli with cup-shaped gemma-receptacles or if receptacles lacking then thallus with a dark line in midrib region   **81. Marchantia**

80. PREISSIA CORDA IN OPIZ, *Naturalientausch*: 647. 1829

One species.

1. P. quadrata (Scop.) Nees, *Naturg. Europ. Leberm.* 4: 135. 1838   (Fig. 139)

Monoecious or dioecious. Plants dull, pale green, often with reddish-purple margin, becoming striate but not parchment-like when dry, in rosettes or patches. Thalli 5–10 mm wide, to 4 cm long, branching dichotomously and ventrally, faintly reticulate and with conspicuous pores, midrib not obvious, margins wavy; ventral surface greenish to purplish. Dorsal epidermal cells without trigones, pores elevated, barrel-shaped, surrounded by 5–6 rings of superposed thick-walled cells. Thallus in T.S. with midrib gradually merging into wings, air-cavities containing large, very thin-walled chlorophyllose cells, chlorenchyma layer narrow. Ventral scales imbricate, purplish, soon

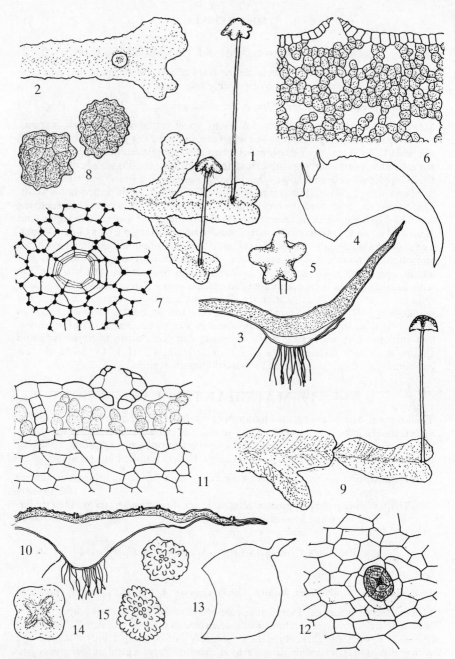

Fig. **139**. 1–8. *Reboulia hemisphaerica*: 1, thallus with female receptacles (× 1.5); 2, thallus with male receptacle (× 1.5); 3, т.s. thallus, chlorenchyma stippled (× 15); 4, ventral scale (× 21); 5, female receptacle (× 4); 6, т.s. pore, chlorenchyma stippled (× 150); 7, pore in surface view (× 84); 8, spores (× 230). 9–15, *Preissia quadrata*: 9, thallus with female receptacle (× 1.5); 10, т.s. thallus, chlorenchyma stippled (× 15); 11, т.s. pore, chlorenchyma stippled (× 150); 12, pore in surface view, cavity cross-hatched (× 150); 13, ventral scale (× 23); 14, female receptacle from above (× 3); 15, spores (× 230).

vanishing, in one row on either side of midrib, ±semicircular with a narrow appendage. Receptacles arising terminally. Male receptacle ± circular, stalk 1–2 cm long. Female receptacles common, ± hemispherical with 3–4 ridges, very shallowly 3–4 lobed stalked to 5 cm. Capsules common, June to November. Spores brown, coarsely papillose, 55–70(–80) μm. $n = 8 + m^*$, 9. On soil, moist or dry rocks and in rock crevices, on cliffs, stream banks, in ravines, dune-slacks and fens, calcicole, ascending to *ca* 900 m, occasional to frequent from Devon, Wales and Lancashire and Yorkshire north to Shetland, very rare elsewhere, occasional in Ireland. 77, H28, C. Europe north to Spitzbergen, Faroes, Iceland, Turkey (?), Kamchatka, Japan, China, Himalayas, Morocco, Azores, Madeira, N. America, Greenland.

When sterile sometimes confused with *Reboulia hemisphaerica* which, however, differs in the inconspicuous pores, the thallus parchment-like when dry, the dorsal epidermal cells with conspicuous trigones and the simple pores.

81. MARCHANTIA L., *Sp. Plant.*: 1137. 1753

Dioecious. Thallus with barrel-shaped pores; air-chambers in a single layer. Cup-shaped gemma-cups with fringed margins, containing lenticular gemmae present. Receptacle stalks with 2 rhizoid furrows, male lobed or rayed; female receptacles rayed with membranous involucres on ventral side, pseudoperianths present. Capsule dehiscing irregularly, wall cells with annular thickenings. Elaters simple. Spores small. Probably about 20 species.

1. M. polymorpha L. *Sp. Plant.*: 1137. 1753   (Fig. 140)
M. alpestris (Nees) Burgeff, *M. aquatica* (Nees) Burgeff

Dioecious. Plants dull, light or yellowish-green, leathery or not in appearance, in rosettes or patches. Thalli prostrate or sometimes ascending in wet habitats, 5–10 mm wide, to *ca* 10 cm long, dark line sometimes present along middle, faintly reticulate or not, pores visible, margins plane to somewhat crisped, cup-shaped, frilled gemma-cups containing lenticular gemmae usually present but sometimes lacking in wet habitats; ventral surface pale green to brownish. Dorsal epidermal cells thin-walled, without trigones, pores barrel-shaped and surrounded by 4–6 rings of superposed cells. Thallus in T.S. with transition from midrib to wings very gradual, air-chambers present or lacking in mid-rib region in some wet habitat forms, containing 2–4-celled, simple or branched filaments of green cells. Ventral scales in 2 rows along midrib and in 2 rows on either side, colourless to purple, sometimes projecting beyond thallus margin. Male receptacles apical, less common than female, peltate, shallowly 8–10-lobed, stalks 2–3 cm. Female receptacles apical, stellate with usually 9–11 rays, hyaline involucres extending to ½ length of rays on ventral side, stalk 2–5 cm. Capsules rare, summer, autumn. Spores ± smooth, 13–14 μm. $n = 9$. Common in man-made habitats, especially as a horticultural weed and on burnt ground, occasional to frequent on soil and rocks by pools, streams and rivers, roadsides, in flushes, fens, bogs and dune-slacks, at low altitudes but ascending to *ca* 1000 m in Scotland. 109, H34, C. Cool to warm temperate regions throughout the world but extending north to Spitzbergen in Europe.

Many authorities regard *M. alpestris* as a distinct species but the only consistent difference between it and *M. polymorpha* is the absence of a dark line in the midrib region and it can only be considered a mere colour variant within a very polymorphic species. Other differences quoted include texture and branching of the thallus, length and thickness of the receptacle stalks, the diameter of the rays of the female receptacles and the extent of the involucre relative to the rays but there is uncorrelated continuous variation. Dr H. Bischler (*in litt.* and various publications) considers there to be so many genetically different strains that the recognition of taxonomic subdivisions within *M. polymorpha* is not a feasible proposition.

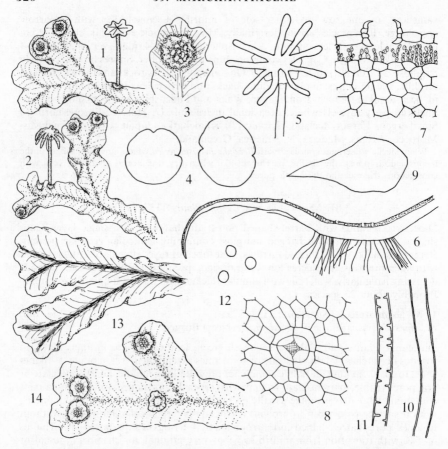

Fig. **140**. *Marchantia polymorpha*: 1 and 2, male and female thalli with receptacles and gemma cups (× 1.3); 3, gemma cup (× 4); 4, gemmae (× 13); 5, female receptacle (× 4); 6, т.s. portion of thallus, chlorenchyma stippled (× 15); 7, т.s. pore, chlorenchyma stippled (× 150); 8, pore in surface view (× 150), 9, ventral scale (× 15); 10 and 11, optical sections or portions of simple and tuberculate rhizoids (× 130); 12, spores (× 230); 13, wet ground '*aquatica*' form with dark midrib region (× 1.3); 14, '*alpestre*' form with leathery thallus (× 1.3).

Plants growing in wet habitats such as fens and flushes are of very distinctive appearance and have in the past been treated as a separate species, *M. aquatica* and are recognised by many authorities as a variety today. The thalli are more elongated with less widely spreading branches, with few or no gemma-cups or receptacles and without pores in the midrib region. I have found, however, that such plants when grown under xeric conditions in a glasshouse for 2–3 years become morphologically indistinguishable from plants from terrestrial habitats and produced abundant gemma-cups and receptacles. They also develop pores, although air-chambers may be only poorly developed in the midrib region. In view of the morphological variability of *M. polymorpha* the distinctive wet-ground forms must be regarded as largely, if not wholly, environmentally induced modifications and *M. aquatica* cannot be retained as a viable taxon at either specific or varietal rank.

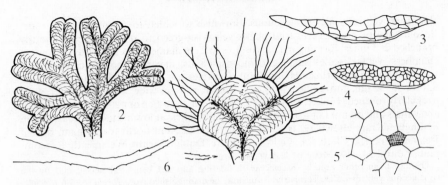

Fig. **141**. *Ricciocarpos natans*: 1, floating thallus (× 5); 2, terrestrial thallus (× 3.5); 3, T.S. floating thallus (× 8); 4, T.S. terrestrial thallus (× 8); 5, epidermal pore (× 210); 6, ventral scales from aquatic thallus (× 6).

SUBORDER RICCIINEAE

Thallus with or without pores, pores where present not usually surrounded by differentiated cells. Thallus containing compact vertical columns of green cells with narrow air-space or plates of green cells forming chambers lacking assimilatory filaments. Antheridia and archegonia embedded in dorsal surface of thallus. Sporophyte at maturity consisting solely of capsule with unistratose wall. Spores large, 40–150 µm. Elaters absent. Two families, the second, the Oxymitriaceae, differing from the Ricciaceae in the presence of tall narrow air chambers in the thallus with stellate pores surrounded by 6 cells with radially thickened walls.

40. RICCIACEAE

Dorsal epidermal pores where present not surrounded by differentiated cells.

> Plants floating, with conspicuous linear or linear-lanceolate ventral scales, or terrestrial with small ventral scales, dorsal epidermis with small distinct pores and scattered oil-cells   **82. Ricciocarpos**
> Plants terrestrial or if aquatic submersed, ventral scales very small or if larger then semilunar, pores if present vestigial, epidermal oil-cells lacking   **83. Riccia**

82. RICCIOCARPOS CORDA IN OPIZ, *Naturalientausch* 12: 651. 1829(–30)

Dorsal epidermis with small but distinct pores and scattered oil-cells, thallus interior composed of unistratose plates of cells forming polygonal chambers. Rhizoids simple, rarely with sparse small tubercles. Ventral scales persistent, linear-lanceolate. One species.

1. R. natans (L.) Corda in Opiz, *Naturalientausch* 12: 651. 1829(–30)   (Fig. 141)

Monoecious or dioecious. Plants floating or terrestrial. Thallus branching dichotomously, grooved. Aquatic plants, green to violet, thallus dichotomously branched, often ± cordate, 4–9 mm long, 4–9 mm wide; in T.S. with usualy two rows of irregular chambers; ventral scales conspicuous, brownish to violet, to 5.5 mm long, serrate; rhizoids few or none. Terrestrial plants yellowish-green to reddish, dichotomously branched 2–3 times, forming partial rosettes to 2 cm diameter; in T.S. with 4–5 rows of chambers; ventral scales small and inconspicuous, linear-lanceolate, colourless to violet, to *ca* 0.6 mm, rhizoids abundant. Capsules not known in Britain or Ireland. Spores 45–55 μm, convex face with *ca* 8 areolae, 6–7 μm wide, across diameter, wing irregularly toothed (Müller, 1954). $n = 9$. Floating in base-rich or neutral ponds, ditches and canals and on mud of dried-out ponds and ditches, at low altitudes, occasional in E. England, rare elsewhere, from N. Somerset and Kent north to Durham, rare in Ireland, Limerick, Clare, S.E. Galway, Kildare, Dublin, Meath, Westmeath, Roscommon, Louth. 34, H9. More or less world-wide.

Distinct from other aquatic Ricciaceae in floating and not being submersed and by the conspicuous ventral scales. Terrestrial forms may be distinguished from *Riccia* section *Ricciella* species by the small dorsal pores and linear-lanceolate ventral scales. Capsules are very rare in Europe suggesting European plants are dioecious but in N. America capsules are frequent (Schuster, 1953) suggesting that at least some plants are monoecious there.

83. RICCIA L., *Sp. Plant.*: 1138. 1753

Monoecious or dioecious. Plants terrestrial or aquatic, often winter ephemerals or annuals. Thalli usually 2–3 times dichotomously branched, composed of ventral portion of compact parenchymatous tissue and a dorsal portion of either vertical columns of green cells and hyaline apical cells (epidermis) and narrow air-spaces, or plates of cells forming chambers, rounded or hexagonal in T.S.; epidermis without oil-cells, pores lacking or if present not surrounded by specialised cells. Ventral scales arising in a single median row, becoming by thallus growth 2 lateral rows of half-scales (loosely referred to below as scales). Antheridia and archegonia (mixed in monoecious species) immersed in dorsal side of thallus. Sporophyte immersed in thallus. Spores large, 40–150 μm, released by decay of thallus tissue. A world-wide genus with 35 European species.

There seems no general agreement about the arrangement of species within subgenus *Riccia*. Grolle (1983a) arranges them in alphabetical order and the arrangements of Müller (1954), Corley & Hill (1981) and Jovet-Ast (1986) differ. I have followed the latter as being the most recent authoritative account.

The species of *Riccia* exhibit considerable phenotypic variation. Species with reddish or purplish coloration may exhibit this more strongly under dry conditions. The development of marginal cilia is affected by moisture conditions and their apparent absence has to be used with care in identification of members of section *Riccia*. Thallus sections are most easily cut using fresh material although satisfactory sections can be obtained from dried specimens provided they are not too old (this applies particularly to section *Ricciella*). Jovet-Ast's account of Mediterranean *Riccia* species (*Cryptogamie Bryol. Lichén.* 7, suppl. to fasc. 3, 287–431, 1986) deals with all British and Irish species except *R. rhenana* but care should be exercised in using sections of the key to the subgenus *Riccia* dealing with presence or absence of marginal cilia.

Unless otherwise stated descriptions of thallus and illustrations of thallus sections are from the ultimate lobes. Older parts of thalli may be thinner and less deeply grooved or channelled. Spore dimensions include the wings and are of the longest diameter where spores are not circular when viewed from above.

Species of *Riccia* may occur mixed so care must be taken when determining mixed gatherings.

1 Thallus in section with ± polygonal chambers, older parts sometimes spongy in
    appearance, plants aquatic or terrestrial   2
  Thallus in section compact, without large air-chambers, not spongy in appearance,
    plants terrestrial   7
2 Plants terrestrial, not strap-like, older parts spongy in appearance, capsules not
    protruding strongly on ventral side of thallus   3

Plants aquatic or terrestrial, strap-like or not, older parts not spongy, capsules
  protruding strongly on ventral side of thallus   5
3 Thallus lobes 0.5–0.8 mm wide, channelled, often reddish or violet-tinged
                                                              3. R. huebeneriana
  Thallus lobes 1–4 mm wide, neither channelled nor reddish to violet-tinged   4
4 Plants yellowish-green, ultimate thallus lobes 1–3 times as long as wide, spores with
  4–6 partial or complete areolae across convex face   **1. R. cavernosa**
  Plants bluish-green, ultimate thallus lobes as wide as or wider than long, spores with
  ca 10 partial or complete areolae across diameter of convex face
                                                              2. R. crystallina
5 At least older parts of thallus channelled especially when dry, not reticulate, 2.0–3.7
  times as wide as thick, capsules common   **6. R. canaliculata**
  Thallus not channelled, lobes often reticulate, 3–6(–8) times as wide as thick,
  capsules very rare
                                                                            6
6 Cells of ventral scales to 45 μm in longest dimension   **4. R. fluitans**
  Cells of ventral scales to 65 μm in longest dimension   **5. R. rhenana**
7 Thallus lobes with ciliate margins, cilia curved over thallus when dry
                                                              13. R. crozalsii
  Thallus margins naked, or if ciliate then cilia not incurved when dry   8
8 Underside of thallus with persistent, overlapping, purplish-black, glossy, ventral
  scales   **11. R. nigrella**
  Ventral scales colourless or not, ephemeral   9
9 Thallus lobes in T.S. 4–6 times as wide as thick, spores 80–96 μm   **8. R. glauca**
  Thallus lobes 1–3 times as wide as thick or if wider then spores 100–128 μm   10
10 Thallus lobes shallowly and broadly channelled, 2–4 times as wide as thick
                                                              12. R. beyrichiana
  Thallus lobes deeply grooved or channelled, 1–3 times as wide as thick   11
11 Subepidermal cells of thallus lobes hyaline, thick-walled   **7. R. sorocarpa**
  Subepidermal cells with chloroplasts, thin-walled   12
12 Older parts of thalli pale brown to orange-brown, epidermal and green cells mostly
  20–40 μm wide, spore wing with thickened margin   **9. R. bifurca**
  Older parts green to purplish, cells wider, spore wing lacking thickened margin   13
13 Marginal cilia* if present not inflexed over thallus when dry, ridges of triradiate face
  of spore not anastomosing to form areolae   **10. R. subbifurca**
  Marginal cilia if present inflexed over thallus when dry, triradiate face of spores
  with complete areolae   **13. R. crozalsii**

*Marginal cilia will develop if plants are cultured under moist conditions (as they may also do in
R. glauca, *R. bifurca* and *R. beyrichiana*).

Subgenus *Ricciella* (A. Braun) Reichenb., *Deutsch. Bot. Herbarienb.* 23: 213. 1841

Thallus with chambers, epidermis breaking down to form lacunae in older parts of
thallus or not; margins without cilia.

Section *Spongodes* Nees, *Naturg. Europ. Leberm.* 4: 391. 1838

Plants terrestrial. Dorsal epidermal cells in older parts of thallus breaking down to
form lacunae, rendering thallus spongy in appearance. Capsules embedded in thallus,
not protruding strongly on ventral side.

1. R. cavernosa Hoffm. emend. Raddi, *Opusc. Sci.* (*Bologna*) 2: 351–3. 1818

(Fig. 142)

Monoecious. Plants yellowish-green, greenish when dry, forming partial or complete rosettes, 1.0–2.5 cm diameter. Thalli ± flat, older parts spongy in appearance, 2–3 times dichotomously branched, ultimate lobes 1.5–3.0 mm long, 1–2 mm wide, ± 1–3 times as long as wide, apex not grooved. Thallus in T.S. 3–5 times as wide as thick, with 2–3 layers of air-chambers. Ventral scales lacking. Capsules very common throughout year, not or scarcely prominent on either surface. Spores reddish-brown to dark brown or blackish, 64–96(–120) μm, convex face with 4–6 partial or complete areolae, 7–20 μm wide, across diameter but sculpturing varying from spore to spore, rounded or truncate tubercles at corners of areolae, triradiate face with irregular ridges anastomosing or not to form partial or complete areolae, triradiate ridges thickened, wing papillose-crenulate. $n = 8$. On mud by ponds and lakes, on damp soil and in dune-slacks, often where basic, rare to occasional in England and Wales, very rare in Scotland and Ireland. 46, H4. Europe north to Fennoscandia and N. Russia, Iceland, India, Mongolia, Africa, Gran Canaria, N., C. and southern S. America, Caribbean, Australia.

All older British and Irish records of *R. crystallina* belong to *R. cavernosa* which is widespread, *R. crystallina* only being known from two vice-counties. *R. crystallina* differs in the wider, relatively shorter lobes of the thallus, bluish-green rather than yellowish-green colour and in spore morphology. *R. huebeneriana* is a smaller, often reddish or violet-tinted plant with the thallus relatively thicker in section; there is also a difference in spore sculpturing. For a comparative account of these three species see Paton, *Trans. Br. bryol. Soc.* 5, 222–5, 1967.

2. R. crystallina L. emend. Raddi, *Opusc. Sci.* 2: 251–3. 1818   (Fig. 142)

Monoecious. Plants bluish green, frequently bluish when dry, forming compact partial or complete rosettes to *ca* 2.5cm diameter. Thalli ± flat, at least older parts spongy in appearance, 1–2 times dichotomously branched, ultimate lobes short, 1.5–4.0 mm wide, as wide as or wider than long, apex not grooved. Thallus in T.S. 3–5 times as wide as thick, with 2–3 layers of air-chambers. Ventral scales lacking. Capsules very common, winter to early summer, not or scarcely prominent on either surface. Spores yellowish-brown, 56–84 μm, convex face with *ca* 10 partial or complete areolae, (5–)7–10 μm wide, across diameter, truncate tubercles at corners of areolae, triradiate face similar but with prominent triradiate ridges, wing toothed. $n = 7 + m^*$, 8. On shaded soil in fields, on paths and on mud by pools, very rare, Scilly Is., Renfrew. 2. S. and C. Europe, France, Baltic Russia, Turkey, Israel, N. and S. Africa, Macaronesia, C. and S. America, Caribbean, Australia.

3. R. huebeneriana Lindenb., *Nova Acta Acad. Leop.* 18: 504d. 1836   (Fig. 142)

Monoecious. Plants green, usually tinged red or violet, forming partial or complete rosettes *ca* 1 cm diameter. Thallus lobes channelled, at least older parts spongy in appearance, 2–3 times dichotomously branched, ultimate lobes 1.5–2.0 mm long, 0.5–0.8 mm wide, 1.5–4 times as long as wide. Thallus in T.S. 1.5–2.0 times as wide as thick. Ventral scales colourless to violet, soon lost. Capsules frequent, autumn, prominent on ventral side of thallus. Spores yellowish-brown, 60–80 μm, convex face with *ca* 8 areolae, 8–12 μm wide, across diameter, truncate tubercles at corners of areolae, triradiate face with partial or complete areolae and resembling convex face, triradiate ridge not obvious, wing entire. $n = 7 + m^*$, 8. On mud by ponds and lakes, rare in England and Wales in scattered localities from E. Cornwall and Sussex north to M.W. Yorks and S. Northumberland, Is. of Bute, Wicklow. 12, H1. Rare in Europe from Spain, Portugal and Italy north to Fennoscandia and N. Russia, India, Japan, N. Africa.

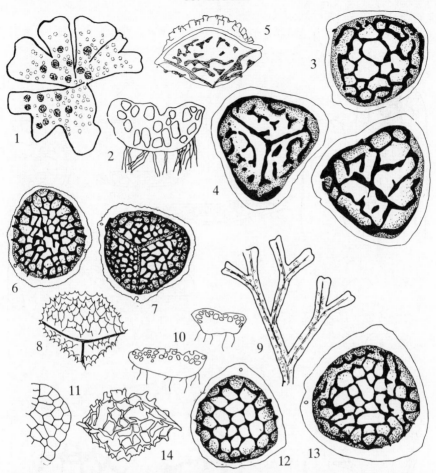

Fig. **142.** 1–5. *Riccia cavernosa*: 1, partial rosette (× 2.5); T.S. thallus lobe (× 21); 3, spores, convex face (× 450); 4, spore, triradiate face (× 450); 5, spore, side view (× 450); 6–8, *R. crystallina*: 6, spore, convex face (× 450); 7, spore, triradiate face (× 450); 8, spore, side view (× 450). 9–14, *R. huebeneriana*: 9, thallus (× 5); 10, T.S. thallus branches (× 21); 11, ventral scale (× 155); 12, spore, convex face (× 450); 13, spore, triradiate face (× 450); 14, spore, side view (× 450).

Section *Ricciella*

Plants aquatic or terrestrial. Epidermal cells on dorsal side of thallus not breaking down, older parts of thallus not spongy in appearance. Capsules protruding very strongly on ventral side.

4. R. fluitans L. emend. Lorbeer in K. Müll., *Hedwigia* 80: 93. 1941   (Fig. 143)

Dioecious. Plants light green, somewhat translucent, submersed or forming partial rosettes on mud. Thalli strap-like, to 3 cm long, regularly dichotomously branched, lobes not narrowing towards apex, apex somewhat grooved, lobes 0.4–1.8(–2.4) mm wide, dorsal side often reticulate at least towards apex, not channelled, rhizoids sparse

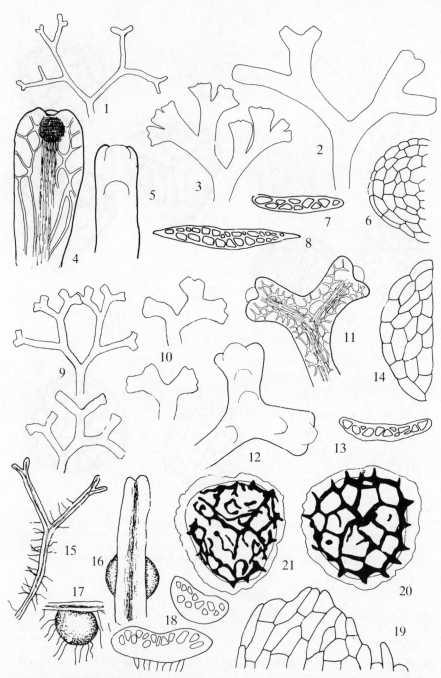

Fig. **143**. 1–6, *Riccia fluitans*: 1 and 2, aquatic thalli (× 3); 3, thallus from mud (× 3); 4 and 5, thallus lobes, dorsal and ventral sides, (× 30); 6, ventral scale (× 155); 7, т.s. aquatic thallus (× 38); 8, т.s. terrestrial thallus (× 23). 9–14, *R. rhenana*: 9, aquatic thalli (× 3); 10, terrestrial thalli (× 3); 11 and 12, thallus lobes, dorsal and ventral sides (× 15); 13, т.s. aquatic thallus (× 38); 14, ventral scale (× 155). 15–21, *R. canaliculata*: 15, thallus (× 3); 16 and 17, thallus with capsule, dorsal and side views (× 15); 18, т.s. thallus (× 38); 19, ventral scale (× 155); 20 and 21, spores, convex and triradiate faces (× 275).

or lacking in aquatic plants, sparse in terrestrial plants. Thallus in T.S. 3–6(–10) times as wide as thick, chambers in 1–2 irregular layers. Ventral scales sparse, soon vanishing, situated along midrib region, hyaline, ± semicircular, cells to 45(–50) μm in longest dimension. Damaged or old thalli sometimes producing propaguliferous shoots. Capsules very rare, summer, 300–370 μm diameter. Spores (50–)56–75(–80) μm, convex face with 5–7 partial or complete areolae, 8–16(–20) μm wide, across diameter, triradiate face with irregular ridges and prominent triradiate ridges. $n = 7 + m^*$. In ponds, canals and ditches, sometimes abundant, and on mud, decaying vegetation and logs in drying-out ponds etc., at low altitudes, occasional to frequent in S.E. England and E. Anglia, decreasing because of drainage, rare elsewhere in England north to Durham and Westmorland, rare in S. and E. Wales and E. Ireland. 56, H13. World-wide except for subarctic and arctic regions.

Data on capsules and spores is taken from Paton, *J. Bryol.* 7, 253–9, 1973, which see also for a detailed comparison with *R. canaliculata*. Although in the past these two species have been confused in Britain they are good species differing in a number of respects. *R. canaliculata* differs from *R. fluitans* in being monoecious with the frequent production of capsules, the markedly channelled thallus with lobes narrowing towards the apex and the ventral scales tearing during the course of growth to form two rows of half-scales.

The situation concerning *R. fluitans* and *R. rhenana* is much more equivocal and whether the two should be regarded as distinct species is very doubtful. Müller (1954) lists several characters distinguishing them but phenotypic variability of the two species renders these characters valueless. This has been shown experimentally by Berrie (*Bryologist* 67, 146–52, 1964) who has also shown that *R. rhenana* is an autodiploid derivative of *R. fluitans*.

5. R. rhenana Lorbeer in K. Müll., *Hedwigia* 80: 94. 1941   (Fig. 143)

Dioecious. Plants light green, scarcely translucent, submersed or forming partial rosettes on mud. Thalli strap-like, regularly dichotomously branched, 0.5–1.0 mm wide in aquatic plants, to 2.8 mm wide in terrestrial plants, to 1.5 cm long, lobes not narrowing to apex, apex strongly grooved, truncate, dorsal side reticulate, not channelled, rhizoids sparse or lacking in aquatic plants, sparse in terrestrial plants. Thallus in T.S. *ca* 5 times as wide as thick, chambers in 1(–2) irregular layers. Ventral scales sparse, soon vanishing, hyaline, ± truncated semicircular, cells to 65 μm in longest dimension. Capsules unknown. In pools and on mud at edges of drying-out ponds, rare, N. Wilts, Surrey (extinct), S. Essex, Herts, Bucks, W. Norfolk, Cambridge. 7. From Belgium, France, and Switzerland north to Russia and S. Sweden, Turkey.

For the first report of *R. rhenana* in Britain see Crundwell, *Trans. Br. bryol. Soc.* 3, 449–50, 1958 and for the differences from *R. fluitans* see Jones (1952).

6. R. canaliculata Hoffm., *Deutsch. Fl.* 2: 96. 1795   (Fig. 143)

Monoecious. Plants terrestrial, light green, forming loosely interwoven patches. Thalli strap-like, dichotomously branched, channelled especially in older parts and when dry, 0.3–0.7 mm wide, to 2 cm long, lobes often becoming narrower towards apex, dorsal side not reticulate. Thallus in T.S. 2–4 times as wide as thick, air-chambers in 1–2 irregular layers. Ventral scales hyaline, semicircular at apex but splitting and becoming lateral and subtriangular. Capsules common, summer to winter, protruding strongly from ventral surface of thallus, 0.7–1.0 mm diameter. Spores 96–130 μm, convex face with 3–5 areolae, 20–34 μm wide, across diameter, triradiate face with irregular ridges, sometimes anastomosing to form partial or complete areolae, wing sinuate-crenulate. $n = 7 + mn^*$, $14 + 2m^*$. On intermittently exposed mud at edges of lakes and reservoirs at low altitudes, very rare, W. Sussex, Shropshire, Merioneth (extinct), Anglesey, Cheshire, Midlothian, Stirling, W. and Mid Perth, Dumbarton. 10. Rare in Europe from Spain and Portugal, Corsica and Greece north to S. Fennoscandia, Turkey, Israel, Morocco, Algeria, Tunisia, N. America.

For the occurrence of this plant in Britain see Jones, *Trans. Br. bryol. Soc.* **2**, 1–10, 1952. A second species, *R. duplex* Lorbeer, has been shown to be a diploid derivative of the haploid *R. canaliculata* (Berrie, *Bryologist* **67**, 146–52, 1964). British material of *R. canaliculata* has two chromosome races with $n = 7 + m$ and $14 + 2m$ and possesses some attributes of *R. duplex* (see Paton, *J. Bryol.* **7**, 253–9), this and the observations of Berrie (loc. cit.) very much calling into question the specific status of *R. duplex*.

Subgenus *Riccia*

Thalli never spongy in appearance, margins ciliate or not. Dorsal epidermal cells often soon lost. Chlorophyllose tissue composed of vertical columns of cells with narrow air-spaces, forming a compact layer of green cells.

7. R. sorocarpa Bisch., *Nova Acta Acad. Leop.* 17: 1053. 1835   (Fig. 144)

Monoecious. Plants glaucous green, green or rarely violet beneath, in partial or complete rosettes, 0.8–2.0 cm diameter. Thalli 2–3 times dichotomously branched, deeply furrowed, lobes 0.5–2.0 mm wide, narrowed to apex, marginal cilia lacking. Thallus in T.S. 2–3 times as wide as thick, margins acute, epidermal cells hyaline, rounded or mamillate, thin-walled, soon vanishing, subepidermal cells also hyaline, thick-walled and persisting. Ventral scales hyaline, soon vanishing. Capsules very common, autumn to early summer. Spores dark brown, 76–90 μm, convex face with 7–8 areolae, 11–15 μm wide, across diameter, truncate tubercles at corners of areolae, triradiate face with irregular ridges but no areolae, wing entire but becoming eroded at maturity. $n = 8$. Fallow and stubble fields, banks, paths, roadsides, quarries and soil amongst rocks, frequent or common in lowland habitats, rarely ascending to *ca* 600 m, extending north to Ross and E. Sutherland. 104, H30, C. Europe north to Spitzbergen, Iceland, Turkey, Cyprus, Iran, Iraq, Lebanon, Israel, Jordan, Siberia, N. Africa, Macaronesia, N. America, Mexico, Greenland, Australia.

R. sorocarpa may be distinguished in the field from *R. glauca* and *R. beyrichiana* by the furrowed thallus and from all other British and Irish *Riccia* species by the hyaline, thick-walled subepidermal cells, these being thin-walled and chlorophyllose in other species. *R. subbifurca* differs in its smaller size, violet-coloured underside and, sometimes, the presence of marginal cilia.

8. R. glauca L., *Sp. Plant.*: 1139. 1753   (Fig. 144)

Monoecious. Plants green or glaucous green, green underneath, in partial or complete rosettes, 1–2 cm diameter. Thalli 2–3 times dichotomously branched, broadly and shallowly channelled towards lobe tips, elsewhere ± flat, lobes (0.5–)1.0–3.0 mm wide, not narrowed at apex, papillose marginal cilia, to *ca* 500 μm sometimes present, spreading when dry. Thallus in T.S. 4–6 times as wide as thick, sides very oblique, margin acute, hyaline epidermal cells soon vanishing. Ventral scales hyaline, soon vanishing. Capsules very common except in summer. Spores dark brown, 80–96 μm, convex face with 7–8 areolae, 10–20 μm wide, across diameter, tuberculate at corners of areolae, triradiate face ± similar, wing entire. $n = 8$. Fallow and stubble fields, pond margins, banks, paths, roadsides and quarries, occasional to common in lowland habitats except where basic, north to Inverness and E. Ross. 84, H23, C. Mediterranean Europe north to S. Fennoscandia and N. Russia, Iceland, Turkey, Lebanon, Japan, Morocco, Algeria, Macaronesia, N. America.

9. R. bifurca Hoffm., *Deutsch. Fl.* 2: 95. 1795   (Fig. 145)
R. arvensis Aust.

Monoecious. Young parts of plants glaucous green to bright green, brownish beneath, older parts pale brown to orange-brown, in partial rosettes 0.5–1.5 cm diameter or

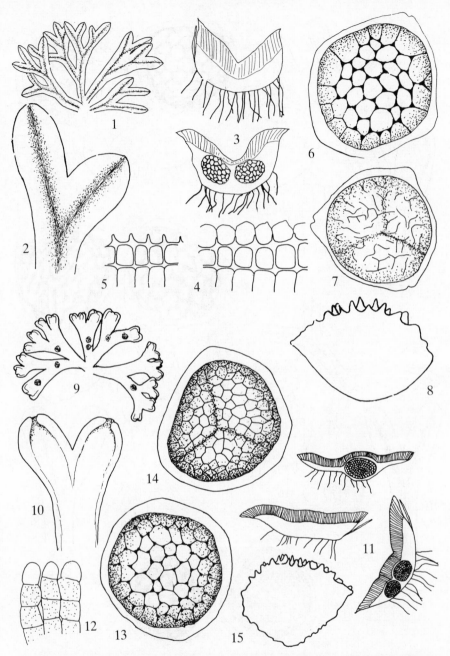

Fig. **144**. 1–8, *Riccia sorocarpa*: 1, partial rosette (× 2.5); 2, thallus lobes (× 7.5); 3, т.s. thalli with and without capsules (× 21); 4, т.s. surface cells near thallus tip showing thin-walled epidermal cells and thick-walled subepidermal cells (× 155); 5, т.s. older part of thallus lacking thin-walled epidermal cells (× 155); 6, spore, convex face (× 385); 7, spore, triradiate face (× 385); 8, spore, side view (× 385). 9–15, *R. glauca*: 9, partial rosette (× 3.3); 10, thallus lobes (× 3.3); 11, т.s. thalli, with and without capsules (× 21); 12, т.s. upper cells of thallus (× 155); 13, spore, convex face (× 385); 14, spore, triradiate face (× 385); 15, spore, side view (× 385).

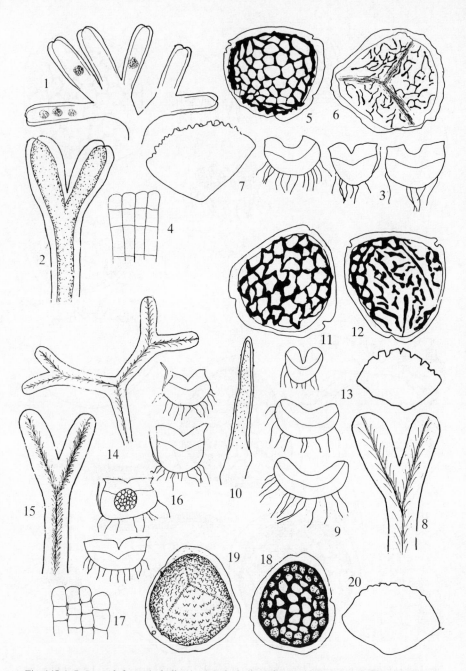

Fig. **145**. 1–7, *Riccia bifurca*: 1, thallus (× 6.6); 2, thallus lobes (× 13.2); 3, т.s. thalli (× 155); 4, т.s. surface of thallus; 5, spore, convex face (× 175); 6, spore, triradiate face (× 175); 7, spore, side view (× 175). 8–13, *R. subbifurca*: 8, portion of thallus (× 16); 9, т.s. thalli (× 38); 10, marginal cilium (× 84); 11, spore, convex face (× 385); 12, spore, triradiate face (× 385); 13, spore, side view (× 385). 14–20, *R. nigrella*: 14, thallus (× 6.6); 15, thallus lobes (× 13.2); 16, т.s. thalli with and without capsules (× 21); 17, т.s. surface cells of thallus (× 155); 18, spore, convex face (× 385); 19, spore, triradiate face (× 385); 20, spore, side view (× 385).

interwoven patches. Thalli irregularly dichotomously branched, furcations distant, deeply channelled at lobe tips, less markedly channelled in older parts, lobes ligulate, 0.4–1.0 mm wide, not narrowed towards apex, margin naked or very rarely with a few papillose cilia. Thallus in T.S. 1.0–1.5 times as wide as thick, margin rounded to acute, epidermal cells thin-walled, mostly 20–40 μm wide. Ventral scales hyaline, soon vanishing. Capsules frequent, winter, spring. Spores dark brown, 84–120 μm, convex face with 7–8 areolae, 10–16 μm wide, across diameter, rounded tubercles at corners of areolae, triradiate face with anastomosing ridges but few complete areolae, triradiate ridges thickened, wing entire to irregularly crenulate, margin thickened. *n* = 8. Damp peaty or gravelly soil on cliff tops and slopes, paths and heaths up to 5 km from coast, Lizard Peninsula, Cornwall. 1. Europe north to Fennoscandia and N. Russia, Iceland, Turkey, Iran, Algeria, Macaronesia, N. America, Greenland, Australia, New Zealand.

In Britain *R. bifurca* has been confused with other *Riccia* species and the only confirmed records are from the Lizard Peninsula, Cornwall (see Paton, *J. Bryol.* **11**, 1–6, 1979). It is readily distinguished by the light brown or orange-brown colour of older parts of the thallus, the epidermal cells usually being 20–40 μm wide (mostly more than 40 μm in other species) and the wing of the spores having a thickened margin.

Spore size in the two British specimens I examined was 84–102 μm and 96–120 μm and Paton gives a size range of 60–110 μm. As with other species spore size is very variable and cannot be used for discriminating species of section *Riccia* as has been done by some authorities (e.g. Macvicar, 1926; Arnell, 1956).

10. R. subbifurca Warnst. ex Crozals, *Rev. Bryol.* 30: 62. 1903   (Fig. 145)
R. commutata Jack ex Levier, *R. warnstorfii* auct. angl.

Monoecious. Plants greenish, violet beneath, in partial or complete rosettes 0.4–1.5 cm diameter. Thalli 2–3 times dichotomously branched, grooved towards lobe tips, grooved to shallowly channelled in older parts, lobes ligulate, 0.4–0.7(–1.0) mm wide, not narrowed towards apex, margin naked or with sparse papillose cilia to 500 μm long on younger parts, cilia ± spreading when dry. Thallus in T.S. 1.5–3.0 times as wide as thick near apex, thinner in older parts, margin rounded or obtuse, epidermal cells thin-walled, soon lost. Ventral scales colourless to violet, soon vanishing. Capsules very common, winter to autumn. Spores dark brown, 72–88(–100) μm, convex face with 7–8(–10) areolae, 8–12 μm wide, across diameter, truncate tubercles at corners of areolae, triradiate face with irregular ridges sometimes anastomosing to form areolae near wing, triradiate ridges thickened, wing entire. Damp soil in arable fields, paths, banks, pond margins and in turf, at low altitudes, occasional in Britain north to the Inner Hebrides and E. Ross, rare in Ireland. 41, H17, C. Sparsely scattered in Europe from Portugal, Spain, Italy and Yugoslavia north to S. Sweden, Algeria, Morocco, S. Africa, Azores, Tenerife.

The smallest *Riccia* species in the British Isles but may be confused with forms of *R. crozalsii* with few or no marginal cilia. Where cilia are present those of *R. crozalsii* become curved over the thallus in the dry state whilst those of *R. subbifurca* remain spreading. In *R. crozalsii* there are areolae on both faces of the spores but in *R. subbifurca* the ridges on the triradiate face only anastomose to form areolae near the wing and the triradiate lines are more heavily thickened and conspicuous. For the identity of the British plant see Paton, *J. Bryol.* 16, 1990.

11. R. nigrella DC., *Fl. Franc.* ed. 3, 6: 194. 1815   (Fig. 145)

Monoecious. Plants dark green, purplish-black and glossy beneath, in rosettes 0.5–1.0 cm diameter. Thalli 2–3 times dichotomously branched, deeply furrowed, lobes 0.5–1.0 mm wide, not narrowed to apex, marginal cilia lacking. Thallus in T.S. 1–2 times as wide as thick, sides steep, sometimes ± vertical, margins acute and sightly recurved, epidermal cells rounded, thin-walled, sometimes persisting. Ventral scales large,

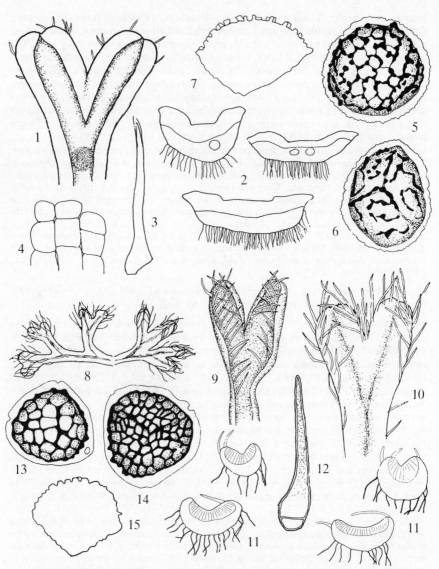

Fig. **146**. 1–7, *Riccia beyrichiana*: 1, thallus (× 6.6); 2, т.s. thalli with and without capsules (× 15); 3, marginal cilium (× 84); 4, т.s. surface cells of thallus (× 155); 5, spore, convex face (× 275); 6, spore, triradiate face (× 275); 7, spore, side view (× 275). 8–15, *R. crozalsii*: 8, thallus (× 5); 9 and 10, dry and moist thallus lobes (× 17.5); 11, т.s. thalli (× 21); 12, marginal cilium (× 155); 13, spore, convex face (275); 14, spore, triradiate face (× 275); 15, spore, side view (× 275).

semilunar, purplish-black, persisting. Capsules common, winter to spring. Spores dark brown, 58–82 μm, convex face with 7–8 areolae, 5–10 μm wide, across diameter, low truncate tubercles at corners of areolae, triradiate face with low irregular tangential ridges, triradiate ridges obvious but not thickened, wing entire. $n = 8$. On exposed soil on paths and slopes, rarely in fields, usually near sea, very rare, W. Cornwall, Worcester (1912), Radnor, Merioneth, Channel Is. 4, C. Spain, Portugal, Mediterranean Europe, W. France (Normandy), Switzerland, Turkey, Israel, Libya, Tunisia, Algeria, Morocco, Macaronesia, N. America, Australia.

Readily recognised by the dark green, deeply furrowed thallus with purplish-black margins and ventral surface and the steep sides.

12. R. beyrichiana Hampe ex Lehm., *Novat. Min. Cogn. Stirp. Pugillus* 7: 1. 1838
(Fig. 146)

Monoecious. Plants glaucous green, of similar colour underneath or more usually with at least margins purplish, in partial rosettes, interwoven mats or scattered. Thalli to 1.5 cm long, 2–3 times dichotomously branched, lobes shallowly and broadly channelled, (0.7–)1.0–2.5(–3.0) mm wide, not narrowed to apex, smooth marginal cilia to 520 μm long often present, spreading when dry. Thallus in т.s. 2.5–4.0 times as wide as thick, margin obtuse or rounded, often slightly reflexed, epidermal cells very thin-walled, soon vanishing. Ventral scales small, colourless or purple. Capsules common, winter to summer. Spores brownish, 104–120(–128) μm, convex face with 6–7 areolae, 12–26 μm wide, across diameter, acute to truncate tubercles at corners of areolae, triradiate face with irregular, sometimes anastomosing ridges but no areolae, triradiate ridges thickened, wing papillose-crenulate. Damp soil on paths, on heaths, in turf, amongst rocks and on pool margins, ascending to *ca* 600 m, occasional in W. and N. Britain, very rare elsewhere, W. Suffolk, rare in Ireland. 35, H11, C. Sparsely distributed through Europe from Portugal, Spain, Italy and Yugoslavia north to Fennoscandia and N. Russia, Faroes, Iceland, Tunisia, Algeria, Azores, N. America.

13. R. crozalsii Levier, *Rev. Bryol.* 29: 73. 1902   (Fig. 146)

Monoecious. Plants green or glaucous green, green to purplish beneath, usually forming partial rosettes. Thalli 2–3 dichotomously branched, to *ca* 5 mm long, furrowed, lobes 0.5–1.0 mm wide, not narrowed to apex, finely papillose, marginal cilia present on younger and sometimes on older thalli, to 560 μm long, erect to spreading when moist, inflexed over thallus in older parts, margin rounded, epidermal cells thin-walled, pyriform. Ventral scales colourless to purplish, soon vanishing. Capsules very common, winter to early summer. Spores dark brown, 80–105 μm, convex face with 6–8 areolae, 12–16(–20) μm wide, across diameter, truncate tubercles at corners of areolae, triradiate face with smaller areolae, triradiate ridges not obvious, wing entire. $n = 7 + m^*$, 8. Banks, paths, arable fields, mud-capped walls and soil amongst rocks, usually near the sea, frequent in southern Cornwall, very rare elsewhere, S. Devon, E. Sussex (1910), Monmouth (1877), Merioneth, Caernarfon, Anglesey, Channel Is. 8, C. Mediterranean Europe, Spain, Portugal, W. France (Normandy), Switzerland, Cyprus, Lebanon, Israel, Tunisia, Algeria, Morocco, Macaronesia, Australia, New Zealand.

For differences from *R. subbifurca* see under that species. Both *R. michelii* Raddi and *R. ciliata* Hoffm. have been reported from Britain. Records of the former appear to have been based upon misidentifications and the only specimen of the latter is probably *R. canescens* Steph. and is probably not a British specimen (see Paton, *J. Bryol.* **11**, 1–6, 1980).

CLASS II. ANTHOCEROTAE

(ANTHOCEROTOPSIDA)

Plants thalloid, forming variously lobed, irregular rosettes. Thallus with or without poorly defined midrib, apical slime papillae lacking, in section 5–30 cells thick, cells of ± uniform size, each usually with a single large chloroplast containing a pyrenoid and surrounding the nucleus, oil-bodies lacking; thallus often with mucilage-filled cavities (lacunae) which may open on the ventral and occasionally the dorsal surface by pores (not stomata), often containing *Nostoc* colonies. Ventral scales lacking. Rhizoids smooth. Antheridia endogenous in origin, few to numerous, broadly ellipsoid, stalked, maturing in succession in antheridial cavities in dorsal surface of thallus. Archegonia borne singly on dorsal surface, pteridophyte-like, consisting only of neck canal cells, ventral canal cell and egg cell embedded in thallus. Sporophyte cylindrical, short, horizontal and immersed in involucre or erect and longly projecting; intercalary meristem present between foot and capsule, seta lacking; capsule epidermis with or without stomata, surrounding layer of chlorophyllose tissue with cells containing two chloroplasts; sporogenous tissue surrounding and overarching well or poorly defined central columella; spore sac at maturity containing spores and geniculate or spherical 1–several-celled pseudoelaters, mature capsules persisting for several weeks, growing from base by meristematic activity and dehiscing by two slits from apex. A small class, commonly referred to as 'hornworts', of 2 families and 6 extant genera and about 330 species. The affinities of the class are totally unknown, either in relation to other bryophytes or to vascular plants.

6. ANTHOCEROTALES

With the characters of the class.

41. ANTHOCEROTACEAE

Capsule erect, longly protruding from perichaetium. Epidermal cells thin-walled, stomata present; columella distinct; pseudoelaters geniculate. Five extant genera.

Thallus usually pale green with crisped margins, in section with cavities, spores blackish **84. Anthoceros**

Thallus usually dark green, margins rarely crisped, lacking cavities, spores yellowish **85. Phaeoceros**

84. ANTHOCEROS L., *Sp. Plant.*: 1139. 1753

Plants terrestrial, thalloid, variously lobed, lobes ± divided with crisped margins, in section with mucilage-filled or *Nostoc*-containing cavities. Antheridial wall of four large regularly arranged rows of cells. Spores dark brown or blackish. About 200 species.

Mature antheridia 100–130(–150) μm long (excluding stalk)   **1. A. punctatus**
Mature antheridia 50–90 μm long   **2. A. agrestis**

1. A. punctatus L., *Sp. Plant.*: 1130. 1753   (Fig. 147)
A. husnotii Steph.

Monoecious. Annual or perennial plants, pale to yellowish-green, forming deeply concave rosettes, 1.5–3.0 cm diameter, deeply divided into crisped lobes, surface of thallus usually without lamellae. Thallus in section 10–20(–30) cells thick, with numerous cavities sometimes containing *Nostoc* colonies visible from dorsal side as dark spots. Antheridial cavities with 6–20 antheridia 100–130(–150) μm long at maturity. Involucres rarely in pairs, cylindrical, slightly narrowed to truncate apex, 2–5 mm long. Capsules 2–10 cm long. Spores 36–48 μm, blackish, bluntly spinose, spines sometimes coalescing at base. Pseudoelaters 3–5 cells long. Capsules common, throughout year. $n = 4 + m^*$, 5, 6*, 8 + m. Damp soil on cultivated land, by paths, quarries, cliffs, banks of ditches and streams, wood rides, occasional in S. and W. Britain, very rare elsewhere, rare in Ireland, 48, H13, C. W. and S. Europe, Mediterranean islands and N. Africa, Macaronesia, Asia, N. America, Caribbean, Andes.

    Mistaken in the past for *A. agrestis*, large plants of which overlap in size small plants of *A. punctatus*. The only reliable distinguishing feature is the size of the antheridia. The antheridial cavities are visible under a lens as small tubercles on the dorsal surface of the thallus and the antheridia may be obtained by sectioning or dissection, preferably of fresh material. Some authorities consider that *A. agrestis* should be regarded merely as a variety of *A. punctatus*. This argument is supported by the occurrence of a variety, var. *crispatus* auct. known from Sardinia and Gran Canaria with antheridia 90–110 μm long but the difference in geographical distributions of *A. punctatus* and *A. agrestis* would justify the recognition of them as subspecies. For a detailed account of the differences between the two species see Paton, *J. Bryol.* **10**, 257–61, 1979.

2. A. agrestis Paton, *J. Bryol.* 10, 257. 1979   (Fig. 147)
A. crispulus auct., *A. punctatus* auct., *A. punctatus* var. *cavernosus* auct.

Similar to *A. punctatus*. Plants pale to yellowish-green, forming concave rosettes, 0.5–1.5 cm diameter, shallowly divided into linear lobes, margins undulate to crisped, surface usually with low lamellae radiating outwards. Thallus in section 5–12(–18) cells thick, with numerous cavities sometimes containing *Nostoc* colonies visible from dorsal side as dark spots. Antheridial cavities with 2–14 antheridia 50–90 μm long at maturity. Involucres often in pairs, cylindrical, slightly narrowed to truncate apex, 1–3 mm long. Capsules 1–2(–3) cm long. Spores blackish, spinose, 38–62 μm. Capsules common, July–December. $n = 4 + m^*$. Damp soil on cultivated land, by paths, rare or occasional, in scattered localities from Cornwall east to Kent and north to Argyll and Angus, very rare in Ireland, W. Donegal and Tyrone. 54, H2. N., E. and C. Europe, Faroes, Morocco, Madeira, Canaries, N. America.

85. PHAEOCEROS PROSK., *Bull. Torrey Bot. Club* 78: 346. 1951

Plants terrestrial, thalloid, lobed, lobes simple or dichotomously branched, margins ± plane to crisped, in section lacking mucilage cavities. Antheridial wall of numerous, small, irregularly arranged cells. Spores yellowish.

1. P. laevis (L.) Prosk., *Bull. Torrey Bot. Club* 78: 347. 1951   (Fig. 147)
Anthoceros laevis L.

Annual or perennial plants, dark green, forming ± plane rosettes 0.5–2.0 cm diameter, divided into broad lobes, margins very rarely crisped, surface of thallus without lamellae, tuber-like thickenings often produced on ventral side of thallus; thallus in section 6–10 cells thick. Antheridial cavities with 2–5 antheridia. Involucres sometimes

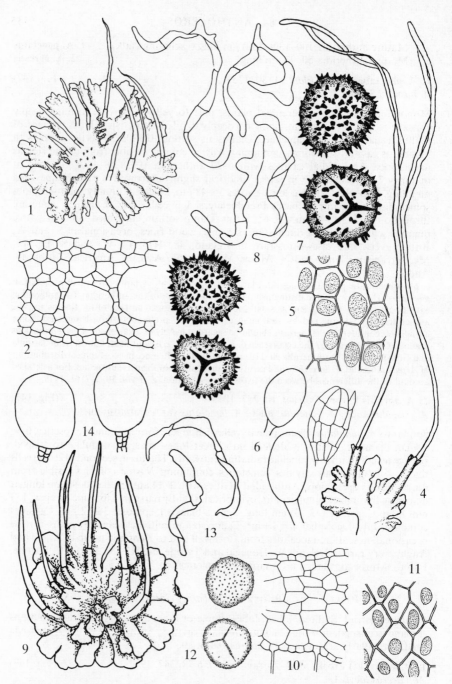

Fig. **147**. 1–3, *Anthoceros agrestis*: 1, thallus with young sporophytes (× 3.5); 2, т.s. thallus (× 275); 3, spores (× 450). 4–8, *A. punctatus*: 4, portion of thallus with sporophytes (× 3.5); 5, dorsal epidermis of thallus, chloroplasts stippled (× 275); 6, antheridia (× 155); 7, spores (× 450); 8, pseudoelaters (× 150). 9–13, *Phaeoceros laevis*: 9, thallus with young sporophytes (× 3.5); 10, т.s. thallus (× 275); 11, dorsal epidermis of thallus, chloroplasts stippled (× 275); 12, spores (× 450); 13, pseudoelaters (× 150); 14, antheridia (× 155).

in pairs, cylindrical, slightly narrowed to truncate apex, 2–4 mm long. Capsules 1.5–4.0 cm long. Spores yellowish, papillose, 40–56 μm.

Ssp. laevis

Dioecious. Capsules common, January–December. $n = 4 + m^*$. Damp soil on cultivated land, by paths, ditch banks, wood rides, occasional in S. and W. England, Wales and Scotland north to Argyll and Angus, very rare elsewhere, occasional in Ireland. 66, H27, C. S. and W. Europe, Turkey, S. and E. Asia, Azores, Madeira, Tenerife, N. and S. Africa, N. and S. America, Caribbean, Australasia.

Ssp. carolinianus (Michx.) Prosk., *Rapp. Comm. VIII Congr. Int. Bot. Paris* 14–16: 60. 1954

P. carolinianus (Michx.) Prosk.

Monoecious. Capsules common, autumn. $n = 5 + 0–5m$. Very rare but probably overlooked, E. Cornwall. W. Sussex, Surrey, S. Essex, Monmouth. 5. C., N. and E. Europe, Iceland, E., S. and W. Asia, Madeira, Canaries, Cape Verde Is, C. Africa, N. America.

The only constant difference between the two subspecies is sex and the only grounds for retaining the two taxa at that rank is the difference, at least in Europe in their geographical distribution. Schuster (1953) regards spp. *carolinianus* merely as a form of more moist conditions. On the other hand, Hassel de Menendez, *J. Hattori bot. Lab.* **62**, 281–8, 1987, reports differences in spore morphology in the type specimens of the two subspecies detectable by scanning electron microscopy. However, in the absence of data from at least several gatherings of each I cannot accept the suggestion that they are distinct species as nothing is known of the degree of variation in spore morphology in *Phaeoceros*. For an account of *P. laevis* in Britain see Paton, *J. Bryol.* **7**, 541–3, 1973. The record of *P. bulbiculosus* (Brotero) Prosk. (as *Anthoceros dichotomus* Raddi) from Devon is based on an erroneous identification.

ADDENDUM

5. Marsupella profunda Lindb., *Rev. Bryol.* 14: 19. 1887

This species, which is closely related to *M. sprucei*, has recently been found to occur in Cornwall. The main characters of *M. profunda*, with those of *M. sprucei* in square brackets, are:

Leafy shoots 0.4–1.0 mm wide [0.3–0.6(–0.8) mm wide], inflorescences 0.6–1.6(–1.8) mm wide [(0.3–)0.5–1.0(–1.2) mm wide]. Rhizoids sometimes purplish-red, especially distally [sometimes reddish-brown but not purplish-red]. Leaves bilobed mostly 1/3–1/2 [mostly 1/5–1/3], lobes mostly obtuse [mostly acute]. Female bracts with sinus sometimes rounded [usually acute], lobes mostly obtuse or subobtuse [acute to obtuse]. Moist acidic crumbling granite or clay, sometimes with *M. sprucei*, in disused china clay works, very rare, W. Cornwall. 1. Portugal.

I am grateful to Mrs J.A. Paton for permission to utilise the manuscript of a paper to be published in *Journal of Bryology* **16** in 1990.

GLOSSARY

acuminate   Longly tapering narrow point (Fig. **150**, 13).

acute   Pointed, with angle at tip less than 90° (Fig. **150**, 14).

adnate   Joined, with reference to structures of different origin.

air-chamber   Cavities in thallus of Marchantiales, usually opening to the surface by pores (Fig. **137**, 15).

alternate   Of leaves, arising at different levels on opposite sides of the stem (e.g. Fig. **70**, 1–3).

amphigastria   Underleaves.

anisophyllous   Of Jungermanniales, with stem with two rows of lateral leaves and with the third row (underleaves) smaller or lacking, rendering stem bilaterally symmetrical and dorsiventral.

antheridium   Male reproductive structure of liverworts, containing sperm, spherical or ellipsoid and with a short or long stalk (Fig. **148**, 12, 13).

antical   Upper or dorsal side or margin relative to substrate.

apiculus   Abrupt projection at apex (*adj.* **apiculate**) (Fig. **150**, 19).

appendage   Of ventral scales, small piece of unistratose tissue appended to scale (Fig. **137**, 17).

appressed   Of leaves, pressed against stem (Fig. **68**, 5).

approximate   Of leaves, close but not overlapping.

archegonium   Female reproductive structure of liverworts, usually flask-shaped, consisting of neck of 4–5 tiers of cells and an egg cell surrounded by a venter (Fig. **148**, 11).

areolae   Small angular-shaped areas, usually bounded by lamellae, on surface of a spore (*adj.***areolate**) (Fig. **136**, 5, 6).

areolation   The network of the cells of a leaf.

ascending   Growing upwards at an angle to the horizontal.

autoecious   Antheridia and archegonia on separate branches on same plant (Fig. **148**, 2, 3, 7).

autoicous   See *autoecious*.

axil   Angle between leaf base and stem (*adj.* **axillary**).

beak   Of perianth, projection formed by strong constriction below mouth (Fig. **50**, 2).

bifid   Deeply divided almost to base into two lobes.

bilobed   Of leaves, divided into two lobes (Fig. **150**, 10–12).

bipinnate   Twice pinnately branched.

bract   Leaf-like but often modified structure subtending antheridia (**male bract**, Fig. **148**, 7) or archegonia or perianth (**female bract**, Fig. **31**, 5, 6; **148**, 5, 6, 8, 9).

bracteole   Usually modified underleaf associated with a male or female inflorescence and sometimes fused to one or both associated bracts (Fig. **31**, 16).

caducous   of propaguliferous branches or leaves, readily falling off and leaving lengths of bare stem (Fig. **96**, 6).

calcicole   Plant growing on a substrate rich in calcium.

calcifuge   Plant intolerant of substrata rich in calcium.

calyptra Structure developed from venter of archegonium and surrounding developing capsule until maturity (Fig. **2**, 1; **148**, 8, 10).

canaliculate Of leaves, sides folded up so that leaf is channelled.

canaliculate-concave Intermediate between canaliculate and concave.

capsule Spore-containing part of sporophyte (Fig. 8; **148**, 7, 8, 10).

chlorophyllose Tissue composed of cells containing chloroplasts.

cilia Hair-like uniseriate structures (*adj*. **ciliate**.) (Fig. **107**, 9).

circinate Of shoot tips, strongly curved (like a young fern frond).

clavate Club-shaped (Fig. **68**, 13).

compressed Of perianths, flattened in one plane, usually dorsi-ventrally (Fig. **75**, 1).

cleistocarpous Of capsules, not dehiscing, spores being released by decay of capsule wall (e.g. *Riccia*, *Sphaerocarpos*).

collateral Side by side.

collenchymatous Of cells having thickened corners, i.e. with trigones.

columella Central mass or column of sterile tissue surrounded by the spore sac in the capsule of Anthocerotae and mosses.

commissure Line of junction of ventral and dorsal lobes of leaf, i.e. the keel region.

complanate Of shoots or perianths, flattened in one plane (Fig. **70**, 1–3; **104**, 1).

complicate Folded.

compound oil-bodies See *oil-bodies*.

compound pore See *pore*.

concave Saucer-shaped when viewed from dorsal side.

conduplicate Of bilobed leaves with the two lobes folded together face to face (Fig. **1**, 3, 4; **149**, 1*b*, *c*, 5, 6).

conduplicate-concave Of leaves, intermediate between conduplicate and concave.

connate Joined with reference to organs of similar origin, e.g. bracts and bracteole (Fig. **31**, 16).

connivent With tips approaching each other (Fig. **21**, 4; **150**, 11).

constricted Abruptly narrowed (e.g. of perianth abruptly narrowed to beaked mouth).

contiguous Of leaves, touching but not overlapping.

convex Inverted saucer-shaped when viewed from dorsal side.

cordate Heart-shaped, usually with reference to leaf base (Fig. **150**, 7).

cortex Outermost layer of stem cells in contrast to inner or medullary cells.

crenulate With small rounded projections formed by projecting cell walls.

crisped Strongly curved and undulate.

cruciate Cross-like (Fig. **137**, 12).

cucullate Of leaves, with margin at apex incurved to form hood-like tip (Fig. **54**, 5).

cuneate Wedge-shaped (Fig. **87**, 3).

cushion ± hemispherical colony formed by shoots radiating from central point.

cuticle Layer on outer cell surface.

cylindrical Cylinder-shaped.

decumbent Prostrate with ascending tips.

decurrent Of leaf base, extending down on one or both sides of stem (Fig. **24**, 1; **150**, 11, 12).

dehisce Opening to releasing spores.

deltoid Shaped like the Greek letter *Δ*.

dentate Toothed, teeth usually composed of one or more cells (Fig. **85**, 6).

denticulate With small teeth usually composed of only part of a cell projecting from margin (Fig. **83**, 7).

GLOSSARY

denuded Eroded away.

depauperate Small plants, poorly developed.

dichotomous Regularly forking with branches of ± equal size (Fig. **145**, 14).

dioecious Antheridia and archegonia on separate plants (Fig. **148**, 1).

dioicous See *dioecious*.

diploid Having two sets of chromosomes (i.e. 2*n*).

discoid Of gemmae, disc-like and thickened in middle.

distal End or portion furthest from point of origin.

distant Of leaves, far apart.

distichous Of leaves, in two ranks, one on either side of stem.

divergent Of leaf lobes, spreading away from each other.

dorsal Side of shoot, lobe of leaf or leaf margin on side away from substrate.

dorsiventral Flattened, growing in two planes as in thallus.

elaters Elongated cells with spiral thickening derived by mitotic division from sporogenous tissue and mixed with spores in capsule (Figs. **125**, **126**).

elaterophore Tuft of elaters attached to inside base or apex of capsule.

ellipsoid Three-dimensional equivalent of elliptical.

elliptical With convex sides, widest at middle (Fig. **150**, 3).

emarginate With small indentation at leaf apex (fig. **150**, 21).

endogenous Of gemmae and branches, arising internally.

entire Of leaves, with margins smooth, not toothed or crenulate (q.v. **simple**).

epidermis Superficial layer of cells of thallus (*adj.* **epidermal**).

epiphyll Plant growing on the leaf of another plant (*adj.* **epiphyllous**).

epiphyte Plant growing on another plant, usually in the bark of a tree or shrub but more loosely a bryophyte growing on another bryophyte (*adj.* **epiphytic**).

erect Of shoots, growing ± at right angles to substrate; of leaves, lying ± parallel to stem.

erecto-patent Of leaves, between erect and patent, at an angle of about 20–45° to stem.

eroded Irregularly worn away, e.g. of leaves by production of marginal gemmae.

evolute Turned inside out.

exogenous Arising on surface or from superficial layer of cells.

exserted Of perianth, projecting beyond bracts (Fig. **148**, 7).

falcate Curved like a sickle (Fig. **4**, 2).

fasciculate In tight bunches.

female Of bracts, uppermost two or more bracts surrounding archegonia or base of perianth (Fig. **148**, 5, 6, 9).

filiform Fine or thread-like.

fimbriate Fringed with long teeth or cilia.

flagelliform Of shoots, thread-like with minute leaves (Fig. **27**, 5); such shoots sometimes referred to as **flagella**.

flexuose Wavy.

foot Basal part of sporophyte embedded in gametophyte tissue.

fruit Lay term for capsule.

fugacious Readily or quickly deciduous.

furcate Dichotomous.

fusiform Tapering at either end, spindle-shaped (Fig. **136**, 3).

gametangia Gamete producing structures, antheridia (male), archegonia (female).

gametophyte The conspicuous leafy or thalloid haploid phase of the liverwort life-cycle, i.e. the 'plant', and which bears the gametangia.

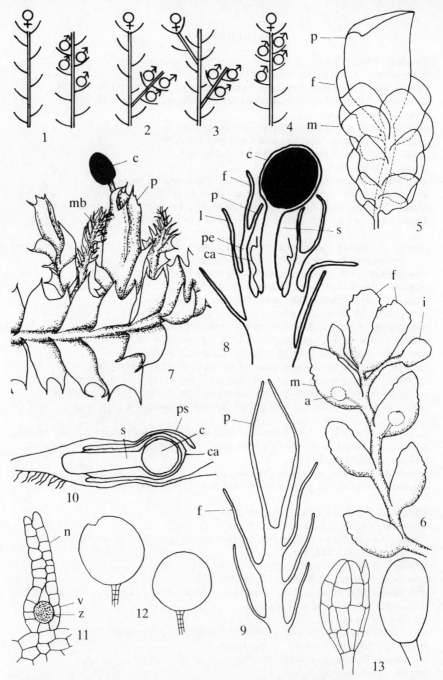

Fig. 148. 1–4, distribution of gametangia in liverworts: 1, dioecious; 2 and 3, autoecious; 4, paroecious. 5, Paroecious inflorescence of *Radula complanata* with perianth and bracts (×44). 6, Paroecious inflorescence of *Aphanolejeunea microscopica* with male and female bracts, antheridia and young subinvolucral innovations (×180). 7, Autoecious inflorescence of *Lophocolea bidentala* with exserted perianths, capsule and male branches (×24). 8, L.S. female inflorescence of *Nardia scalaris* showing perigynium and immersed perianth (×45). 9, L.S. female inflorescence of *Jungermannia pumila* (without perigynium) (×30). 10, L.S. mature sporophyte of *Pellia epiphylla* (×30). 11, Archegonium of *Pellia epiphylla* with fertilised egg cell (zygote) (×760). 12, Antheridia of *Fossombronia angulosa* (×125). 13, mature and dehisced antheridia of *Anthoceros punctatus* (×300). *a* = antheridium; *c* = capsule; *ca* = calyptra; *f* = female bract; *i* = innovation; *m* = male bract; *mb* = male branch; *n* = neck; *p* = perianth; *pe* = perigynium; *ps* = pseudoperianth; *s* = seta; *v* = venter; *z* = zygote.

gemma(e) Vegetative propagule(s) borne on leaves, stems or thalli of many liverwort species.

gemma-cup Receptacle containing gemmae (Fig. **137**, 10).

gemmiferous Of leaves or stems bearing gemmae.

geniculate With a knee-like bend.

genotype The genetic or hereditary constitution of an organism (*adj.* **genotypic**).

gibbous With a swelling on one side; of a sinus, with a pouting lip.

glaucous With whitish, greyish or bluish bloom.

globose Spherical.

gregarious Plants growing close together but not in mats or tufts.

habit Aspect or appearance of a plant.

haploid Having a single set of chromosomes (i.e. *n*).

heterochromatin Substance in the nucleus which shows maximum staining at interphase (*adj.* **heterochromatic**) (Fig. **128**, 7, 8).

heteroecious Of variable sexuality, e.g. sometimes monoecious, sometimes dioecious, or autoecious and paroecious.

hyaline Transparent and colourless.

hyaline papilla See *papilla*.

imbricate Appressed and overlapping like roof tiles (Fig. **149**, 4*a*).

incised With a narrow sinus or slit.

incrassate Of cells, with thick walls (Fig. **150**, 26).

incubous With dorsal leaf margin nearer to stem apex than the ventral margin so that forward margin of one leaf overlies rear margin of leaf in front (Fig. **149**, 1*a, b*, 2).

incurved Curved upwards and inwards (Fig. **17**, 7).

inflated Swollen.

inflexed Turned upwards and inwards.

innovation New shoot, usually from below perianth (Fig. **148**, 6).

innovating Producing new shoots (**innovations**), usually from beneath perianth (Fig. **148**, 6).

insertion Line of attachment of leaf to stem; **transverse insertion**, at right angles to axis (Fig. **149**, 1*c*); **oblique insertion**, diagonal to axis (Fig. **149**, 1*a, e*); **longitudinal insertion**, parallel to axis.

intercalary Between base and apex of stem, usually of male inflorescences, originally terminal, but becoming so positioned by further growth of the stem apex (Fig. **32**, 11).

involucre Strictly, a sheath growing from a thallus and protecting a single antheridium, archegonium or sporophyte as in the Marchantiales or Sphaerocarpales. Frequently also loosely applied to the pseudoperianth of the Metzgeriales and the female bracts and bracteoles of the Jungermanniales (*adj.* **involucral**).

isophyllous Of shoots of Jungermanniales or Calobryales, having three ranks of leaves of ± similar size, rendering stem radially symmetrical.

julaceous Cylindrical and worm-like in appearance because of imbricate arrangement of leaves (Fig. **68**, 1).

keel Commissure or line of fold separating dorsal and ventral conduplicate lobes of a leaf (*adj.* **keeled**).

laciniate Deeply and irregularly divided.

lamella(e) Low ridge(s) (*adj.* **lamellate**).

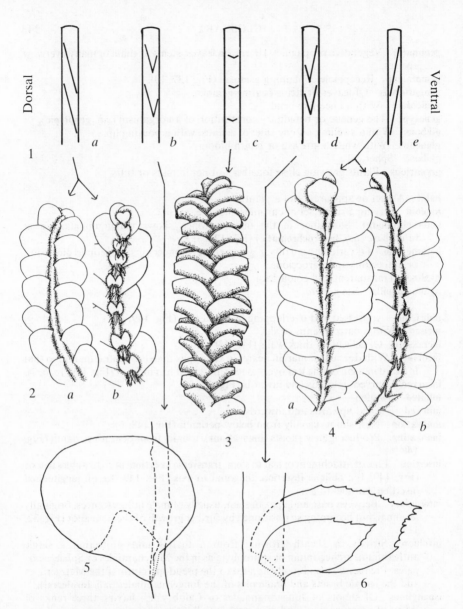

Fig. **149**. Leaf insertions and derivative shoot and leaf morphologies. 1, Diagram of side view of stem, dorsal and ventral sides indicated; *a* oblique insertion giving rise to incubous leaves; *b* J-shaped insertion giving rise to incubous conduplicate leaves; *c* transverse insertion; *d* J-shaped insertion giving rise to succubous conduplicate leaves; *e* oblique insertion giving rise to succubous leaves. 2*a*, *b*, *Calypogeia suecica*, dorsal and ventral sides, leaves incubous (× 30). 3, *Marsupella emarginata*, leaves transversely inserted (× 30). 4 *a*, *b*, *Chiloscyphus polyanthos*, dorsal and ventral sides, leaves succubous (× 16). 5, Conduplicate leaf of *Radula complanata*, dorsal side, (× 46). 6, Conduplicate leaf of *Scapania umbrosa*, dorsal side (× 76).

lamina   Blade of leaf.

lanceolate   Lance-shaped, about three times as long as wide (Fig. **150**, 5).

leaf   Unistratose structure borne on stem functionally analogous to the leaf of vascular plants.

ligulate   Strap-shaped, narrow and parallel-sided (Fig. **137**, 17).

linear   Long, narrow and parallel-sided (Fig. **143**, 2, 3).

lingulate   Tongue-shaped, broad and ± parallel-sided (Fig. **150**, 6).

lobule   Term applied to small ventral lobe of leaf in Lejeuneaceae and Radulaceae.

lumen   Cell cavity.

lunate   Shaped like a cresent moon.

male   Of bracts, in male inflorescence subtending antheridia (Fig. **148**, 5, 6).

mamilla   Hollow projection from cell surface (*adj*. **mamillose**).

marsupium   A down-growth of stem apex into the substrate, carrying with it and forming a protective structure round the developing sporophyte (Fig. **16**, 4; **93**, 4).

medulla   Central cells of stem (*adj*. **medullary**).

meristem   Region of growth by active cell division.

microphyllous   With minute leaves.

middle lamella   Layer of pectic substances between cellulose walls of adjacent cells.

mid-leaf   Position approximately halfway up leaf and midway between margins.

midrib   Thickened central part of thallus.

monoecious   With both antheridia and archegonia on same plant, i.e. autoecious, paroecious or synoecious, or more strictly, autoecious only.

monogeneric   With only one genus.

monoicous   See *monoecious*.

monotypic   With only one species.

mucilage hair   Hair producing mucilage, usually at or near thallus apex as in *Metzgeria* (Fig. **134**, 8).

mucro   Short projection from leaf apex (adj. **mucronate**).

mycorrhizal   Associated with a fungus.

n   Functionally haploid chromosome number, as opposed to x, the basic chromosome number; numbers such as $n = 8 + m$, $9 + m$ denote the presence of 8 or 9 chromosomes plus a microchromosome (a small chromosome half or less the size of the next smallest member of the complement). Numbers with an asterisk (e.g. $n = 9*$) are based upon British or Irish material.

ob-   Of a shape, inverted (e.g. obcordate, obovate, obpyriform).

oblong   Shortly rectangular, but with reference to leaves, with rounded corners.

obscure   Indistinct, difficult to see.

obtuse   Bluntly pointed, angle at apex more than 90°; more loosely, blunt (Fig. **150**, 16).

ocelli   Enlarged and/or discoloured leaf cells, usually in groups or rows (Fig. **111**, 10).

oil-body   Translucent or opaque oil-containing structure; **simple oil-bodies**, each composed of a single oil-containing structure (Fig. **63**, 4); **compound oil-bodies**, each consisting of two or more oil-containing structures (Fig. **57**, 12).

opaque   Not transparent or translucent.

opposite   Of two leaves arising at the same level on opposite sides of the stem (Fig. **94**, 1).

orbicular   More or less circular in outline (Fig. **150**, 1).

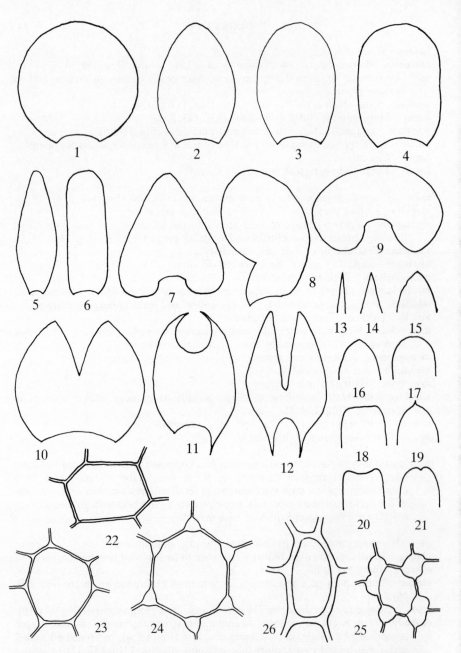

Fig. **150**. 1–12, leaf shapes: 1, orbicular; 2, ovate; 3, elliptical; 4, oblong; 5, lanceolate; 6, lingulate; 7, cordate; 8 and 9, reniform; 10, orbicular, bilobed to $\frac{1}{2}$, sinus acute; 11, ovate, decurrent on one side, bilobed to $\frac{1}{3}$, lobes connivent, sinus rounded; 12, ovate, decurrent on both sides, bilobed to $\frac{2}{3}$, sinus obtuse. 13–21, apex shapes: 13, acuminate; 14, acute; 15, subacute; 16, obtuse; 17, rounded; 18, truncate; 19, obtuse and apiculate; 20, retuse; 21, emarginate. 22–26, cell types: 22, thin-walled and without trigones (*Chiloscyphus polyanthos*); 23, thin-walled with small trigones (*Lophocolea bidentata*); 24, thin-walled with large trigones (*Nardia geoscyphus*); 25, thin-walled with very large bulging trigones (*Pleurozia purpurea*); 26, thick-walled (incrassate) without trigones (*Anthelia julacea*). All cells × 630.

ovate Egg-shaped in outline and widest below middle, about twice as long as wide (Fig. **150**, 2).

ovoid Egg-shaped, widest below middle, about twice as long as wide.

palmate Lobed in a hand-like fashion.

papilla Solid projection from cell surface (*adj.* **papillose**). **Hyaline papilla**, very thin-walled ephemeral cell situated near base of apical tooth of ventral leaf lobe in the Lejeuneaceae, the exact position of which in relation to the tooth is of great taxonomic importance.

parenchymatous Of thin-walled cells lacking thickened corners (i.e. without trigones).

paroecious With antheridia in axils of bracts immediately beneath female inflorescence (Fig. **148**, 4, 5, 6).

paroicous See *paroecious*.

patent Spreading at an angle of about 60° to stem.

pectinate Comb-like (Fig. **37**, 7).

pellucid Transparent, clear.

perianth Protective tubular structure formed from the fusion of two leaves and, if present, an underleaf often but not always following fertilisation of an archegonium (Fig. **148**, 5, 7, 8, 9).

perigynium Fleshy tubular structure developing at apex of female stem after fertilisation of an archegonium and, often surmounted by a perianth, protecting the developing sporophyte (Fig. **148**, 8).

phenotype The sum total of observable properties of an organism, resulting from the interaction of genotype and environment (*adj.* **phenotypic**).

pinnate With spreading branches produced ± regularly on either side of stem.

plicate With longitudinal pleats (Fig. **23**, 2).

pore Opening to exterior in surface of thallus; **compound pore**, pore surrounded by 2 or more rings of superposed cells (Fig. **140**, 7, 8); **simple pore**, pore surrounded by one or more concentric but not superposed cells (Fig. **138**, 11, 12).

porose Of cell walls, perforated, with pits (Fig. **3**, 4).

postical Lower or ventral side relative to substrate.

procumbent Prostrate.

proliferating Producing usually numerous branches.

propagule Vegetative reproductive structure (i.e. gemma, caducous leaf, etc.).

protonema(ta) Filiform or thalloid structure produced by germinating spore and from which the gametophyte shoots grow, characteristic of mosses but very small or lacking in most liverworts.

proximal End or part nearest point of origin.

pseudo-elaters Irregular-shaped sterile cells produced from sporogenous tissue of capsule in Anthocerotae and Sphaerocarpales which lack true elaters (Fig. **147**, 8, 13).

pseudolateral Of an inflorescence, terminal but becoming apparently lateral because of continued growth by a branch from immediately below inflorescence (Fig. **120**, 10).

pseudoperianth Strictly, a sheath or protective structure arising from thallus and protecting two or more archegonia or a developing sporophyte; often loosely referred to as an involucre (q.v.), characteristic of many Metzgeriales.

pyriform Pear-shaped.

receptacle Specially modified sexual branches of the Marchantiales.

recurved Curved down.

remote   Of leaves, very far apart.
reniform   Kidney-shaped (Fig. **150**, 8, 9).
reticulate   With net-like pattern.
retuse   Broad apex with slight indentation in middle.
revolute   Of margin, rolled back and under.
rhizoids   Elongate, unicellular hair-like structures arising from stem or thallus, usually from ventral surface.
rhizome   A horizontal underground stem analogous to the rhizome of vascular plants (*adj.* **rhizomatous**).
rhomboidal   Narrowly diamond-shaped.

saccate   Swollen at base, sac-like (Fig. **148**, 5).
saxicole   Plant growing on rock surface (*adj.* **saxicolous**).
scale   Flat, usually unistratose, ±translucent structure, either associated with gametangia on dorsal surface of some thalloid Metzgeriales or (**ventral scales**) on ventral surface of thallus of Marchantiales and some Metzgeriales.
scarious   Thin and papery in texture.
secund   Turned to one side.
serrate   Toothed like a saw with teeth consisting of one or more cells (Fig. **85**, 6).
serrulate   Finely toothed with teeth consisting of only part of a cell (Fig. **83**, 7).
seta   Stalk between foot and capsule of sporophyte, very short until maturity when it usually elongates very rapidly and markedly by cell elongation (Fig. **148**, 8, 10).
simple   Of leaves, not lobed, with margins entire, crenulate or toothed.
simple oil-bodies   See *oil-bodies*.
simple pore   See *pore*.
sinuose   Of margin, wavy outline.
sinus   Indentation between lobes of a leaf (Fig. **150**, 10–12).
slime papilla   Thin-walled, ephemeral cell, usually near stem apex, with swollen tip exuding mucilage, and protecting growing point.
spathulate   Spoon-shaped.
spicate   Of male inflorescence, closely imbricate bracts forming a spike-like structure (Fig. **104**, 9).
spinose   With sharply pointed teeth (Fig. **86**, 2).
spinulose   With minute finely pointed teeth.
sporogenous tissue   Tissue within the walls of a liverwort capsule or spore-sac of a hornwort capsule giving rise to elaters or pseudoelaters by mitotic division and spores by meiosis.
sporophyte   Diploid phase of life-cycle consisting of foot embedded in gametophyte tissue, seta and capsule. In the Anthocerotae there is no seta.
spreading   Of leaves, more or less at right angles to stem.
squarrose   Upper part of leaf curved back at an angle of 90° or more from the lower part of the leaf.
stellate   Star-shaped.
sterile   Without gametangia.
stoloniform   Stems arching and rooting at points touching substrate, analogous to stolons of higher plants.
stoma(ta)   Pore(s) with two guard cells in wall of capsule of Anthocerotae.
stylus   Small, narrow or flattened structure associated with ventral leaf lobe (Fig. **122**, 3).
subinvolucral bracts   Bracts between uppermost leaves and female bracts of the inflorescence, in paroecious species constituting the male bracts below the female bracts.

subulate Awl-shaped.

succubous With ventral margin nearer to the stem apex than dorsal margin so that rear margin of one leaf overlies the forward margin of the leaf behind (Fig. **149**, 1*d*, *e*, 4).

synoecious Antheridia and archegonia mixed in same inflorescence, a situation very rarely encountered in liverworts.

synoicous See *synoecious*.

terete Rounded and smooth in transverse section (perianth in Fig. **5**, 1).

terrestrial Growing on soil, or more loosely in a non-aquatic habitat.

tetrad Of spores, the products of meiosis remaining united in a group of four at maturity.

thallus Flattened shoot not differentiated into stem and leaves.

trigone Thickening at cell angle (Fig. **150**, 23–25).

trigonous Three-sided in transverse section.

triploid Having three sets of chromosomes (i.e. 3*n*).

triradiate ridges Three rays radiating from centre of ventral face of a spore resulting from the formation of spore tetrads following meiosis.

tristichous In three ranks.

truncate Cut-off; ending abruptly (Fig. **150**, 18).

tubercle Small blunt protuberance.

tuberculate Of rhizoids, with papilla-like projections from inner wall (Fig. **140**, 10).

underleaves The third row of leaves on the ventral side of the stem in the Jungermanniales.

undulate Wavy.

uniseriate With cells in a single row.

unistratose Cells in a single layer only.

urceolate Urn-shaped; narrow below mouth then enlarged.

valve One of the sections of capsule wall after regular dehiscence.

venter Layer of cells surrounding egg cell in swollen part of archegonium (Fig. **148**, 11).

ventral Side of shoot, lobe of leaf or leaf margin closest to substrate.

ventral scale See *scale*.

vermicular Worm-like; long, narrow and wavy.

verrucose Of cuticle, with wart-like papillae, roughened.

vitta Band of elongated cells in a leaf (Fig. **70**, 4).

wing (1) Lamina of thallus (Fig. **135**, 1); (2) extension of keel of leaf (Fig. **82**, 7); (3) extension of longitudinal angle of perianth.

xerophytic Of dry habitats.

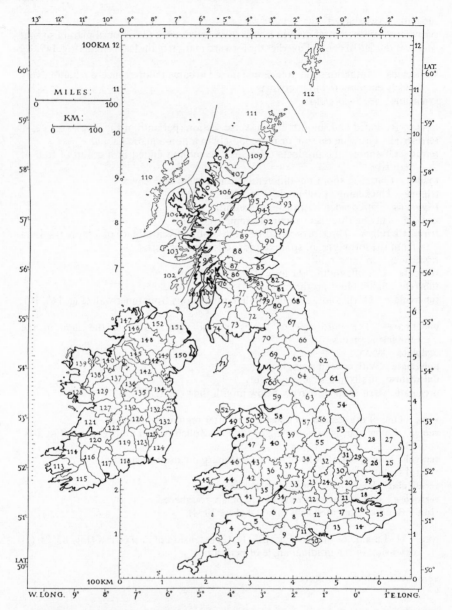

Fig. **151**. British and Irish vice-county map based upon the New Naturalist vice-county map of the British Isles. For key to numbering see facing page.

BRITISH AND IRISH VICE-COUNTIES

The British and Irish vice-counties or botanical districts mentioned in the text are listed below and figured on the facing page. The British vice-counties were delimited in 1852 and the Irish in 1901 but because of political boundary changes many of these bear little relation to current administrative counties.

ENGLAND AND WALES
1 W. Cornwall
2 E. Cornwall
3 S. Devon
4 N. Devon
5 S. Somerset
6 N. Somerset
7 N. Wilts.
8 S. Wilts.
9 Dorset
10 Isle of Wight
11 S. Hants.
12 N. Hants.
13 W. Sussex
14 E. Sussex
15 E. Kent
16 W. Kent
17 Surrey
18 S. Essex
19 N. Essex
20 Herts.
21 Middlesex
22 Berks.
23 Oxford
24 Bucks.
25 E. Suffolk
26 W. Suffolk
27 E. Norfolk
28 W. Norfolk
29 Cambridge
30 Bedford
31 Hunts.
32 Northampton
33 E. Gloucester
34 W. Gloucester
35 Monmouth
36 Hereford
37 Worcester
38 Warwick
39 Stafford
40 Shropshire
41 Glamorgan
42 Brecon
43 Radnor
44 Carmarthen
45 Pembroke
46 Cardigan
47 Montgomery
48 Merioneth
49 Caernarfon
50 Denbigh

51 Flint
52 Anglesey
53 S. Lincoln
54 N. Lincoln
55 Leicester
56 Nottingham
57 Derby
58 Cheshire
59 S. Lancashire
60 W. Lancashire
61 S.-E. York
62 N.-E. York
63 S.-W. York
64 Mid-west York
65 N.-W. York
66 Durham
67 S. Northumberland
68 N. Northumberland
69 Westmorland
70 Cumberland
71 Isle of Man

SCOTLAND
72 Dumfries
73 Kirkcudbright
74 Wigtown
75 Ayr
76 Renfrew
77 Lanark
78 Peebles
79 Selkirk
80 Roxburgh
81 Berwick
82 E. Lothian
83 Midlothian
84 W. Lothian
85 Fife
86 Stirling
87 W. Perth
88 Mid Perth
89 E. Perth
90 Angus
91 Kincardine
92 S. Aberdeen
93 N. Aberdeen
94 Banff
95 Moray
96 E. Inverness
97 W. Inverness
98 Argyll
99 Dumbarton

100 Clyde Isles
101 Kintyre
102 S. Ebudes (Islay, etc.)
103 Mid Ebudes (Mull, etc.)
104 N. Ebudes (Skye, etc.)
105 W. Ross
106 E. Ross
107 E. Sutherland
108 W. Sutherland
109 Caithness
110 Outer Hebrides
111 Orkney
112 Shetland

IRELAND
113 S. Kerry
114 N. Kerry
115 W. Cork
116 Mid Cork
117 E. Cork
118 Waterford
119 S. Tipperary
120 Limerick
121 Clare
122 N. Tipperary
123 Kilkenny
124 Wexford
125 Carlow

126 Laois
127 S.-E. Galway
128 W. Galway
129 N.-E. Galway
130 Offaly
131 Kildare
132 Wicklow
133 Dublin
134 Meath
135 Westmeath
136 Longford
137 Roscommon
138 E. Mayo
139 W. Mayo
140 Sligo
141 Leitrim
142 Cavan
143 Louth
144 Monaghan
145 Fermanagh
146 E. Donegal
147 W. Donegal
148 Tyrone
149 Armagh
150 Down
151 Antrim
152 Derry

BIBLIOGRAPHY

A list of works referred to in the text freely consulted in the compilation of the flora. References dealing with specific taxa are given in the text where appropriate.

Arnell, S. *Illustrated Moss Flora of Fennoscandia*. I. Hepaticae. Lund: Gleerup, 1956.

Buch, H. Die Scapanien Nordeuropas und Siberiens II. Systematische Tiel. *Soc. Sci. Fennica Comm. Biol.* **3** (1), 1–177. 1928.

Bryologist, The. Journal of the American Bryological and Lichenological Society, 1898–.

Bulletin of the British Bryological Society. Cardiff: British Bryological Society, 1975–.

Corley, M.F.V. & Hill, M.O. *Distribution of Bryophytes in the British Isles. A Census Catalogue of their Occurrence in Vice-Counties*. Cardiff: British Bryological Society, 1981.

Cryptogamie: Bryologie, Lichénologie. Paris: Museum National d'Histoire Naturelle, 1980–.

Düll, R. Distribution of the European and Macaronesian Liverworts (Hepaticophytina). *Bryol. Beitr.* **2**, 1–114, 1983.

Eggers, J. Artenliste der Moose Makaronesiens. *Crypt. bryol. lichén.* **3**, 283–335, 1982.

Fritsch, R. *Index to Plant Chromosome Numbers – Bryophyta*. Utrecht/Antwerp: Scheltema & Holkema, 1982.

Grolle, R. Verzeichnis der Lebermoose Europas und benachbarter Gebeite. *Feddes Repertorium* **87**, 171–279, 1976.

Grolle, R. Hepatics of Europe including the Azores: an annotated list of species with synonyms from the recent literature. *J. Bryol.* **12**, 403–59, 1983a.

Grolle, R. Nomina generica Hepaticarum: references, types and synonymies. *Acta bot. Fennica* **121**, 1–62, 1983b.

Jones, E.W. Advances in the knowledge of British Hepatics since 1926. *Trans. Br. bryol. Soc.* **2**, 1–10. 1952.

Journal of Bryology. Oxford: Blackwell Scientific Publications, 1972–.

Lindbergia. Arrhus & Leiden: Nordic and Dutch Bryological Societies, 1971–.

Macvicar, S.M. The distribution of Hepaticae in Scotland. *Trans. Proc. Bot. Soc. Edinburgh* **25**, 1–332, 1910.

Macvicar, S.M. *The Student's Handbook of British Hepatics*, 2nd edn. Eastbourne, Sumfield: 1926.

Müller, K. Die Lebermoose Europas. In Rabenhorst's *Kryptogamen-Flora von Deutschland, Osterreich und der Schweiz* IV, 1, 3. Leipzig: Geest & Portig, 1954.

Müller, K. Die Lebermoose Europas. In Rabenhorst's *Kryptogamen-Flora von Deutschland, Osterreich und der Schweiz* VI, 2, 3. Leipzig: Geest & Portig, 1957.

Revue Bryologique et Lichénologique. Paris: Museum National d'Histoire Naturelle, 1928–79.

Schuster, R.M. Boreal Hepaticae. A manual of the liverworts of Minnesota and adjacent regions. *Am. Midl. Nat.* **49**(2), 1953.

Schuster, R.M. *The Hepaticae and Anthocerotae of North America East of the Hundredth Meridian*, Vol. 1. New York: Columbia University Press, 1966.

Schuster, R.M. *The Hepaticae and Anthocerotae of North America East of the Hundredth Meridian*, Vol. 2. New York: Columbia University Press, 1969.

Schuster, R.M. *The Hepaticae and Anthocerotae of North America East of the Hundredth Meridian*, Vol. 3. New York: Columbia University Press, 1974.

Schuster, R.M. *The Hepaticae and Anthocerotae of North America East of the Hundredth Meridian*, Vol. 4. New York: Columbia University Press, 1980.

Smith, A.J.E. (ed.). *Provisional Atlas of the Bryophytes of the British Isles*. Natural Environment Research Council, Huntingdon, 1978.

Transactions of the British Bryological Society. London: Cambridge University Press, 1947–71.

Watson, E.V. *British Mosses and Liverworts*, 3rd edn. Cambridge: Cambridge University Press, 1981.

INDEX

Synonyms and excluded taxa are in italics; page numbers of illustrations are in italics.

Acrobolbaceae, 234
Acrobolbus Nees, 234
  wilsonii Nees, *233*, 234
Adelanthaceae, 44
Adelanthus Mitt., 44
  decipiens (Hook.) Mitt., 44, *45*
  dugortensis Douin & Lett, 46
  lindenbergianus (Lehm.) Mitt., *45*, 46
  unciformis (Hook. f. & Tayl.) Mitt., 46
Alicularia breidleri Limpr., 152
  compressa (Hook.) Nees, 150
  geoscyphus De Not., 152
  scalaris (S.F. Gray) Corda, 150
Anastrepta (Lindb.) Schiffn., 98
  orcadensis (Hook.) Schiffn., 98, *99*
Anastrophyllum (Spruce) Steph., 123
  donnianum (Hook.) Steph., 126, *127*
  hellerianum (Nees ex Lindenb.) Schust.,
    124, *125*
  joergensenii Schiffn., 126, *127*
  minutum (Schreb.) Schust., 124, *125*
  saxicola (Schrad.) Schust., *125*, 126
Aneura Dum., 297
  incurvata Steph., 300
  latifrons Lindb., 302
  multifida (L.) Dum., 300
  palmata Dum., 302
  pinguis (L.) Dum., 297, *298*
  sinuata Dum., 300
Aneuraceae, 297
Anomylia cuneifolia (Hook.) Schust., 202
Anthelia (Dum.) Dum., 88
  julacea (L.) Dum., 88, *89*
  juratzkana (Limpr.) Trev., *89*, 90
Antheliaceae, 88
Anthoceros L., 334
  agrestis Paton, 335, *336*
  crispulus auct., 335
  dichotomus Raddi, 337
  husnotii Steph., 335
  laevis L., 335
  punctatus auct., 335
    var. *cavernosus* auct., 335
  punctatus L., 335, *336*

  var. *crispatus* Mont., 335
Anthocerotaceae, 334
Anthocerotae, 334
Anthocerotales, 334
Anthocerotopsida, 334
Aphanolejeunea Evans, 278
  microscopica (Tayl.) Evans, *277*, 278
Aplozia atrovirens (Dum.) Dum., 139
  caespiticia (Lindenb.) Dum., 145
  cordifolia Dum., 142
  crenulata (Mitt.) Lindb., 144
  lanceolata sensu Dum., 137
  pumila (Tayl.) Dum., 140
  riparia (Tayl.) Dum., 139
  schiffneri Loitl., 140
  sphaerocarpa (Hook.) Dum., 143
Apometzgeria Kuwah., 308
  pubescens (Schrank) Kiwah., *307*, 308
Arnelliaceae, 230
Aytoniaceae, 316

Barbilophozia Loeske, 92
  atlantica (Kaal.) K. Müll., *94*, 95
  attenuata (Mart.) Loeske, 96, *97*
  barbata (Schmid ex Schreb.) Loeske, 98,
    99
  floerkei, (Web. & Mohr) Loeske, *94*, 95
  gracilis (Schleich.) K. Müll. 96
  hatcheri (Evans) Loeske, 96, *97*
  kunzeana (Hüb.) Gams, 93, *94*
  lycopodioides (Wallr.) Loeske. *97*, *98*
  quadriloba (Lindb.) Loeske, 93, *94*
Bazzania S.F. Gray, 32
  pearsonii Steph., *31*, 34
  triangularis (Lindb.) Pears., 34
  tricrenata (Wahlenb.) Lindb., *33*, 34
  trilobata (L.) S.F. Gray, 32, *33*
Blasia L., 296
  pusilla L., *293*, 296
Blasiaceae, 296
Blepharostoma (Dum. emend. Lindb.) Dum.,
    18
  trichophyllum (L.) Dum., 20, *21*

Calobryales, 10
Calypogeia Raddi, 35
    arguta Nees & Mont., 43, *43*
    azurea Stotler & Crotz, 41, *42*
    fissa (L.) Raddi, 36, *37*
    integristipula Steph., 38, *39*
    meylanii Buch, 38
    muelleriana (Schiffn.) K. Müll., 36, *37*
    neesiana (Mass. & Carest.) K. Müll., 38, *39*
        var. *meylanii* (Buch) Schust., 38
    sphagnicola (H. Arn. & J. Perss.) Warnst. & Loeske, *39*, 41
    submersa (Arn.) C. Müll., 38, 41
    suecica (H. Arn. & J. Perss.) K. Müll., *39*, 41
    trichomanis auct., 41
Calypogeiaceae, 35
Cephalozia (Dum.) Dum., 47
    affinis Lindb. ex Steph., 55
    ambigua Mass., 48, *52*
    bicuspidata (L.) Dum., 49
        ssp. *ambigua* (Mass.) Schust., 48
        ssp. *lammersiana* (Hüb.) Schust., 49
        var. *atra* Arnell., 48
        var. bicuspidata, 49, *50*
        var. lammersiana (Hüb.) Breidl., 49, 50
    catenulata (Hüb.) Lindb., 51, *52*
    connivens (Dicks.) Lindb., 59, *60*
    fluitans (Nees) Spruce, 63
    francisci (Hook.) Dum., 63
    hibernica Spruce ex Pears., *60*, 61
    lammersiana (Hüb.) Spruce, 49
    leucantha (Schiffn.) K. Müll., *52*, 53
    loitlesbergeri Schiffn., 58, *59*
    lunulifolia (Dum.) Dum., 55, *56*
    macrostachya Kaal., 53
        var. macrostachya, 53, *54*
        var. spiniflora (Schiffn.) K. Müll., 53, 54
    media Lindb., 55
    pleniceps (Aust.) Lindb., 56, *57*
        var. *macrantha* (Kaal. & Nichols.) K. Müll., 56
Cephaloziaceae, 46
Cephaloziella (Spruce) Schiffn., 72
    baumgartneri Schiffn., 78, *79*
    byssacea sensu Schust., 79
    calyculata (Durieu & Mont.) K. Müll., *87*, 88
    dentata (Raddi) Migula, 84, *85*
    divaricata (Sm.) Schiffn., 79, *80*
    elachista (Jack ex Gott. & Rabenh.) Schiffn., *75*, 76
    hampeana (Nees) Schiffn., *77*, 78
        var. *pulchella* Jens., 76
    integerrima (Lindb.) Warnst., 86, *87*

massalongi (Spruce) K. Müll., 82, *83*
    var. *nicholsonii* (Douin & Schiffn.) Jones, 84
    myriantha (Lindb.) Schiffn., 76
    nicholsonii Douin & Schiffn., *83*, 84
    pearsonii (Spruce) Douin, 122
    rubella (Nees) Warnst., 76, *77*
    spinigera (Lindb.) Joerg., 74, *75*
    starkei auct., 79
    stellulifera Schiffn., 81, *81*
    striatula (C. Jens.) Douin, 74
    subdentata Warnst., 74
    turneri (Hook.) K. Müll., *85*, 86
Cephaloziellaceae, 72
Cephaloziopsis pearsonii (Spruce) Schiffn., 122
Chandonanthus setiformis (Ehrh.) Lindb., 92
Chiloscyphus Corda, 211
    bispinosus (Hook. f. & Tayl.) Engel & Schust., 205
    cuspidatus (Nees) Engel & Schust., 205
    fragrans (Moris & De Not.) Engel & Schust., 210
    latifolius (Nees) Engel & Schust., 205
    pallescens (Ehrh. & Hoffm.) Dum., 213
    polyanthos (L.) Corda, 211
        var. *fragilis* (Roth.) K. Müll., 213
        var. pallescens (Ehrh. ex Hoffm.) Hartm., *212*, 213
        var. polyanthos, 211, *212*
        var. *rivularis* (Schrad.) Nees, 211
    profundus (Nees) Engel & Schust., 208
    semiteres (Lehm.) Lehm. & Lindenb., 208
Cladopodiella Buch, 62
    fluitans (Nees) Buch, 63, *64*
    francisci (Hook.) Joerg., 63, *64*
Codoniaceae, 281
Cololejeunea (Spruce) Schiffn., 273
    calcarea (Libert) Schiffn., 274, *275*
    microscopica (Tayl.) Evans, 278
    minutissima (Sm.) Schiffn., 276, *277*
    rossettiana (Mass.) Schiffn., *275*, 276
Colura (Dum.) Dum., 272
    calyptrifolia (Hook.) Dum., 272, *273*
Conocephalaceae, 316
Conocephalum Hill corr. Wiggers, 316
    conicum (L.) Underw., *315*, 316
Crossocalyx hellerianus (Nees ex Lindenb.) Meyl., 124
Cryptothallus Malmb., 299
    mirabilis Malmb., *298*, 299

Diplophyllum (Dum.) Dum., 168
    albicans (L.) Dum., *169*, 170
    gymnostomophilum (Kaal.) Kaal., 176
    obtusifolium (Wahlenb.) Dum., *171*, 172
    taxifolium (Wahlenb.) Dum., 170, *171*

Douinia (C. Jens.) Buch, 167
  ovata (Dicks.) Buch, 167, *168*
Drepanolejeunea (Spruce) Schiffn., 262
  hamatifolia (Hook.) Schiffn., *261*, 262
Dumortiera Nees, 314
  hirsuta (Sw.) Nees, 314, *315*

Eremonotus Lindb. & Kaal. ex Pears., 122
  myriocarpus (Carringt.) Lindb. & Kaal. ex
    Pears., *120*, 122
Eucalyx hyalinus (Lyell) Carringt., 146
  obovatus (Nees) Carringt., 149
  paroicus (Schiffn.) Macv., 147
  subellipticus (Lindb. ex Kaal.) Breidl., 147

Fossombronia Raddi, 281
  angulosa (Dicks.) Raddi, 282, *284*, *288*
  caespitiformis De Not. ex Rabenh., 282,
    284
  crozalsii Corb., *285*, 286
  dumortieri Hüb. & Genth. ex Lindb., 282
  fimbriata Paton, 287, *288*
  foveolata Lindb., 282, *284*
  husnotii Corb., 283, *284*
    var. *anglica* Nich., 283
  incurva Lindb., *285*, 286
  loitlesbergeri Schiffn., 286
  mittenii Tindall, 286
  pusilla (L.) Nees, 283
    var. *decipiens* Corb., 283
    var. maritima Paton, *285*, 286
    var. pusilla, 283, *285*
  wondraczekii (Corda) Dum., *285*, 286
Frullania L., 251
  dilatata (L.) Dum., 257, *257*
  fragilifolia (Tayl.) Gott. et al., 255, *256*
  germana (Tayl.) Gott. et al., 252, *252*
  maritima Steph., 255
  microphylla (Gott.) Pears., *254*, 255
  tamarisci (L.) Dum., 253, *253*
  teneriffae (Web.) Nees, 252, *252*
Frullaniaceae, 250

Geocalycaceae, 201
Geocalyx Nees, 215
  graveolens (Schrad.) Nees, 216, *216*
Gongylanthus Nees, 232
  ericetorum (Raddi) Nees, 232, *233*
Gymnocolea (Dum.) Dum., 119
  acutiloba (Schiffn.) K. Müll., *120*, 121
  inflata (Huds.) Dum., 119, *120*
    var. *acutiloba* (Schiffn.) Arnell, 121
    var. heterostipa (Carringt. & Spruce)
      K. Müll., 121
    var. inflata, 121

Gymnomitriaceae, 153
Gymnomitrion Corda, 163
  adustum Nees emend. Limpr., 161
  alpinum (Gott. ex Limpr.) Schiffn., 162
  apiculatum (Schiffn.) K. Müll., *165*, 166
  concinnatum (Lightf.) Corda, 164, *165*
  corallioides Nees, 164, *165*
  crenulatum Gott. ex Carringt., *165*, 166
  obtusum (Lindb.) Pears., 164, *165*
  varians (Lindb.) Schiffn., 162

Haplomitriaceae, 10
Haplomitrium Nees, 10
  hookeri (Sm.) Nees, 10, *11*
Harpalejeunea (Spruce) Schiffn., 260
  ovata (Hook.) Schiffn., 260, *261*
Harpanthus Nees, 213
  flotovianus (Nees) Nees, *214*, 215
  scutatus (Web. & Mohr) Spruce, *214*, 215
Hepaticae, 9
Hepaticopsida, 9
Herbertaceae, 16
Herbertus S.F. Gray, 16
  aduncus auct., 16
  aduncus (Dicks.) S.F. Gray ssp. hutchinsiae
    (Gott.) Schust., 16, *17*
  borealis Crundw., 18, *19*
  hutchinsiae (Gott.) Evans, 16
  stramineus (Dum.) Lett, 16, *17*
Hygrobiella Spruce, 65
  laxifolia (Hook.) Spruce, 65, *66*

Isopaches bicrenatus (Schmid. ex Hoffm.)
    Buch, 112
  hellerianus (Nees & Lindenb.) Buch, 124

Jamesoniella (Spruce) Carringt., 131
  autumnalis (DC.) Steph., 131, *132*
  carringtonii (Balfour) Schiffn., 220
  schraderi (Mart.) Schiffn., 132
  undulifolia (Nees) K. Müll., *132*, *132*
Jubula Dum., 258
  hutchinsiae (Hook.) Dum., 258, *259*
Jubulaceae, 258
Jungermannia L., 135
  atrovirens Dum., *138*, 139
  borealis Damsh. & Váňa, 140, *141*
  caespiticia Lindenb., 145, *145*
  confertissima Nees, 142, *143*
  exsertifolia Steph. ssp. cordifolia (Dum.)
    Váňa, 141, *142*
  gracillima Sm., 144, *145*
  hyalina Lyell, 146, *148*
  lanceolata auct., 137
  leiantha Grolle, 137, 138
  longiflora Nees, 105

Jungermannia L. (*Cont.*)
  obovata Nees, *148*, 149
  paroica (Schiffn.) Grolle, 147, *148*
  polaris Lindb., 140, *141*
  pumila With., *138*, 140
  sphaerocarpa Hook., 143, *143*
  subelliptica (Lindb. ex Kaal.) Levier, 147, *148*
Jungermanniaceae, 131
Jungermanniales, 12
Jungermanniales acrogynae, 12
Jungermanniales anacrogynae, 280

Kurzia v Martens, 24
  pauciflora (Dicks.) Grolle, 25, *26*
  sylvatica (Evans) Grolle, 25, *27*
  trichoclados (K. Müll.) Grolle, *27*, 28

Leiocolea Buch, 112
  alpestris (Schleich. ex Web.) Isov., 116, *117*
  badensis (Gott.) Joerg., *117*, 118
  bantriensis (Hook.) Joerg., 116, *117*
  collaris (Nees) Schljak., 116
  gillmanii (Aust.) Evans, 114, *115*
  heterocolpos (Thed. ex Hartm.) Buch, *115*, 118
  muelleri (Nees ex Lindenb.) Joerg., 116
  rutheana (Limpr.) K. Müll., 114, *115*
  turbinata (Raddi) Buch, *117*, 118
Lejeunea Libert corr. Hampe, 262
  azorica Steph., 264
  cavifolia (Ehrh.) Lindb., 263, *266*
    var. *planiuscula* Lindb., 264
  diversiloba auct., 268
  flava (Sw.) Nees, 265, *267*
  hibernica Bischl., 268, *269*
  holtii Spruce, 268, *269*
  lamacerina (Steph.) Schiffn., 264, *266*
    var. *azorica* (Steph.) Greig-Smith, 264
  macvicari Pears., 270
  mandonii (Steph.) K. Müll., 270, *271*
  patens Lindb., 265, *267*
  planiuscula (Lindb.) Buch, 264
  ulicina (Tayl.) Gott. et al., 270, *271*
Lejeuneaceae, 259
Lepicoleaceae, 15
Lepidozia (Dum.) Dum., 28
  cupressina (Sw.) Lindenb., 30, *31*
  pearsonii Spruce, *29*, 30
  pinnata (Hook.) Dum., 30
  reptans (L.) Dum., 28, *29*
  setacea auct., 25
  sylvatica Evans, 25
  trichoclados K. Müll., 28
Lepidoziaceae, 21

Leptoscyphus Mitt., 202
  anomalus (Hook.) Mitt., 135
  cuneifolius (Hook.) Mitt., 202, *203*
  taylori (Hook.) Mitt., 133
Lophocolea (Dum.) Dum., 204
  alata Mitt. ex Larter, 205
  bidentata auct., 205
  bidentata (L.) Dum., 204
    var. bidentata, 205, *206*
    var. rivularis (Raddi) Warnst., 205, *206*
  bispinosa (Hook. f. & Tayl.) Gott. et al., 205, 207
  cuspidata (Nees) Limpr., 205
  fragrans (Moris & De Not.) Gott. et al., *209*, 210
  heterophylla (Schrad.) Dum., 208, *209*
  latifolia Nees, 205
  semiteres (Lehm.) Mitt., *207*, 208
Lophozia (Dum.) Dum., 99
  alpestris auct., 105
  alpestris (Schleich. ex Web.) Evans, 116
  atlantica (Kaal.) Schiffn., 95
  attenuata (Mart.) Dum., 96
  badensis (Gott.) Schiffn., 118
  bantriensis (Hook.) Steph., 116
  barbata (Schmid. ex Schreb.) Dum., 98
  bicrenata (Schmid. ex Hoffm.) Dum., 112, *113*
  capitata (Hook.) Macoun, 111, *113*
  confertifolia Schiffn., 104
  collaris (Nees) Dum., 116
  elongata (Lindb.) Steph., 106
  excisa (Dicks.) Dum, 106, *107*
    var. *cylindracea* (Dum.) Müll., 106
  floerkei (Web. & Mohr) Schiffn., 95
  gillmanii (Aust.) Schust., 114
  guttulata (Lindb.) Evans, 105
  hatcheri (Evans) Steph., 96
  heterocolpos (Thed. ex Hartm.) Howe, 118
  herzogiana Hodgs. & Grolle, 108, *109*
  incisa (Schrad.) Dum., *110*, 111
  kaurinii (Limpr.) Steph., 114
  kunzeana (Hüb.) Evans, 93
  longidens (Lindb.) Macoun, *101*, 102
  longiflora sensu Macvicar, 104
  longiflora (Nees) Schiffn., *103*, 105
  lycopodioides (Wallr.) Cogn., 98
  muelleri (Nees ex Lindenb.) Dum., 116
  obtusa (Lindb.) Evans, 109, *110*
  opacifolia Culm. ex Meylan, *110*, 111
  perssonii Buch & S. Arn., 106, *107*
  porphyroleuca (Nees) Schiffn., 105
  quadriloba (Lindb.) Evans, 93
  quinquedentata (Huds.) Cogn., 128
  rutheana (Limpr.) Howe, 114
  schultzii (Nees) Schiffn., 114

var. *laxa* Schiffn. ex Burrell, 114
silvicola Buch, 104
sudetica (Nees ex Hüb.) Grolle, *101*, 105
turbinata (Raddi) Steph., 118
ventricosa (Dicks.) Dum., 102
  var. confertifolia (Schiffn.) Husn., *103*, 104
  var. longiflora auct. non (Nees) Macoun, 104
  var. silvicola (Buch) Jones ex Schust., *103*, 104
  var. ventricosa, 102, *103*
  wenzelii (Nees) Steph., *101*, 105
Lophoziaceae, 90
Lunularia Adans., 312
  cruciata (L.) Dum. ex Lindb., 312, *313*
Lunulariaceae, 312

Madotheca cordaeana (Hüb.) Dum., 248
  laevigata (Schrad.) Dum., 246
  platyphylla (L.) Dum., 250
  porella (Dicks.) Nees, 246
  thuja auct., 248
Marchantia L., 319
  alpestris (Nees) Burgeff, 319
  aquatica (Nees) Burgeff, 319
  polymorpha L., 319, *320*
Marchantiaceae, 317
Marchantiales, 310
Marchantiopsida, 9
Marchesinia S.F. Gray, 279
  mackaii (Hook.) S.F. Gray, 279, *279*
Marsupella Dum., 153
  adusta (Nees emend. Limpr.) Spruce, 161, *162*
  alpina (Gott. ex Limpr.) H. Bern., *162*, 163
  apiculata Schiffn., 166
  aquatica (Lindenb.) Schiffn., 156
    var. *pearsonii* (Schiffn.) E.W. Jones, 156
  boeckii (Aust.) Kaal., 159
    var. boeckii, 159, *160*
    var. stableri (Spruce) Schust., 159, *160*
  brevissima (Dum.) Grolle, 162, *162*
  condensata (Ångstr. ex Hartm.) Lindb. ex Kaal., 160, *161*
  emarginata (Ehrh.) Dum., 154
    var. aquatica (Lindenb.) Dum., *155*, 156
    var. emarginata, 154, *155*
    var. pearsonii (Schiffn.) Corley, *155*, 156
  funckii (Web. & Mohr) Dum., *157*, 158
  hungarica Boros & Vajda, 158
  jörgensenii Schiffn., 156
  pearsonii Schiffn., 156
  profunda Lindb., 159, 338

pygmaea (Limpr.) Steph., 158
robusta (De Not.) Evans, 156
sparsifolia (Lindb.) Dum., *157*, 159
sphacelata (Gies. ex Lindenb.) Dum., 156, *157*
  var. *media* (Gott.) E.W. Jones, 156
sprucei (Limpr.) H. Bern., *157*, 158
stableri Spruce,
sullivantii (De Not.) Evans, 156
ustulata Spruce, 158
varians (Lindb.) K. Müll., 162
Mastigophora Nees, 15
  woodsii (Hook.) Nees, *11*, 15
Metzgeria Raddi, 302
  conjugata Lindb., 306, *307*
  fruticulosa (Dicks.) Evans, *305*, 306
  furcata (L.) Dum., 304, *305*
    var. *fruticulosa* (Dicks.) Lindb., 306
  hamata auct., 306
  leptoneura Spruce, 306, *307*
  pubescens (Schrank) Raddi, 308
  temperata Kuwah., 304, *305*
Metzgeriaceae, 302
Metzgeriales, 280
Microlejeunea diversiloba auct., 268
  ulicina (Tayl.) Evans, 270
Microlepidozia setacea auct., 25
  sylvatica (Evans) Joerg., 25
  trichoclados (K. Müll.) Joerg., 28
Moerckia Gott., 294
  blyttii (Moerch.) Brockm., *295*, 296
  flotoviana (Nees) Schiffn., 294
  hibernica (Hook.) Gott., 294, *295*
Mylia S.F. Gray, 133
  anomala (Hook.) S.F. Gray, *134*, 135
  cuneifolia (Hook.) S.F. Gray, 202
  taylorii (Hook.) S.F. Gray, 133, *134*

Nardia S.F. Gray, 149
  breidleri (Limpr.) Lindb., *151*, 152
  compressa (Hook.) S.F. Gray, 150, *151*
  geoscyphus (De Not.) Lindb., *151*, 152
  scalaris S.F. Gray, 150, *151*
Nowellia Mitt., 61
  curvifolia (Dicks.) Mitt., 61, *62*

Odontoschisma (Dum.) Dum., 67
  denudatum (Mart.) Dum., 69, *70*
  elongatum (Lindb.) Evans, 69, *71*
  macounii (Aust.) Underw., 69, *71*
  sphagni (Dicks.) Dum., 68, *68*
Orthocaulis atlanticus (Kaal.) Buch, 95
  attenuatus (Mart.) Evans, 96
  floerkei (Web. & Mohr) Buch, 95
  kunzeana (Hüb.) Buch, 93
  quadriloba (Lindb.) Buch, 93

Pallavicinia S.F. Gray, 294
  lyellii (Hook.) Carruth., *293*, 294
Pallaviciniaceae, 292
Pedinophyllum (Lindb.) Lindb., 218
  interruptum (Nees) Kaal., *217*, 219
    var. *pyrenaicum* (Spruce) Kaal., 219
Pellia Raddi, 289
  borealis Lorbeer, 290, *291*
  endiviifolia (Dicks.) Dum., 292, *293*
  epiphylla agg., 289, *291*
  epiphylla (L.) Corda, 290, *291*
    ssp. *borealis* (Lorbeer) Messe, 290
  fabbroniana Raddi, 292
  neesiana (Gott.) Limpr., *291*, 292
Pelliaceae, 289
Petalophyllum Nees & Gott. ex Lehm., 287
  ralfsii (Wils) Nees & Gott. 287, *288*
Phaeoceros Prosk., 335
  bulbiculosus (Brotero) Prosk., 337
  carolinianus (Michx.) Prosk., 337
  laevis (L.) Prosk., 335, *336*
    ssp. carolinianus (Michx.) Prosk., 337
    ssp. laevis, 337
Plagiochila (Dum.) Dum., 219
  ambagiosa auct., 228
  asplenioides auct., 223
  asplenioides (L. emend. Tayl.) Dum., 223, 225
    ssp. *porelloides* (Torrey & Nees) Schust., 223
    var. *major* Nees, 223
  atlantica F. Rose, 228, *229*
  britannica Paton, 225, 226
  carringtonii (Balfour) Grolle, 220, *221*
  corniculata auct., 229
  exigua (Tayl.) Tayl., *227*, 229
  killarniensis Pears. 221, *222*
  major (Nees) S. Arn., 223
  porelloides (Torrey ex Nees) Lindenb., 223, *224*
  punctata Tayl., *227*, 227
  spinulosa (Dicks.) Dum., *222*, 226
    var. *inermis* Carringt., 221
    var. *killarniensis* (Pears.) Macv., 221
  stableri Pears., 229
  tridenticulata auct., 229
Plagiochilaceae, 218
Plectocolea hyalina (Lyell) Mitt., 146
  obovata (Ness) Lindb., 149
    var. *minor* (Carringt.) Schljak., 147
  paroica (Schiffn.) Evans, 147
  subelliptica (Lindb. ex Kaal.) Evans, 147
Pleuroclada albescens (Hook.) Spruce, 65
  islandica (Nees) Pears., 65
Pleurocladula Grolle, 65
  albescens (Hook.) Grolle, 65, *66*

  islandica (Nees) Grolle, 65
Pleurozia Dum., 235
  purpurea Lindb., 235, *235*
Pleuroziaceae, 234
Porella L., 245
  arboris-vitae (With.) Grolle, 246, *247*
    var. *killarniensis* (Pears.) Corley, 248
    var. *obscura* (Nees) Corley, 248
  cordaeana (Hüb.) Moore, *149*, 150
    var. *faroensis* (C. Jens.) E.W. Jones, 250
    var. *simplicior* (Zett.) Arn., 250
  laevigata (Schrad.) Pfieff. 246
  obtusata (Tayl.) Trev., 248, *249*
  pinnata L., 246, *247*
  platyphylla (L.) Pfeiff., *249*, 250
  thuja auct., 248
Porellaceae, 245
Potamolejeunea holtii (Spruce) Greig-Smith, 268
Preissia Corda, 317
  quadrata (Scop.) Nees, 318, *319*
Pseudolepicoleaceae, 18
Ptilidiaceae, 243
Ptilidium Nees, 243
  ciliare (L.) Hampe, 243, *244*
  pulcherrimum (Web.) Vainio, *244*, 245

Radula Dum., 236
  aquilegia (Hook. f. & Tayl.) Gott. et al., *240*, 242
  carringtonii Jack, *241*, 242
  complanata (L.) Dum., 237, *238*
    ssp. *lindbergiana* (Gott.) Schust., 287
  holtii Spruce, 239, *241*
  lindbergiana Gott., 237
  lindenbergiana Gott. ex Hartm. f., 237, *238*
  voluta Tayl. ex Gott. et al., *239*, 240
Radulaceae, 236
Reboulia Raddi, 317
  hemisphaerica (L.) Raddi, 317, *318*
Riccardia S.F. Gray, 299
  chamedryfolia (With.) Gott., 300, *301*
  incurvata Lindb., 300, *301*
  latifrons (Lindb.) Lindb., 302, *303*
  multifida (L.) S.F. Gray, 300, *301*
  palmata (Hedw.) Carruth., 302, *303*
  pinguis (L.) S.F. Gray, 297
  sinuata (Hook.) Trev., 300
Riccia L., 322
  arvensis Aust., 328
  beyrichiana Hampe ex Lehm., *332*, 333
  bifurca Hoffm., 328, *330*
  canaliculata Hoffm., *326*, 327
  canescens Steph., 333
  cavernosa Hoffm. emend. Raddi, 324, *325*

ciliata Hoffm., 333
commutata Jack ex Levier, 331
crozalsii Levier, *332*, 333
crystallina L. emend. Raddi, 324, *325*
fluitans L. emend. Lorbeer, 325, *326*
glauca L., 328, *329*
huebeneriana Lindenb., 324, *325*
michelii Raddi, 333
nigrella DC., *330*, 331
rhenana Lorbeer, *326*, 327
sorocarpa Bisch., 328, *329*
subbifurca Warnst. ex Crozals, *330*, 331
warnstorfii auct. angl., 331
Ricciaceae, 321
Ricciocarpos Corda, 321
natans (L.) Corda, 321, *321*

Saccobasis polita (Nees) Buch, 130
Saccogyna Dum., 218
viticulosa (L.) Dum., *217*, 218
Scapania (Dum.) Dum., 172
aequiloba (Schwaegr.) Dum., 195, *196*
aspera M. & H. Bern., *196*, 197
bartlingii (Hampe) Nees, 178
calcicola (H. Arn. & J. Perss.) Ingham, *177*, 178
compacta (H. Roth) Dum., 174, *175*
crassiretis Bryhn, 193
curta (Mart.) Dum., *180*, 181
cuspiduligera (Nees) K. Müll., *174*, 178
degenii Schiffn. ex K. Müll., 185, *187*
dentata Dum., 188
gracilis Lindb., 198, *199*
gymnostomophila Kaal., 176, *177*
intermedia (Husn.) Pears., 188
irrigua (Nees) Nees, 185, *186*
lingulata Buch, 183, *184*
microphylla Warnst., 183
mucronata auct. angl., 183
mucronata Buch ssp. *praetervisa* (Meyl.) Schust., 183
   var. *arvernica* (Culm.) K. Müll., 183
   var. *praetervisa* (Meyl.) Buch, 183
nemorea (L.) Grolle, 193, *194*
nemorosa (L.) Dum., 193
nimbosa Tayl. ex Lehm., *200*, 201
obliqua (Arn.) Schiffn., 189
ornithopodioides (With.) Waddell, 199, *200*
paludicola Loeske & K. Müll., 186, *187*
paludosa (K. Müll.) K. Müll., 191, *192*
parvifolia Warnst., 179, *180*
praetervisa Meyl., 183, *184*
scandica (H. Arn. & Buch) Macv., 181, *182*

subalpina (Nees) Dum., 189, *191*
uliginosa (Sw. ex Lindenb.) Dum., 189, *192*
umbrosa (Schrad.) Dum., *194*, 195
undulata (L.) Dum., 188, *190*
Scapaniaceae, 166
Solenostoma atrovirens (Dum.) K. Müll., 139
caespiticium (Lindenb.) Steph., 145
cordifolium (Dum.) Steph., 142
crenulatum (Sm.) Mitt., 144
gracillimum (Sm.) Schust., 144
hyalinum (Lyell) Mitt., 146
levieri (Steph.) Steph., 142
oblongifolium auct., 140
obovatum (Nees) Mass., 149
ontariensis Schust., 146
paroicum (Schiffn.) Schust., 147
pumilum (With.) K. Müll., 140;
   ssp. *polaris* (Lindb.) Schust., 140
schiffneri (Loitl.) K. Müll., 140
sphaerocarpoideum (De Not.) Paton & Warburg, 139
sphaerocarpum (Hook.) Steph., 143
subellipticum (Lindb. ex Kaal.) Schust., 147
triste (Nees) K. Müll., 139
Southbya Spruce, 230
nigrella (De Not.) Henriques, *231*, 232
stillicidiorum (Raddi) Lindb. 230
tophacea (Spruce) Spruce, 230, *231*
Sphaerocarpaceae, 308
Sphaerocarpales, 308
Sphaerocarpos Boehmer, 309
michelii Bellardii, 309, *309*
texanus Aust., *309*, 310
Sphenolobopsis Schust. & Kitag., 122
pearsonii (Spruce) Schust., *120*, 122
Sphenolobus exsectiformis (Breidl.) Steph., 128
exsectus (Schrad.) Steph., 128
hellerianus (Nees ex Lindenb.) Steph., 124
minutus (Schreb.) Berggr., 124
pearsonii (Spruce) Steph., 122
politus (Nees) Steph., 130
saxicola (Schrad.) Steph., 126

Targionia L., 312
hypophylla L., 312, *313*
Targioniaceae, 311
Telaranea Spruce ex Schiffn., 22
murphyae Paton, *23*, 24
nematodes (Gott. ex Aust.) Howe, 22, *23*
sejuncta auct., 22
setacea auct., 25
sylvatica (Evans) K. Müll., 25
trichoclados (K. Müll.) K. Müll., 28
Tetralophozia (Schust.) Schljak., 91
setiformis (Ehrh.) Schljak., 92, *92*

Trichocolea Dum., 20
　tomentella (Ehrh.) Dum., *19*, 20
Trichocoleaceae, 20
Tritomaria Schiffn. ex Loeske, 127
　exsecta (Schrad.) Loeske, 128, *129*

exsectiformis (Breidl.) Loeske, 128, *129*
polita (Nees) Joerg., *129*, 130
quinquedentata (Huds.) Buch, 128, *129*

Wiesnerellaceae, 314